# II Conference on Plasma & Laser Research Technologies 2016



Journal of Physics: Conference Series Volume 747

Moscow, Russia
25 – 27 January 2016

**Editors:**

**V.A. Kurnaev**
**E.I. Dodulad**

ISBN: 978-1-5108-3058-5
ISSN: 1742-6588

**Printed from e-media with permission by:**

Curran Associates, Inc.
57 Morehouse Lane
Red Hook, NY 12571

**Some format issues inherent in the e-media version may also appear in this print version.**



Printed by Curran Associates, Inc. (2016)

For permission requests, please contact the Institute of Physics
at the address below.

Institute of Physics
Dirac House, Temple Back
Bristol BS1 6BE UK

Phone: 44 1 17 929 7481
Fax: 44 1 17 920 0979

techtracking@iop.org

**Additional copies of this publication are available from:**

Curran Associates, Inc.
57 Morehouse Lane
Red Hook, NY 12571 USA
Phone: 845-758-0400
Fax: 845-758-2633
Email: curran@proceedings.com
Web: www.proceedings.com

# Table of contents

## Volume 747

**II Conference on Plasma & Laser Research and Technologies**
**25–27 January 2016, Moscow, Russia**

**Accepted papers received: 1 September 2016**
**Published online: 21 September 2016**

### Preface



### Papers

### Plasma Physics

## Modern aspects of solid state matter physics and nanophysics

## Laser Physics

## Charged particle accelerators

Journal of Physics: Conference Series **747** (2016) 011001 doi:10.1088/1742-6596/747/1/011001

# II International Conference on Plasma and Laser Research and Technologies

**V A Kurnaev and E I Dodulad**
National Research Nuclear University MEPhI (MEPhI), Kashirskoe highway 31, Moscow, 115409 Russian Federation

E-mail: VAKurnaev@mephi.ru, EIDodulad@mephi.ru

II Conference on Plasma and Laser Research and Technologies took place on January 25th until January 27th , 2016 at National Research Nuclear University "MEPhI" (NRNU MEPhI). It was organized by the Institute of Laser and Plasma Technologies and was supported by the Competitiveness Program of NRNU MEPhI.

The Conference consisted of four sections: Plasma physics and controlled nuclear fusion, Laser physics, Modern aspects of solid state matter physics and Charged particle accelerators. The Conference provided participants an opportunity to present their research results for the consideration of a wide audience from the sidelines of science.

The main topics of the Conference were:

- Controlled nuclear fusion with magnetic and inertial confinement;
- Low-temperature plasma and its application in modern technology;
- Laser physics and technologies for industry, environmental control and precise measurements;
- Optical information control, holography, spintronics and photonics;
- Modern aspects of solid state matter physics and nanophysics;
- Charged particle accelerators.

More than 200 specialists on plasma, laser and solid state physics took part in the II Conference. They represented leading Russian scientific research centres and universities (such as Troitsk Institute of Innovative and Thermonuclear Research, Institute of Crystallography, National Research Centre "Kurchatov Institute", Institute of Physical Chemistry and Electrochemistry and others) and universities from Belarus, Ukraine, Germany, USA, Canada, Belgium, and Sweden.

All report presentations were broadcasted online on the NRNU MEPhI official site. The translation was watched by viewers from Moscow, Prague, St. Petersburgh and other cities, who could not attend the Conference.

We would like to thank heartily all of the speakers, participants and organizing committee members for their contribution to the conference.

# Peer review statement

All papers published in this volume of *Journal of Physics: Conference Series* have been peer reviewed through processes administered by the proceedings Editors. Reviews were conducted by expert referees to the professional and scientific standards expected of a proceedings journal published by IOP Publishing.

Journal of Physics: Conference Series **747** (2016) 012001 doi:10.1088/1742-6596/747/1/012001

# Analysis of main regularities of periodic discharge passing in fluid stream

**B Yu Bogdanovich[1] and A V Nesterovich**
National Research Nuclear University MEPhI, 115409 Russia, Moscow, Kashirskoe highway, 31

E-mail: BYBogdanovich@mephi.ru

**Abstract.** This paper describes the basic processes in the gas discharge plasma accompanying the jet of liquid flowing between the high-voltage electrode in the form of a tubular tip and the plane grounded electrode. The process is quasi-periodic due to a jet break caused by the "core" discharge on the surface of the jet and the arising shock wave. The main discharge is accompanied by one or more pinch discharges. At the bottom of a pinch in the metal there occurs a zone with ~ (100 ÷ 300) micron size and with ultra-high energy release, where the metal is hardened. The paper analyzes the cause of formation of pinches and of formation of channels in the matter of the anode which are similar to channels generated in gas during the passage of intense beams. The paper concludes that there is a self-focusing mechanism not previously studied.

## 1. Introduction

A periodic discharge in a flow of liquid (PDFL) was first implemented in NRNU MEPhI in 1995 [1] and since then has found application in a number of technological processes used in aircraft industry [2, 3], heavy engineering [4, 5], in the manufacture of endoprostheses [3], tools, in the manufacture of rails [6], and others. In the study of PDFL its potential has not been fully identified. One of the constraints in this area is the difficulty in interpreting the physical processes that determine its flow. It is easiest to examine the PDFL mechanism, starting with the operation of a processing system in the original version.

## 2. Description of a processing system

The physical nature of a high-voltage discharge in a flow of liquid is illustrated by to the given principal diagram of a system (Figure 1). The processed material (3) is placed on the stage (5) which is movable both automatically and manually. Above the specimen there is a hollow metal tip (1) attached to a reservoir of liquid (7) by a dielectric flow line (8). On the tip (1) a constant high potential (generally negative, with the declining current-voltage characteristic of around 7 kV) is supplied from the power unit (9). The specimen (3) with the stage (5) is located in the dielectric chamber (6). The excess of liquid is drained through the drain system (4). In case of a gap 3 ... 15 mm, a periodic discharge (2) with a repetition frequency f ≈ 50 ... 300 Hz appears between the specimen and the tip. The arc discharge with duration of a few tens of nanoseconds (a crater diameter of 5 ... 20 mm) has "point" effect on the treated surface. The displacement speed of the stage typically varied from 0.25 to 2.0 mm/sec.

[1] To whom any correspondence should be addressed.

A jet of liquid flowing out of the container on a flat surface does not retain its cylindrical shape. It is known [8], herewith there appears a system of capillary waves fixed with respect to this obstacle. As the electric current passes through a jet from the constant rechargeable high-capacity capacitor, the normal instability in the form of wavy distortion of the smooth cylindrical jet [9] is significantly worsened by electrical charges [10] accumulated on its raised surface. Due to the presence of charges in a liquid jet the expansion pressure of Faraday-Maxwell will arise [11].

## 3. Photometric measurements of a discharge

The process of discharge leads to the disintegration of the continuity of a liquid mass and the emergence of shock waves accompanied by a powerful crash and emissions of some parts of a substance on the remote distance in relation to the scale of the experiment (up to tens of cm) (Figure 2).

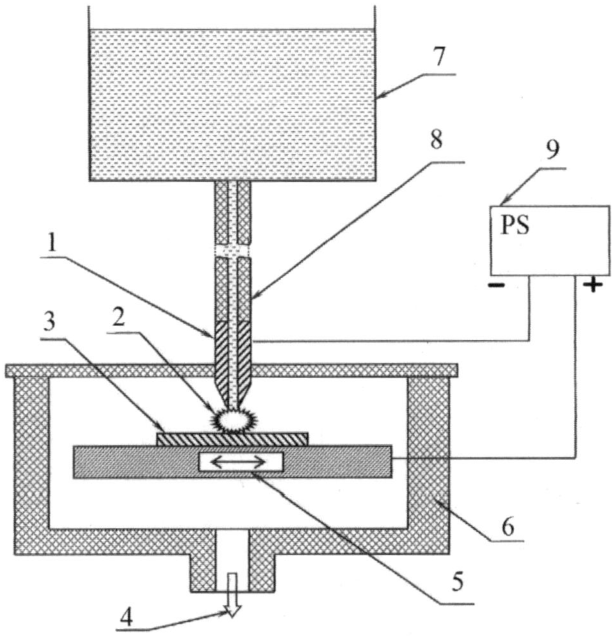


**Figure 1.** The diagram of a system for processing the specimens in the PDFL: 1 - the tip; 2 - the area of a discharge; 3 - the specimen; 4 - the drain system; 5 - the stage; 6 - the dielectric chamber; 7 - a reservoir of liquid; 8 a dielectric flow line; 9 - the power unit

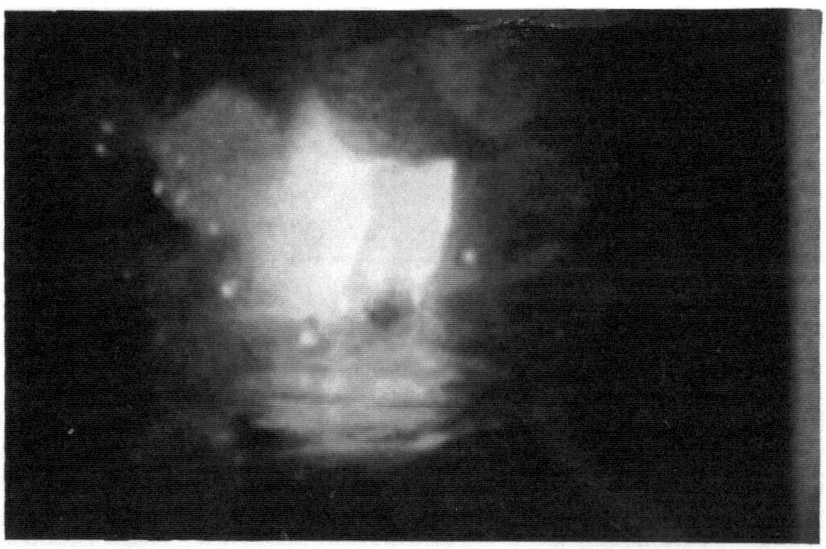

**Figure 2.** A jet break by a discharge and formation of clusters in the form of spheres.

In close proximity to a jet of liquid clearly distinguishable objects are formed taking the shape of spheres or polyhedrons, and having a "granular" structure like a peeled pomegranate, according to the authors, the most likely explanation for the appearance of these objects, in case of long-lived (up to several seconds), could be the formation, which is known from physical electronics and physics of a gas discharge, of hydrated ions of oxonium, whose structure at the molecular level is shown in Figure 3. It has the character of a dodecahedron.

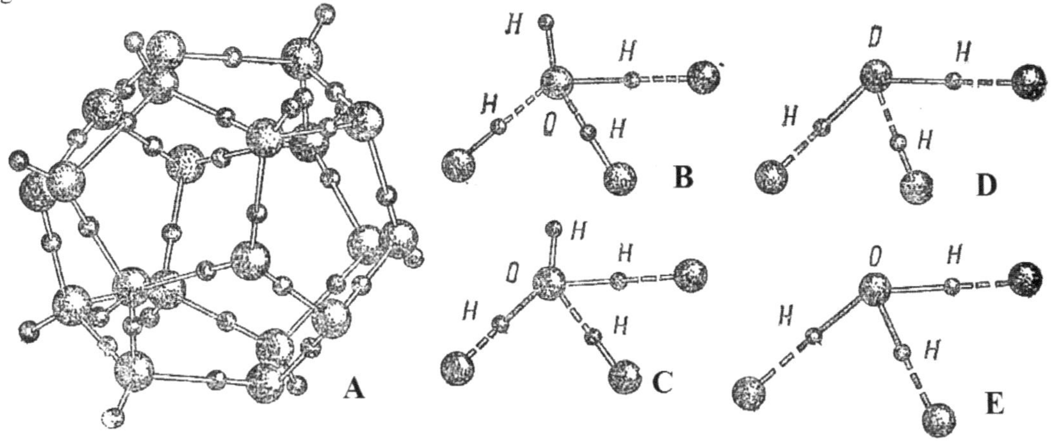


**Figure 3.** Ion $H^+ (H_2O)_{21}$:
a – the structure of ion; b, c – the top of the first type, where a molecule of water and oxonium ion is located respectively; d, e – the top of the second type with a molecule of water and oxonium ion; solid lines – valence bonds; a dashed line – a hydrogen bond.

This hypothesis is one of the key hypotheses explaining the nature of ball lightning [12]. This structure has a high capacitance and a lifetime.
The presence of such clusters on the basis of hydrated ions causes the appearance of the discharges of completely different type, in particular, the pinch discharges, in the localization area of the main discharge in a liquid jet (Figure 4).

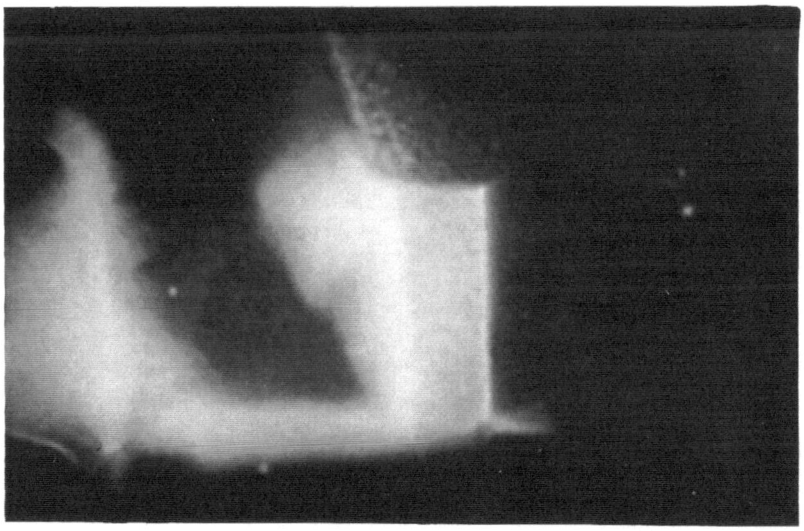

**Figure 4.** A pinch discharge (at the left) accompanying the appearance of the main one (at the center).

Journal of Physics: Conference Series **747** (2016) 012001 doi:10.1088/1742-6596/747/1/012001

From the analysis of photo and filmed materials it is possible to make an unambiguous conclusion about the cluster origin of the pinch discharges. It is clearly seen that there may be several clusters and pinches on one "main" discharge along the jet (Figure 5). Clusters have substantially the same size and, consequently, the same capacitance and charge. This accounts for almost the same size of traces on the anode which remained at the bottom of the pinch discharges. Noteworthy is the fact that all types have in the bottom the typical structure which is characteristic of the so-called θ - pinch, when the flow of electrons moves on a narrowing spiral, like a tornado.

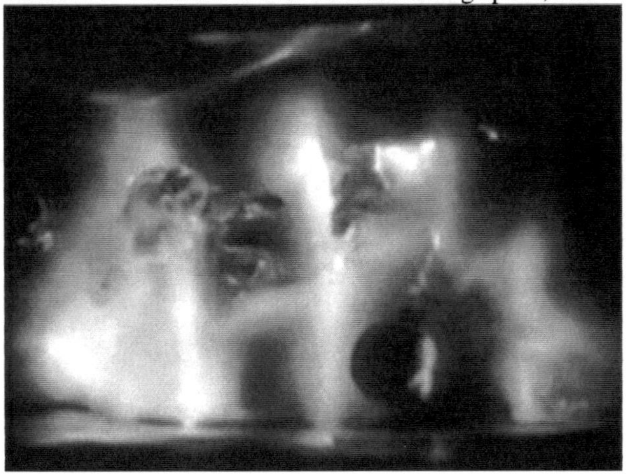

**Figure 5.** A few pinches simultaneously and a black object on their background.

## 4. The physical model of the process

Along a jet of liquid (water and etc.) on its surface there is a pinch discharge contributing to the formation of a hydrodynamic wave which breaks a jet. The process is periodic with a frequency of up to several tens of Hz. In the area of the jet localization (Figure 6), a region of enhanced ionization and low pressure caused by the flows of counter ions and electrons ($e_0$ and $i^+$) appears as in any electric discharge pump (the power unit of a processing unit - standard, intended for supplying magnetic discharge pumps such as NORD-100 ). In the presence of the emitting surface in the discharge area (E), or simply a charged region (as in a thundercloud) and conditions arise for the pinch discharges, especially if preionization and lowering of the gas (air) density have occurred. Electrons moving initially in parallel (vertically) form a pinching flow, since their space charge has been neutralized by the positive ions, and "parallel currents" are known to be attracted. A well-known «Z» -pinch is formed. But the process does not end here. The «Z» - pinch when approaching the anode becomes the «θ» - pinch, as indicated by the results of the analysis of photo and filmed materials. The author of [13] managed to find out the cause of this phenomenon. Its essence lies in the fact that only one of the directions of electrons rotation corresponds to the stable motion of electrons, namely, that coincides with the direction of magnetic induction vector of the azimuthal magnetic field $\overrightarrow{B_\varphi}$ formed by the current of 'primary' electrons and ions in a flow. This practically means a "twisting" of the particle flux (electrons) in this direction when approaching the anode. The process is self-supporting, a so-called "collapse" is formed, and at the bottom of a pinch the current density may reach several tens of kiloAmperes per square millimeter (up to 5 kA at the pinch bottom not more than 30 microns in diameter).

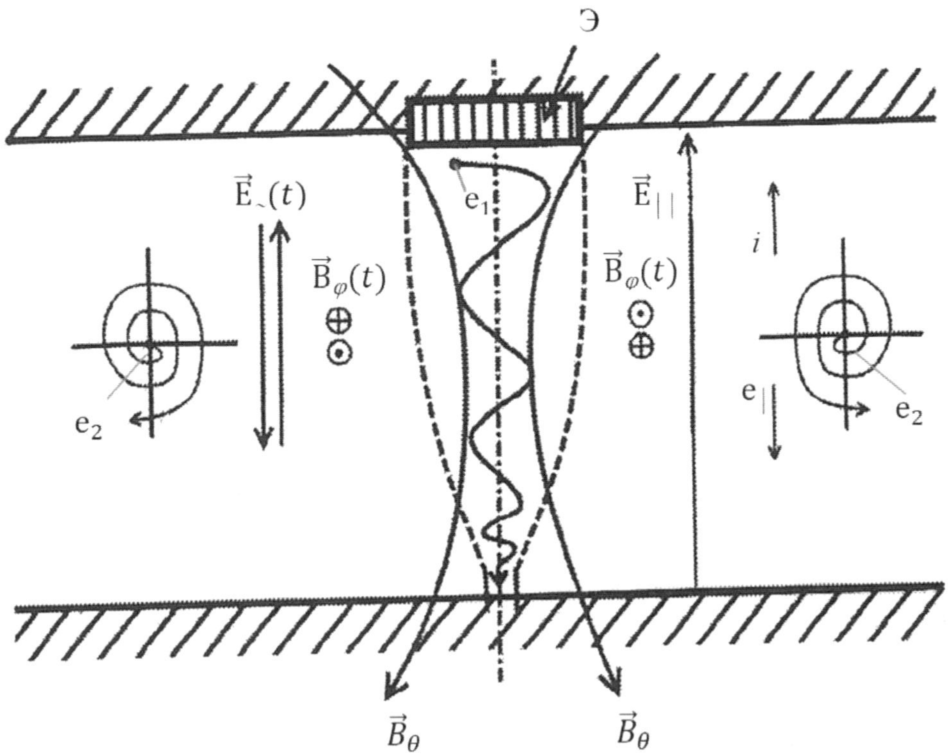


**Figure 6.** The diagram explaining the emergence of electric and magnetic fields and particle dynamics in the process of PDFL.

The azimuthal component of the electrons velocity generates the magnetic field B, which is essentially the field of a magnetic dipole. It can have an impact on the dynamics of particles in the surrounding area, but mostly - by collapsing a jet. Near a pinch the field $\vec{B_\varphi}$ can accelerate ionization electrons in the betatron mode, stimulating the formation of plasmoids of ball and cylinder type, similar to the super high frequency (SHF) systems type - injectors based on the cyclotron resonance. Strengthening of the formation effect of plasmoids may be due to the presence of SHF - fields generated by the pinch discharges. The presence of a powerful SHF radiation has been detected by the authors, including in the frequency range corresponding to the electron cyclotron resonance, which corresponds to values of magnetic induction $|\vec{B_\varphi}|$ in close proximity to the pinch discharges with currents up to 5 kA.

The presence of such plasmoids also was detected using photo and filmed materials.

Special attention is paid to the presence of absolutely black (non-transparent for radiating a discharge or laser light) spherical objects with the size of about 0.5 - 1 mm, sometimes with a central spherical part having presumably its own glow. The experimenters [14] encountered the objects on the facilities for the detection of an air spray glow. It has been suggested that the internal environment is formed by the radiation of the excited components of air and water ionized by particles (protons) with a short path. The outer sphere is separated from the inner sphere of black color where the light emission is suppressed. In the direction of the potential gradient in the outer sphere has been a modest glow trail.

During the unsteady evolution of a cold plasma experiencing the accelerating effect of Coulomb forces and the wave capture of light particles (electrons, protons) at the initial stage of its inception, there is a process of free charges pairing, which results in both the emitting plasma states and the states of self-absorption of electromagnetic energy that are characteristic of relaxation phases of the Josephson generation [16].

Taken into account the fact that the source of the gas-kinetic and plasma excitation in the collapse of a particle is the point source, in the nature of the spatial distribution of localization areas of the particles energy the extrema of density are observed. The emergence of such spatially localized extrema of density (detected by light emission of excited components of air) owes its origin to the effect of a "virtual" electrode [15].

We cannot disregard the appearance of glowing regions in the form of hemispheres on the "grounded" electrode surface (in this case - the anode) (Figure 7) both in the discharge process and in its absence. According to the authors, this mysterious effect can be explained as follows.

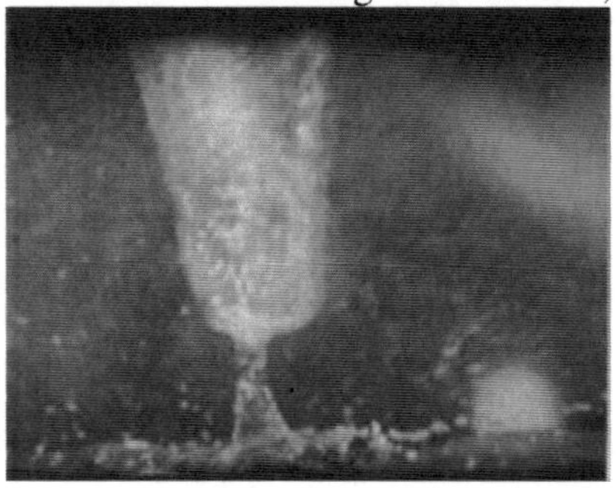

**Figure 7.** A glowing region in the form of a hemisphere on the anode surface after the end of discharge.

Similarly to the "step potential" known to specialists in electrical engineering and related in particular to the presence of leakage currents on the insulators of high voltage lines, in this case, the leakage current on a liquid jet, which in the immediate area generates the voltage other than zero, may be responsible for a glow. So, if the voltage on the power unit with an incomplete discharge capacity is an order of magnitude smaller, for example, 1 kW, on the surface of the water droplet with a diameter of 1 mm, which lies on the anode, it may be the electric field intensity ~ 20 kV/cm sufficient to the emergence of the so-called corona discharge in the humid air. This hypothesis is supported by the fact that such glows are displaced radially from the axis of a jet of water, most likely under the influence of a shock wave in the air.

Studies of the pinch discharges are of interest not only for solving technological applications mentioned in the paper introduction, but also for the problems of a fundamental nature. As the calculations and materials research of electrodes exposed to the pinch effects have shown, in the area of the bottom of pinches at the surface layer the zones up to 0.3 mm deep and with a diameter of about 0.7 mm are formed, in which a substance becomes amorphous instead of crystalline. Properties of a metal subjected to such an impact indicate that it has stayed in the stage of ultra-high pressure (up to ~$10^7$ kg/cm$^2$) and temperature of about several thousand degrees. From the theory of shock waves is known that at pressures in the range of $10^4$ ... $10^5$ kg/cm$^2$ significant deviations from the linear law of compressibility are already observed and polymorphic phase transitions are found which are associated with changes in the atomic structure. At pressures of $10^6$ ... $10^7$ kg/cm$^2$ electron transitions occur; the destruction of electron shells and partial loss of the individual properties by atoms are also possible. A further pressure increase up to $10^8$ ... $10^9$ kg/cm$^2$ leads to the complete destruction of electron shells and the degeneration of the properties: the substance is converted into electron-nuclear gas. High pressures are accompanied by high temperatures. The relationship between pressure $p$, density 〉 (specific volume $V$) and temperature $T$ in the substance is expressed by the equation of state.

As a result of the materials research of the substance subjected to the PDFL, it was found that there are specific features associated, according to the authors, with the ultra-high densities of current and low pulse width with the pinch discharges. As can be seen from Figures 8 and 9, in the cross-section there are channels (up to a few cm) with the cross sectional dimensions of about (0.5 ÷ 1) mm, inside which the metal has an amorphous structure and color characteristic of the superheated metal and rotating around its own axis. This behavior of the charged particles flow is characteristic of the gas medium, when the beam is "self-focused" by its own magnetic field, in particular, when electrons passing through the plasma. Herewith, the repulsive effect of the electrons space charge is compensated or overcompensated by the space charge of positive ions.

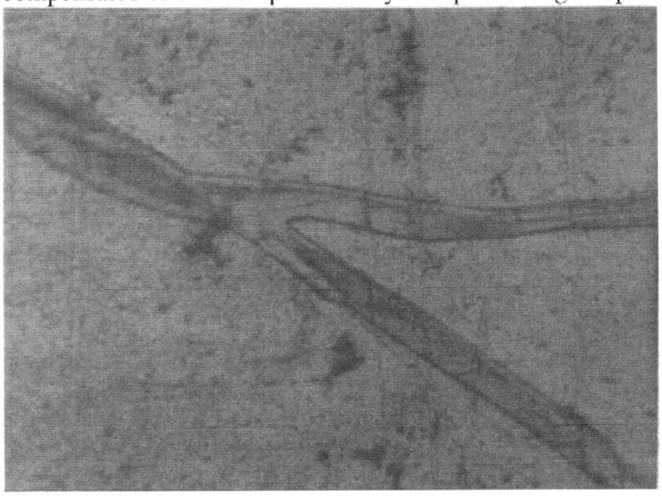

**Figure 8.** Pictures of channels in the substance, ×350.

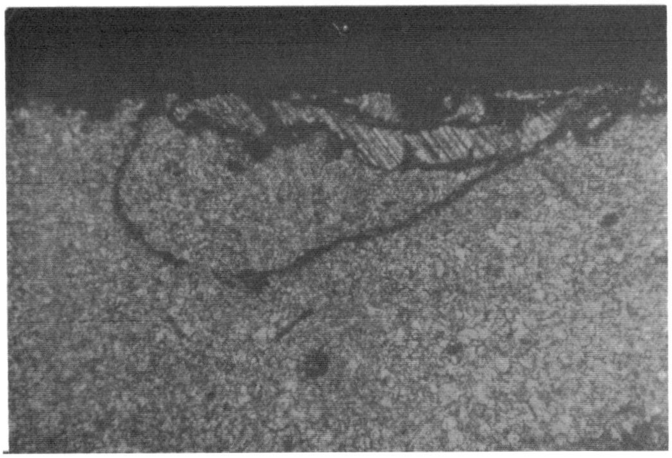

**Figure 9.** Pictures of channels in the substance, ×350.

When electrons passing through the anode (metal or semiconductor) in the localization area of the conduction electrons flow (near the bottom of a pinch or the main discharge), a considerable number of "holes" is formed, the mobility of which is significantly lower than that of the conduction electrons. In particular, in Ge it is more than 2 times lower and in InSb and InAs it is about 70 times lower. This gives reason to believe that at some point, "the accumulated holes" are like a fixed background that compensates or overcompensates the charge of the electron flow focused by its own azimuthal magnetic field. The cross section shows that the trace of the flow narrows and then widens again. Most likely, there is a situation similar to the Hall effect in metals and semiconductors, particularly since we do not exclude the effect of the cross sectional magnetic field from "neighboring" currents, for example, the main discharge on the pinch discharge (in the space above the surface we can see the curvature of pinch axes being close to the main discharge - Figure 4).

The rotation of the current channel around its own axis can be attributed to reasons similar instabilities such as "snakes" or "sausages"-type during the passage of the particles flow in the gas.

## 5. Conclusions.

Results of the study of this process may be of practical importance for electrical engineering, in particular for the proper design of high-current pulse mode facilities to which the future of high-power electronics is ascribed.

Journal of Physics: Conference Series **747** (2016) 012001 doi:10.1088/1742-6596/747/1/012001

The work was performed as a part of the Grant Agreement # 14-19-00022 between the Russian Science Foundation, "National Research Nuclear University "MEPhI"" and B.Yu. Bogdanovich, Head of the project, to conduct fundamental and pilot studies.

**References.**


[1] Bogdanovich B Yu, Kalin B A, Nesterovich A V and Puchkov AM 1996 A periodic arc discharge in a flow of liquid as a means of increasing microhardness and "etching" of the metal surface *Abstracts of IV All-Russian Conf. on the modification of the properties of structural materials by beams of charged particles* (Tomsk: TPU) p 102

[2] Fetisov G P, Sokolova N H, Sokolov V S, Len' E N and Nesterovich A V 2001 A study of the structure and properties of steel specimens after exposure to a periodic discharge in a flow of liquid (PDFL) *Metals technology* **12** 23–29

[3] Akhmetova E R and Tazetdinov R G. Modification of titanium alloys for the friction pairs by a periodic discharge in a flow of liquid *Bulletin of MAI* **1** 73–83

[4] Gavze A L, Matevosyan A P, Nesterovich A V and Bogdanovich B Yu 2001 Heat treatment using highly concentrated energy sources *Metal science and heat treatment* **9** 34–38

[5] Gavze A L Shestakov I I Nesterovich A V Bogdanovich B Yu and Alferov G V 2003 Hardening of the surfaces of metallic materials using a periodic high-voltage discharge in a flow of liquid *Metal working* **3** 8–11

[6] Bogdanov V M *et al.* Rail reinforcement device *Patent RU* **2253712** 16.04.2004

[7] Bogdanovich B Yu and Nesterovich A V 2015 Improving corrosion resistance of technological and structural metals using a periodic discharge in a flow of liquid *Defence equipment enginery* **7** 2–7

[8] Andronov A A, Vitt A A and Khaikin S E 1959 *Theory of vibrations* (Moscow: Physmathlit)

[9] S.T. Kalashnikov 1985 *Electricity* (Moscow: Nauka, Physmathlit) p 547

[10] 1971 *High-speed deformation: Collection of scientific articles* (Moscow: Nauka) p 128

[11] Zolotyh B N 1966 *Physical fundamentals of spark processing of metals* (Moscow: Engeeneering) p 235

[12] Stakhanov I P 1985 *On the physical nature of lightning* (Moscow: Energoatomizdat) p 208

[13] Nesterovich A V A method of transporting the beams of charged particles *Patent* **1281144** 08.04.1985

[14] Pakhmanin Yu A *et al.* 2002 *Water is a cosmic phenomenon* ed Yu A Pakhmanin (Moscow: RANS) p 368

[15] Naumov N D 1998 Unsteady model for formation of a virtual cathode *Doklady Physics* **359** 323-325

[16] Nezlin M B 1982 *Dynamics of beams in plasma* (Moscow: Energoizdat)

# Investigation of electrolyte electric discharge characteristics

**D L Kirko[1] and A S Savjolov**
National Research Nuclear University MEPhI (MEPhI), Kashirskoe shosse 31, Moscow, 115409 Russian Federation

E-mail: kirko@plasma.mephi.ru

**Abstract**. The most important electrical characteristics of electrolyte electric discharge were investigated. The electric burning discharge was obtained with the help of different electrolytes. The spectral composition of the electric discharge electromagnetic radiation was determined, the plasma temperature was determined. The spectrum of the electric discharge high-frequency oscillations was calculated in the region ν=10 kHz–80 MHz. The most appropriate modes of the electric burning discharge in different electrolytes were proposed.

## 1. Introduction

If there are electric discharges in electrolytes, the intensely radiating plasma area, which is situated inside the liquid medium, appears. This results in intensive boiling of electrolyte near the electric discharge and in the formation of liquid flows [1,2]. Electrolyte discharges are used for processing and modification of different metal complex surfaces and in more efficient methods of hydrogen production [1-4]. The burning discharge is caused by broad spectrum of high-frequency oscillations [5,6]. If the discharge plasma is placed in the magnetic field, all the processes of the electric discharge are stabilized. The electrolyte discharge characteristics were investigated in this work and different working mediums were used.

## 2. The investigation of the electric discharge voltage-current characteristics

Cylindrical and conical chambers were used in these experiments. Chambers were made of plexiglass with volume 120-310 cm$^3$. The experimental setup is shown in figure 1. Electrolytes contained sodium carbonate, sodium hydro carbonate and potassium hydroxide. The discharge supply was implemented with the help of full-wave rectifier: voltage 0-250 V, frequency 100 Hz. Tungsten rods or titanic rods were used as cathodes (diameter 2-3 mm). The plates, made form stainless steel, were used as anodes (plate thickness 0.1-0.2 mm).

Let's analyze the main patterns of the electrolyte discharge development when the sodium carbonate is used. The concentration of the sodium carbonate $Na_2CO_3$ was equal to $C$=0.01-1 M or $m$=1.06-106 g in every 10$^3$ cm$^3$ of distilled water (1 M – mass of 1 mole of matter in grams). The voltage-current characteristic of the electric discharge is shown in figure 2(a) there the concentration of the sodium carbonate is equal to 0.5 M. At first we have a linear relationship. The cathode arcing and discharge ignition occur near the point 1. After that the plasma area appears around the cathode surface. The color of discharge glow was mainly yellow-red. Previously it was found that the ignition voltage increase occurs as the content of the working medium decreases [5,6]. The cathode region is situated near the cathode surface; this region has the biggest radiating capacity with thickness 1-2 mm.

[1] To whom any correspondence should be addressed.

The next middle plasma area is rarefied and is surrounded by steam-gas layer with thickness 1-2 cm. The voltage-current relationship becomes smaller in the 1-2 region, because of the medium resistance increase. Then the discharge combustion intensifies and the voltage-current relationship becomes linear in the 2-3 section. The 2-3 section is the main working area, inside of which the discharge combustion is stable and normal. The high electric oscillations occur in the 2-3 section. The gradual decrease of the discharge glow takes place inside the 3-4 section. The discharge combustion ceases in point 4.

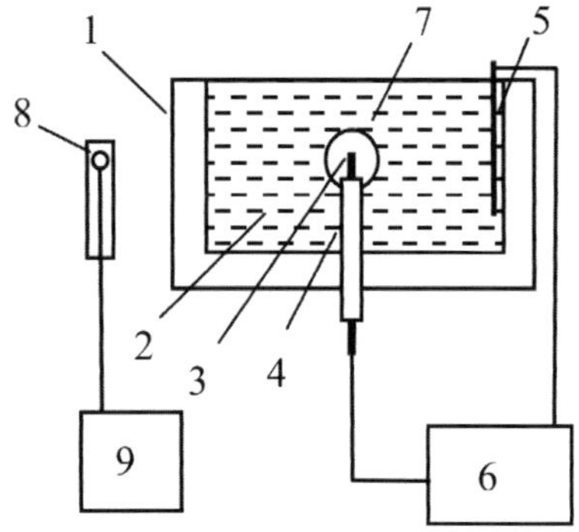


**Figure 1.** Experimental setup: 1-chamber, 2 – electrolyte, 3 – cathode, 4 – ceramic tube, 5 – anode, 6 – power supply, 7 – electric discharge, 8 – magnetic probe, 9 – frequency analyzer

Previously it was found that the sodium carbonate content in electrolyte affects the shape of the discharge voltage-current characteristic [6]. The following sodium carbonate concentration values were used for the investigation of this relationship: 1M, 0.75 M, 0.5 M, 0.25, 0.1 M, 0.05 M, 0.01 M. The voltage-current characteristics for 0.5 M и 0.1 M values are shown in figure 2. If the working medium content decreases, the initial linear section of the voltage-current characteristic becomes straighter in figure 2(b). The ignition voltage value increases in point 1. The 1-2 and 2-3 sections also become straighter. The next points 3-4-5 points come nearer to the initial points 3-2-1 section. As a result, the total area of the closed voltage-current characteristics becomes smaller.

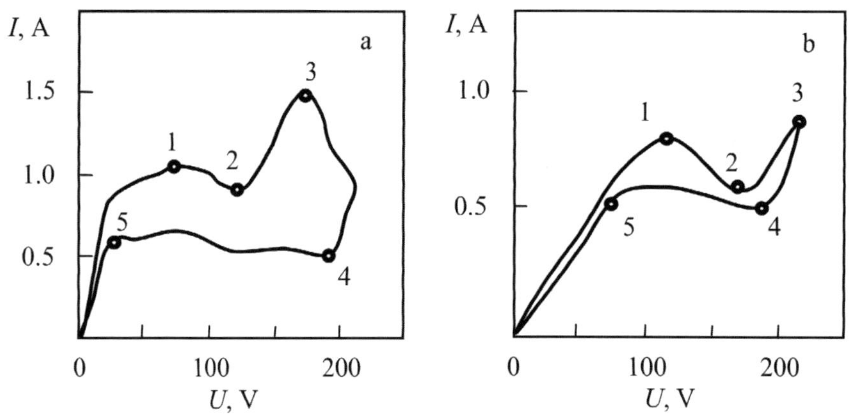


**Figure 2.** The relationships between the electrolyte discharge voltage-current characteristics and the sodium carbonate concentration: (a) the concentration is equal to 0.5 M, (b) the concentration is equal to 0.1 M

If the concentration of the working medium is minimum 0.01 M, the voltage-current characteristic looks like a practically linear section without twists. The voltage-current characteristic shapes for all contents of working medium were investigated in these experiments. If we want to get the optimum shape of the voltage-current characteristic, we need to choose the following concentration of the sodium carbonate: M=0.5-0.75.

## 3. The investigation of electric oscillations

Magnetic probes were used for the investigation of electric oscillations in electrolyte. All signals were recorded with the help of TDS 2024B oscilloscope and processed due to Origin software. The probe signals were also sent to C4-25 analyzer (working range 10 kHz-70 MHz). Magnetic probes are small coils with diameter 2-4 mm and number of coil blocks 80-170. The potassium hydroxide KOH was used as a working medium. The external constant magnetic field $B$=200 Gs was used for stabilization of the burning discharge. The following fundamental oscillation frequencies were determined: (36±2) kHz, (51±3) kHz, (203±10) kHz, (274±13) kHz, (6.4±0.3) MHz, (12.3±0.5) MHz, (25±1) MHz, (32±2) MHz, (43±2) MHz. If we use 190-310 kHz range, the continuous frequency distributions occur. The spectrum of the discharge electric oscillations is shown in figure 3.

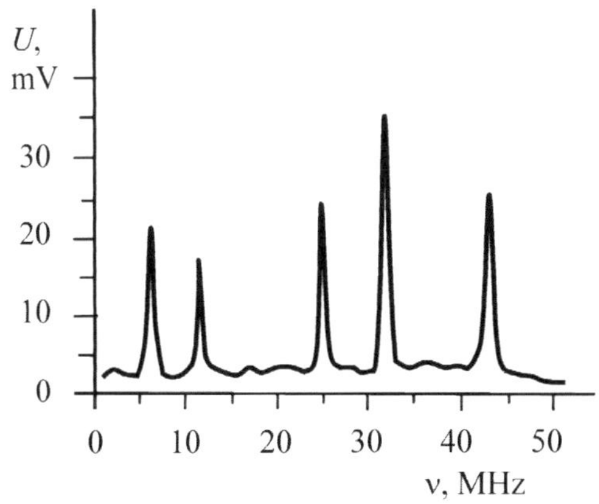


**Figure 3.** The spectrum of the electrolyte discharge high-frequency oscillations

Let's analyze all possible oscillation processes which can occur in the discharge plasma of the electrolyte. If we have an external constant magnetic field, cyclotron frequencies can appear. If the field magnetic induction is equal $B$=200 Gs, we will have the following cyclotron frequencies: $\omega_{Bi}=eB/m_p=1.92\cdot10^6$ s$^{-1}$ and $\omega_{Be}=eB/m_e=3.52\cdot10^9$ s$^{-1}$. These frequencies can correspond with ion-cyclotron plasma waves or electron-cyclotron plasma waves. The wave lengths for different frequencies of electric oscillations were calculated in accordance with dispersion relation for low-frequency waves, which occur in plasma [7]: if the frequency is ν~205 kHz, the wave length is λ~5 cm; if the frequency is ν~35 kHz, the wave length is λ~2 mm [5,6]. These wave lengths are practically equal to the characteristic of anode and cathode, which are used in the discharge system.

Debye radius as the parameter of the plasma screening was $r_d$ ~$5\cdot10^{-6}$ cm at all standard characteristics of plasma. This value is much smaller than standard Debye radius of plasma structures ($l$~$10^{-3}$-$10^{-2}$ cm), which occur during the electrolyte discharge.

The spectra of the electrolyte discharge electromagnetic radiation were investigated by the concentration M=0.5 of the sodium hydroxide KOH. The following lines had the most intensive radiation: K I 770 nm, K II 394 nm, Na I 589.5 nm, Na I 588.9 nm. The sodium atoms were used as electrolyte impurities. The following lines had the less intensive radiation: K II 423 nm, K I 404 nm, $H_\alpha$ 656 nm, $H_\beta$ 486 nm, Hγ 434 nm. The plasma temperature inside the cathode region with the help of the method of relative intensities [8] due to lines of atomic hydrogen $H_\alpha$ and $H_\beta$ was calculated: $T$=2400±200 K. The working range was used by the current: $I$=1.25±0.15 A. We also calculated the plasma concentration for the same current with the help of Stark broadening of $H_\alpha$, $H_\beta$ lines: $n_e$=(2.8±0.2)·$10^{15}$ $cm^{-3}$. The oscillation processes in the region 4-7 kHz were recorded for time dependencies of hydrogen line intensity.

## 4. The investigation of the electrolyte working medium characteristics

Earlier experiments with composites showed that if we fill crystalline cells with electrolyte, the voltage occurs in these cells are up to 1.5 V [9]. We investigated the change of the discharge characteristics after the addition of crystalline powders to the electrolyte. One electrode the aluminum plate with thickness 0.1-0.2 mm and total area 80-120 $cm^3$ was placed inside the plexiglass vessel. The calcite fine-dispersed powder $CaCO_3$ was placed on the vessel bottom (medium size of each granule 1-5 μm). The electrolyte contains distilled water and sodium carbonate $Na_2CO_3$ with concentration 0.5 M. The thickness of electrolyte layer inside the vessel was bigger that the thickness of the powder layer. The second electrode - copper rod with diameter 3-4 mm was vertically fixed in the center of the vessel. This electrode was fixed at the distance from the vessel bottom and contacted the calcite powder. The voltage and the current were measured with the help of micro voltmeter (measurement accuracy: $\Delta U$=±1 μV, $\Delta I$=±0.1 μA) after the discharge ceasing. The thickness of the calcite powder layer inside the vessel was equal to $h_l$=3-40 mm. The negative potential was recorded on the aluminum electrode; the positive potential was recorded on the central copper electrode. Let's analyze the measurement results for the following cylindrical vessel: diameter $d$=40 mm, height $h_3$=50 mm, the depth of the central electrode submersion $h_2$=40 mm. The relationship between the electrode voltage and the thickness of the calcite layer is shown in figure 4.

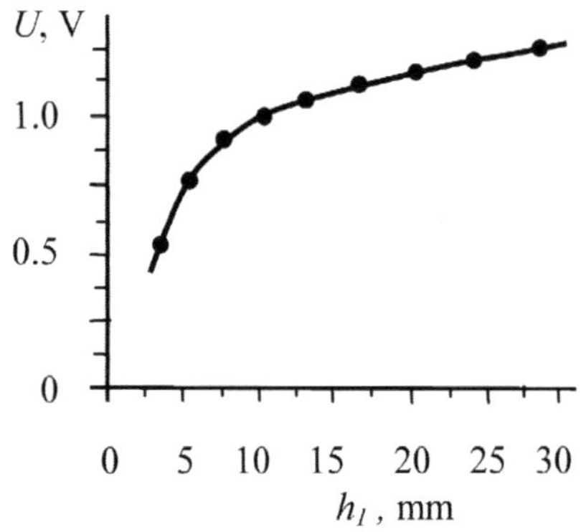


**Figure 4.** The relationship between the electrode voltage and the thickness of the calcite layer

If the range of the load was 100 kΩ -300 kΩ, the current was equal to 45 μA-10 μA. If the calcite layer thickness and its volume increase, the voltage increases too. This phenomenon is apparently caused by the increase of the number of charge carriers in calcite medium and in the electrolyte medium. Such phenomena in new materials can be represented as plasmon waves, which occur in near-surface layer of the medium [10]. This means that additional potentials can appear in the medium, which contains electrolyte and fine-crystalline component.

## 5. Conclusion

The following results were obtained during the course of this work. The main patterns of the voltage-current characteristic shape change in accordance with the working medium content in the electrolyte were investigated. If the working medium content increases, the total area of the voltage-current characteristic also increases. The optimum concentration of sodium carbonate, which is used as the working medium, is M=0.5-0.75.

The fundamental oscillation frequencies for ν=10 kHz-80 MHz working range were obtained during the investigation of the discharge oscillation processes, so the potassium hydroxide KOH was used as a working medium. Our calculations showed that ion-cyclotron plasma waves and electron-cyclotron plasma waves can occur in such discharge.

The electrolyte elemental composition and the presence of impurities in the electrolyte were calculated with the help of spectral methods. The electrolyte, which contains potassium hydroxide, has the intensive lines of potassium ions and potassium atoms. The plasma temperature inside the cathode region of the discharge was calculated: $T$=(2400±200) K.

**References**

[1] Gaisin Al F, Nasibullin R T 2011 *Plasma Phys. Rep.* **37** 896-903
[2] Kayumov R R, Gaysin Al F , Son E E, Gaysin Az F, and Gaysin F M 2010 *Physica Scripta* **T142** 014038-014041
[3] Suminov I V, Epel'fel'd A V, 2001 *Pribory* **9** 13-23 (in Russian)
[4] Kanarev F M 2010 *Low temperature electrolysis of water* (Krasnodar: Krasnodarsk. Univ.) (in Russian)
[5] Kirko D L, Savjolov A S, and Vizgalov I V, 2011 *Vestnik Kazan. Technol. Univ.* **15** 86–89 (in Russian)
[6] Kirko D L, Savjolov A S, Vizgalov I V 2013 *Russian Phys. J.* **55** 1243-1247
[7] Kroll N A, Trivelpiece A W 1973 *Principles of plasma physics* (New York: McGrawHill)
[8] Huddlstone R, Leonard S 1965 *Plasma diagnostic techniques* (New York: Academic Press)
[9] Reid M 2009 *Raum und zeit* **162** 76-81 (in German)
[10] Klimov V V 2010 *Nanoplasmonics* (M: Fizmatlit) (in Russian)

Journal of Physics: Conference Series **747** (2016) 012003 doi:10.1088/1742-6596/747/1/012003

# Behavior specificities of the plasma in the REB – polymeric anode interactions.

**S S Ananyev, S A Dan'ko, E D Kazakov, Yu G Kalinin, A A Kurilo and M G Strizhakov**[1]

National Research Center "Kurchatov Institute", Academician Kurchatov Square 1, Moscow 123182, Russia

E-mail: electric2003@inbox.ru

**Abstract**. To study Relativistic Electron Beam (REB) interaction with a set of polymeric targets we have made experiments for the investigation of the plasma dynamics in high-voltage diode of high-current generator. It is necessary for applied goals, such as correction of mathematical models of high power density deposition in the matter, for testing materials reaction under the powerful impulse of ionizing radiation, etc. An electron-optical streak camera (EOC) was used to measure the velocity of the visible light border of the plasma glowing. It was estimated to be from 10 to 35 km/s at the energy-release density in range of 200 – 800 J/cm². In a few of experiments, besides regular motion of contrast border in the direction from electrodes to the middle of the diode, the motion from the middle to the periphery of the high-voltage diode with velocity up to 500 km/s was observed. It was by an order of magnitude greater than regular one.

## 1. Introduction

Investigation of the various material properties is necessary for a wide number of scientific and technical purposes. These studies are especially important for material technologies, radiation resistance of the electronics. Experimental data in this field is also essential to obtain parameters and coefficients in equations of state for polymeric materials and for mathematical models verification [1], [2]. Moreover, the experiment reveals interesting details are to be introduced in models. For example, we demonstrated that materials with similar chemical composition may have very different response to the same powerful pulse impact [3], [4].

One of the most important parameters, which can be measured, is the velocity of ablated target material. The glow of expanding matter was registered in visible range by streak-camera.

The choice of material was based on importance in different applications and on existence of worked out equations of state for the polymeric materials such as epoxy. The second type of targets was made of polystyrene. View of a set of polymeric targets is presented in figure 1.

[1] To whom any correspondence should be addressed.

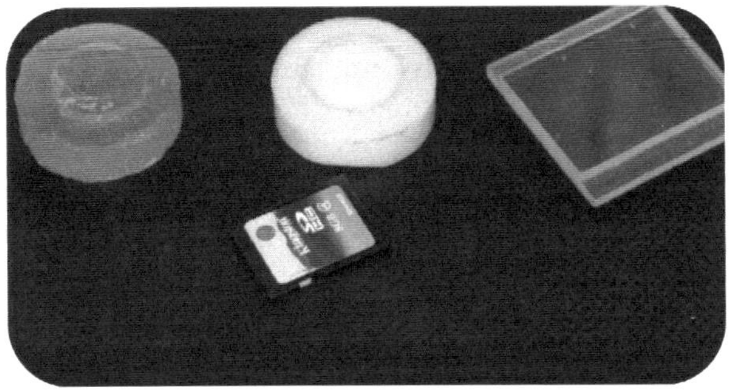

**Figure 1**. Types of targets. From left to right: amber-colored epoxy resin; milk-colored epoxy; polystyrene. SD-card is given for scaling.

## 2. Experimental set-up

This experiment was carried out on high-current generator "Calamary" with the following parameters: current was above 25 kA; output voltage up to 300 kV; maximal electron's energy was 350 keV (median energy of Calamary potential); pulse duration on FWHM was 100 ns [5].

The Calamary is completely simple in construction. Basic element of device is double forming line (DFL) with electrical length 70 ns. DFL's forming spark-gap switch is powered by the 20 stage Marx generator, containing 140 capacitors of 0.1 μF in each. There are a lot of electromagnetic interference (EMI) on the Calamary machine caused by air spark-gap switches using in the generator. Distilled water is used in DFL as a dielectric: its specific resistance is 2×106 Ω×cm.

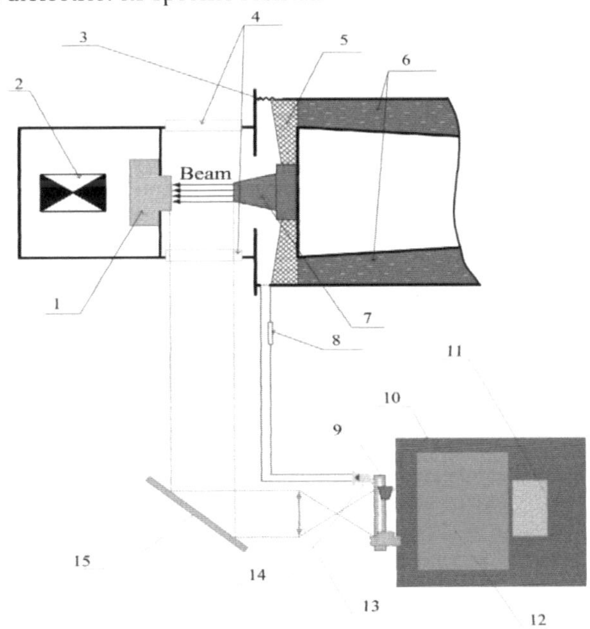


**Figure 2**. Diagnostics scheme: 1 – target (anode); 2 – pinhole camera; 3 – current instrument shunt; 4 – glass windows for observation; 5 – insulator; 6 – distilled water; 7 – brass cathode; 8 – resistor; 9 – entering slit of EOC; 10 – EMI shield; 11 – photo camera; 12 – EOC; 13 – lens; 14 – LED; 15 – mirror.

Diagnostics, which scheme is sketched in figure 2, consists of X-ray pinhole camera 2 (see in figure 3) and electron-optical streak camera 12 equipped by mirror 15, focusing lens 13 and placed in shielding box 10 against EMI. An image of EOC screen is captured by photo camera 11. Light emitting diode 14 (LED) – novelty in our registration system. LED is used for synchronization of generator current with EOC streak and powered from current shunt 3 via the resistor.

Energy-release area on the sample surface was measured by the double-film pinhole X-ray camera. The second X-ray film was situated behind the amplification luminophor inset [6]. The density of energy release was measured in range from 200 to 800 J/cm² [7].

**Figure 3.** Diode unit: 1 – viewing hole; 2 – target place; 3 – pinhole camera.

## 3. Experimental results

In figure 4 EOC images from experiments with different sample materials are seen. Images were given as typical, to present common form of illuminated area. In these images plasma from the anode arises a bit later then from the cathode, and its velocity is rather constant.

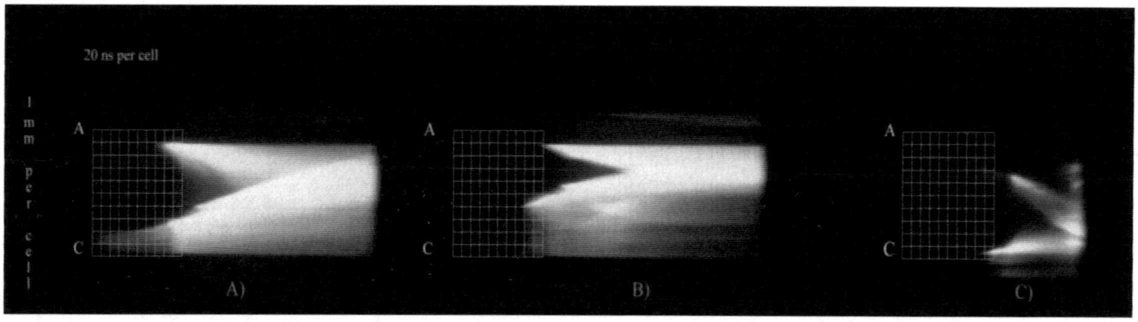


**Figure 4.** Examples of EOC records: A – two aluminum foils are as a target; B – two Al foils with smaller gap; C – epoxy target.

However, when the diode gap was of 7-8 mm the phenomena presented in figure 5(a, b) in experiment with polystyrene anode was observed. Weakly lighting matter moves regularly up to the reversal of diode voltage polarity. Then we observe the flashout near the diode axis and glow expansion from the axis to the electrodes with the velocity approximately of 500 km/s.

This effect is also observed on epoxy targets (in figure 5 (c, d)). We called it as "Cobra's teeth" – that's why EOC images looks like it. It should be noted that "Cobra's teeth" for epoxy targets is reproduced more stable than for polystyrene anodes.

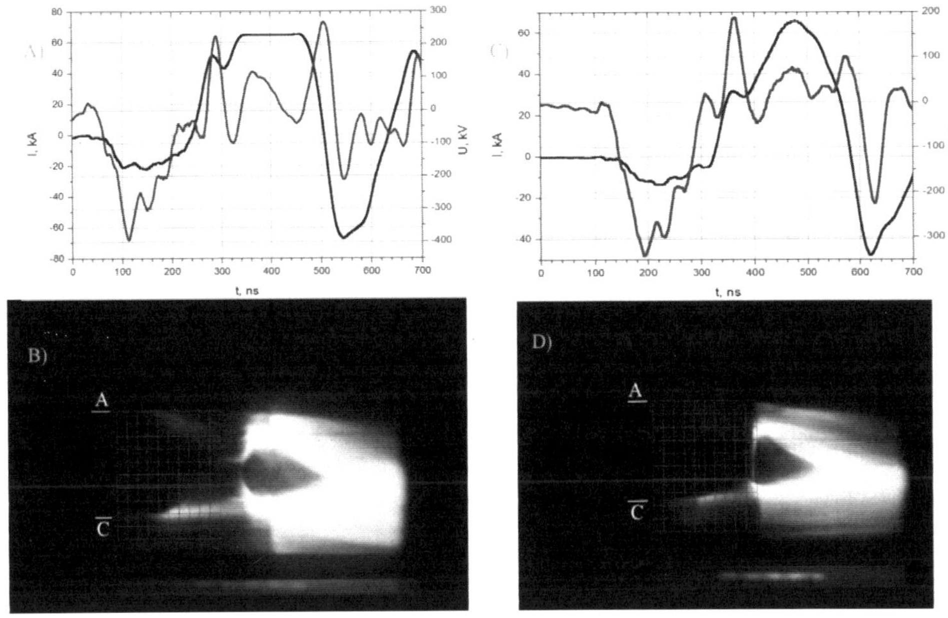


**Figure 5.** Experimental results: A and C – waveforms of the generator current (black graph) and voltage (blue graph) for polystyrene (A) and epoxy (C) targets; graphs were synchronized by the time with located below EOC images; B and D – "Cobras's teeth" on polystyrene and epoxy anodes respectively.

## 4. Conclusion

In several experiments we observed not only the typical motion of glowing matter from electrodes to center of high-voltage diode, but also we saw the motion from center to the periphery of the diode with velocity by an order greater than typical one – hundreds of kilometers per second. It occurs after the diode polarity reversal.

We suppose that the cause of this phenomenon consists in the non-radiating matter collection near the diode axis during the first half-period of the discharge and initiation of the discharge glow through well conductive plasma after the diode polarity reversal at high current magnitude. For checking hypothesis, we planned to enrich diagnostics by the laser shadow photography and by the second pinhole camera (perpendicularly to the first camera).

## Acknowledgements

This work was supported by grant of RFBR No. 15-02-03544-A.

## References

[1] Demidov B A *et al* 1997 *J. Tech. Phys. The Russian J. of Applied Phys.* **42** 11 pp 1264-1269.
[2] Akkerman A F *et al* 1986 *J. of Exp. and Theor. Phys.* **64** 1043
[3] Demidov B A *et al* 2008 *J. Surf. Inv. X-ray, Synch. Neutr. Tech.* **8** 55
[4] Demidov B A *et al* 2009 *J. Surf. Inv. X-ray, Synch. Neutr. Tech.* **9** 1
[5] Demidov B A *et al* 1979 *Atomnaja Energija* **46** pp 100-104
[6] Demidov B A, Efremov V P, Kalinin Yu G, Kazakov E D, Metelkin S Yu, Petrov V A and Potapenko A I 2015 *J. of Phys.: Conf. Series* **653** (2015) 012009
[7] Ananyev S S, Dan'ko S A, Kazakov E D, Kalinin Yu G, Kurilo A A, Mineeva T A, Potapenko A I and Strizhakov M G 2015 *Zvenigorod Int. Conf. on Plasma Phys. and Controlled Fusion* collection of abstracts p 187

Journal of Physics: Conference Series **747** (2016) 012004 doi:10.1088/1742-6596/747/1/012004

# Peculiarities of proton microscope development for the aims of proton radiography

**H Y Barminova**[1]
Plasma Physics Department, National Research Nuclear University MEPhI, 115409, Russian Federation, Moscow, Kashirskoe highway, 31

E-mail: barminova@mephi.ru

**Abstract**. Proton radiography facility based on a GeV-energy beam is expected to be useful to study the dynamic processes in matter under extreme conditions. The main factors affecting pRad facility spatial and temporal resolutions and general peculiarities of proton microscopy are described. The results of the microscope simulation for GeV-energy facility are presented.

## 1. Introduction

Proton radiography is one of numerous radiographic methods used for the research of matter. Well-known methods such as X-ray radiography, electron and positron radiography, neutron radiography and muon radiography as well as proton radiography have a specific range of their applicability. For particular physical tasks the applicability of a specific radiographic method is mainly determined by the task requirements as well as by the experimental conditions and the costs of the implementation of the method. The method of proton radiography is useful for dense and superdense objects study, both static and dynamic [1-3] because of the specific interaction of energetic proton with matter, namely because the curve of the energy release during the complex proton-matter interaction has a significant "plateau" range and the final peak (Bragg's peak). For instance, such a shape of the energy release distribution allows to carry out the experiments for cylindrical shock wave observation and investigation. The existence of the Bragg's peak allows to carry out some static experiments when the energy release should be in a localized area. This property is especially useful in medicine for proton therapy and tomography coupled.

Another benefit of proton radiography are the excellent proton transmission characteristics, for instance, the proton penetration through the substance having an optical thickness 350 g/cm is ~15%, in contrast to γ-ray transmission estimated as $\sim 10^{-7}$, so such a tool will be useful for extensive objects study.

The applicability of proton radiography is restricted due to resolution limitations. The spatial resolution limit appears to be due to a few reasons, such as the scattering of the particles inside the object investigated, the interaction of transmitted protons with the detector substance (the "device" function) and accelerator technology limitations. The respective impacts of these reasons are different and may be reduced by means of additional technologies and treatment. One of the methods to improve spatial resolution is to increase the beam energy.

Proton radiography with high energy beam requires full accelerator technique involvement instead of low-energy proton radiography devices, working at tens of MeV energies. The GeV-energy facility

[1] To whom any correspondence should be addressed.

should include telescopic ion-optical system or proton microscope [4] to obtain the best spatial resolution.

Spatial resolution is of great importance to study both the dynamic processes and static objects. The study of the dynamic processes requires additionally the appropriate temporal resolution. Temporal resolution doesn't depend on the properties of the microscope. The factors affecting the facility temporal resolution are temporal structure of the beam and temporal characteristics of the registration system (detector).

In this paper the properties of proton microscope and their influence on spatial resolution as well as the peculiarities of the microscope development in the GeV-energy range are described.

## 2. Peculiarities of proton microscope development

To develop an appropriate ion-optical system serving as proton microscope one should fulfill the following basic principles. Proton microscope should consist of the "forming" part and the "measuring" part, and the target chamber should be placed between them.

The main mission of the "forming" optics is to make an effective illumination of the target (or the object under investigation) with possible intensity profile uniformity. The mission of the "measuring" optics is to form an image corresponding to the object studied at the plane of registration (detector plane). Note that in proton imaging experiments typically the radiochromic film (one or more) is used as a detector, placed in the plane of the image. The detector placement with respect to the object position corresponds to the optical system magnification.

One of the possible proton microscope schemes is shown in Figure 1, where the magnetic quadrupole lenses in triplet configuration serve as a "forming" microscope part, and four magnetic quadrupole lenses in "Russian quadruplet" configuration serve as a "measuring" part.

To operate high energy proton beam the microscope scheme should consist of quadrupole magnetic lenses, providing the sufficient dynamic aperture to decrease the possibility of the image distortion.

The optical scheme of the proton microscope should be based on electromagnetic lenses. The microscope based on the electromagnetic lenses has flexible tuning and is easily operated in contrast with the scheme based on permanent magnets. The ion-optical system built on the basis of electromagnetic lenses has the highest long-term stability. Such a scheme doesn't require the mechanical displacement of the optical elements to tune the optics for the specific object, so it avoids the probable element shifts, turns and following optics mismatches, hence it doesn't require the additional time during the experiment or before it for the mismatch elimination. Such a scheme does not require the replacement of the lenses caused by the lens parameter degradation under the radiation exposure during the long period due to the easy tuning ability. In contrast, the scheme based on the permanent magnets requires lens re-magnetization after a rather short operation time. Recent investigations have demonstrated that strong-field permanent magnets of pRad microscope are significantly degraded in a short time [5]. It was found that under the exposure of equivalent dose of $10^{14}$ protons the decrease of the field level achieves up to 20%. Besides the level diminution the field shape became distorted with significant deviation from the linear character. In addition, the field inside the permanent quadrupole lenses became significantly asymmetric, that may be caused by the specific lens construction and the technological principle of its manufacturing. In the case of the proton microscope the lens field nonlinearity appeared not to be a dominant or even significant factor deteriorating the image. The dominant factor is the field diminution leading to the optical system focus displacement and requiring the perpetual tuning. This is a strong reason to avoid the permanent magnet application for the aims of radiography facility development.

Let us consider the possible optical solutions for GeV-energy proton microscopy based on electromagnetic lenses. The first scheme proposed has the "forming" part that consists of the quadrupole lenses in triplet configuration. The "measuring" part consists of four electromagnet quadrupoles in configuration called "Russian quadruplet" [6]. Such quadruplet configuration has reflection symmetry and allows the same particle excursion in both transverse coordinate planes. It is a preferable optical scheme for such a telescopic system. The layout of the scheme is shown in Figure 1.

The simulation of the scheme in the case of the beam with 1 GeV-energy was carried out with the help of G4Beamline [7]. In Figure 2 the particle trajectories are presented. Some scheme elements such as the microscope entrance window, the target chamber, the collimators as well as the virtual detectors are illustrated on probation, without their real dimensions, to simplify the illustration.

The scheme simulated has the "measuring" part with magnification coefficient equal approximately to "4".

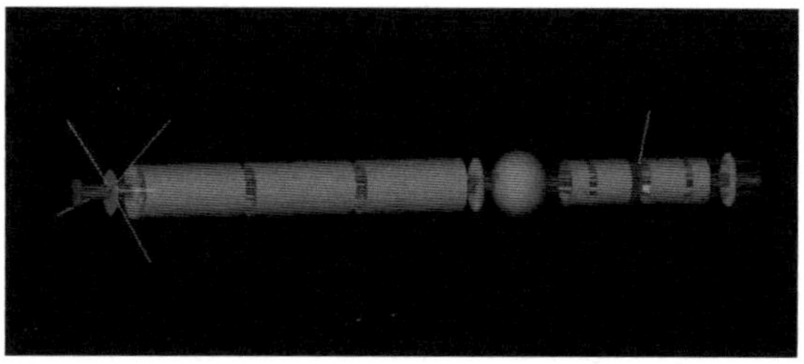

**Figure 1.** Microscope with the "forming" triplet and the "measuring" Russian quadruplet.

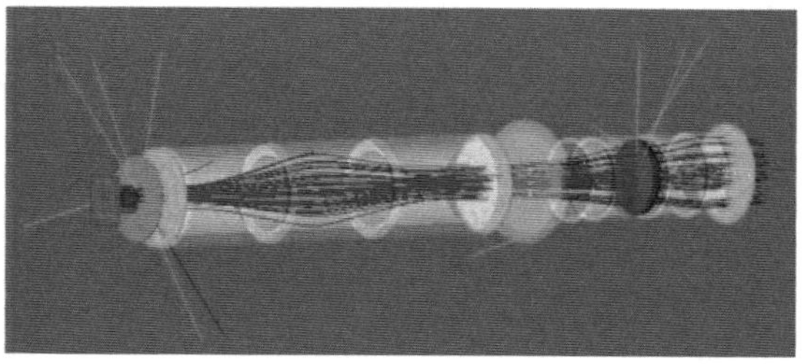

**Figure 2.** Particle trajectories for a beam with 1 GeV-energy simulated in the case of the scheme "triplet + Russian quadruplet".

The geometry of the beamline depends on the initial experiment conditions, such as the delivered beam energy, the accessible length of the beamline or the accessible experimental area dimension, the size of the chamber which must contain the investigated object. The microscope design depends on the costs too, which affect the choice of the accessible lenses.

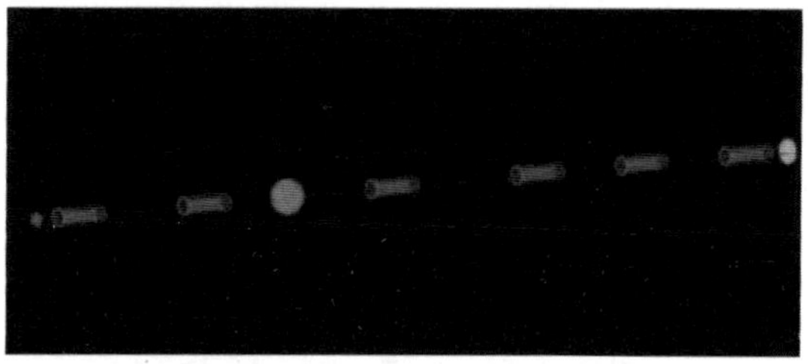

**Figure 3.** Microscope with the "forming" doublet and the "measuring" quadruplet.

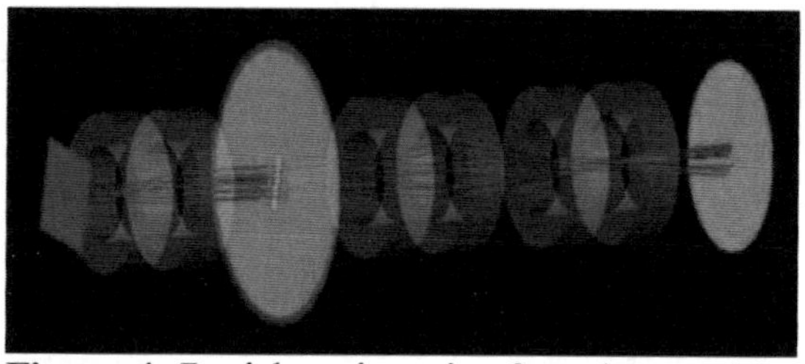

**Figure 4.** Particle trajectories for a beam with 9 GeV-energy simulated in the case of the scheme "doublet + quadruplet".

In Figure 3 another microscope scheme is presented. The scheme consists of the "forming" doublet and the "measuring" quadruplet [8]. Figure 4 corresponds to the case of 9 GeV energy beam simulation for the scheme illustrated in Figure 3. In this case the microscope with magnification coefficient "-1" is realized.

To simulate the beam dynamics in both the schemes the same initial beam parameters are considered. The following parameters are used during the calculation: the beam divergence at the entrance window of the microscope 5 mrad, the beam normalized rms emittances 32 $\pi$ mm mrad and 11 $\pi$ mm mrad for horizontal and vertical plane respectively, the beam momentum spread 1%, the beam intensity $10^7$ ppb, the beam size at the entrance of the microscope 3-10 mm.

Spatial resolution of the schemes is estimated with the help of the sharp-edge method and is used for the scheme optimization. The best spatial resolution in the cases considered above corresponds to second scheme.

Generally, the coefficient of the microscope magnification should be chosen proceeding from the basic experiment conditions, namely the accessible lens parameters and the accessible experimental area size. The increase of the coefficient improves the resolution in whole [4], but it is a subject of careful choice to satisfy all the requirements of the task. In any case, the results of preliminary calculations [8] have shown the possibility to obtain the scheme spatial resolution not worse than 10 μm in the case of the magnification coefficient "-1".

The significant factor which affects the effectiveness of the microscope scheme is the angular acceptance of the system, so the shortest systems have the best spatial resolution under the same conditions. The angular acceptance is an important factor for the "forming" part of the scheme too, because it affects the microscope field-of-view.

Finally, let us estimate qualitatively the influence of the general task parameters on the image blur. The image distortion caused by the proton multi-scattering in the object may be characterized by the expression $\sigma \sim l^{3/2}/p$, where σ is the dispersion, $l$ – the length of the object investigated in the direction of the beam inlet, $p$ – the beam momentum. The distortion caused by the chromatic aberrations of the microscope may be presented as $\sigma \sim (l/p^3)^{1/2}$, and the blur caused by the processes inside the detector may be characterized as $\sigma \sim l_{\det} l^{1/2}/p$, where $l_{det}$ – is the conditional "depth" of the detector. While fixing the distortions, caused by the Coulomb and nuclear scattering inside the object and in the detector, the external factors defining the image blurring are the beam energy and the size of the object studied. While the beam energy fixed, the best spatial resolution could be achieved for the microscope scheme with minimum chromatic length.

## 3. Conclusions

The applicability of proton radiography for the aims of the study of matter depends on the facility resolution. Spatial facility resolution is improving with increasing beam energy, so GeV-energy range is preferable to study dense objects and fast dynamic processes, and particularly the matter extreme states. Proton microscope is needed to operate the high energy beam for the object image creation. The microscope should be built with quadrupole electromagnets and consist of an "illuminating" or "forming" part and a "measuring" part. The microscope significantly contributes to the spatial resolution limitation. To avoid the image errors caused by the optical scheme the microscope should be carefully designed. Depending on the specific experimental conditions one should choose the ion-optical scheme with admissible highest angular acceptance, appropriate dynamic aperture, reflection symmetry of the "measuring" optics and minimum chromatic length.

**Acknowledgements**

This work was supported by the Competitiveness Program of NRNU MEPhI.

## References

[1] Borghesi M *et al* 2003 *Appl Phys Lett* **82** 1529
[2] Borghesi M *et al* 2010 *Laser Part Beams* **28** 277
[3] Kar S *et al* 2008 *High Energy Density Phys* **4** 26
[4] LANL 1997 *Tech. rep. LA-UR-974172*
[5] Danly C R, Merrill F E, Barlow D and Mariam F G 2014 *Rev. Sci. Instrum.* **85** 083305
[6] Dymnikov A D and Yavor S Y 1963 *Techn. Phys.* **33** 851
[7] Roberts T J, Beard K, Huang D, Ahmed Sh, Kaplan D M and Spentzouris L 2008 *Proc. EPAC'08 (Genoa, Italy)* ed I Andriann, O Brünning, Ch Petit-Jean-Genaz and P Pierini (EPS-AG) p 2776
[8] Barminova H Y 2014 *Proc. IEEE Int.Workshop on Beam Dynamics and Optimization BDO'14 (S.-Petersburg, Russia)* ed D A Ovsyannikov (IEEE Inc) art no 6890004

Journal of Physics: Conference Series **747** (2016) 012005 doi:10.1088/1742-6596/747/1/012005

# Plasma potential and geodesic acoustic mode evolution with Helium puffing in the ECRH regime on the T-10 tokamak

**V N Zenin[1,2,4], G F Subbotin[1,2], L A Klyuchnikov[1] and A V Melnikov[1,3]**
[1] National Research Centre 'Kurchatov Institute', Moscow 123182, Russia
[2] Moscow Institute of Physics and Technology, Dolgoprudny 141700, Russia
[3] National Research Nuclear University MEPhI, Moscow 115409, Russia

E-mail: vitalyzenin@mail.ru

**Abstract.** The evolution of the Geodesic Acoustic Mode (GAM) and mean plasma electric potential were examined in the regime with short (5 ms) Helium puffing into Electron Cyclotron Resonance heated discharge of the T-10 tokamak. It was shown that a Helium pulse leads to temporal perturbation of the plasma electron temperature and density and concomitant evolution of the mean potential, happening in the diffusive time-scale ~ 30 ms. Afterwards, the potential restores to the new stationary level with the same time-scale. On top of that GAM amplitude reduces sharply (within 2-5 ms) and GAM frequency also decreases within 30 ms after Helium puffing. Afterwards GAM amplitude and frequency relax to a new stationary level within about 50-70 ms. The evolution of electron density, electron and ion temperatures, total stored energy and plasma density turbulence is discussed in order to clarify their links with potential and GAM evolution.

## 1. Introduction
Physical mechanisms of the turbulent energy transport present an open area in fusion research. One of the important phenomena in this field are Zonal Flows and their high-frequency counterpart GAMs. They have been intensively studied as a possible mechanism of the plasma turbulence self-regulation [1]. They were first introduced in [2] within the ideal Magneto Hydro Dynamic (MHD) model. GAM is a perturbation with $m = n = 0$ for electrostatic potential, which is coupled to pressure perturbation $m = 1$, $n = 0$. GAM frequency is proportional to the square root of temperature [2]. The main features of GAM on T-10 were described in [3]. The recent observations show that periphery plasma parameters play important role in the GAM behavior [4]. Besides, the mean electric field (or potential) plays a role in plasma confinement and turbulent transport [5]. This paper presents the evolution of the GAM and mean electric potential induced by the perturbation of the periphery due to the Helium puffing into Electron Cyclotron Resonance Heated (ECRH) plasmas.

## 2. Experimental setup
T-10 is a circular tokamak (major radius $R = 1.5$ m, minor radius $a = 0.3$ m, $B < 2.5$ T) with a rail carbon limiter. Plasma potential oscillations in the hot plasma were studied with Heavy Ion Beam Probe (HIBP), a unique method of the direct measurement of plasma electric potential. The basic principles of HIBP were described in [6], while the T-10 HIBP with $Tl^+$ probing beam was described in [5, 7].

Discharge scenario and the evolution of the main plasma parameters are presented in figure 1.

[4] To whom any correspondence should be addressed.

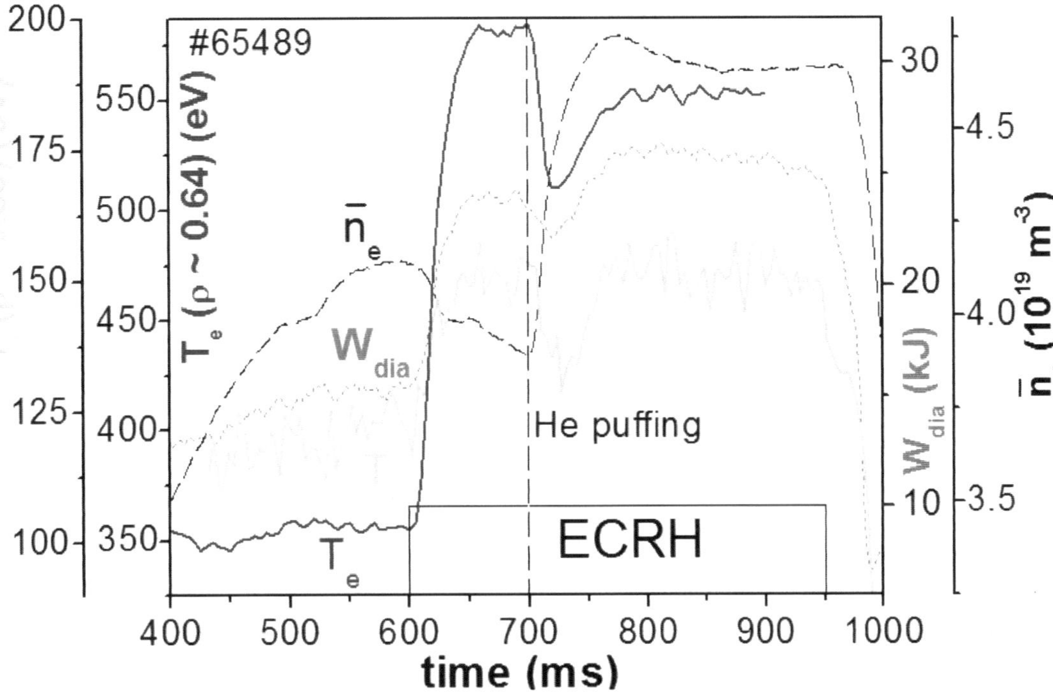


**Figure 1.** Scenario of the discharge with He puffing and the evolution of the main plasma parameters. Here ($\overline{n}_e$ - central chord line-averaged plasma density, $T_e$ (r/a = 0.64) electron temperature, and $T_i$ (r/a = 0.80) ion temperature, $W_{dia}$ – total stored energy.

The high density ($\overline{n}_e = (3.5\text{-}4.5)\times10^{19}$ m$^{-3}$) ohmic plasma is auxiliary heated by Electron Cyclotron Resonance Heating (ECRH) from 600 to 950 ms. A short pulse (~5 ms) of He puffing is performed at 700 ms. After puffing the central chord line-averaged plasma density, measured by microwave interferometer stops the decay due to the pump-out and starts to raise. On top of that both $T_e$, retrieved from the database of the Electron Cyclotron Emission, and $T_i$, measured by CHERS [8] decreases for about 25 ms, then relax to a new steady state level, which appear to be somewhat lower than the initial one before He pulse. At this new steady state the electron density rises up by about 15% with respect to the initial level. After He pulse the energy content ($W_{dia}$, as measured by diamagnetic loop) decreases for about 25 ms due to the temperature decrease then reaches a new steady state. New level of energy content is higher by ~ 10% than the level before puffing due to the density increase.

## 3. Experimental results

### *3.1. Evolution of the GAM properties due to the Helium pulse.*

The fine details of the main plasma parameters and GAM evolution, as measured in the plasma periphery r = 19 - 30 cm are presented in figure 2. Fourier spectrogram of potential oscillations, measured by HIBP is shown in figure 2 (a). GAM is clearly seen as a dominating frequency peak, evolving in time with the ECRH and Helium pulse. GAM amplitude, which is calculated over the wide frequency region (5-25 kHz), with a pronounced peak indicated by dark grey color in figure 2 (a), is shown in figure 2 (b). Right after the Helium pulse GAM amplitude abruptly falls down from 25 V almost to the noise level during 2-5 ms. GAM amplitude reaches minimum values and then evolves slowly to the new steady-state with a characteristic time of about 70 ms, which is close to the density and temperature relaxation time. The time trace of the GAM frequency or the mean frequency of the pronounced GAM peak is shown in figure 2 (c). The GAM frequency increases from its ohmic level of ~ 12 kHz to its ECRH level ~ 17 kHz according to the $T_e$ raise due to the ECRH. Then it decreases from ~ 17 kHz down to ~12-13 kHz during ~ 25 ms after the Helium pulse, which is in a sharp contrast to the dynamics of the GAM amplitude.

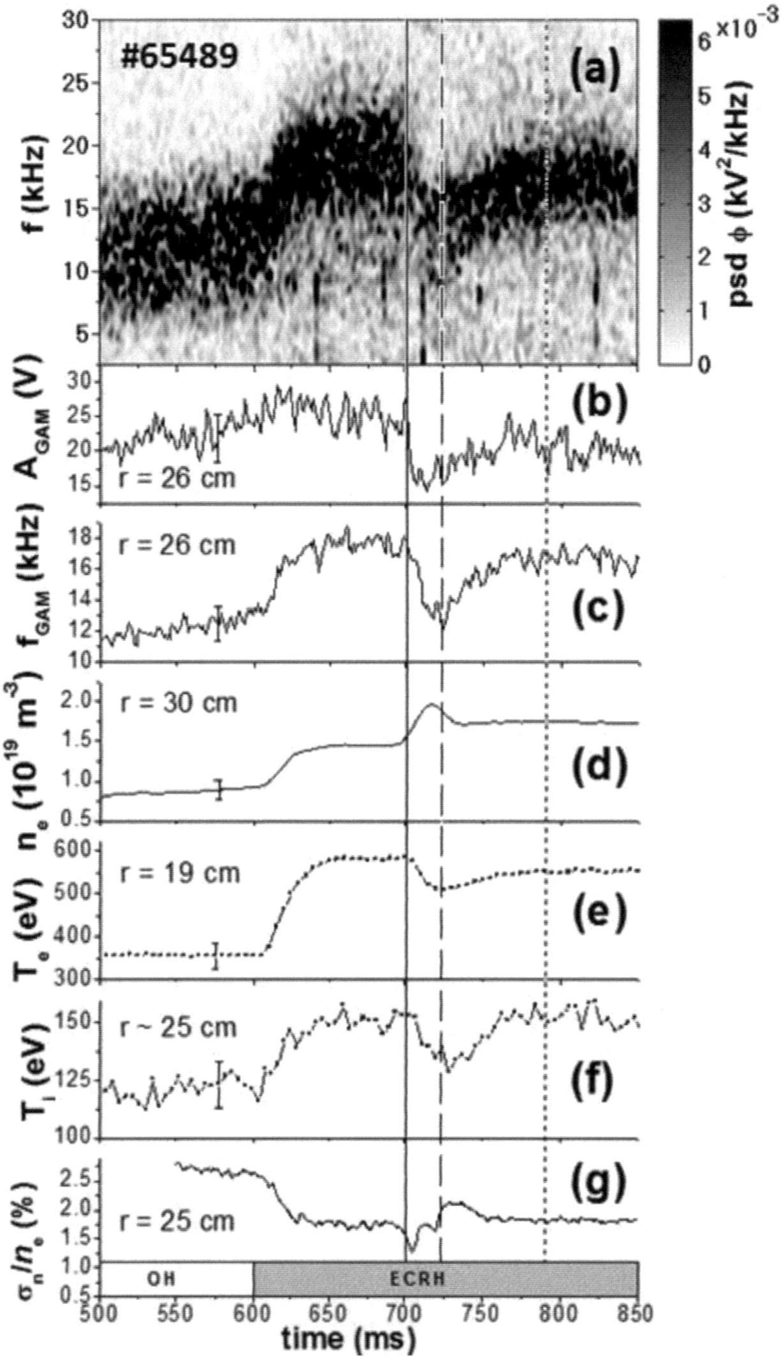


**Figure 2.** (a) — power spectrogram (power spectral density, psd) of potential oscillations, measured by HIBP at r = 26 cm, (b) — GAM amplitude, (c) — GAM frequency, (d) — electron density, measured by microwave interferometry at the edge chord, (e) — electron temperature, retrieved from the database of the Electron Cyclotron Emission, (f) — ion temperature, measured by CHERS, (g) — density turbulence level, measured by Correlation Reflectometry. Both (f) and (g) were averaged over the series of identical shots (##65488 — 65492). Solid vertical line indicates Helium puffing, dash line — minimum of the GAM frequency, coinciding with a minimum of $T_e$, dot line — the steady state for GAM frequency.

After reaching the minimum value GAM frequency increases to the new steady-state. It relaxes to this new steady state during about 50 ms, mainly in line with evolution of $T_e$ and $T_i$. Other plasma parameters also evolve after the impurity pulse. Figure 2 (d) shows the time-trace of the local (edge) density, measured by the outermost chord of the microwave interferometer. Figure 2 (e) presents electron temperature, derived from the T-10 database measured by Electron Cyclotron Emission according to [9], while figure 2 (f) is for the ion temperatures measured by Charge Exchange

Recombination Spectroscopy [8]. Figure 2 (g) shows the evolution of the broadband turbulence level obtained by Correlation Reflectometry.

### *3.2. Evolution of the mean potential due to the Helium pulse*

In the high-density regime under study the mean potential level has negative absolute values all over the radius [5, 10]. The fine details of the global plasma parameters and mean potential evolution are shown in figure 3. The time evolution of the line-averaged density, measured by the central chord of microwave interferometer is shown in figure 3 (a), while the energy content, measured by diamagnetic signal is shown in figure 3 (b). After ECRH switch-on the absolute value of plasma potential decreases in consistency with earlier observations [10]. Contrary, after the Helium pulse its absolute value increases. This increase lasts much longer than Helium pulse (5 ms) and takes about diffusion time (~ 25-30 ms).

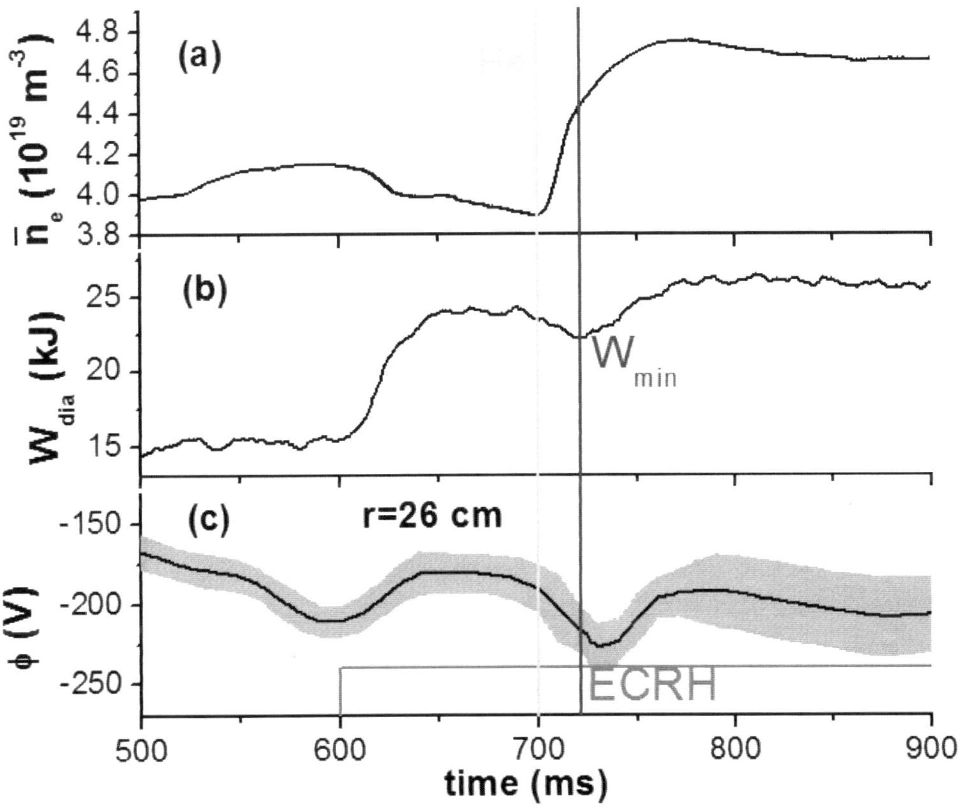


**Figure 3.** Evolution of electron density (a), diamagnetic energy (b) and plasma potential (c), grey shade represents experimental errors.

Afterwards mean potential starts to recover, the absolute value of its average level falls during 30-35 ms. Then the potential reaches a new quasi steady-state level for t=800-900 ms. This new level of absolute potential appears to be slightly higher than the initial stationary level before Helium puffing t=630-670 ms, which is consistent with the general link between mean potential and stored plasma energy [5]. Note that the minimum of energy content coincides in time with change in sign of the density time derivative. The time instant of potential minimum at the periphery is delays with respect to the minimum of total stored plasma energy by about 10 ms, which can be caused by the transient processes in the bulk plasma particle transport or Helium distribution over the plasma column.

## 4. Summary

Helium puffing dramatically changes the GAM behavior: GAM frequency decrease temporarily after the He pulse. GAM frequency evolution is consistent with the evolution of $T_e$ and $T_i$ at the plasma periphery. Unlike GAM frequency GAM amplitude reacts on Helium pulse abruptly (2-5 ms). Such sharp decrease can be probably linked with global character of GAM [11]. The further increase of GAM amplitude is consistent with collisional damping, caused by edge density evolution. On top of that Helium pulse affects the mean plasma potential. Potential evolution mainly follows the density one and reveals its links with energy content.

**Acknowledgements**. This work was carried out by Russian Science Foundation, Project 14-22-00193. The authors are grateful to Dr. D.A. Shelukhin for providing correlation reflectometry data.

## References

[1] Fujisawa A *et all.* 2007 *Nucl. Fusion* **47** S718–26
[2] Winsor N, Johnson J L and Dawson J M 1968 *Phys. Fluids* **11** 2448
[3] Melnikov A V *et all.* 2006 *Plasma Phys. Control. Fusion* **48** S87–110
[4] Melnikov A V *et all.* 2015 *Nucl. Fusion* **55** 063001
[5] Melnikov A V *et all.* 2013 *Nucl. Fusion* **53** 093019
[6] Dnestrovskij Y N, Melnikov A V, Krupnik L I and Nedzelskij I S 1994 *IEEE Trans. Plasma Sci.* **22** 310–31
[7] Melnikov A V *et all.* 1995 *Rev. Sci. Instrum.***66** 317–19
[8] Krupin V A, Klyuchnikov L A, Korobov K V, Nemets A R, Nurgaliev M R, Gorbunov A V, Naumenko N N, Troynov V I, Tugarinov S N and Fomin F V 2014 *Probl. At. Sci. Technol. Ser. Thermonucl. Fusion* **37** 60–70
[9] Kirneva N A, Martynov A A, Pavlov Y D, Trukhin V M and Trukhina E V 2014 *Probl. At. Sci. Technol. Ser. Thermonucl. Fusion* **37** 56–61
[10] Melnikov A V *et all.* 2011 *Nucl. Fusion* **51** 083043
[11] Melnikov A V, Eliseev L G, Lysenko S E, Perfilov S V, Shurygin R V, Krupnik L I, Kozachek A S and Smolyakov A I 2015 *JETP Lett.***100** 555–60

Journal of Physics: Conference Series **747** (2016) 012006 doi:10.1088/1742-6596/747/1/012006

# Magnetic discharge accelerating diode for the gas-filled pulsed neutron generators based on inertial confinement of ions

**K I Kozlovskij[1], A E Shikanov, E D Vovchenko, V L Shatokhin, A A Isaev and A S Martynenko**

National Research Nuclear University MEPhI (Moscow Engineering Physics Institute), Kashirskoe shosse 31, 115409, Moscow, Russian Federation

E-mail: cozlowskij2013@yandex.ru

**Abstract**. The paper deals with magnetic discharge diode module with inertial electrostatic ions confinement for the gas-filled pulsed neutron generators. The basis of the design is geometry with the central hollow cathode surrounded by the outer cylindrical anode and electrodes made of permanent magnets. The induction magnitude about 0.1−0.4 T in the central region of the discharge volume ensures the confinement of electrons in the space of hollow (virtual) cathode and leads to space charge compensation of accelerated ions in the centre. The research results of different excitation modes in pulsed high-voltage discharge are presented. The stable form of the volume discharge preserveing the shape and amplitude of the pulse current in the pressure range of $10^{-3}$-$10^{-1}$ Torr and at the accelerating voltage up to 200 kV was observed.

Nowadays compact pulsed neutron generators for the study and monitoring of oil and gas deposits are applied. The basic element of such pulsed neutron generators is usually a neutron accelerating tube, which characteristics largely determine the diagnostic capabilities of geophysical instruments. However, such neutron tubes have some disadvantages which are caused by a number of physical processes during their work and this reduces the competitiveness of neutron generators based on them. In particularly pulsed neutron generators based on accelerating tubes of direct action have limited resource (determined by the degradation of solid-state neutron targets), not high efficiency of energy transfer to accelerated nuclide of hydrogen and problems with radiation safety in operation and utilization [1]. Therefore, along with improvement of traditional neutron accelerating tube, there is an active search and development of new types of accelerating systems.

Such new developments include a gas-filled discharge-plasma diode with oscillation mode of ions between the electrodes, and interaction of ions flows with the structural elements is minimized. Abroad the similar systems are referred to Inertial Electrostatic Confinement (IEC) devices, as their work is based on the inertial-electrostatic confinement of deuterons [2]. There is no traditional solid target in IEC diodes filled with gaseous deuterium. The interaction of accelerated hydrogen nuclide is performed with plasma or with an opposite flow of nuclide (plasma target). In this case, the working resource of the pulsed neutron generator can be increased. The efficiency of neutrons generation also is increased using the special geometry of electrodes with high transmittance of ions in the center of discharge gap [3].

In this paper, we consider a new improved design of magneto discharge module of IEC diode and the results of different excitation modes in pulsed high-voltage discharge. The basis of this design, as

[1] To whom any correspondence should be addressed.

in the previous version [3], is the electrode system with a hollow central cathode surrounded by the outer cylindrical anode. However, to improve the efficiency of ion generation in the electrode gap, the new version of diode module for acceleration of charged particles in crossed ExH fields is organized. This leads to additional electron drift and a substantial lengthening of their trajectories. Moreover, the magnetic field confines the electrons in space of hollow cathode (virtual cathode) and provides the space charge compensation of accelerated ions in the center of the system. This is important, because it allows to increase the ionic current in comparison with the current limited by the "law 3/2".

The scheme of the new design of diode module is show in the figure 1. A pulse of accelerating voltage comes into a vacuum chamber through a high-voltage dielectric input. Insulator is designed for operating voltage up to 200 kV. The anode is made in the form of a hollow cylinder (inner diameter is 6 cm, the height is 4 cm). In addition, it is made from permanent magnet, magnetized in the direction of the axis of the diode. The cathode consists of two coaxial magnetic rings (external diameter 3 cm), magnetized along the axis of the diode. The magnitude of magnetic field in the central region of discharge volume is about 0.4 T.

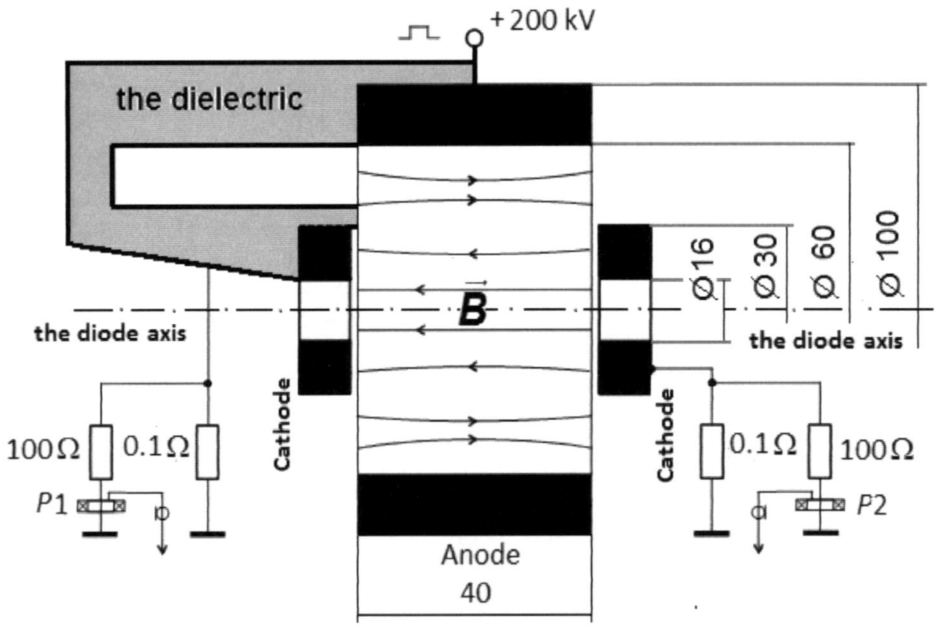


**Figure 1.** The scheme of a gas-filled diode module with magnet electrodes
($P_1$ and $P_2$ – Rogowski coil

Diode module is located in a dielectric vacuum chamber, pumped up to $10^{-4}$ Torr. Vacuum system has the ability to control and vary the pressure and composition of working gas. In our experiments, the pressure is varied from $10^{-1}$ to $10^{-3}$ Torr. As a source of accelerating high voltage, unlike previous developments of IEC-diode, pulsed Marx generator was used. It provides the amplitude of the accelerating voltage up to 350 kV with duration less than 1 μs in the idle current and the pulse current up to 1 kA in the short circuit mode. To control the discharge current Rogowski coils (*P1* and *P2*) are applied. In addition, probe for measuring the pulse voltage are included in primary diagnostic tools.

Experiments have shown that electrical breakdown of the insulator between the anode and the cathode in a vacuum limits the maximum amplitude of the accelerating voltage. Its value is equal to 200 kV for our geometry of electrodes and the pressure range $10^{-1}$–$10^{-3}$ Torr. For comparison, it should be noted that for gas-filled neutron tubes the maximum of accelerating voltage is usually about

100–120 kV. Thus, a new diode module with the magnetic electrodes provides the increase of the accelerating voltage in comparison with the previous variant without magnets.

According to existing concepts, the reduction of deuterium pressure in the discharge chamber reduces the ionization probability and falls the current. As a result it leads to a decreasing number of interactions between deuterons. This process defines the lower limit of the working pressure. However, on the other hand, a rapid increase of the current leads to a breakdown at high pressures. It's also possible to shunt he main discharge current due to the parasitic currents. These processes affect the upper limit of the operating pressure.

A stable discharge for the pressure range $10^{-2}$ to $10^{-1}$ Torr was observed without magnetic field and only in the presence of pre-ionization [3]. In this case, the amplitude of the accelerating voltage increased from 60 to 100 kV and the range of pressure was markedly decreased. In the presence of magnetic field, as we can see, the current pulse shape and current amplitude did not change in a wider pressure range from $5{\cdot}10^{-4}$ to $10^{-1}$ Torr, because in the region of smaller pressures the gas ionization takes place effectively. In addition to this, in the magnetic field it is expected to increase the neutron yield due to the better gas ionization and the increasing number of interactions acts between deuterons and the increasing number of oscillations at use lower pressure.

In order to optimize the magnetic system we made computer simulation of the magnetic field distribution in the electrode gap. The results are shown in figure 2 (a), (b).

(a) (b)

**Figure 2.** The distributions of the magnetic field in the electrode gap.
In the first case (a) the magnetization direction of the anode and the cathode coincides and in the second one (b) it has an opposite direction.

One can see from figures, in the case of the first configuration (a) the magnetic lines of the anode and the cathode are partially closed against each other. In the second configuration (b) it is expressed much less. Consequently, the second configuration is more optimal, because of better inhibiting the electrons from their drift to anode. However, neither configuration is capable to completely suppress the emission electron current. It is planned to consider various screens systems and the use of magnetic coils instead of permanent magnets.

This work is supported by the Ministry of education and science of Russian Federation according to the agreement № 14.575.21.0049 (RGMEFI 57514X0049).

References

[1] Bogdanovich B Yu, Nesterovich A V, Shikanov A E, Vorogushin M F and Svistunov Yu A 2009 *Remote Radiation Control with Linear Accelerators* vol. 1.: *Linear Accelerators for Bremsstrahlung and Neutron Generation* (Moscow: Energoatomizdat) p 272

[2] Miley G H and Murali S K 2014 *Inertial Electrostatic Confinement (IEC) Fusion. Fundamentals and Applications* (New York: Springer) p 400

[3] Shikanov A E, Vovchenko E D, Kozlovskii K I and Shatokhin V L 2015 *Journal of technical physics* **1** (85) 47-51

# Project of the borehole neutron generator for the direct determination of oxygen and carbon by activation method

**B Yu Bogdanovich[1], E D Vovchenko[1], A V Iliinskiy[2], A A Isaev[1], K I Kozlovskiy[1,3], A V Nesterovich[1], V A Senyukov[1] and A E Shikanov[1]**

[1] National Research Nuclear University MEPhI (Moscow Engineering Physics Institute), Kashirskoe highway 31, 115409, Moscow, Russia

[2] Institute of geophysical and radiation technologies of the International Academy of Sciences of Higher School, Atarbekova street 4/B, 107076, Moscow, Russia

E-mail: cozlowskij2013@yandex.ru

**Abstract**. The paper deals with application features of borehole neutron generator (BNG) based on the vacuum accelerating tube (AT) with laser-plasma ion source for determination of oxygen isotope $^{16}O$ and carbon isotope $^{12}C$ by direct activation. The project of pulsed BNG for realization of an activation method in the conditions of natural presence of productive hydrocarbons is offered. The diode system with radial acceleration, magnetic electron insulation and laser-plasma source of deuterons at the anode in a sealed-off vacuum accelerating tube is applied. The permanent NdFeB magnet with induction about 0.5 T for produce the insulating magnetic field in the diode gap is proposed. In the experiments on the model of BNG with the accelerating voltage source (≈350 kV), performed by the scheme of Arkadiev-Marx generator, the output of (d, d) neutrons was ~$10^7$ pulse$^{-1}$.

Nowadays in nuclear-geophysical research of oil borehole great attention is paid to direct determination of carbon and oxygen by using of high-frequency neutron generator (frequency is about 10-20 kHz) with multi-channel gamma-ray spectrometer for inelastic scattering and radiation capture (C/O - logging) [1,2]. Such method has the advantage over classical neutron method of indirect identification of productive fluids based on determination of chlorine deficiency in which result efficiency depends on the salinity degree of water accompanying the oil in the geological layer. However, from the viewpoint of mathematical processing and interpretation of data such method is very difficult to develop for realization by specialists.

The report proposes a project of borehole neutron generator (BNG) on the base of a sealed accelerating tube (AT). With its help you can successfully implement an alternative method of the direct determination of the oxygen isotope $^{16}O$ and $^{12}C$ carbon isotope in the oil deposits by the use of activating effect as a result of such nuclear reactions as $^{16}O(n, p)^{16}N$ and $^{12}C(n, p)^{12}B$. These reactions are endothermic with following energies: $Q_1 = -9.63$ MeV and $Q_2 = -12.59$ MeV, respectively [3]. Threshold energies of neutrons $T_1$ and $T_2$ thus can be determined using equation, which follows from conservation laws of energy and momentum:

$$T_1 = (17/16)\,|Q_1| = 10.2 \text{ MeV}, \quad T_2 = (13/12)\,|Q_2| = 13.6 \text{ MeV}.$$

[3] To whom any correspondence should be addressed.

The dependence of the nuclear cross sections of these reactions $\sigma_{1,2}$ (in barns) from the neutron's kinetic energy $T_n$ are easy to calculate, using a root approximation of these dependences near the threshold [4], as well as table data [3]. These cross sections are as follows:

$$\sigma_1 \cong 0.0215\,(T_n - 10.23)^{1/2}, \quad \sigma_2 \cong 0.032\,(T_n - 13.64)^{1/2}.$$

As we can see from these formulas for the neutron energy in range 14 MeV generated by the nuclear reaction T(d, n)$^4$He cross section is increasing with rising of the kinetic energy of deuterons $T_d$. In the case of a direct neutron flight, when its motion direction coincides with the motion direction of deuteron, the following formula for cross section of activation can be obtained, if we use the known dependence $T_n = f(T_d)$ [5]

$$\sigma_1 \cong 0.0215\,\{[(14.08 + 0.62T_d)^{1/2} + (0.02T_d)^{1/2}]^2 - 10.23\}^{1/2},$$
$$\sigma_2 \cong 0.032\{[(14.08 + 0.62T_d)^{1/2} + (0.02T_d)^{1/2}]^2 - 13.64\}^{1/2}.$$

Evaluations made with using these expressions showed that the activation effect would be sufficient for the implementation of these logging methods if $T_d \geq 0{,}3$MeV and neutron output is more than $10^{10}$ neutrons per pulse. In addition, the deuterons acceleration should be in the direction perpendicular to the wellbore.

Nowadays the experimental samples of AT with vacuum-arc source of deuterons are developed [6]. They can generate in a full solid angle up to $10^{10}$ neutrons per pulse. The maximum value of $T_d$ in such AT does not exceed 0.15 MeV and deuterons are accelerated along the tube and in parallel the axis of wellbore. The $T_d$ increase due to increasing of the amplitude of the acceleration voltage leads to the insurmountable problems of providing electrical isolation in such neutron devices. In addition to this, the experience of such AT operation showed the significant instability of pulsed neutron yield associated with the deuteron source peculiarities. This instability can reach 50%. In turn, this requires add in the BNG a monitor of the pulse neutron flux, which significantly complicates the logging methods.

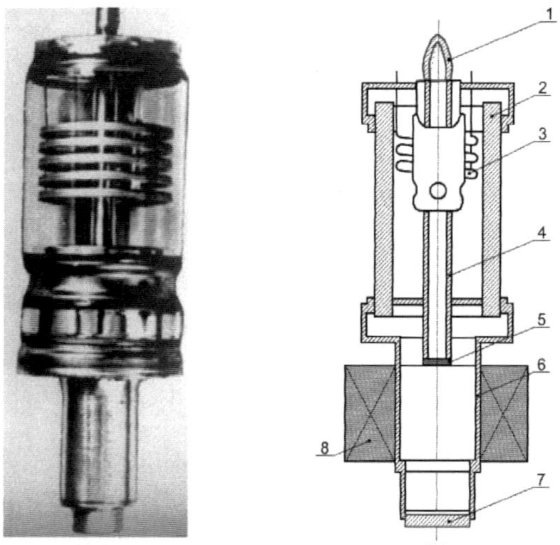


(a) (b)

**Figure 1.** The accelerating neutron tube based on the laser-plasma source of neutrons: visual appearance (a); schematic cross section (b)
1 – exhaust tube for AT separation, 2 – insulator, 3 – getter, 4 – anode electrode, 5 – plasma-generating target, 6 – cathode metal-tritium target, forming neutrons, 7 – optical window, 8 – magnetic system.

The above disadvantages can be eliminated and the neutron yield is saved in AT diode system with radial acceleration [7], magnetic isolation of the accelerating gap [8] and laser-plasma source of deuterons. Figure 1 shows a visual appearance and a schematic cross section of such AT. The insulating magnetic field is stationary and is formed by solenoids or annular permanent magnets with the induction vector directed along the axis of AT symmetry for the suppression of electronic conductivity in the diode gap. The magnitude of induction of 0.5 T is the estimation for the critical magnetic field in which the electrons are not to reach the anode and for the amplitude of accelerating voltage about 150 kV and the cathode radius about 3 cm. Modern technologies based on the compound of NdFeB allow creating the permanent magnets with induction in the end face of up to 1 T. The schematic cross section of the existing laboratory model of BNG based on AT with deuteron laser source is shown in figure 2.

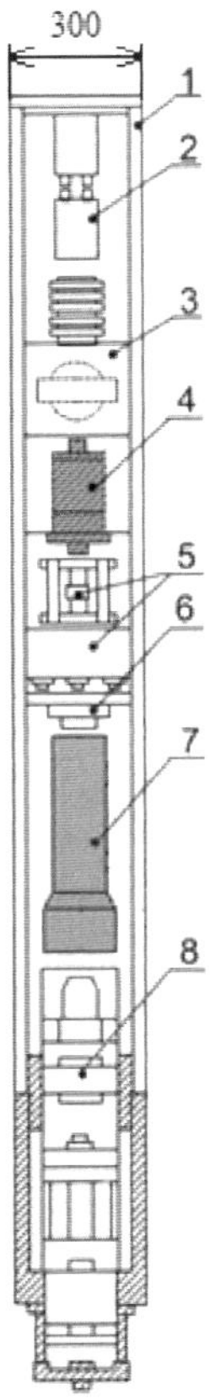


**Figure 2.** The block diagram of the borehole neutron generator

1 – the body, 2 – the capacitive energy storage, 3 – high voltage unit on the basis of the pulse HV transformer or Arkadiev-Marx generator, 4 – the neutron tube, 5 – system to scan the laser beam, 6 – laser-gap switch, 7 – pulse YAG laser, 8 – the module of capacitive energy storage

Laser beam is focused on the plasma-generating target of AT. This target is made in the tablet form from the metal which can dissolves hydrogen and can absorb the deuterium. Plasma containing deuterons falls into the area surrounded by the cylindrical cathode, which is the neutron-producing target. During this time, a short pulse of high voltage is supplied to electrodes of diode synchronously with the laser pulse.

Under the influence of the electric field, deuterons were extracted from plasma and were accelerated perpendicularly the target surface. In the result of bombing by deuterons and the nuclear reaction T*(d, n)*$^4$He the neutron flux was produced. In this case, the maximum energy of neutrons corresponds to the emission perpendicularly to the axis of AT symmetry. Figure 3 shows the experimental dependence of the neutron flux $F$ on the amplitude of accelerating voltage $U$ for the different frequency of the working pulses in AT. This result was received on the experimental model of BNG with the power system based on a high-voltage transformer.

In further experiments, in order to increase the discharge current and accelerating voltage the Arkadiev-Marx generator was suggested. This generator is designed for up to 400 kV amplitude and generates pulses of nanosecond duration. It consists of 30 stages, each with two capacitor type K5-4 connected in parallel with the total capacitor about of the $C_0 = 2 \times 4700 = 9400$ pF, which stores maximum energy about 1.5 J during a charging voltage of 15–18 kV. Respectively the equivalent capacitance is $C_0/30 = 310$ pF. The generator's design feature is the use of a spiral geometry for the serial connection of capacitors and the spark switch. Such geometry ensures a reliable electric strength between input and output of Marx generator without increase of the overall device size.

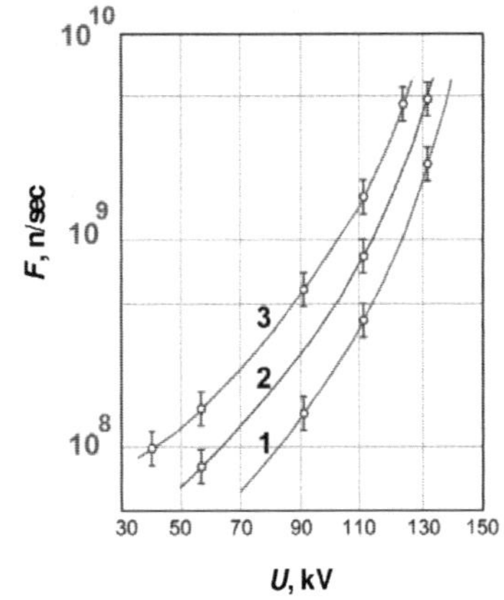


**Figure 3.** The dependence of the neutron flux *F* on the amplitude of accelerating voltage *U*(1-12.5 Hz; 2 – 25 Hz; 3 – 50 Hz)

Furthermore, by forming a short high voltage pulse in conjunction with the magnetic isolation of electron we believe that the requirements for high-voltage isolation system of diode can be significantly reduced. In experiments on BNG model we used laser with the energy about 50 mJ [9] and forming the single voltage pulse with amplitude up to 350 kV on the accelerating gap of AT. For these experimental conditions, the expected yield of neutrons (*d, d*) will be ~$10^7$ 1/pulse.

In conclusion, should be noted that the increase of laser energy and the decrease of dimensions of laser sources may be achieved in a new series of YAG: Nd lasers and CO2 lasers. These divices can be able to generate pulses with energies about one joule at frequencies up to 100 Hz. New perspective laser and its combination with magnetic isolation of electrons in the diode gap, as well as new designs of the compact high-power pulsed sources of accelerating voltage clearly show that the creation of an effective BNG is really. In turn, this will allow to use the simple technique of neutron activation with the direct determination of carbon and oxygen.

This work was performed under Agreement № 15-19-00151 between the Russian Research Foundation, National Research Nuclear University MEPhI, and the project manager A.V. Nesterovich on the financing of fundamental and pilot research studies.

**References**

[1] Tropin A N and Enikeeva F H 2002 *Scientific and Practical Conference "Nuclear Geophysics 2002"* (Tver) p 15-16
[2] Amurskiy A G, Bogolyubov E P and Titov I A 2003 *Proceedings of the conference "The portable neutron generators and based on them technologies"* (Moscow: VNIIA) p 17
[3] Maslov I A and Luknickiy V A 1971 *Reference book with neutron activation analysis* (Leningrad: Nauka) p 312
[4] Landau L D and Lifshits E M 1963 *Quantum mechanics, non-relativistic theory* (Moscow: GIFL) p 702
[5] Bekurts K and Virtc K 1968 *Neutron physics* (Moscow: Atomizdat) p 456
[6] Bobilev V T, Bogolyubov E P, Ryzhkov V I, Kurdyumov I G and Kuznetsov Yu P 2003 *Proceedings of the conference "Portable neutron generators and based on them technologies"* (Moscow: VNIIA) p 19
[7] Bespalov D F, Mints A Z, Pleshakova R P and Shikanov A E 1980 *Pulsed neutron generator.* USSR Author's Certificate № 457 406 (priority: 25.12.1972)
[8] Bespalov D F, Kozlovskiy K I, Tsybin A S and Shikanov A E 1980 *Pulsed neutron tube* USSR Author's Certificate № 766 048 (priority: 27.03.1979)
[9] Dydychkin V N and Shikanov A E 1991 *Atomic energy* **2** (70) 41-42

# The form of the free surface of hydrogen isotopes in the spherical shell

**V M Izgorodin, E Y Solomatina, A P Pepelyaev, E I Osetrov[1] and M A Rogozhina**

FSUE "RFNC-VNIIEF", Sarov, the Nizhniy Novgorod region

E-mail: evgeniy-osetrov@mail.ru

**Abstract**. Initial study of hydrogen isotopes distribution on inner surface of a hollow spherical shell under cryogenic conditions is given. Comparison of theoretical and experimental surfaces of ice layers of various hydrogen isotopes is performed.

Laser confinement fusion experiments needs a cryogenic target representing a polymeric spherical shell with deuterium-tritium (D-T) fuel frozen on its inner surface in the form of the homogeneous cryogenic layer.

The basic problem in cryogenic targets production technology is producing of smooth thermonuclear solid fuel layer of the given thickness in the shell. The three-dimensional problem of D-T fuel cryogenic layer formation in the spherical container has been formulated in order to solve the problem.

The problem represents a complex of equations of heat conduction of five-layer medium together with an equation of gas dynamics and an equation of state for a gas phase in an inner part. An outer part is the spherical container with the spherical shell of polystyrene or beryllium in the center. Helium functions as heat exchange gas between them. Solid layer of D-T mixture with the nonuniformity of thickness is located on an inner surface of the shell. In the central part there is gaseous phase of hydrogen isotopes. The outer surface of the container is exposed to nonuniform thermal radiation. The exchange gas is considered to be fixed and the gas pressure is 0.1 to 10 Torr. Diameter of the container is 5-10 mm, diameter of the shell is 1.5-3 mm.

[1] To whom any correspondence should be addressed.

Journal of Physics: Conference Series **747** (2016) 012008 doi:10.1088/1742-6596/747/1/012008

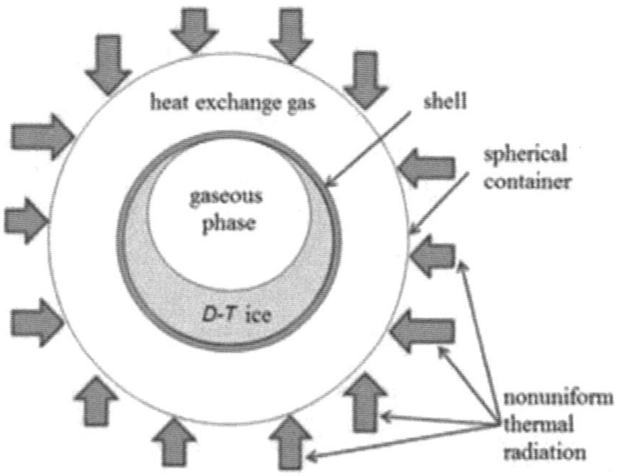


**Figure 1.** Scheme of a three-dimensional problem of cryogenic layer formation in the spherical container.

In order to receive entry conditions for solution of the three-dimensional problem of D-T fuel cryogenic layer formation in the spherical container forms of free surfaces of liquid have been calculated at various situations of container filling. Calculation was yielded by means of solving the hydrostatic balance equation for axially symmetrical problems by a method of finite differences [1]. The given method of the solution of the differential equations is based on the replacement of derivatives by differential schemes. Contours of section of interfaces "liquid-gas" for three cases of filling of the shell are presented in figure 2.

The nondimensional form of equation looks like [2]:

$$\frac{1}{r}\frac{d}{dr}(r\sin\Phi) = \pm z + C_0,$$

where $C_0 = \frac{2}{R_0}$ ($R_0$ - curvature radius in the point z=0).

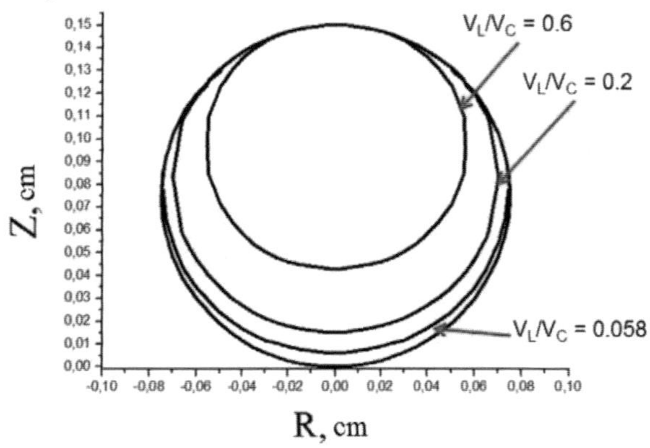


**Figure 2.** Contours of section of interfaces "liquid-gas" for three cases of filling of the shell (1 - $V_L/V_C = 0.058$; 2 - $V_L/V_C = 0.2$ and 3 - $V_L/V_C = 0,6$). $V_L$ and $V_C$ - volume of the condensed gas and the shell accordingly.

The apparatus for targets research at low temperatures consists of research cryostat, systems of simultaneous pumping out of gas channels, helium and hydrogen isotopes supplying system, the equipment for temperature measurement and control, experimental cells and optical system of visual control. The necessary temperature for experiment carrying out (10K - 20K) is reached on a thermoadjustable object stage mounted at the second stage of the cryocooler.

Several approaches of manufacturing and assembly of experimental cells in which the polystyrene shell is disposed have been realized. The primary goal during assembly is maintenance of tightness of cell and fill tubes for helium and hydrogen isotopes supplying. Now two most technological variants providing simplicity of installation and durability of mounting: a cell (figure 3) and the spherical cell with semishells sealed with indium are used.

Experiments on hydrogen isotopes inlet through a fill tube in the spherical polystyrene shell with subsequent freezing have been made. In figure 4 process of deuterium liquefaction with subsequent freezing at temperature decreasing below triple point is given.

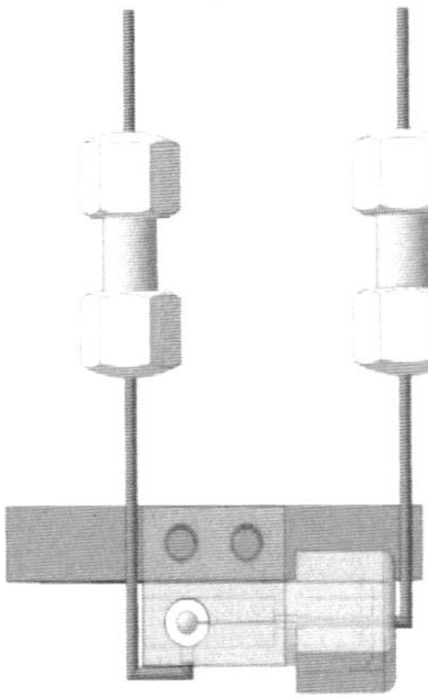

**Figure 3.** Experimental cell.

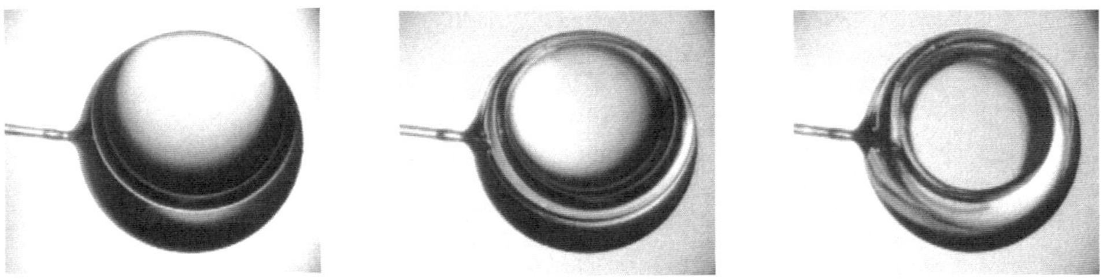

| Beginning of condensation | Beginning of crystallization | Solid phase |
|---|---|---|
| $T_A$ =18,6 K, $T_B$ =19,4 K. | $T_A$ =18,47 K, $T_B$ =19,27 K. | $T_A$ =18,37 K, $T_B$=19,17K. |

**Figure 4.** Process of deuterium liquefaction with subsequent freezing.

Stages of carrying out experiments on hydrogen isotopes inlet through the fill tube in the spherical polystyrene shell with subsequent freezing:

1) pumping out of cryostat volume and gases (helium, hydrogen isotopes) supplying systems with the help of backing and turbomolecular pumps up to $10^{-5}$mbar;
2) the procedure of experimental assemblage cooling down to 30K;
3) the puffing of helium as heat exchange gas into the cell;
4) the puffing of hydrogen isotopes into the spherical polystyrene shell;
5) shell cooling down to temperature below triple point with video recording of «gas-liquid-solid» phase transitions process.

Comparison of theoretical and experimental surfaces of ice layers of various hydrogen isotopes (figure 5) has been performed. Good correlation observed between theoretical and experimental equilibrium surfaces of liquid in the spherical container.

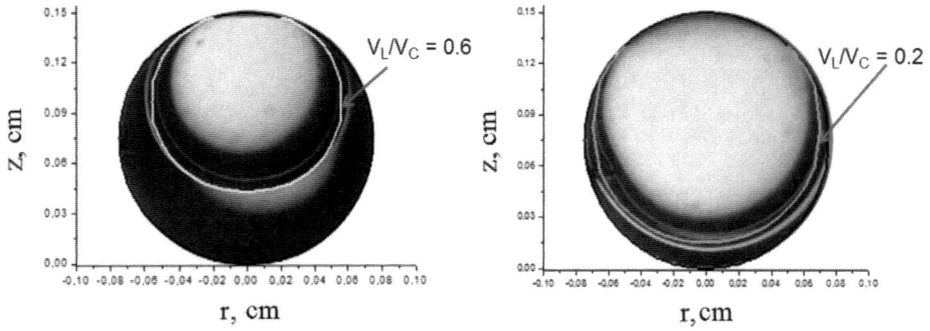


**Figure 5.** Comparison of theoretical and experimental data for two cases of the shell filling. $V_L$ and $V_C$ – volume of condensed gas and the shell accordingly.

**References**

[1] Myshkis A D 1976 *Hydromechanics of weightlessness* (M: Publishing house Science) p 504

[2] Labuntsov D A and Jagov V V 2000 *Mechanics of two-phase systems: the Manual for high schools* (M: Publishing house MEI) p 374

# The polystyrene microsphere filling with hydrogen isotopes through the fill tube with consequent freezing

**V M Izgorodin, E Y Solomatina, A P Pepelyaev, M A Rogozhina and E I Osetrov**
FSUE "RFNC-VNIIEF", Sarov, the Nizhniy Novgorod region

E-mail: evgeniy-osetrov@mail.ru

**Abstract.** Process of spherical polystyrene capsules filling with hydrogen isotopes through the fill tube for the purpose of a cryogenic target building is described. The scheme of the stand for researches and a technique of carrying out of experiments is represented. Results of capsules filling and subsequent freezing for protium, deuterium and protium- deuterium mixture are shown.

## 1. Introduction
Nowadays megajoule energy laser facility for carrying out inertial confinement fusion experiments is being built in RFNC-VNIIEF. The idea of inertial confinement fusion is based on possibility of symmetric compression and heating up to thermonuclear temperatures of small amount of the fuel contained in the target in nanoseconds. The elementary fuel target represents a polymeric spherical capsule with DT fuel cryogenic layer of the specified thickness, generated on its internal surface. Development of cryogenic targets production technology includes some steps one of which is working out of spherical capsule filling procedure with hydrogen isotopes through a fill tube [1].

## 2. Research stand
The stand for targets research at low temperatures consists of research cryostat, systems of simultaneous pumping out of gas channels, helium and hydrogen isotopes supplying system, the equipment for measurement and control of temperature, experimental boxes and optical system of visual control.

### *1.1. The research cryostat*
Optical cryostat with system of cooling on the basis of two-stage pulse tube cryo-refrigerator with refrigeration capacity 1Watt at 4.2 K is used for carrying out researches. Main parts of research cryostat are cryo-refrigerator, helium compressor, helium supplying lines [2].

The cryo-refrigerator produces different temperatures: 45K at the first stage and 4.2K at the second stage. The necessary temperature for liquefaction process research with the subsequent freezing of hydrogen isotopes in a spherical capsule (from 10K to 20K) is reached on a thermoadjustable object stage mounted at the second stage cold station.

### *1.2. Gas channels pumping out system, helium and hydrogen isotopes supplying system*
The system of simultaneous pumping out of cryostat working volume and gas channels, with the subsequent delivery of gases into the box and the polystyrene capsule [3, 4] has been developed and installed. The system arrangement is shown in figure 1.

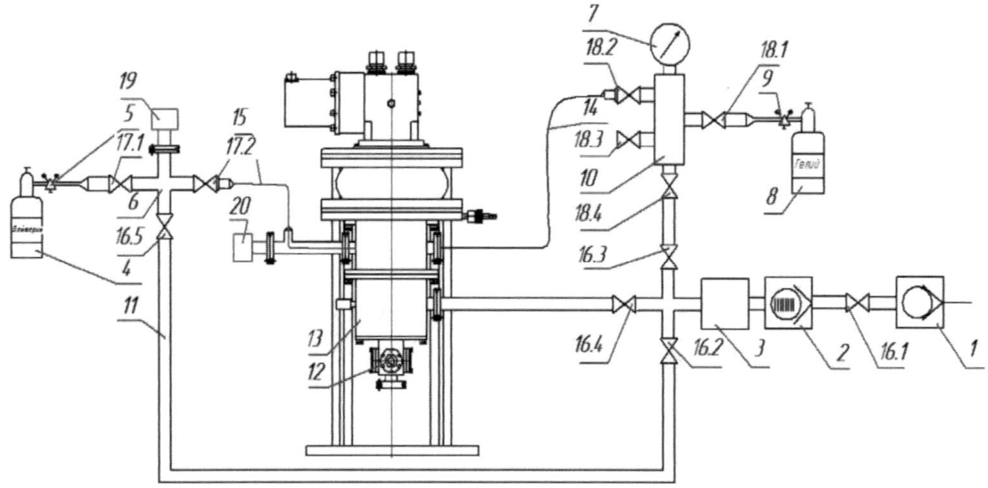


**Figure 1.** Arrangement of the stand.

1. Backing pump; 2. Turbomolecular pump; 3. Buffer volume of the pump; 4. Hydrogen isotopes cylinder; 5. Reducer P1; 6. Cross fitting in hydrogen isotopes supply line; 7. The pressure-and-vacuum gage; 8. Helium cylinder; 9. Reducer P2; 10. Cross fitting in helium supply line; 11. Braided bellow; 12. Box with polystyrene capsule; 13. The cryostat; 14. Helium supply fill tube; 15. Hydrogen isotopes supply fill tube; 16.1-16.5. Vacuum valve; 17.1-17.2 Fine control leak valve; 18.1-18.4. Valve; 19. Pressure gauge D1; 20. Pressure gauge D2.

System pumping out is perfomed with the help of backing and turbomolecular pumps in three stages: after pumping out of buffer volume hydrogen isotopes supply channels, heat exchange gas (helium) supply channels and cryostat working volume are evacuated accordingly. The given scheme of pumping out allows to reach necessary vacuum at the minimum quantity of the used pump equipment for pumping out of three systems isolated from each other. Besides, the pumping out system is integrated with helium and hydrogen isotopes supply systems that allows to carry out corresponding gases puffing in experimental box and polystyrene capsule at overlapping of certain valves.

### *2.3. Temperature control*

Control of temperature of experimental assemblages is carried out by means of two sensors. Sensor DT1 is mounted within a stage 11 mm far from the capsule center (its indications are denoted by TA). The second sensor DT2 is mounted directly on the body of experimental box (TB). Sensors run in a range of temperatures from 1,4K to 325K with accuracy ±0,009K and ±0,001K at T=20K accordingly. Time delay of sensors data reading by the controller is 0,5s. Data are displayed on a personal computer.

It is necessary to maintain defined temperature for a long time at zone of research object for carrying out of experiments. This is carried out by a heater installed at a distance of 11 mm from the centre of a capsule. Temperature control is carried out by the controller (directly from the control panel or the remote personal computer), the average dispersion of temperature data values thus does not exceed 0,002K.

Journal of Physics: Conference Series **747** (2016) 012009 doi:10.1088/1742-6596/747/1/012009

*2.4. Experimental box*
The experimental box represents a system of isolated gas volumes with different pressure (figure 2). The polystyrene capsule is fixed in the box–container which is mounted at cryostat working zone, pumped out to high vacuum. The box is filled with helium that equalizes polystyrene capsules surface temperature. The capsule is in turn filled with gaseous hydrogen isotopes under pressure of about 105 Pa. The primary goal during assembling is maintenance of tightness of box and fill tubes for supplying of operating gases (helium, hydrogen isotopes) taking into account their use at cryogenic temperatures [5]. Thus simplicity of box installation and reliability of its connections should be provided as the size of experimental assemblages are approximately 30x20x10 mm.

Some ways of manufacturing and assembly of experimental boxes have been developed and realized. Now the most technological variant in a form of box-cell which is presented in figure 3 is used.

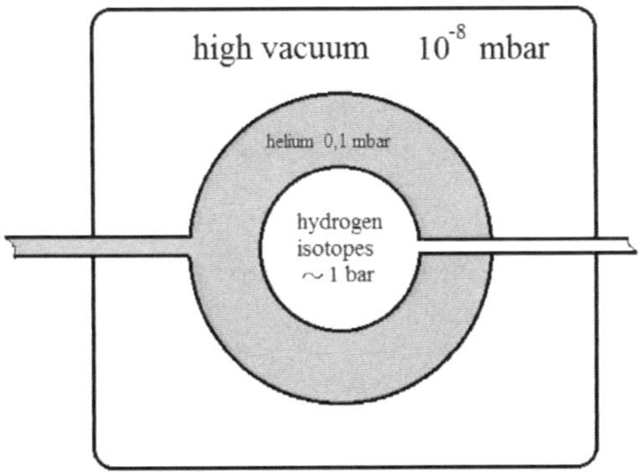


**Figure 2.** The scheme of experimental box.

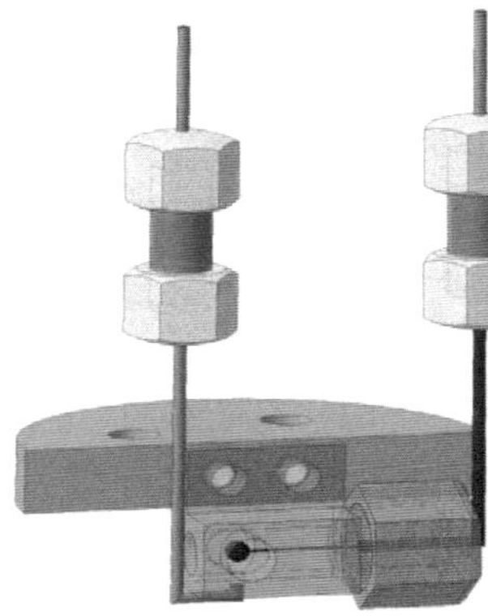

**Figure 3.** The experimental box-cell.

*2.5. Visual control system*
In order to get high quality image diode illumination microscope (the aperture 0.11, working distance 90 mm) is used allowing to observe an accurate contour of polystyrene capsules and subsequent liquefied and frozen hydrogen isotopes borders. The image is registered by a digital camera and is transferred to the personal computer.

## 3. Technique of carrying out experiments
Stages of carrying out experiments on hydrogen isotopes inlet through a fill tube in spherical polystyrene capsule with the subsequent freezing:

- cryostat volume and gases (helium, hydrogen isotopes) supplying systems pumping out with the help of backing and turbomolecular pumps to $10^{-3}$Pa;
- the procedure of experimental assemblage cooling to 30K;
- the puffing of helium as heat exchange gas into the box cavity;
- the puffing of hydrogen isotopes into the spherical polystyrene capsule;
- capsule cooling down to temperature below triple point with «gas-liquid-solid» phase transitions recording process on video.

In experiments on hydrogen isotopes liquefaction with the subsequent freezing hydrogen ($H_2$), deuterium ($D_2$) and mixture of hydrogen-deuterium (HD) are used [6]. Gas filling occurs in the capsule already mounted in the cryostat, fixed in the copper box. Temperature data is obtained by means of the sensor placed on a surface of the box. Before gas puffing required temperature (T=30K) is achieved and is maintained by the controller. Then there is heat exchange gas puffing into the box and hydrogen isotopes puffing into the capsule. Gradual decrease of temperature (1K in 3-4 minutes), set by the controller program, prior to the beginning of liquefaction process in the capsule occurs. Process of liquid phase increase in the polystyrene capsule as a result of filling through a fill tube is presented in Figure 4.

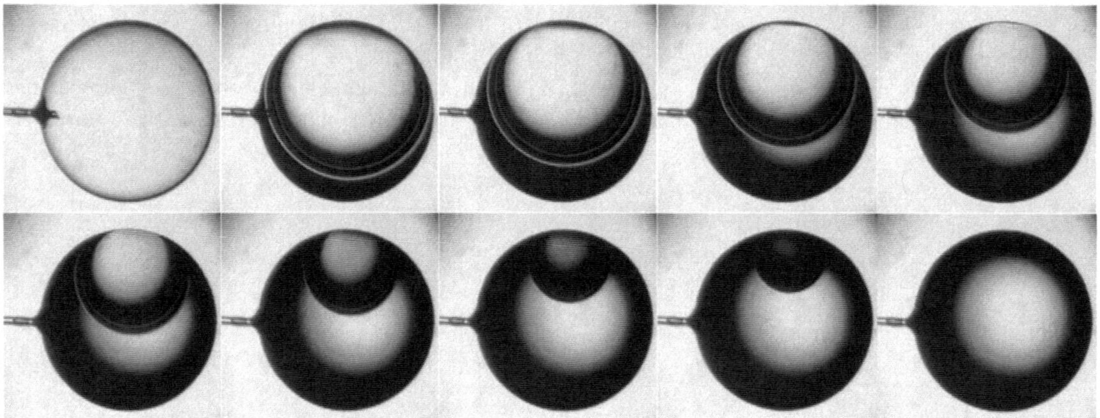

**Figure 4.** – Process of spherical polystyrene capsule filling with liquid phase hydrogen isotopes.

## 4. Results of experiments

Dynamics of hydrogen cryogenic layer behavior at different temperature near triple point is shown in Figure 5.1-5.3 (hydrogen), 6.1-6.3 (deuterium), 7.1-7.3 (hydrogen-deuterium mixture). Experiments were conducted in "box-cell" experimental assemblage. Polystyrene capsule diameter is Ø1,325 mm, diameter of the glass fill tube is 56 microns.

Hydrogen liquefaction process in the capsule is shown in figure 5.1. Gas is condensed at $T_B$=14,7K that corresponds to saturated vapor pressure value P=10,6kPa in the capsule (the temperature of triple point of hydrogen is 13,96K). Rough estimate gives us that mass of condensed gas is $m_{H2}$=0,07 mg. At temperature $T_B$=14,6K we can observe transition of liquid phase in a solid one (figure 5.2) which has completely come to the end at $T_B$=14,2K (figure 5.3).

The experiment with deuterium was carried out in the same way. The temperature of deuterium triple point is 18,7K and it is higher than that of hydrogen, therefore deuterium liquefaction process

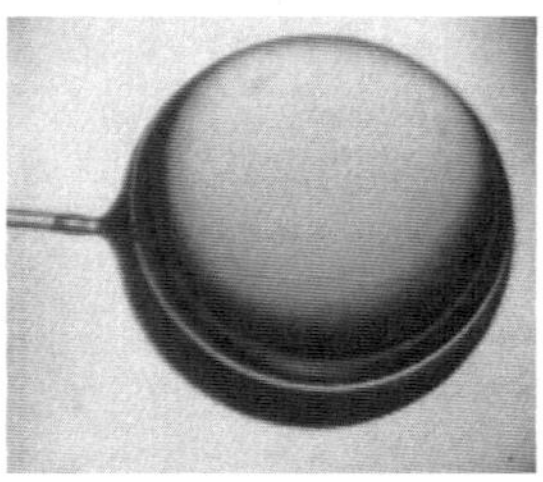

**Figure 5.1.** Beginning of condensation $T_A$=13,9K, $T_B$ =14,7K.

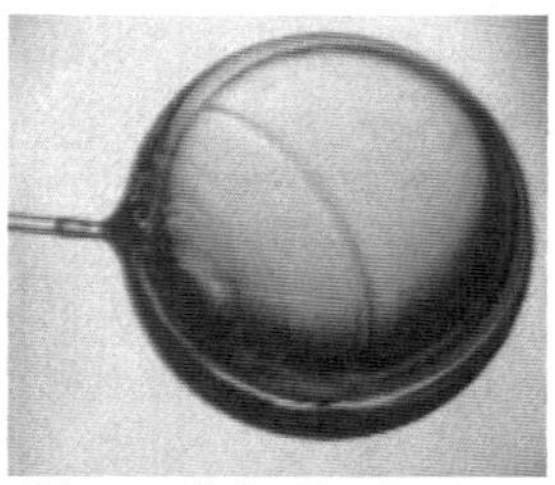

**Figure 5.2.** Beginning of crystallization $T_A$=13,8K, $T_B$ =14,6K.

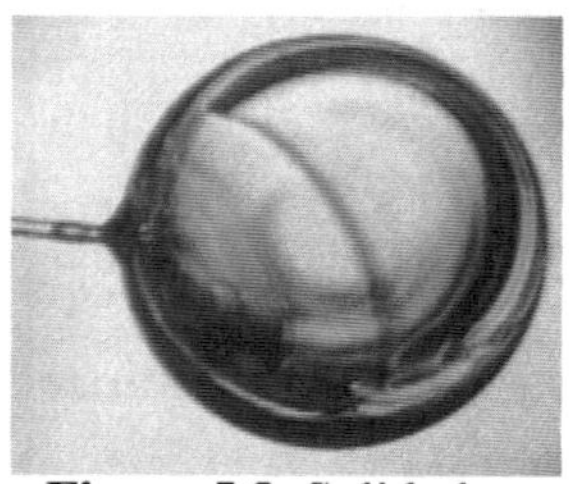

**Figure 5.3.** Solid phase $T_A$=13,4K, $T_B$ =14,2K.

has begun at higher temperature $T_B$=19,4K, $P_{nas.p.}$ = 23,4kPa (figure 6.1). Calculated condensed gas mass value is $m_{D2}$=0,168mg. With decrease of temperature to $T_B$=19,27K deuterium freezing process on capsule internal surface has come to the end at $T_B$=19,17K (figure 6.2, 6.3).

Journal of Physics: Conference Series **747** (2016) 012009 doi:10.1088/1742-6596/747/1/012009

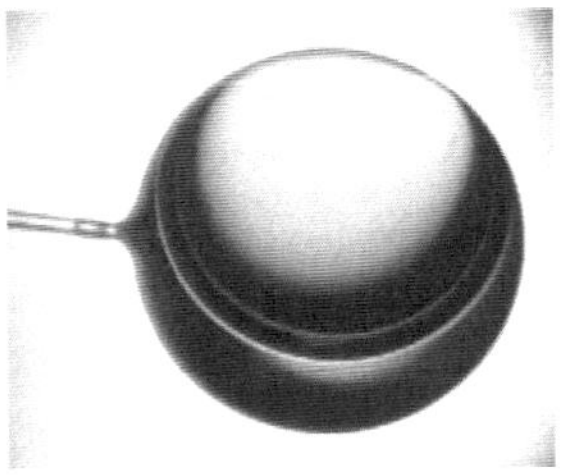

**Figure 6.1.** Beginning of condensation $T_A$=18,6K, $T_B$=19,4K.

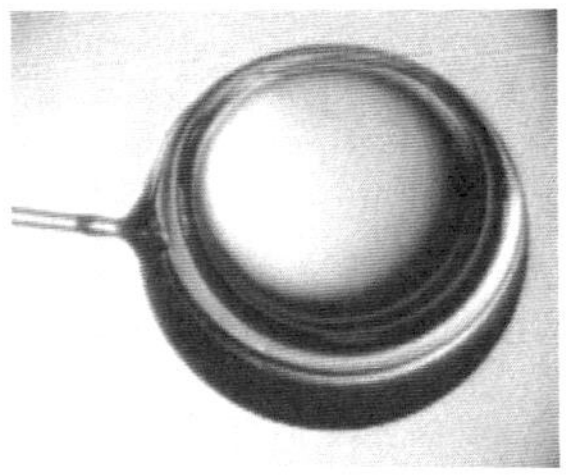

**Figure 6.2.** Beginning of crystallization $T_A$ =18,47K,$T_B$ =19,27K.

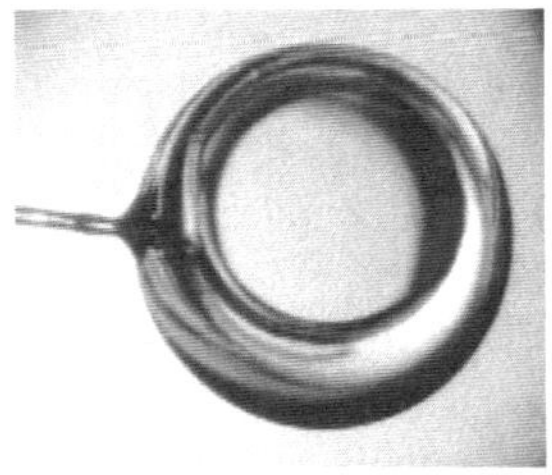

**Figure 6.3.** Solid phase $T_A$=18,37K, $T_B$=19,17K.

In a series of the experiments distinctions in protium and deuterium behaviour have been noticed. Freezing of heavy hydrogen isotope occurred all around on sphere inner surface while in a case of protium rather the most part of ice was formed in the bottom of microsphere. It was revealed that in experimental conditions uniformity of deuterium distribution in triple point between the top and the bottom pole of the spherical capsule is considerably higher. It is connected, possibly, with higher than of protium viscosity, more magnitude of surface tension coefficient near triple point ($\sigma_{D2}(18,5\text{K}) = 4{,}006 \cdot 10^{-3}\frac{\text{H}}{M}$, $\sigma_{H2}(13{,}95 = 3{,}004 \cdot 10^{-3}\frac{\text{H}}{M}$).

Temperature of hydrogen-deuterium mixture triple point is 16,6K, thereby phase transformations occur around that temperature. In figure 7.1 condensation beginning is shown at $T_B$ =16,97K, with temperature decrease to 16,74K process of crystallization has begun and at $T_B$=16,6K all the liquid has transformed into solid phase (figure 7.2, 7.3). Calculated mass of condensed mixture is $m_{HD}$=0,025 mg.

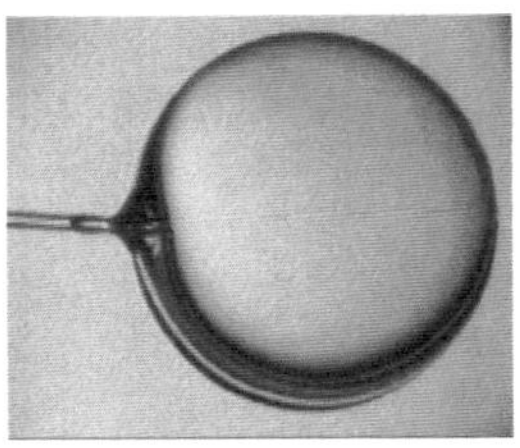

**Figure 7.1.** Beginning of condensation $T_A$=16,05K, $T_B$=16,97K.

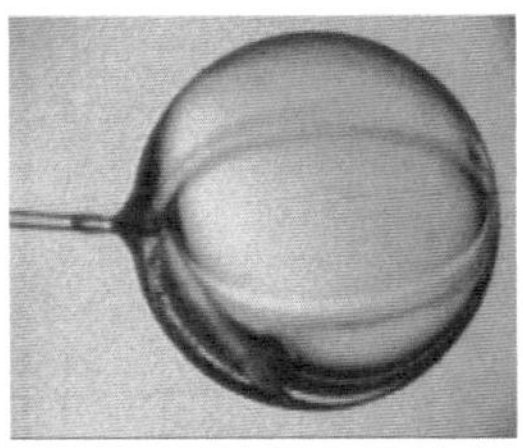

**Figure 7.2.** Beginning of crystallization $T_A$ =15,81K, $T_B$=16,74K.

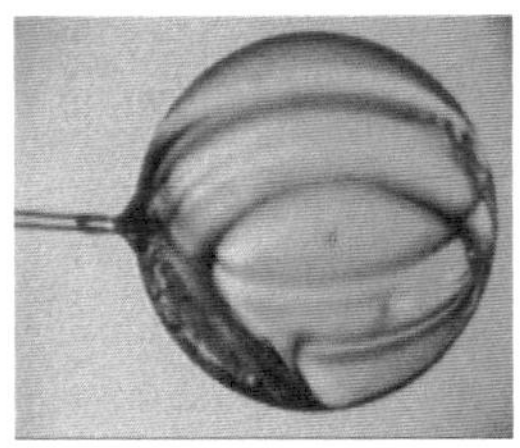

**Figure 7.3.** Solid phase $T_A$=15,7K, $T_B$ =16,6K.

## 5. Conclusion

The technology of spherical polysterene capsules filling by hydrogen isotopes through a fill tube has been developed.

Experiments on liquefaction of protium, deuterium and protium-deuterium mixture with the subsequent freezing on internal surface of the capsule are made. The technics of carrying out the experiment is worked through.

The given work is the initial stage of developing the technology of cryogenic targets development with smooth defined thickness homogeneous layer of solid fuel for carrying out experiments on laser confinement fusion.

Journal of Physics: Conference Series **747** (2016) 012009 doi:10.1088/1742-6596/747/1/012009

**References**

[1] Cryogenic Targets: Current Status and Future Development. LLEReview, Volume 114.
[2] Scott R B 1962 *Technics of low temperatures* Publishing house of the foreign literature,.
[3] Pipko A I, Plisovsky V Y and Penchko E.A. 1970 *Designing and calculation of vacuum systems* (M: "Energy")
[4] Heinz B 1960 *Introduction in the vacuum technics* Gosenergoizdat
[5] Solntsev J P and Stepanov G A 1982 *Materials in the cryogenic technics: the Directory* (L: Mechanical engineering)
[6] *Hydrogen. Properties, reception, storage, transportation, application* 1989 Under the editorship of Hamburg D Y and Dubovkina N F (M.: Chemistry)

Journal of Physics: Conference Series **747** (2016) 012010 doi:10.1088/1742-6596/747/1/012010

# Carbon dust particles in a beam-plasma discharge

**O A Koval [1], I V Vizgalov and A V Shalpegin**
National Research Nuclear University MEPhI, 115409, Russia, Moscow, Kashirskoe highway 31

E-mail: helgakoval@gmail.com

**Abstract.** This paper focuses on dynamics of micro-sized carbon dust grains in beam-plasma discharge (BPD) plasmas. It was demonstrated that injected dust particles can be captured and transported along the discharge. Longitudinal average velocity of the particles in the central area of the plasma column was 17 m/sec, and 2 m/sec in the periphery. Dust injection caused a decrease of emission intensity of metastable nitrogen molecular ion. This effect is suggested for a spectroscopy method for particles' potential measurements. Five-micron radius carbon dust grains obtained potential above 500 V in the experiments on PR-2 installation, proving the feasibility of BPDs for the charging of fine dust particles up to high potential values, unattainable in similar plasma conditions.

## 1. Introduction

The presence of dust particles in fusion plasma is recognized as a serious issue for safe and efficient operation of the ITER tokamak [1]. This paper focuses on dynamic of micro-sized carbon dust grains in plasma relevant to presheath fusion plasma in tokamaks. The purpose of these experiments was to define conditions of dust particles capture into discharge, to investigate their trajectories, and to define their potential in BPD. Additionally, dust particles charging to a high potential values can have broad industrial applications in such areas as microelectronics and photonics, where dust particles subjected to plasma ions implantation [2] can be used for creation of new unique micro- and nano-disperse composite materials. One of the main features of the BPD is the presence of thermal and energetic electron components, moreover energetic electron component is often in a non-equilibrium state and can consist of more than one energy population. Ionization degree of such plasmas can be close to unity. Dust charging experiments have been performed in the BPD device for the first time and a special spectral method for potential estimation has been developed.

## 2. Experimental setup

Research was carried out on the linear plasma device PR-2 (MEPhI) with magnetic field of a mirror machine configuration (0.16 T at the axis). An electron beam driven discharge is implemented to generate and heat plasma. It is initiated by a high power electron gun with directly heated tungsten cathode. Mixture of argon and nitrogen was used as a working gas in the experiments, residual gas pressure in the chamber was $10^{-3}$ Pa, working pressure was $2{\cdot}10^{-1}$ Pa. Typical plasma parameters obtained at PR-2 are: $n_e$ up to $10^{13}$ cm$^{-3}$, $T_e = 5 - 25$ eV [3]. In addition, plasma device is equipped with Langmuir probes, optical spectroscopy system, static mass-spectrometer of plasma ions and quadrupole residual gas analyser (Figure 1). Experiment took place in the collector part of the PR-2 (pointed by arrow). Basic experiment scheme is shown in Figure 2a. Dust injector consists of two main elements – piezoelectric crystal and a container with carbon dust grains attached to it. Vibrations

[1] To whom any correspondence should be addressed.

generated by the piezo crystal spread to the container causing dust to drop through a narrow slit at the bottom of the container. The injected mass flow rate was varied between $2 \cdot 10^5$ and $5 \cdot 10^5$ particles/s by changing the amplitude of the voltage applied to the piezo crystal. Fast camera (1200 fps) Nikon J1 was used for observation of particles motion, and a blue laser (450 nm, up to 5W) was installed for extra illumination. In addition, an optical spectrometer AvaSpec-2048x14 was used for studies of influence of dust presence on plasma emission spectrum. Pre-characterized carbon dust particles with linear size of 10 microns were injected (Fig 2b).

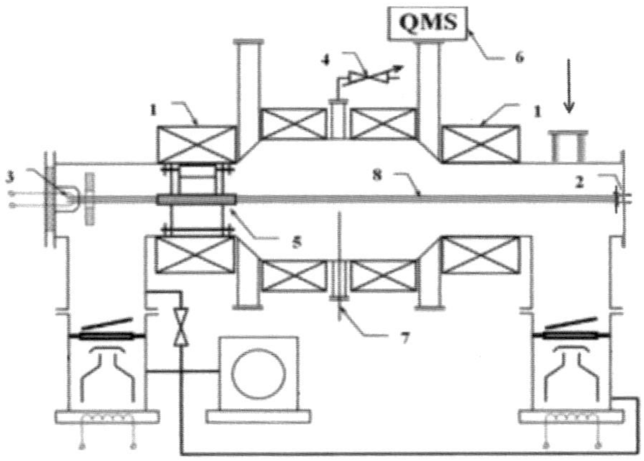


**Figure 1.** Scheme of the PR-2: 1 - magnetic field coils, 2 - collector, 3 - electron gun, 4 - gas inlet system, 5 - static mass-spectrometer, 6 - quadrupole mass-spectrometer, 7 - Langmuir probes, 8 – plasma column.

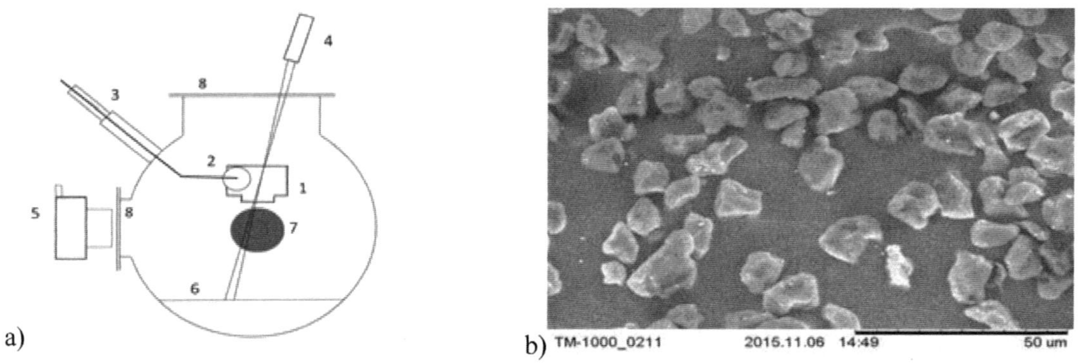


a) b)

**Figure 2.** a) Scheme of dust injection experiment: 1 injector, 2 – flexible fixture, 3 – vacuum inlet, 4 – laser, 5 – fast camera (or optical spectrometer), 6 - plate for dust collection, 7 – plasma column, 8 – quartz windows; b) SEM image of carbon dust particles

## 3. Results

In the first experiments Ar was used as working gas, discharge power was set to 200 W – dust particles fell down through the plasma column without any changes in their trajectories. It was found that with the increase of discharge power up to 1kW, dust particles were captured into the BPD. Majority of the dust particles were captured in the central area of the discharge and moved directly to the collector (Figure3 a). Average value of the longitudinal velocity component ($v_z$) for such particles was 17 m/sec. Others moved in the periphery area of plasma column, trajectories had helical shape ($v_z$ ~ 2 m/sec) (b). Some particles fell out after a half of a turn in the gradient area of the discharge (c, d) ($v_z$ ~ 6 m/sec). When an agglomeration of several dust particles appeared in plasma, analog of Coulomb explosion was observed (e, f). We could observe particles motion only from a single viewpoint, thus part of the trajectory behind plasma column wasn't detected. After the experiments, dust particles, which were captured and transported by plasma, were found on the collector plate ((2) in Figure1)

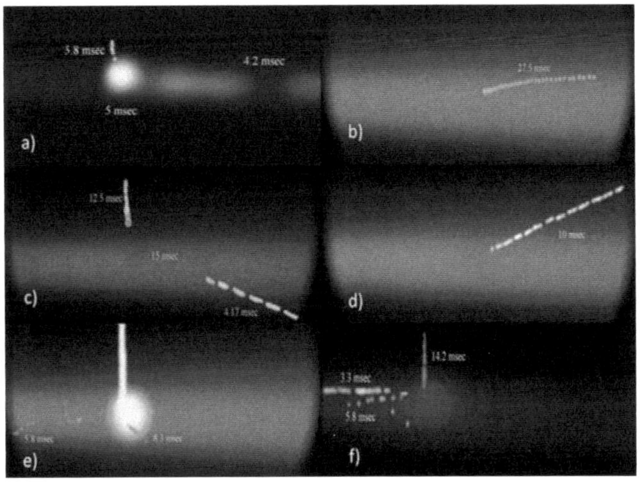


**Figure 3.** Recorded trajectories of 5-micron radius dust grains in a plasma-beam discharge.

Plasma radiation spectra have been recorded in the discharges with and without dust injection. Ion spectral lines are dominant over the neutral radiation in the BPD spectra since the intense ionization processes take place in the central region of the discharge due to the presence of the energetic electron beam. No new lines or any significant increase of continuum radiation has been observed during the dust injection as shown in the averaged difference spectrum (Figure 4b). We interpret the absence of new lines, especially C-lines, as an indication that dust particles did not evaporate and therefore modify plasma elemental composition in the area of observations. In contrary, the presence of dust particles clearly decreases the emission intensity of nitrogen molecular ion (0–0) band of the first negative system between $B^2\Sigma_u^+$ and $X^2\Sigma_g^+$ levels at 391.4 nm. This is assumed to be caused by the collection of the positive metastable ions on the surface of the negatively charged dust particles. Therefore, the decrease of the ions concentration is proportional to the concentration and potential of the dust particles.

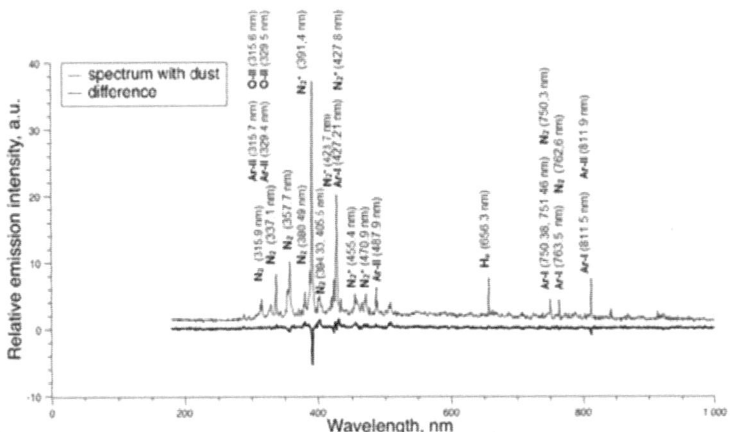


a)

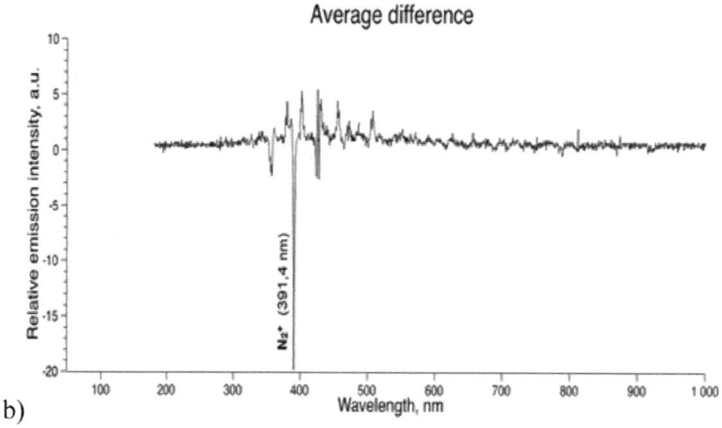


b)

**Figure 4.** Plasma spectral lines: a – beam-plasma discharge with dust particles and difference (this spectrum minus spectrum without dust at the same conditions); b – averaged over eight experiments difference spectrum.

## 4. Discussion

Observed types of trajectories can be explained by the structure of the plasma-beam discharge and dust particles charging. Positive ions overcompensate the discharge axial zone with generation of a radial electrostatic field and a drift plasma beam rotation. Since the Lorentz force on charged dust particles from magnetic field is negligible, the helical shape of their trajectories can be caused by the radial electrostatic force upon negatively charged dust particles along with tangential drag upon them from the azimuthal plasma drift. In longitudinal direction particles move with constant velocity. In the axial central region due to the presence of high-energy primary electron beam dust particle potential could be much higher than in the gradient region. Therefore, the electrostatic force countervails the gravity force. Moreover, ions accelerate to the collector and drag dust particles in the same direction. In fact, large centrifugal force and gravity force cause drop-out of grains. All these assumptions about types of dust particles trajectories were confirmed by the computer simulations (Figure 5). Two coaxial cylinders represent central and periphery area of plasma column. This model allows to deduce trajectories dependence on initial conditions, grain's radius, potential and plasma parameters.

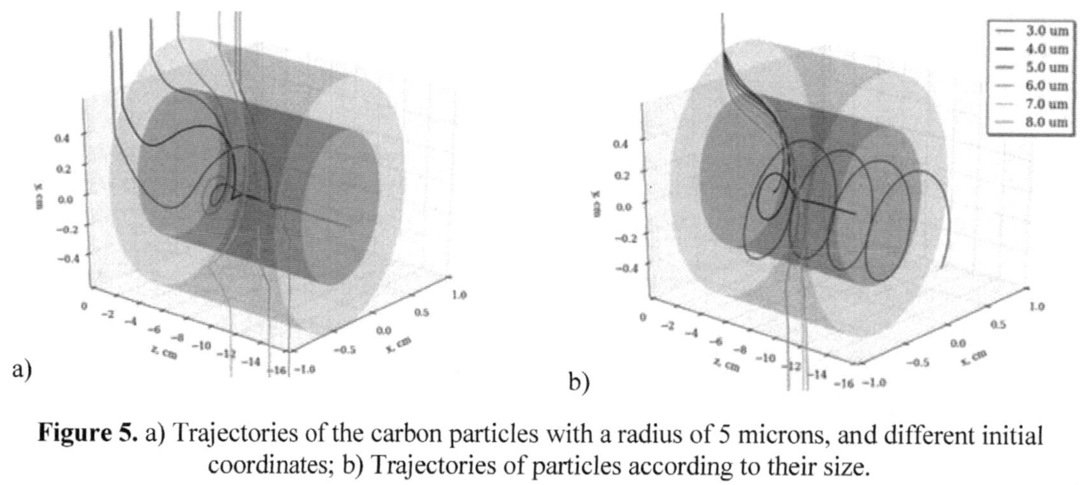

**Figure 5.** a) Trajectories of the carbon particles with a radius of 5 microns, and different initial coordinates; b) Trajectories of particles according to their size.

An average value of the 391.4 nm line intensity relative decrease $\Delta\Phi/\Phi$, and thus the density of $N_2^+$ was 10%. It is essential that the upper level of the first negative system is metastable with a lifetime τ = 65 ns [4]. Based on this effect, knowing plasma parameters it is possible to determine dust grains potential. Thus, spectrum and probes measurements are required to use this method. We assume, that excited ions $N_2^+$, entering the field of negatively charged dust particle, get to its surface and non-radiatively relax on it during their lifetime. This effect causes a change in their number, and therefore decreases of the spectral line intensity

$$\frac{\Delta\Phi}{\Phi} \cong \frac{4/3\,\pi r_{abs}^3 N_d}{1}, \tag{1}$$

where $N_d$ – dust particles concentration, $r_{abs}$ – ion collection radius.

Concentration of dust particles at the injection area was estimated to be $\leq 10^3$ cm$^{-3}$ based on the dust injection rates and recorded dust velocities. Dust particles ion collection radius was calculated taking into account OML theory [5] and expressions for spherical Langmuir probes [6]:

$$r_{abs} \cong \left(\frac{e\varphi}{T_e}\right)^{3/4} \left(\frac{M}{m}\right)^{1/4} r_D, \tag{2}$$

where $\varphi$ - probe potential, $T_e$ - electron temperature, $M/m$ ion to electron masses ratio, $r_D$ - Debye radius. This This resulted in the typical value of the ion collection radius $r_{abs}{\sim}100r_D$ According to these assumptions dust grains obtained potential higher than 500V.

Further research can be continued for both simulation of the dust particles behaviour in tokamak edge plasmas and application of dust in plasma methods in technology. For instance, dust particles manipulation can be used for blanket deposition inside narrow tubes. The possibility of dust particles charging to an extra high potential values is interesting in perspective of immerse ion implantation into dust grains or fusion reactions on them. Several upgrades of PR-2 device are being designed to allow such experiments, namely spatial and temporal focusing of the secondary emission electron beam [7], and reconfiguration and increase of magnetic field, in order to further increase the charge of injected dust particles.

**References**

[1] Krasheninnikov S I, Smirnov R D and Rudakov D L 2011 *Plasma Phys. Control. Fusion.*, 53.083001

[2] Filippov A V, Vasil'ev M N, Gavrikov A V, Pal' A F, Petrov O F, Starostin A N and Fortov V E Ultrahigh charging of dust particles in nonequilibrium plasma 2007 *JETP Letters*. **86** 14-19

[3] Gutorov K M, Vizgalov I V and Kurnaev V A 2012 *Plasma Phys. Reports* **38** 1050-1052

[4] Lofthus A and Krupenie Paul H 1977 *J. Phys. Chem.* **6**

[5] Allen J E Probe Theory - The Orbital Motion Approach 1992 *Physica Scripta* **45** 497-503

[6] Demidov V I, Kolokolov N B and Kudrjavcev A A 1996 *Zondovye metody issledovanija nizkotemperaturnoj plazmy [Probe methods for low temperature plasmas]* (M.:Jenergoatomizdat) (in Russian)

[7] Koval O A and Vizgalov I V Using an auto-oscillating beam-plasma discharge to implant ions intodust particles 2016 *Bulletin of the Russian Academy of Sciences: Physics* **80** 133-137

# Effect of radiation refraction on the characteristics of a cw optical discharge

**A F Glova[1,2,3], A Yu Lysikov[1], D D Malyuta[1], S S Nelyubin[1], P I Peretyatko[1] and Yu F Ryzhkov[1]**

[1]State Research Center of Russian Federation Troitsk Institute for Innovation and Fusion Research (SRC RF TRINITI), Moscow, Troitsk, 142190 Russia

[2]National Research Nuclear University Moscow Engineering Physics Institute (NRNU "MEPhI"), Moscow, 115409, Russia

E-mail: afglova@triniti.ru

**Abstract.** One-dimensional model of a continuous optical discharge is proposed, allowing the estimation of its parameters in a converging laser beam taking into account the refraction of the radiation on the discharge. The influence of spherical aberrations of a focusing lens on the measured thresholds of maintenance of the discharge in the laser plasmatron mode has been discussed.

## 1. Introduction

The continuous optical discharge supported by a $CO_2$-laser radiation was extensively studied in the last quarter of last century. At the present time, these studies are continued at the new stage [1] when a possibility was shown of effective synthesis of diamond coating using a continuous optical discharge in the laser plasmatron mode [2]. One of the main characteristic of an optical discharge in this mode is the threshold power of laser radiation that is necessary to maintain the discharge in the dependence on the gas flow rate. It was shown in the experiments [3] that this dependence has a minimum for long-focusing lenses and is characterized by monotonic decrease in the power as the gas flow rate decreases up to zero as short-focus lenses are used. The numerical calculations [4] confirmed these features and for discharge in Ar under atmospheric pressure made it is possible to find a conventional boundary between long-focus and short-focus lenses with respect to the value of the half-angle of focusing the radiation in the vicinity of $\varphi \approx 0.125$ rad.

The optical discharge plasma is characterized by large transverse gradients of the electron density and the gas density and, because of this, can cause the radiation refraction on the discharge. A qualitative indication of the existence of noticeable laser beam refraction on an optical discharge is presented in [5]. Authors [6] experimentally proved for the first time the possibility of refraction broadening of a laser beam on the optical discharge plasma that was supported by this laser beam. The value of the broadening was estimated in recent work [7] and corresponds to the experimental data [6]. Some of the experimentally observed features of discharge burning in a laser beam at the weak focusing [8] are also related with the radiation refraction.

[3] To whom any correspondence should be addressed.

The results of experimental and theoretical studies devoted to determination of the discharge boundaries in a convergent laser beam depending on the conditions of its maintenance are presented in [9-12].

The aim of this work is to study the influence of the radiation refraction and the lens spherical aberrations on the boundaries position and the conditions of maintenance of a continuous optical discharge.

## 2. Focusing of radiation

Figure 1 shows the scheme of radiation focusing. Here $r_0$ is the radius of the focused laser beam, $P$ is the laser radiation power, $v_0$ is the gas flow rate, $r_1$, $r_2$ are radii of the front and back discharge boundaries, $z_1$, $z_2$ are the coordinates of the discharge boundaries at the optical axis, $\varphi_p$ is the radiation refraction angle, $r_f$ is the focal spot radius, $\varphi = (r_0-r_f)/f$ is the angle of radiation focusing (here $f$ is the lens focal length).

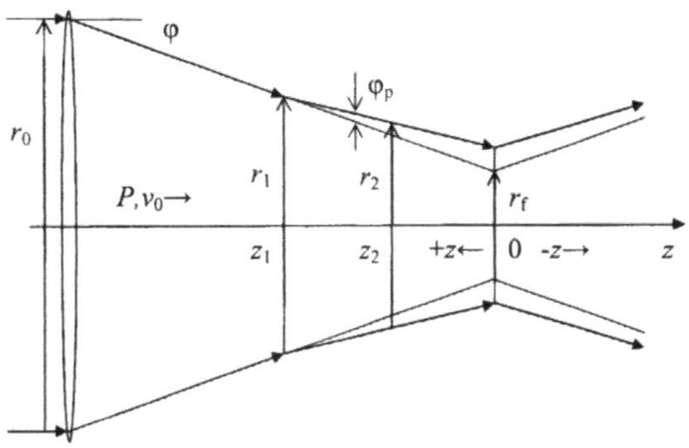


**Figure 1.** Scheme of radiation focusing.

Taking into account the lens spherical aberrations, focal spot radius $r_f$ is

$$r_f = \frac{a(2r_0)^3}{f^2} + \gamma_0 f, \tag{1}$$

here $r_0=rH$ and $\gamma_0=\gamma/H$ is the beam radius and the radiation divergence after telescoping the beam exited from the laser, $r$ and $\gamma$ are the radius and the divergence in the laser output, $H$ is the telescope magnification. For convex-flat cylvite lenses used in the experiments and in calculations, coefficient $a$=0.044, the values $r$=10 mm and $\gamma=3.5\cdot10^{-4}$ rad correspond to the parameters of the laser used.

Figure 2 presents the dependences $r_f$ on $f$ calculated by formula (1) at $H$=1, 1.5 and 2. At constant $f$, radius $r_f$ can both increase and decrease with increasing $H$. The position of the minimum in the dependences shown in figure 2 corresponds to focal length $f_{opt}$ and spot radius $r_{fmin}$ calculated by formulas $f_{opt}= (2a/\gamma_0)^{1/3}2r_0$ and $r_{fmin}=1.5f_{opt}\gamma_0$.

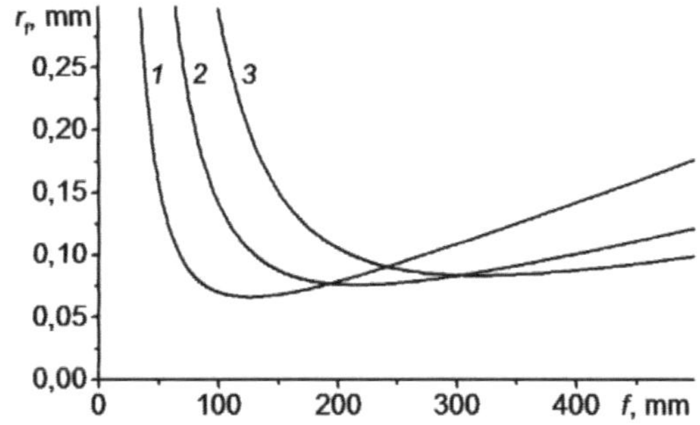


**Figure 2.** Dependences $r_f$ on $f$ for $H$=1 (*1*), 1.5 (*2*) and 2 (*3*).

### 3. Model of discharge and calculation results

In a paraxial approximation, coordinates $z_1$ and $z_2$ are (see figure 1)

$$z_1 = \frac{r_1 - r_f}{\varphi}, \tag{2}$$

$$z_2 = \frac{r_2 - r_f - z_1\varphi_p}{\varphi - \varphi_p}. \tag{3}$$

In a stationary discharge, the radiation intensity at the back boundary must be equal to the intensity at the front boundary. Based on this, we take into account the increase in the intensity at the back boundary due to the beam compression and simultaneously its decrease because of radiation absorption. From the obtained balance relationship, we find

$$r_2 = \frac{r_1}{[1 + \exp(-\alpha(z_1 - z_2))]^{1/2}}, \tag{4}$$

where $\alpha$ is the radiation absorption coefficient.

The angle of radiation refraction we find using the equation of a light beam in the paraxial approximation

$$\frac{d}{dz}\left(N_{\mathrm{p}}\frac{dr}{dz}\right) = \nabla N_{\mathrm{p}}, \tag{5}$$

where $N_p=\beta N_e+(1-\beta)N_g$ is the discharge refractive index, $r$ is the transverse coordinate, $\beta=[n_eT/(n_0T_0)]$ is the degree of gas ionization, $N_e$ and $N_g$ are the refractive indexes of the electron and gas components, $T$ is the discharge temperature, $n_e$ is the electron density, $n_0$ is the normal gas density, $T_0$=300 K. Neglecting the dependence of $N_p$ on $z$ and taking into account $N_g\approx1$, $N_e\approx1$ and replacing $\delta r$ for $(r_1+r_2)/2$, we obtain for $\varphi_p=dr/dz$

$$\varphi_p = \frac{[\beta|\delta N_e| + (1-\beta)|\delta N_g|](z_1 - z_2)}{0.5(r_1 + r_2)}, \tag{6}$$

here $|\delta N_e|\approx n_e/2n_{ec}$, $n_{ec}\approx1.2\cdot10^{19}$ cm$^{-3}$ is the critical electron density for $CO_2$-laser, $|\delta N_g|=(N_{g0}-1)(1-T_0/T)$, $N_{g0}$ is the refractive index of the gas at normal conditions.

Now we will use the expression from [13] for the propagation velocity $v_0$ of thin optical discharge in laser beam through a cold gas

$$v_0 = \frac{1}{c\rho_0}\left(\frac{k\alpha}{T}\right)^{1/2}\left(I - \frac{\Phi}{\alpha} - \frac{\eta kT}{\alpha R^2}\right)^{1/2}, \tag{7}$$

here $c$ is the specific heat capacity, $k$ is the thermal conductivity coefficient, and $\Phi$ is the specific radiation loss power. Term $\eta kT/\alpha R^2$ approximately describes the lateral losses of a discharge with radius $R$ for thermal conduction, $\eta$=2. At $v_0$=0, it follows from (7) that the threshold intensity at the discharge front is

$$I_t = \frac{\Phi}{\alpha} + \frac{\eta kT}{\alpha R^2}. \tag{8}$$

Let expression (8) is also valid for the discharge with a finite thickness. Then, taking into account $R=(r_1+r_2)/2$, we find the radius for the front discharge boundary

$$r_1 = \left(\frac{0.86P}{\pi I_t}\right)^{1/2} = \left(\frac{0.86P}{\pi\left(\frac{\Phi}{\alpha} + \frac{4\eta kT}{\alpha(r_1 + r_2)^2}\right)}\right)^{1/2}, \tag{9}$$

here factor 0.86 reflects the fraction of the total power $P$ in the laser beam.

Equations (2)-(4), (6) and (9) form a system of equations from which we can find $r_1$, $r_2$, $z_1$, $z_2$, $\varphi_p$ using known $\Phi$, $\alpha$, $n_e$ and $T$.

The calculation were performed for Ar at pressure $p$=1 atm. The values $\Phi$, $\alpha$, $n_e$ as functions of $T$ were calculated using the formulas from the monograph [14]. The thermal conductivity coefficient of Ar was linearly extrapolated with respect to temperature using the data [15]. This system of equations is valid for short-focus lenses, when the discharge is localized in the beam and exists in a steady state at $v_0$=0. Under these conditions, the measurement results [9] can be used, according to which the discharge temperature in Ar at $p$=1 atm and $v_0$=0 is 17 kK and it is independent on $P$.

Figure 3 shows the calculated dependences of $z_1$, $z_2$, $\varphi_p$ on $P$ for the short-focus lens with $f$=130 mm ($H$=2.06, $\varphi\approx$0.16 rad); the dashed line shows the calculations for $\varphi_p$=0. The power varies in the range beyond of which discharge does not exist because of violation of the condition of the intensity equality at the boundaries. For the chosen lens, the relative decrease in $z_2$ caused by the radiation refraction is insignificant, since inequality $\varphi_p$<<$\varphi$ is fulfilled over entire range of the power change.

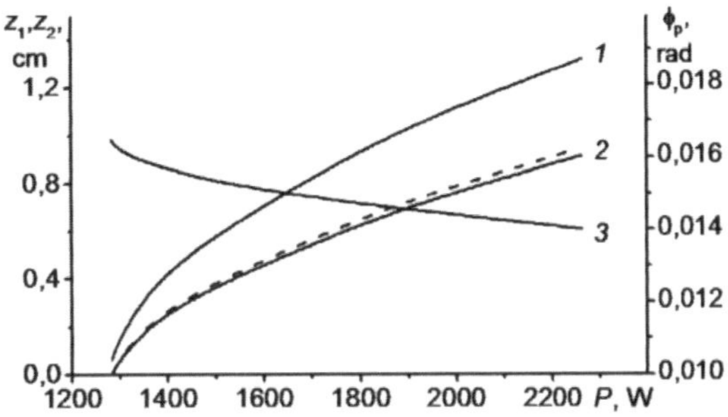


**Figure 3.** Calculated dependences of $z_1$ (*1*), $z_2$ (*2*), $\varphi_p$ (*3*) on $P$ for $f$=130 mm. Ar, $p$=1 atm, $v_0$=0, $T$=17 kK. Dashed line is the calculation for $\varphi_p$=0.

This model allows one to calculate, for short-focus lenses, the radii, the boundary coordinates and the refraction angle also for the discharge in the laser plasmatron mode. In this mode, as $v_0$ increases, the discharge shifts to the focal plane, and the radiation intensity $I_1$ at its front boundary increases according to the expression

$$I_1 = I_{10} + (v_0 c \rho_0)^2 \frac{T}{k\alpha}. \quad (10)$$

At given value of $v_0$, the intensity $I_1$ is a known value, since the expression contains the known intensity $I_{10}$=0.86$P$/($\pi r_{10}^2$) at the known radius $r_{10}$ calculated for $v_0$=0. Because of this, the values $r_1$ and $z_1$ are also known at $v_0 \neq 0$ and are determined by formula $r_1=[0.86P/(\pi I_1)]^{1/2}$ and by equation (2). As a result, the number of unknown parameters decreases to three ($r_2$, $z_2$, $\varphi_p$), and these parameters can be found by solution of the system consisting of equations (3), (4), (6).

Figure 4 shows the results of calculations of the threshold power $P_{th}$ in dependence on $v_0$ for short-focus lenses with $f$=100 mm (curves *1*, $\varphi\approx$0.21 rad) and $f$=130 mm (curves 2, $\varphi\approx$0.16 rad) and the same $H$=2.06 at $\varphi_p \neq 0$ (solid curves) and $\varphi_p$=0 (dashed curves). Insignificant increase in the discharge temperature with $v_0$ [4] was neglected, and its value used in the calculations was 17 kK. The threshold power $P_{th}$ was found for $z_2$>0 near the focal plane, although this model allows one to calculate also a negative value of $z_2$ behind the focal plane. However, at $z_2$<0, the balance of the intensities at the boundaries is violated fast, and the choice of near-zero positive values of $z_2$ provided adequate accuracy of the calculations. It is seen from figure 4 that the inclusion of the refraction for $v_0$ at the right of the characteristic kink in the dependences presented leads to the increase in $P_{th}$ as compared to the case $\varphi_p$=0, and this increase becomes more significant with increasing $f$. We can't to confirm such behaviour of $P_{th}$ on $v_0$ in our conditions due to the features of the plasmatron construction not allowing to use the short-focus lenses.

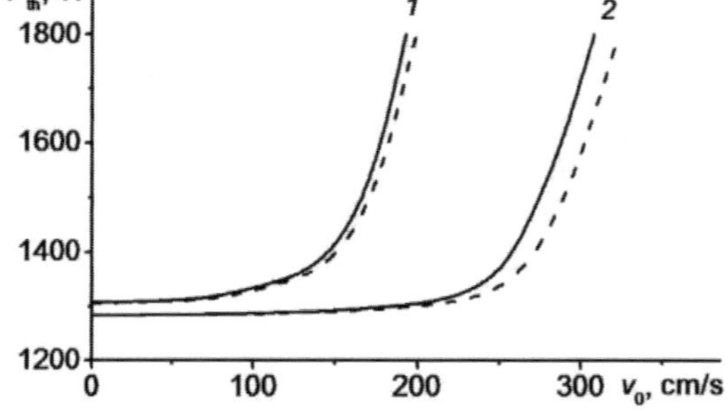


**Figure 4.** Calculated dependences of $P_{th}$ on $v_0$ for *f*=100 (*1*) and 130 mm (*2*) at $\varphi_p \neq 0$ (solid lines) and $\varphi_p=0$ (dashed lines). Ar, *p*=1 atm, *T*=17 kK.

## 4. Measurement of the threshold power

A comparison of the values of $P_{th}$ in figure 4 for both lenses at $v_0$~100 cm/s shows that $P_{th}$ for the lens with *f*=100 mm is higher. It can be a result of the fact that radius $r_f$=0.32 mm at *f*=100 mm is noticeably larger than $r_f$=0.2 mm at *f*=130 mm (figure 2). Because the increase in the flow rate fast shifts the discharge to a near-focal region, the threshold power for the lens with smaller focal length must increase to provide the unchanged threshold radiation intensity at the discharge front. This conclusion does not reflect increasing difference between the calculated values of the threshold power with the increase in the flow rate related to the necessity of fulfilling the intensity balance at the discharge boundaries. Based on this conclusion, let us compare the results of measuring the threshold power for two long-focus lenses shown in figure 5. The lens with *f*=170 mm has a smaller radius $r_f$, since its focal length, as well as the focal length of the lens with *f*=152 mm belong to the descending branch of dependence $r_f$ on *f* in figure 2. Because of this, the threshold power for the lens with *f* =170 mm must be lower.

As was noted above, the existence of the minimum in dependence of $P_{th}$ on $v_0$ is typical of long-focus lenses when the discharge decays at $v_0$=0. The model under consideration does not describe the appearance of the minimum, since $P_{th}$ is calculated using the results of the calculations for $v_0$=0.

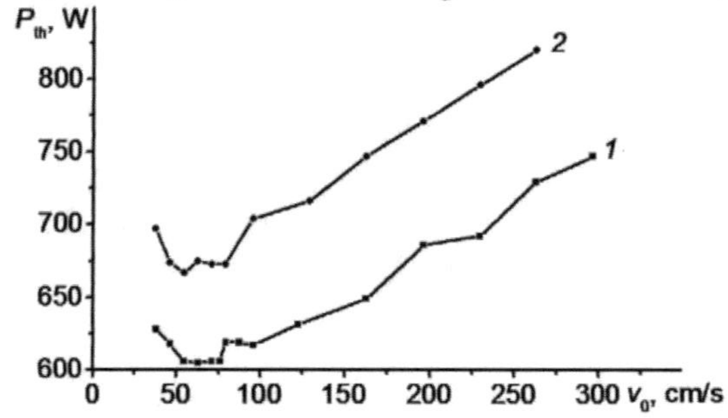


**Figure 5.** Measured dependences of $P_{th}$ on $v_0$ for Ar at *p*=1 atm, *f*=170 (*1*) and 152 mm (*2*), *H*=2.06.

## 5. Conclusions

In this work it was shown that the radiation refraction on the optical discharge noticeably influences the threshold power depending on the gas flow rate as short-focus lenses were used near the boundary of transition to long-focus lenses; the criteria for the short-focus lenses were chosen on the base of previous studies. Spherical aberrations of lenses can also strongly affect the threshold power.

## Acknowledgments

This work was supported by State Corporation “Rosatom”, State contract No. N.4x.44.90.13.1090 from 02.04.2013.

**References**

[1] Bolshakov A, Konov V, Vostrikov V, Dubrovskii V, Kosyrev F, Naumov V, Shachkin L 2008 *Quant. Electron.* **38** 165
[2] Konov V, Uglov S 1998 *Quant. Electron.* **28** 291
[3] Gerasimenko M, Kozlov G, Kuznetsov V and Masyukov V 1979 *Tech. Phys. Letters* **5** 954
[4] Raizer Yu, Surzhikov S 1984 *Quant. Electron.* **14** 1526
[5] Fowler M, Smith D 1975 *J. Appl. Phys.* **46** 138
[6] Glova A, Lebedev F and Yartsev V 1985 *Sov. J. Quant. Electron.* **15** 1633
[7] Galushkin M and Lebedev F 2015 Modern Laser-Information Technologies ed V Panchenko and F Lebedev (Moscow: Nauka) pp 455-472
[8] Kozlov G and Kuznetsov V 1994 *Tech. Phys. Letters* **20** 46
[9] Gerasimenko M, Kozlov G and Kuznetsov V 1983 *Plasma Physics Reports* **9** 1269
[10] Gladush G and Yavokhin A 1983 *Sov. J. Quant. Electron.* **13** 908
[11] Raizer Yu 1984 *Sov. J. Quant. Electron.* **14** 40
[12] Burmistrov V, Glova A, Lebedev F and Yartsev V 1987 *Sov. J. Quant. Electron.* 17 381
[13] Boni A, Su F 1974 *Phys. Fluids* **17** 340
[14] Raizer Yu 1974 Laser Spark and Discharges Propagation (Moscow: Nauka)
[15] Kikoin I 1976 Tables of Physical Quantities (Moscow: Atomizdat)

# Experimental facility for reactor experiments on study of spectral-luminescent characteristics of nuclear-excited plasma

**Yu N Gordienko[1,3], E G Batyrbekov[1], M K Skakov[1], Yu V Ponkratov[1], M U Khasenov[2], Zh A Zaurbekova[1], N I Barsukov[1], T V Kulsartov[1] and Ye Yu Tulubayev[1]**

[1] National Nuclear Center of the Republic of Kazakhstan, Krasnoarmeyskaya str., 2, 071100, Kurchatov, Kazakhstan

[2] National Laboratory Astana, Kabanbay batyr av., 53, 010000, Astana, Kazakhstan

E-mail: gordienko@nnc.kz

**Abstract.** The description of experimental facility and reactor ampoule device for carrying out the experiments on study of spectral-luminescent characteristics of nuclear-excited plasma formed by products of 6Li(n,α)T nuclear reaction under conditions of neutron irradiation is given in paper.

## 1. Introduction

Direct conversion of nuclear reactions' energy is a promising trend in the field of producing large amounts of light energy, including its most perfect form - coherent light. Exceptional features of nuclear power sources such as high power density, small size, the ability to effectively excite and ionize large amounts of active gaseous media at high pressures, give a certain potential advantages over other traditional methods of active media excitation [1]. There are two possible ways of nuclear energy conversion into the energy of coherent light emission: direct and combined nuclear pumping [2]. The efficiency of nuclear energy conversion into laser emission depends on the choice of the active gaseous media excited by the products of nuclear reactions. In this connection, the spectral study of optical emission of nuclear-excited plasma is of interest to solve the problems associated with finding of gaseous media with high efficiency of nuclear reactions energy conversion into energy of laser or spontaneous emission. This paper describes the ampoule device and experimental facility for study of spectral-luminescent characteristics of gaseous media, excited by the products of nuclear reactions in the IVG.1M research reactor. In the experiments on studies of the gaseous mixtures luminescence efficiency for ionization and excitation of gas mixtures was chosen $^{6}Li(n,\alpha)T$ reaction which was not used before.

## 1. Experimental part

### *1.1. Methods of gas excitation by nuclear reactions' products*

For the effective excitation of gas media it is essential that isotopes interacting with neutrons must be in direct contact with gas medium. There are two possible methods of nuclear reactions energy use for ionization and excitation of plasma – use of surface sources of charged particles (inner surface of laser chamber is coated by thin layer of isotope $^{10}B$, $^{6}Li$, $^{235}U$ or its compound $^{235}UO_2$, $^{235}U_3O_8$) and volume

[3] To whom any correspondence should be addressed.

sources of charged particles (gaseous isotope or its compound: $^{3}$He, $^{235}UF_6$, $^{10}BF_3$ include in composition of working mixture).

Figure 1 shows the scheme of experiment to study optical radiation of nuclear-excited plasma formed by nuclear reaction products. For the ionization and excitation of gas media in experiments performed at stationary reactor IVG.1M, was selected a reaction $^{6}$Li(n,α)T that was not used earlier:

$$(^{6}\text{Li} + \text{n} \rightarrow (^{4}\text{He}\ (2.73\ \text{MeV}) + (^{3}T\ \ (2.05\ \text{MeV})) \tag{1}$$

The cross-section of this reaction is 945 barns [3, 4]. In addition, surface method using charged particles was selected for gas media excitation.

*1.2. The scheme of reactor experiments on study of luminescence of gaseous mixtures*

Figure 1 shows the scheme of experiments on study of optical emission of nuclear-excited plasma formed by nuclear reaction's products during thermal neutrons interaction with charged particle's sources.

Light radiation (4) from the ampoule, occurred in result of studied gas mixture excitation by nuclear reaction products is removed from the ampoule through quartz window (6) and, reflecting by hinged mirror (7) it gets to focusing lens (8). Then through optical fiber (9) the light directs to entrance of optical spectrometer (13) and entrance slit of monochromator (10) with photoelectric multiplier (11), operating in photon-counting mode. Luminescence spectrum obtained by optical spectrometer is recorded to computer (14).

The IVG.1M research reactor used as a source of neutron flux for nuclear reactions and excitation of gaseous mixture.

*1.3. IVG.1M research reactor*

The IVG.1M reactor – a water cooled heterogeneous research nuclear reactor on thermal neutrons with water moderator and beryllium neutron reflector.

The main parameters of the IVG.1M reactor:

- thermal power – 30 MW
- effective core size – 548 mm
- core height – 800 mm
- uranium-235 content in core – 4.6 kg
- thermal neutron flux density in center of experimental channel – $3.5\cdot10^{14}$ n/cm$^2\cdot$s.

In Table 1 the neutron flux at the reactor power 6 MW is presented.

**Table 1.** Neutron flux in core center of the IVG.1M reactor at the reactor power 6 MW.

| Energy group | from 0 to 0.67 eV | from 0.67 to $0.1\cdot10^{6}$ eV | from 0.1 to 10 MeV | Integral flux |
|---|---|---|---|---|
| Neutron flux, 1/(cm$^2\cdot$s) | $10.87\cdot10^{14}$ | $0.42\cdot10^{14}$ | $0.22\cdot10^{14}$ | $1.5\cdot10^{14}$ |

*1.4. Experimental bench*

The LIANA experimental facility is designed to carry out studies on hydrogen isotopes mass-transfer processes in structural materials of fission and fusion reactors under neutron irradiation in the IVG1.M research reactor [5]. For reactor investigations of spectral-luminescent characteristics of nuclear-excited plasma the modernization of experimental facility was made. The modernization consists of: finalization of the exhaust system and a complete set of information measuring complex (IMC) with equipment and devices, which convert the measured optical characteristics of the light emission of nuclear-induced plasma in records of registration devices. Figure 2 shows a schematic diagram of the LIANA experimental facility after modernization. The developed additional systems marked with dotted lines.

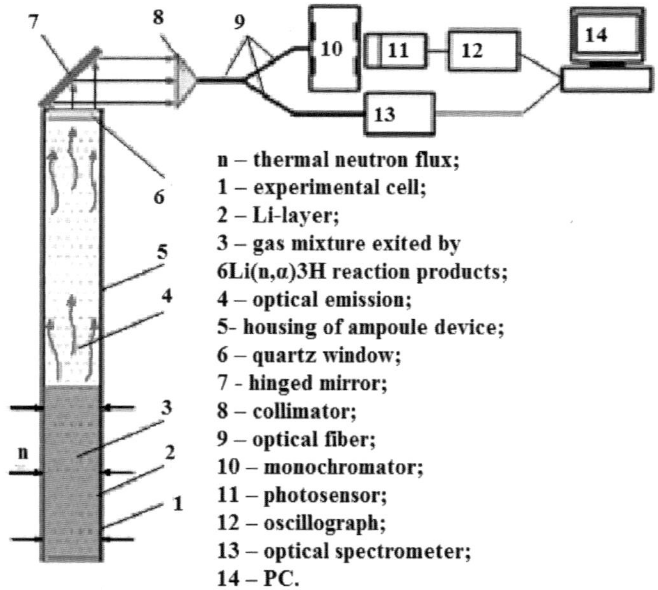


**Figure 1**. Scheme of experiments on study of optical radiation of nuclear-excited plasma.

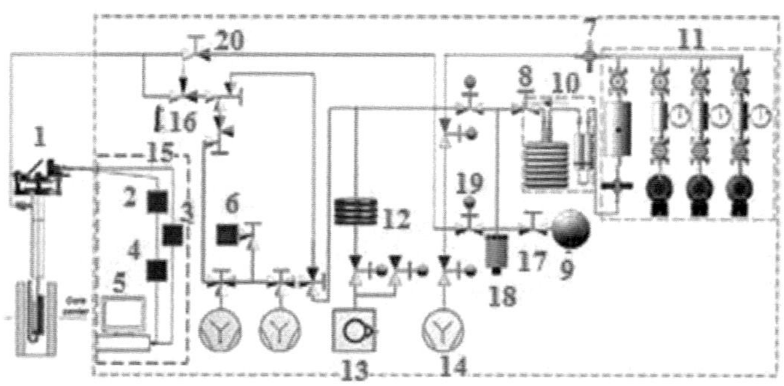

**Figure 2.** Scheme of LIANA experimental facility for experiments on study of spectral-luminescent characteristics of gaseous mixtures.
1 – experimental ampoule device; 2 – monochromator with PEM; 4 – digital oscillograph; 5 – computer; 6 – mass-spectrometer; 8, 17, 20 – manual operated valves; 9 – tank; 10 – gas mixture purification system; 11 – gas mixture preparation system; 12 – nitric trap for for-vacuum pump; 13-for-vacuum pump; 14 – high-vacuum pump; 15 – pressure sensor; 16 – gas regulator; 18 – vacuum-meter VDG-1; 19 – mechanical vacuum valve.

The LIANA experimental facility consists of ampoule device with experimental cell (where the investigated sample of gas mixture is placed) and operational unit (marked with red line in Figure 3). Operational unit consists of vacuum system, ampoule device gas feeding system and informational and measuring complex. Vacuum system allows prepare the facility to work, to create the necessary conditions for experiments and prepare the investigated gaseous mixture of three gases of different concentrations and pressures. Exhaust system provides pumping of investigated gaseous mixture from tank to experimental cell of AD through filtration system. Filtration system was included in exhaust system especially to remove impurities ($N_2$, CO2, $O_2$, $H_2$, etc.). These impurities appear in gaseous mixture during experiments preparation and lead to decreasing of luminescent properties of gaseous mixture. Purification includes a blowing of gases through absorbing device (titanium and copper oxide with temperature of 923 K). Informational and measuring complex provides control and operation of facility's parameters during experiments (see Figure 3).

*1.5. Experimental reactor ampoule device*

To perform reactor experiments upon the scheme presented in Figure 2, the ampoule device (AD) was specially designed with surface source of charged particles (see Figure 3). As source of charged particles as was stated above the natural lithium coating by thin layer to the inner surface of experimental cell of the ampoule device was used (natural composition of lithium isotopes $^6$Li (7.5 %) + $^7$Li (92.5 %). Given below are technical characteristics of ampoule device (Table 2).

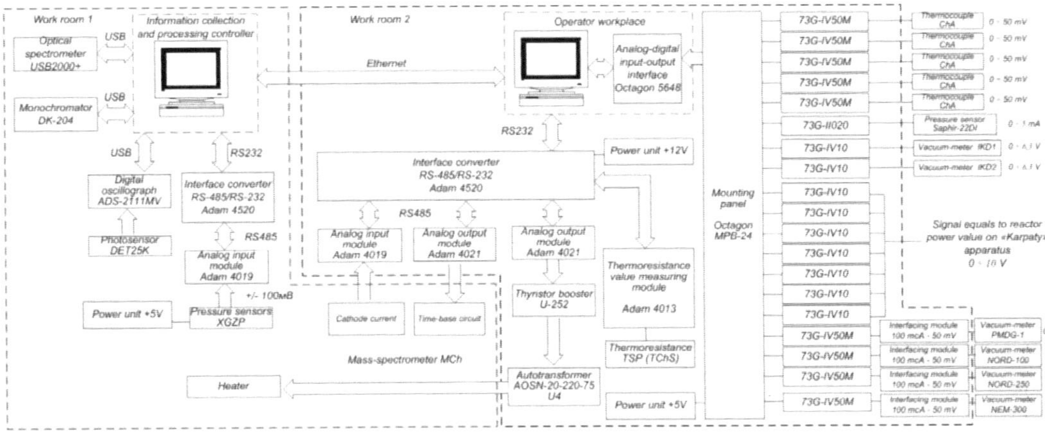


**Figure 3**. Scheme of modernized IMC of LIANA facility.

**Table 2.** Technical characteristics of AD.

| Item | Value |
|---|---|
| AD length, mm | 2300 |
| AD diameter, not more, mm | 40 |
| Experimental cell's length, mm | 200 |
| Experimental cell's diameter, mm | 20 |
| Fissionable material | natural Li |
| Lithium layer thickness, μm | 50 |
| Analyzed gases and mixtures | He, Ar, Ne, Kr, Xe |
| Electric heater capacity, from W | 400 |
| Thermocouple's material (ChA) | chromel-alumel |
| Analyzed pressure of gases in volume of AD, atm | from 0.05 to 1 |
| AD, experimental cell body's material | stainless steel 12Ch18Ni10Ti |

*1.6. Technique of reactor experiments*

Reactor experiments to study spectral luminescent characteristics of gaseous mixtures excited with nuclear reaction $^6Li(n,\alpha)T$ products will be carried out using the designed ampoule device according to the following scheme:

- Ampoule device is loaded in experimental channel of IVG.1M reactor. Thus, AD experimental cell was placed at the level of reactor core center.
- Hereafter, the estimation of reactor reactivity change made by ampoule is carried out and thereafter all necessary technological procedures on preparation of loaded irradiated ampoule device and LIANA bench to carry out reactor experiments were conducted.

- Then, prior to reaching the targeted level of power with a reactor, the studied premixed gas accumulated in the reserve storage (9) (see Figure 2) is supplied into the ampoule device capacity up to required pressure.
- After this, the reactor is set on the targeted level of power and the measurements of light fluxes are carried out. The light radiation occurred as a result of excitation of studied gas mixture by nuclear reaction products influenced by neutron flux, withdraws from the ampoule through quartz window and is fed to optical spectrometer input and entrance slit of monochromator with PMT operating in photon counting mode using a collimator and optical fiber cable (with two light valves).
- Then, luminescent spectrum are recorded at stationary level of power, whereafter IVG.1M reactor is set on the succeeding stationary level of power and the following measurements of light fluxes are carried out.
- When luminescent spectrum of gas mixture at all studied stationary levels of reactor power are measured, the planned reactor shutdown will be done.

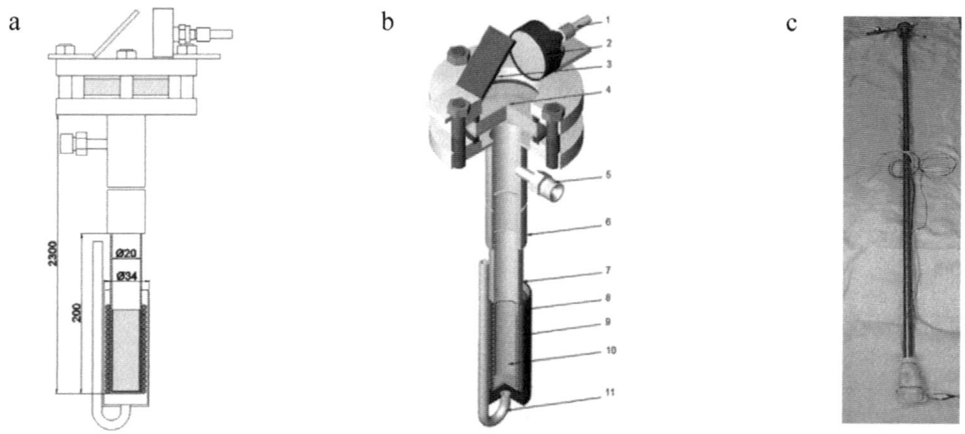


**Figure 4**. Scheme (a) and view (b) of reactor ampoule device for investigations of optical radiation of nuclear-excited plasma. 1 – fiber-optic lightguide; 2 – collimator; 3 – hinged mirror; 4 – quartz window; 5 – gas mixture pumping and supply tract; 6 – ampoule device body; 7 – experimental cell body; 8 – cooling casing; 9 – heater; 10 – (lithium) fissile materials thin layer; 11 – cooling tract.

## Conclusion

Conducted out-of-pile experiments with designed ampoule device showed the reliable operation of temperature regulation and stabilization systems and also systems of studied gaseous mixtures supply and pressure control in ampoule device. Thus, the designed ampoule device completely meets all requirements of reactor experiments and is ready to conduct experiments for studying optical radiation of nuclear-excited plasma produced by nuclear reaction products in neutron radiation field of IVG.1M reactor.

## Acknowledgements

The work supported by the Ministry of Education and Science of the Republic of Kazakhstan, Grant #2068/GF4.

## References

[1] Batyrbekov E 2013 Converting nuclear energy into the energy of coherent optical radiation *Laser and Particle Beams* **31** 673-687
[2] Batyrbekov E G 1989 Electric-discharge xenon laser with weak ionization by an external source

*Kvantovaia Elektronika* **16** 2165-2169
[3] Mis'kevich A I 2011 *Direct nuclear energy conversion into light and laser radiation: Training manual* (M.: NRNU MEPhI) p 164
[4] Khasenov M U 2011 *Nuclear-excited plasma of gaseous mixtures and nuclear pumped lasers: Monograph* (Almaty) p 187
[5] Gordienko Yu N 2014 Application of hydrogen permeability method in reactor experiments on study of hydrogen isotopes interaction with structural materials *Tomsk Polytechnic University Bulletin* **324** 149-162

# The role of the heating of the vacuum chamber on the water content in plasma and gas

**A V Bernatskiy[1, 3], V N Ochkin[1] and R N Bafoev[2]**
[1] P.N. Lebedev Physical Institute of the Russian Academy of Sciences, 53 Leninskiy Prospekt, Moscow, 119991, Russia
[2] Moscow Institute of Physics and Technology (State University), 9 Institutskiy per., Dolgoprudny, Moscow Region, 141700, Russia

E-mail: berav@sci.lebedev.ru

**Abstract**. The behavior of the H2O concentration in stainless steel chamber is compared for the cases of gas with and without plasma discharge at different temperatures of chamber wall. The experiments were performed with the use of mixtures of rare gases and water vapor. In absence of plasma discharge the dynamics of the density of the water molecules was determined by their interaction with the surface. In the presence of a gas discharge the concentration of water molecules in the volume of the chamber was determined by optical actinometry method. The role of the chamber wall temperature is discussed.

## 1. Introduction

Monitoring the content of water molecules in the gas phase in vacuum and plasma discharge installations is well known problem [1]. In vacuum systems it is primarily associated with the adsorption of molecules on vessel surfaces. In this case for the relatively low temperature the chemical transformations of molecules are negligible so the adsorption can be considered as purely physical mechanism. But situation changes dramatically in the presence of the plasma discharge inside the vessel, even if it's not located close to the wall. In this case one has to take into account the non-equilibrium plasma reactions and heterogeneous phase reaction, not only between water molecules, but also between their reaction products. Up until recently, there were no methods of measurement of concentration of water molecules directly in plasma. Modern demand for those caused by the development of technology, especially in powerful electric devices, including the ITER project [2]. The reactor working conditions are limited by water leakage at the rate of $10^{-7}$ $Pa{\cdot}m^3{\cdot}s^{-1}$ [2]. Such small flows can occur both because of the defects of the camera shell or due to the water self release from the surface when the temperature changes. Spectral methods for control of water vapor content in the cold plasma of glow discharge for searching and eliminating the source of subsequent leaks, have been recently proposed in [3, 4].

In the study of dynamics of the water vapor density in the model installation [3], it was observed that the concentration of water molecules in the plasma during the initial period (<1 min) decreases by several times with respect to the original, and then reducing of the concentration slows down. In this paper these stages are studied more detailed.

[3] To whom any correspondence should be addressed.

## 2. Experimental setup

The device and its experimental setup were described in [3, 4]. The vacuum chamber, flanges and gas valves are made of stainless steel 12X18H10T. The profiled sealing copper gaskets are used as a vacuum tight connectors. Unlike [3], the volume of the vacuum chamber is reduced to 22000 cm$^3$ due to the shutdown of the sorption pump. The square of the inner surface of the vacuum chamber is 5900 cm$^2$. The hollow cathode, which volume is 50 cm$^3$, was installed to the center of chamber.

Experiments were carried out at different temperatures of the chamber wall. High temperature were obtained by using an external electric heater in the form of a flat tape with a heat-resistant insulation, which allowed to heat chamber to 295-450 K. Heating and temperature consistency (±5 K at 420 K) were controlled by thermocouples in various zones of the chamber surface.

The concentration of water molecules in the plasma discharge was measured using optical actinometry method [4] by comparing the intensities of the lines of the vibrational-rotational spectra of "hot" hydroxyl [5] in the 306 nm and line of Xe (823.2 nm). The gas mixture for discharge consisted of 99%He + 1%Xe + additive of water vapor. The gas temperature in the plasma discharge was determined from the relative intensities of lines from the "cold" hydroxyl vibrational-rotational band (also in the area of 306 nm) [3, 5].

In all cases the vacuum chamber was pumped out for several hours keeping the walls temperature $T$=(400-420) K before the gas injecting. Measurements were carried out either with pre-heated walls or after their cooling those to room temperature. Results are shown for the case of a mixture gas He:$H_2O$:Xe=99:33:1 at a total pressure of 0.5-1 mbar, the discharge current of 0.28 A.

## 3. Experimental results and discussion

The results of measurements are shown in figures. 1.a, 1.b for the wall temperatures 295±3 K and 395±5 K correspondingly. The initial concentration of water molecules was $n_0$=3.5·10$^{15}$ cm$^{-3}$.

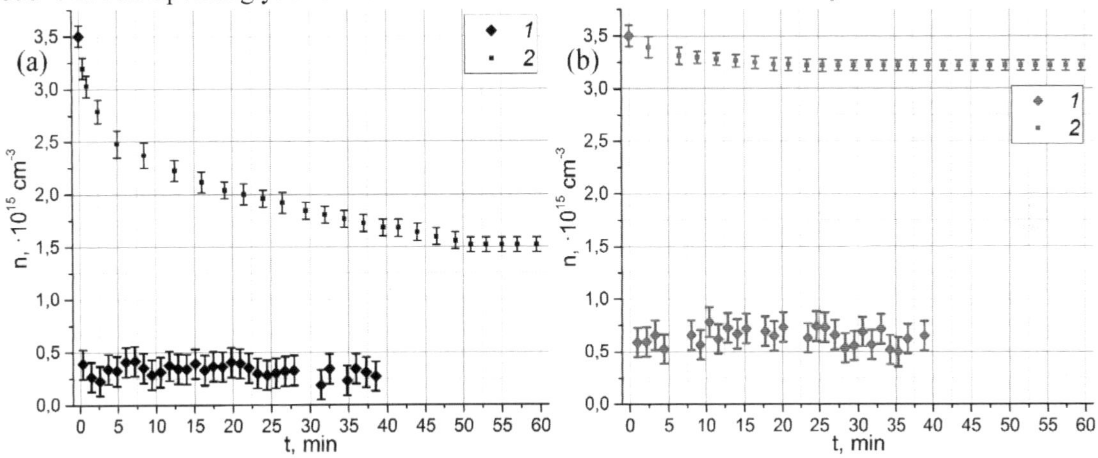


**Figure 1.** Concentration of water molecules changes with time. The initial concentration of water molecules was $n_0$=3.5·10$^{15}$ cm$^{-3}$. (a) chamber without heating; (b) chamber is warmed to 395 K. *1* – with discharge, actinometrical measurements; *2* – without discharge, manometeric measurements.

Manometric measurements were carried out without the discharge. The required number of water molecules were injected to the chamber and the dynamics of water vapor density with time was investigated. The concentration of water molecules in the vacuum chamber was determined and controlled by pressure sensor. Figures 1.a.2, 1.b.2 shows that the molecular density is decreasing with time and the drop saturates after 10 minutes in warm case and after 40 min in case of cold walls. For cold case the reduction of molecules is deeper (~45%) than for the heated one (~8%). The rate and magnitude of water molecules loss depends on the wall temperature, which agrees with the assumption that the physical mechanism of adsorption of molecules on surfaces is taking place.

The measurement results confirm the results of calculations [6, 7], compared with those given in table 1. Wherein: $T$ – the temperature of the vacuum chamber walls; $N$ – concentration of water molecules in a state of dynamic equilibrium with the adsorbed molecules; $P$ – partial pressure of water molecules; $N_A$ – number of water molecules adsorbed on the chamber walls; $\theta$ – number of monolayers of adsorbed particles in the saturation stage (degree of coating); $T^*$, $P^*$, $\theta^*$ – relevant parameters from [6, 7].

**Table 1.** A comparison of the measurements with the calculations of the work [6, 7], no discharge in chamber. The partial initial density of water vapor $n_0$=3.5·10$^{15}$ cm$^{-3}$.

| $T$, K | $N \cdot 10^{19}$ | $P$, Pa | $N_A \cdot 10^{19}$ | $\theta$ | $T^*$, K | $P^*$, Pa | $\theta^*$ |
|---|---|---|---|---|---|---|---|
| 295 | 4.8 | 7 | 2.9 | 3.9 | 295 | 7 | 4 |
| 395 | 7 | 12.1 | 0.7 | 0.9 | 395 | 12 | 0.7 |

In actinometrical measurements we take into account the thermal pushing of actinometer particles out of the discharge area, whose role in this case played a xenon [4]. The gas temperature in the discharge was measured by the intensity distribution in the rotational structure of the spectrum of "cold" hydroxyl [3]. Additional heating of the surface of the hollow cathode was controlled by radiometer, and taken into account during analysis of the results.

The behavior of water molecules in time with the discharge switched on changes significantly compared to the previous case. The concentration of molecules drops much faster with characteristic time t1 and deeper in initial phase (figures 1.a.1, 1.b.1). By the times of more than tens of minutes concentration of molecules is practically maintained in the second phase. The wall temperature does not affects on water concentration essentially on the first phase. In our opinion electron impact is responsible for the fast drop of water concentration in the first phase the dissociation. In the work [3] the rate constant $<v_e \cdot \sigma_{e,diss}>$ of the process of dissociative excitation of the $H_2O$ molecule ($v_e$ – electron velocity, $\sigma_{e,diss}$ – cross section of dissociative excitation of the molecule $H_2O$) was measured under similar conditions and the corresponding characteristic dissociation time was found equal $\tau_{diss}=(<v_e \cdot \sigma_{e,diss}>)^{-1}$~5 s. This time up to a factor 2-3 coincide with those that we observed in the first phase $t_1 \approx \tau_{diss}$.

The second quasi stationary phase also reveals some features which does not agree with predictions of thermodynamic adsorption theory. In the framework of this theory the total balance of particles remain constant and the amount of particles split up between the free volume molecule and the molecules captured by the surface. It becomes obvious if one compare the data from figures 1.a.1, 1.b.1 and from the table 1. In the absence of discharge the change of the wall temperature from 295 K to 395 K leads to releasing ~2·10$^{15}$ cm$^{-3}$ molecules from the surface to the bulk. In the presence of discharge the same wall heating leads to releasing ~0.3·10$^{15}$ cm$^{-3}$ molecules which about 10 times less. It definitely means that the structure of surface layers get modified by discharge despite the fact that the low power (50 W) discharge in our experiments is located quite far from the walls. The probable explanation is that the discharge generate the active particles which reacts with surface particles thus transform the adsorption mechanism from pure physical to more complicated one so formally $\theta_T \neq \theta_P$, where $\theta_T$, $\theta_P$ – numbers of absorbed monolayers in case of thermal equilibrium and in case of plasma discharge influence respectively.

## 4. Conclusion

The behavior of water vapor concentration with time in vacuum chamber with stainless steel walls was studied. The measurement results are correspond to the physical adsorption theory [6, 7] at different temperature conditions. The local plasma source in vacuum chamber far from its wall significantly changes the dynamics of the density of the water vapor. That is caused by the process of emergence of a new active particles in a plasma-chemical decomposition of the starting molecules which is faster than the adsorption process. Measurements lead to the conclusion that the presence of these particles can significantly change the particle desorption mechanism on the surface.

**Acknowledgements**
This work was supported by the Russian Science Foundation, grant no. 14-12-00784.

**References**

[1] Klyuev V V 2003 *Nerazrushayushchij kontrol. Spravochnik* (Moscow: Mashinostroenie) (in Russian)
[2] ITER Final Design Report No. G 31 DDD 14 01.07.19 W 0.1, Section 3.1: Vacuum Pumping and Fuelling Stems, 2001.
[3] Bernatskiy A V, Ochkin V N, Afonin O N and Antipenkov A B 2015 *Plasma Physics Reports* **41** 705
[4] Bernatskiy A V and Ochkin V N 2015 *Bulletin of the Lebedev Physics Institute* **42** 273
[5] Ochkin V N 2009 *Spectroscopy of Low Temperature Plasma* (New York: WILEY-VCH) p 609
[6] Rozanov L N 2011 *Vakuumnaya tekhnika i tekhnologiya* **21** 143 (in Russian)
[7] Rozanov L N 2012 *Vakuumnaya tekhnika i tekhnologiya* **22** 197 (in Russian)

Journal of Physics: Conference Series **747** (2016) 012014 doi:10.1088/1742-6596/747/1/012014

# Simulation of the dynamics of the PF-3 device's plasma

**S S Ananyev, A M Kharrasov and S V Suslin**[1]

National Research Centre "Kurchatov institute", 1, Akademika Kurchatova pl., Moscow, 123182, Russia

E-mail: SuslinSV@yandex.ru

**Abstract.** We have simulated the motion of the current-drive plasma shell (CDPS) in the PF-3 facility – from discharge start to appearance of plasma focus. For this purpose, we modified one-component MHD model, taking Hall effect into account. The extended simulation matches CDPS "runaway" mode, obtained in the PF-3 facility.

## 1. Introduction

Plasma focus (PF) is a kind of pinch phenomena. Dense plasma focus appears as a result of plasma constriction, that takes place on the axis of facility. Through the cumulative heating and high plasma density, PF is the source of powerful X-ray, neutron and other emissions. Withal, the supersonic axial plasma jet was launched from PF in many experiments. The same jet may be useful for purpose of surface alloying [1], or material testing [2].

## 2. Model

The simulation is based on the model previously used in work [3], which assumes single-fluid ideal two-dimensional MHD approximation. It means using equations noted below:

$$\frac{\partial n}{\partial t} + \mathrm{div}(n\vec{V}) = K(r, z, t)$$

$$\frac{\partial B_\vartheta}{\partial t} = -\frac{\partial}{\partial r}(V_r B_\vartheta) - \frac{\partial}{\partial z}(V_z B_\vartheta)$$

$$\frac{\partial (nV_r)}{\partial t} + \frac{\partial (rnV_r^2)}{r\partial r} + \frac{\partial (nV_r V_z)}{\partial z} = -\frac{2}{rm_i}\frac{\partial}{\partial r}(rnT) + \frac{1}{4\pi m_i}\frac{\partial rB_\vartheta}{r\partial r}B_\vartheta$$

$$\frac{\partial (nV_z)}{\partial t} + \frac{\partial (nV_z^2)}{\partial z} + \frac{\partial (rnV_r V_z)}{r\partial r} = -\frac{2}{m_i}\frac{\partial}{\partial z}(nT) + \frac{1}{4\pi m_i}\frac{\partial B_\vartheta}{\partial z}B_\vartheta$$

$$\frac{3}{2}\left(\frac{\partial nT}{\partial t} + \mathrm{div}(nT\vec{V})\right) = -nT\mathrm{div}(\vec{V})$$

where $n$ — concentration of ions; $V$ — velocity of plasma flow; $B_\vartheta$ — toroidal magnetic field ; $r$ and $z$, — radial and axial coordinates correspondingly; $T$ — temperature of electrons, $m_i$ — ion mass, $K$ –function of three variables, that indicates ionization.

We presumed that plasma is singly ionized in terms of plasma dynamics. Ionization was considered in thin area near the insulator through $K(r, z, t)$ function of three variables. The set of equations was

[1] To whom any correspondence should be addressed.

Journal of Physics: Conference Series **747** (2016) 012014 doi:10.1088/1742-6596/747/1/012014

completed with equation of power supply circuit, and boundary conditions – zero speed on the surfaces, temperature equals to 3 eV, and magnetic field strength is continues:

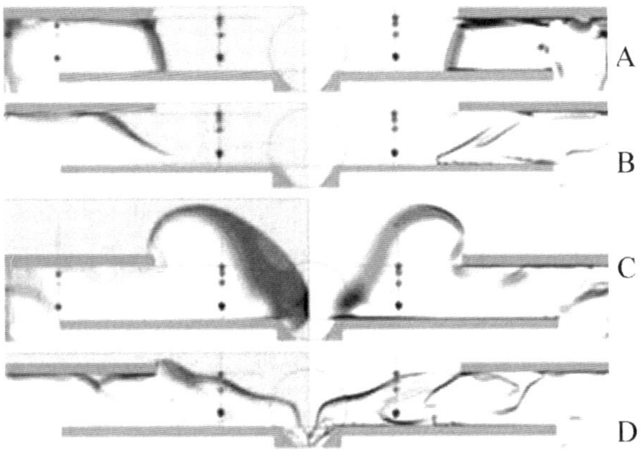


**Figure 1.** Simulated density map (left side) and current map (right side). a, c – without Hall effect; b, d – with Hall effect. Color circles mark magnetic probe's position.

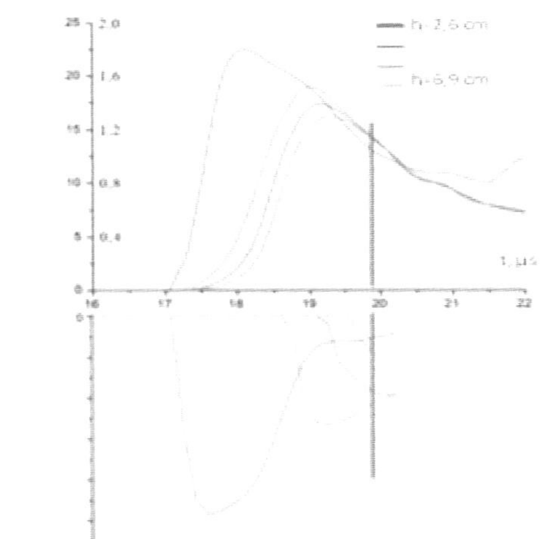

**Figure 2.** Simulated signals of magnetic probes, placed at a distance of 16 cm from the axis and height of h above the anode, without Hall effect (top curves); and with Hall effect (bottom curves)

$$V_r = 0; V_z = 0; T = 3\ \text{eV};$$

Magnetic field strength on the insulator surface is equal to:

$$B_\vartheta = \frac{\mu_0 I}{2\pi r};$$

*I*– discharge current.

The set of equations doesn't contain dissipation. It means infinite conductivity to calculate the energy balance and plasma motion.

All simulations were built for Ne plasma with initial pressure $P_0$ = 1-1.5 Torr, initial voltage $U$ =8-9 kV, stored energy $W$ =290-370 kJ

We varied pressure and parameters of power supply (initial voltage, capacitance and inductance); and we have reached the match of simulated plasma dynamics with one observed by optoelectronic transducer in the vicinity of pinch (within central circle in figure 1). On the contrary, we were unable to match the simulated signals of magnetic probes with the observations.

Magnetic probes show the difference of axial and radial plasma velocities. In the experiment radial propagation is faster than axial one. It means that an additional phenomenon should be taken into account.

There is another feature in simulations – simulated CDPS is more cylindrical with respect to experimental observed CDPS (figure1). It means that its opening angle is noticeably less than one observed with optoelectronic transducer. Furthermore, there is the mode with extremely high opening angle (so called "runaway" mode) [4]. This feature isn't explained by the ideal MHD only. Authors of the papers [5] and [6] explain "runaway" phenomenon through the Hall effect, caused by electrode evaporation. It results to lower gas sweeping; and then to faster radial propagation of CDPS. Axial propagation saves its normal speed, so we observe high opening angle.

We tried to build simulation based on extended MHD (xMHD) model.

First, we estimated how much is influence of the Hall effect.

Equation of magnetic field evolution (in terms of xMHD):

$$\frac{\partial B_\vartheta}{\partial t} = -\frac{\partial}{\partial r}(V_r B_\vartheta) - \frac{\partial}{\partial z}(V_z B_\vartheta) + B_\vartheta\left(\frac{2\varkappa}{r} - \frac{\partial \varkappa}{\partial r}\right)\frac{\partial B_\vartheta}{\partial z} + B_\vartheta\frac{\partial \varkappa}{\partial z}\left(\frac{B_\vartheta}{r} + \frac{\partial B_\vartheta}{\partial r}\right)$$

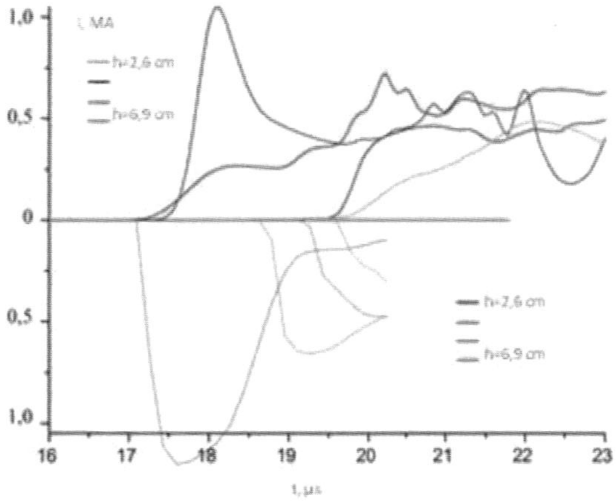


**Figure 3.** Measured signals of magnetic probes, placed at a distance of 16 cm from the axis and height of h above the anode (top curves); and simulated with Hall effect signals (bottom curves).

- $\varkappa = \frac{c^2\chi}{4\pi}$; $\chi = \frac{1}{n_e e c}$; – Hall parameter.

Hall effect preferably appears in areas of density drops. Low electron density and magnetic field drops are also important conditions. These conditions are most possible behind of CDPS. Let's examine how important is the Hall effect. For this purpose we've compared current speed with flows one:

$$V_{cur} = \frac{j}{en_e} = \frac{c\left|rot(\vec{B})\right|}{4\pi en_e} \approx \frac{cB_\vartheta}{4\pi en_e \Delta_{CDPS}} \approx \frac{10^{22}}{n_e} \gtrsim V_{flow} \approx 10^7\frac{cm}{s};$$

This inequality is true when $n_e \lesssim 10^{15} cm^{-3}$;

In practice, most of current is concentrated in back surface of CDPS. It means $\Delta_{CDPS}$ is less; so $n_e$ might be more than $10^{15} cm^{-3}$. Under this condition we need to take the Hall effect into account.

It's interesting to calculate evolution rate of the magnetic field strength caused by plasma flow and Hall effect:

$$\left(\frac{\partial B_\vartheta}{\partial t}\right)_{flow} = -\frac{\partial}{\partial r}(V_r B_\vartheta) - \frac{\partial}{\partial z}(V_z B_\vartheta) \approx \frac{V_{CDPS} B_\vartheta}{\Delta_{CDPS}} \approx \frac{10^7 \cdot 10^4}{3} = 0{,}33 \cdot 10^{11} G/s;$$

- $V_{CDPS}$ is CDPS's velocity and $\Delta_{CDPS}$ is its thickness. Both experiments and numerical simulation show $\Delta_{CDPS}$ =3 cm [7].

$$\left(\frac{\partial B_\vartheta}{\partial t}\right)_{Hall} = B_\vartheta \frac{\partial \varkappa}{\partial z}\left(\frac{B_\vartheta}{r} + \frac{\partial B_\vartheta}{\partial r}\right);$$

At 10 cm radii, Hall effect results to:

$$\left(\frac{\partial B_\vartheta}{\partial t}\right)_{Hall} = -B_\vartheta \frac{c}{4\pi n_{e,min} eh}\left(\frac{B_\vartheta}{\Delta_{\text{ТПО}}}\right) = 10^4 \frac{3 \cdot 10^{10}}{4\pi \cdot 10^{15} \cdot 1 \cdot 4{,}8 \cdot 10^{-10}}\frac{10^4}{3} = 0{,}17 \cdot 10^{12} \text{Гс/с};$$

$n_{e,min}$ is least concentration of electrons, which reaches to $10^{15}$см$^{-3}$. As CDPS moves to axis, $\frac{B_\vartheta}{r}$ increases. Thus we have to consider this member of equation at the late stage of discharge.

Unfortunately, it's too hard to simulate Hall effect influence – it makes calculation unstable. But we have built simplified model, which allows us to calculate plasma dynamics through all the discharge. So we used the equation showed below for calculate field magnitude:

$$\frac{\partial B_\vartheta}{\partial t} = -(k_x + \frac{k_x^*}{r} + 1)\frac{\partial}{\partial r}(V_r B_\vartheta) - \frac{\partial}{\partial z}(V_z B_\vartheta);$$

- $k_x$ and $k_x^*$ - parameters of Hall effect rate. We had observed that $k_x = 2{,}7$; and $k_x^* = 13{,}2$ provides model matches experiments.

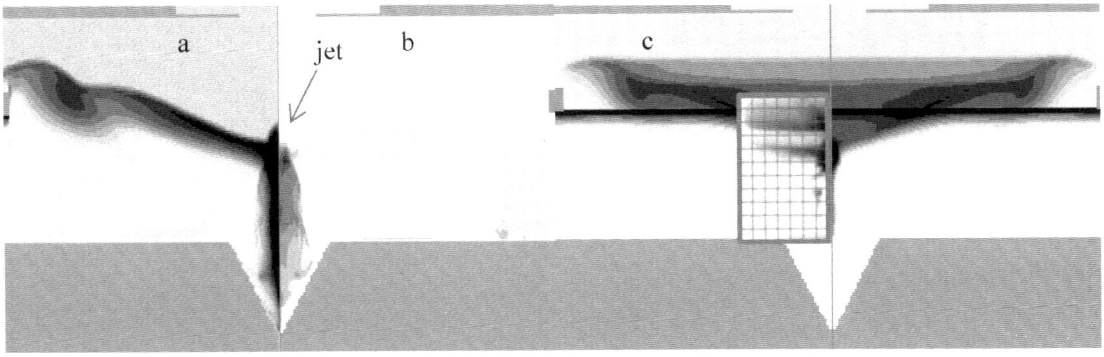


**Figure 4.** Simulation: a – density map and b – magnetic strength map at jet's generation; c – integrated (along line of probing) of density square; optoelectronic image is inserted.

This model is extremely simplified, but it simulates plasma dynamics with high consistency to measurements.

## 3. Discussion

Figure 1 contains results of the simulation (both with Hall effect and without one). It shows strong difference between the cases. When Hall effect is applied, the "runaway" phenomena occurs. The axial spreading of CDPS is slower than radial one – because of Hall effect provides faster magnetic field radial advance (it transports magnetic field in addition to the plasma flow).

So Hall effect influences CDPS shape. But it also affects the temporal parameters of discharge. Figure 2 shows the difference of magnetic probes signals simulated. In the case of Hall effect is neglected, field was registered by four probes with slight time delay - about one microsecond. But with Hall effect, registration delays was more significant – about 3 μs from first to fourth signal. It's noticeable, the same delays also observed experimentally (figure 3) [7].

The forming of plasma focus is shown in figure 4. For purpose of comparison with experiment, we have built image of integral distribution of density square (along a line of probing). The simulated image has the same shape as measured. It also has the same dimensions.

The simulation well corresponds with the real discharge plasma dynamics [7]. After the CDPS collision, the focus compression starts. At this moment, the current breaks rapidly (figure 3). The pinch formed near anode and then it lengthens due to zipper-effect. After that plasma axially leaves a plasma focus in the axial direction. And it forms the plasma jet (figure 4).

It's noticeable, the jet contains strong magnetic field. So it's possible, jet is magnetic-accelerated. In this case, jet's radial propagation is slowly than axial – because of magnetic field is approximately proportional to 1/r (within the jet). Then magnetic pressure is highest on the edge of the jet.

Note, plasma jet contains dense "core" (on the axis), "magnetic bubble" (surrounds core), and dense sheath (which cover other parts). The core determines the upper limit of field magnitude (due to 1/r law). But the core is held by surrounded field. So it is subject of instabilities. It leads to some "clumps" (relatively dense plasma formation) appearance.

## References

[1] Soto L *et. al.* 2014 *Physics of Plasmas* **21** 122703
[2] Ivanov L I *et.al* 2008 *Physics and chemistry of materials treatment* **1** 32-37 [in Russian]
[3] Baronova E O *et al.* 2012 *Plasma Physics Reports* **38** 751-760
[4] Gureev K G *et. al.* 1975 *Fizika plasmy* **1** 120-126 [in Russian]
[5] Vikhrev V V and Gureev K G 1977 *Nuclear Fusion* **172** 291-295
[6] Filippov N V *et. al.* 1998 *ICPP & 25-EPS CCFPP*
[7] Mitrofanov K N *et al.* 2014 *JETP*. **5**. 910-923

Journal of Physics: Conference Series **747** (2016) 012015 doi:10.1088/1742-6596/747/1/012015

# The research of interaction of the capillary discharge with metal foils

**D L Kirko[1] and I D Egorov**
National Research Nuclear University MEPhI (MEPhI), Kashirskoe shosse 31, Moscow, 115409 Russian Federation

E-mail: kirko@plasma.mephi.ru

**Abstract.** The properties of capillary discharge under its interaction with various metal foils are analysed. Spectral composition of capillary discharge jet is investigated. Upon jet interaction with metal foils, plasma domains occur. The properties of glowing plasma domains, which occur in a constant magnetic field, are analysed. The possible internal structure of plasma domains is analysed.

## 1. Introduction
Upon adding organic and inorganic matters to the erosive discharge, glowing plasma domains occur. The lifetime of these domains is up to 1 sec [1-3]. In this version of capillary discharge, a construction with a thin channel in the dielectric and one electrode closing the capillary is used [4,5]. Resulting from the current, which passes through the capillary, intense metal evaporation, takes place and intense plasma jet emerges in the outlet of the capillary. Upon interaction of the capillary discharge with aluminum foil, plasma domains separated from the discharge area occur. The lifetime of these domains is up to 0.5 sec [6]. Taking into account these circumstances, we decided to investigate the properties of these plasma domains and their possible internal structure.

## 2. Experimental setup
In order to obtain the capillary discharge, we used a dielectric arrester (plexiglass, 3-4 mm thick). The diameter of little hole - the capillary was equal to 1-3 mm. A capacitor battery was used as a power supply unit (voltage 200-250 V, energy 100-250 J). The discharge was ignited by a high voltage pulse. As a result, a plasma jet with cylindrical form occurred. The jet length was equal to 8-12 cm, the jet diameter was equal to 3-5 cm. The discharge current was equal to 80-140 A, the pulse duration was equal to 6-10 ms. Nikon 1 J1 camera was used for video- and photo recording of discharge characteristics (temporal resolution 1 ms).

Interaction of the capillary discharge with metal foils was investigated with the help of experimental setup, which is shown in figure 1. Near the discharger (1), at a distance of 1.0-3.0 cm from the capillary, a thin metal foil (4) was located in the holder (material: aluminum, tantalum, copper, brass, 0.05-0.1 mm thick). The jet (2) interacted with this foil. Initial shape of jet is shown as dotted line. As a result of interaction of jet with metal foil, the jet length decreases. Resulting from the interaction, plasma domains (5) occurred and maintained horizontal motion. The dimensions of plasma domains were equal to 0.1-3 mm, their lifetime was equal to 0.02-0.5 s. There were about 1-20 plasma

[1] To whom any correspondence should be addressed.

domains during one shot. The velocity of plasma domain motion was equal to 0.5-8 m/s. Their glowing characteristic color was white or blue. As a rule, they usually moved in the direction of jet propagation in a cone with an angle of about $120^0$.

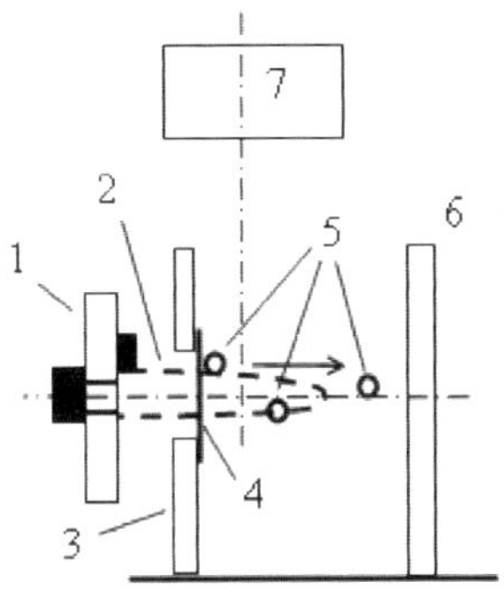


**Figure 1.** Experimental setup, which was used for analysis of interaction of the capillary discharge with metal foils: 1- discharger, 2- capillary discharge jet, 3- holder, 4- metal foil, 5- plasma domains, 6- barrier, 7- spectrometer

## 3. Spectral measurements

We decided to measure temperature and plasma concentration in the area of interaction between capillary discharge jet and metal foil. The spectral methods were implemented. Ava Spec 2048 spectrometer (spectral range 200-1000 nm, spectral resolution 0.3 nm) was used for obtaining of electromagnetic radiation panoramic spectrum. The temporal characteristics were investigated with the help of MUM monochromator (spectral range 200-800 nm, spectral resolution 0.2 nm) and FEU-85 photoelectron multiplier (time resolution 5 ns). The image in figure 2(a) shows the discharge spectrum obtained where the jet interacts with tantalum foil. The discharge spectrum was registered perpendicularly to the jet axis at the distance $x \approx 2.2$ cm from the capillary.

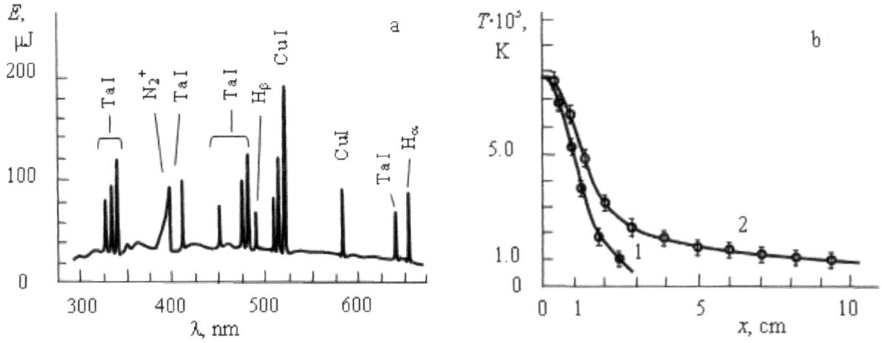


**Figure 2.** Results of spectral measurements, conducted during the interaction of capillary discharge jet with tantalum foil: (a) discharge spectrum; (b) temperature distribution along the jet axis (line 1 – interaction occurs; line 2 - without the interaction)

Of the highest intensity are the lines of atomic copper, Cu I 521.8 nm; atomic tantalum, Ta I 482.5 nm and 335.8 nm; $N_2^+$ molecular nitrogen ion, 378…390 nm. The atomic hydrogen lines of Balmer series $H_\alpha$ 656 nm and $H_\beta$ 486 nm have the lower intensity. We supposed that local thermal equilibrium takes place in the jet plasma. In this case the method of relative intensities of spectral lines can be used. By means of the atomic hydrogen lines $H_\alpha$ and $H_\beta$, we calculated the temperature in the area of interaction between capillary discharge jet and metal foil. The foil was located at the distance $x$=1.8-2.4 cm from the capillary, temperature range was equal to $T$=1000-2200 K. We obtained two temperatures (figure 2(b)): under the jet interaction with the tantalum foil (line 1) for the foil located at the distance $x$=1.8-2.4 cm from the capillary; without the interaction (line 2). The maximum plasma temperature, $T$=7800±300 K, takes place near the capillary. The hydrogen lines, $H_\alpha$ and $H_\beta$, have essential broadening $\Delta\lambda$=5-12 nm that might be attributed to Stark effect. We calculated the plasma concentration in the interaction area according to the Stark $H_\alpha$ and $H_\beta$ line broadening. At the distance $x\approx 2.1$ cm, the plasma concentration was equal to $n_e=(1.4\pm0.2)\cdot10^{16}$ cm$^{-3}$.

## 4. The investigation of plasma domain characteristics

At first we registered plasma domain characteristics which occurred as a result of interaction between jet and aluminum foil. As a result of interaction between plasma domains and vertical plexiglass barriers, the plastic domain elastic reflection occurred. In some cases drastic damping of plasma domain motion was registered. We assumed that plasma domains have totally formed structure and elastic shell. The occurrence of plasma domains, which took place as a result of interaction with tantalum foil, had some special characteristics. The number of plasma domains greatly increased, their dimensions used to increase too. In some cases up to 3 spherical plasma domains occurred (2-3 mm diameter). In other cases domains with smaller dimensions (0.1-1 mm diameter) and irregular shape occurred. The image in figure 3 shows the motion of plasma domains, which have dimensions in the range of 3 mm and are surrounded by a halo, and the motion of 17 plasma domains, which have smaller dimensions.

**Figure 3.** Photo of plasma domains ($t$=6.8 ms after discharge initiation, time resolution is equal to 1 ms)

The process of plasma domain splitting in the smaller domains is quite unusual. This process occurs spontaneously or as a result of plasma domain interaction with barriers. The new small domains move independently and exist for some time. The process of mutual interaction of two domains, which are located at a distance that is equal to several domain dimensions, is also very interesting. This interaction is complex and cannot be reduced to the predominance of attraction or repulsion.

In the area, inside of which moving plasma domains occurred, magnetic field was formed. The magnet with magnetic induction $B$=500-700 Gs was located at a distance $l$=7-15 cm from capillary discharge, at an angle $\alpha=0$-$90^0$. Plasma domains, which occurred as a result of interaction between the jet and the foil, moved toward the magnet. The velocity of plasma domain motion $v$ was equal to 0.5-1.5 m/s. Some domains interacted with magnet surface. The change of plasma domain motion trajectory under the influence of magnetic field also took place. The results of series of experiments, which were conducted under different trajectories of plasma domain motion towards magnetic field, were investigated. Obtained results show that there is no volume positive charge or volume negative charge inside the plasma domain. Taking into account these circumstances, we can assume that plasma domain should have a plasma structure.

Now we have to analyze which phenomena can occur as a result of plasma domain occurrence. Plasma domains occur during an interaction between the capillary discharge jet and metal foil (tantalum or aluminum). If we use tantalum foil, temperature in the interaction area is $T$=1000-2200 K, electron concentration is $n_e=(1$-$2)\cdot10^{16}$ см$^{-3}$. Degree of ionization of the capillary discharge jet is $\alpha=(1.1$-$1.8)\cdot10^{-3}$. Let's assume that plasma domain characteristics are close to characteristics of the capillary discharge jet. Let's propose that aluminum oxidation reaction, which takes place inside the domain, is an energy source. The physical model of plasma near-spherical domain with added aluminum atoms is presented [7]. This model shows that decrease of near-spherical domain charge is caused by drift motion of ions through the plasma domain surface. Energy loss occurs when ions and electrons of plasma domains and air molecules are recombined. The above-mentioned experiments show that plasma domains have a stable internal structure. That is why we can assume that plasma vortex can be generated. During the interaction of the capillary discharge jet with the foil, the most likely areas of plasma domain occurrence are edges of the hole, which was made in the foil. Some studies show that toroidal vortices and other stable vortices can be generated on the edges of holes or on the edges of plates. Plasma domain can have lower viscosity that is why we can assume that such vortex can be generated inside such domain. So we can assume that vortical plasma structure can be generated inside the plasma domain volume.

## 5. Conclusion

The following data were obtained as a result of this study. As a result of interaction of the capillary discharge with metal foils, plasma domains occur. The dimensions of these domains are up to 3 mm, their lifetime is up to 0.5 sec. Plasma domains have an elastic structure. This phenomenon was confirmed during their interaction with vertical and horizontal dielectric barriers. The velocity of plasma domain motion is equal to 0.1-5 m/s. Trajectories of plasma domain motion are usually parabolic, in some cases vertical motions can occur. The characteristics of interaction of plasma domain with barriers and the characteristics of plasma domain motion in magnetic field show that plasma domain has a totally formed internal plasma structure. This structure can be vortical.

## References

[1] Emelin S E, Semenov V S, Bychkov V L, Belisheva N K and Kovshik A P 1997 *Techn. Phys. J.* **67** 19-28 (in Russian)

[2] Bychkov V L, Bychkov A V and Timofeev I B, 2004 *Techn. Phys. J.* **74** 128-133 (in Russian)

[3] Pirozerski A L, Mikhailovskii V Yu and Lebedeva E L, 2014 *Int. J. Appl. Engin. Res.* **9** 24047-24055

[4] Avramenko R F, Bachtin B I, Nikolaeva V I and Poskacheeva L P 1990 *Techn. Phys. J.* **60** 57-64 (in Russian)

[5] Dimitrov S K, Zdanov S K, Kirko D L, Lutchko A S, Mihin S G, Smirnov V M and Telkovsky V G 1992 *Siberian Phys. Techn. J.* **2** 57-60 (in Russian)

[6] Kirko D L, Savjolov A S and Egorov I D 2014 *Proc. 4th Int. Conf. on Atmosphere, Ionosphere, Safety* (Kaliningrad: Imanuil Kant University) pp 179-180

[7] Bychkov V L 2013 *Proc. 10th Int. Conf. on Wave Electrodynamics of a Conductive Liquid* (Yaroslavl, Yaroslavl State University) pp 27-34

Journal of Physics: Conference Series **747** (2016) 012016 doi:10.1088/1742-6596/747/1/012016

# Cathode configuration influence on low-inductance vacuum spark plasma dynamics

**S A Sarantsev[1], Ya M Dvoeglazov, E I Dodulad, I F Raevsky and A S Savjolov**
National Research Nuclear University MEPhI (Moscow Engineering Physics Institute), Kashirskoe highway 31, 115409, Moscow, Russia

E-mail: Sarsa1305@mail.ru

**Abstract.** The results of studies of cathode configuration influence on high current low-inductance vacuum spark (HLVS) plasma dynamics are presented in this work. The research was carried out on "PION" installation using shadowgraphy method. Molecular nitrogen laser ($\lambda$=337 nm) was used as a radiation source. It was determined that the HLVS behavior changes with the increase of number of discharges. In a fresh electrode system (less than 200 discharges) the dependence of constriction position on the discharge trigger position is observed during HLVS development. Also, high gradients of plasma density and secondary constrictions are observed. In a previously exploited electrode system (more than 300 discharges) HLVS behavior changes: plasma density gradients become less expressed, secondary constrictions disappear. In electrode systems with highly developed cathode surface plasma density gradient distribution pattern only slightly changes from discharge to discharge, e.g. the discharge becomes more stable.

## 1. Introduction

Installations based on HLVS discharge plasma source draw researchers' attention for more than 50 years. Such qualities as high plasma parameters (density more than $10^{20}$ cm$^{-3}$, temperature ~ several keV), simple design, low cost and relatively small dimensions make these installations potentially perspective for wide range of technological purposes. One of the possible purposes is its use in controlled fusion problems solving, for example: modelling of current disruption in tokamaks, as a dense, pulsed, high temperature plasma source for various diagnostic techniques testing and development. For purposes of technological application HLVS-based plasma generators can be of interest as point source of soft and hard X-ray and EUV radiation [1-4]. Also there is a possibility to use pulsed plasma of HLVS discharge for irradiation of metal surfaces in order to develop layers with modified structural-phasic states with altered physical, chemical and mechanical properties [5-6], but this area needs more detailed research.

One of the key goals of HLVS studies is development of stable source of electromagnetic radiation, because of that complete understanding of all factors that influence plasma parameters stability is needed. Such instability is a significant aspect that limits HLVS discharge utilization in technological tasks, because providing plasma parameters stability from discharge to discharge is rather difficult problem.

The results of studies of cathode configuration influence on high current low-inductance vacuum spark plasma dynamics are presented in this work. Shadowgraphy technique was used for plasma

[1] To whom any correspondence should be addressed.

dynamics research. Spatial distribution of X-ray radiation was registered by four-channel pinhole camera (hole diameter ~50 μm). Temporal dispersion of X-ray emission was determined by use of X-ray pin-diode.

## 2. Experimental setup and experiment description

Experiments were carried out on PION installation [5] on Plasma Physics Department NRNU MEPhI. Electrode system of the installation is made as pointy anode 3 mm in diameter, and plane cylindrical cathode 22 mm in diameter with 3 mm hole in the center. Electrodes were made from iron Fe (steel 45). Interelectrode gap was 5 mm wide. The discharge was triggered by erosional-type plasma source. Operating voltage was 15 kV, discharge period 5,5 μs, stored electrical energy – 1350 J, maximum of discharge current – 180 kA.

It is important to utilize short-wave lasers, because HLVS discharge plasma shows high parameters (plasma density in peripheral area of discharge is $\geq 10^{17}$ cm$^{-3}$, density in pinch ~ $10^{21}$ cm$^{-3}$, temperature $\geq$ 1 keV). In the considered research molecular nitrogen laser LGI-21 was used, its wavelength is $\lambda$=337 nm. Laser radiation power is 12 μJ.

Studied discharge is powerful source of electromagnetic radiation in wide spectral range – from SHF to hard X-ray radiation. Significant portion of this radiation energy falls at visible and UV light ranges. In such case, during shadowgramm registration it is important to deal with high level of backlight noise from the discharge area. This greatly complicates shadowgramm obtaining process. Backlight noise falling at spectral range close to laser wavelength appears to be comparable to illumination of the laser itself. Therefore in order to minimize backlight noise in the registration technique diaphragm was user, CCD-camera was allocated 6 m away from the plasma source, and image registration was carried through narrow-range (6 nm) interference filter. Experimental setup is presented on Figure 1.

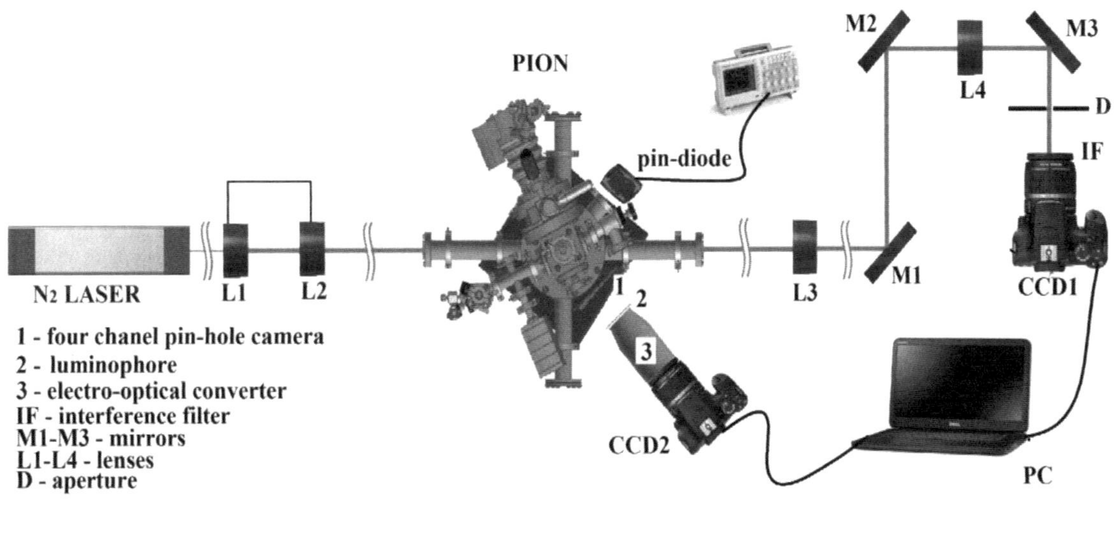


**Figure1.** Experimental setup.

With the rising of number of discharges electrode system suffers significant changes. The hole in the center of cathode becomes welded, cathode surface itself exposed to interaction with plasma flows transforms, sharp brim and weld bed are formed [5]. Such cathode configuration development leads to changes in discharge behavior and has an impact on its stability (repeatability of emission characteristics from discharge to discharge). The goal of the research was to determine how cathode surface development with the growth of number of discharges influences on HLVS discharge plasma dynamics.

Journal of Physics: Conference Series **747** (2016) 012016 doi:10.1088/1742-6596/747/1/012016

As it was mentioned above, temporal dispersion of X-ray emission was determined by use of X-ray silicon-based pin-diode, and spatial distribution was registered by four-channel pinhole camera. Pinhole channels were covered by aluminum filters of different thickness: 60 μm thick – which allowed registration of >6 keV photons, 120 μm thick (>8 keV), 240 μm thick (>10 keV) and 480 μm thick (>13 keV).

Two similar electrode systems were used in the experiments, each was exposed to the same number of discharges (600). Shadowgramms, pinhole images, pin-diode and Rogovsky coil signals were registered in the process. Simultaneous registration of pin-diode and Rogovsky coil signals allowed to determine the precise moment of micropinch formation, this moment was taken as zero-point on the sequences of shadowgramms. The second electrode system was used for the validation of the results of preceding research.

**Figure 2**. Pinhole images of the discharge. (The thinnest filter is on the right) Discharge trigger is on the left side of the anode.

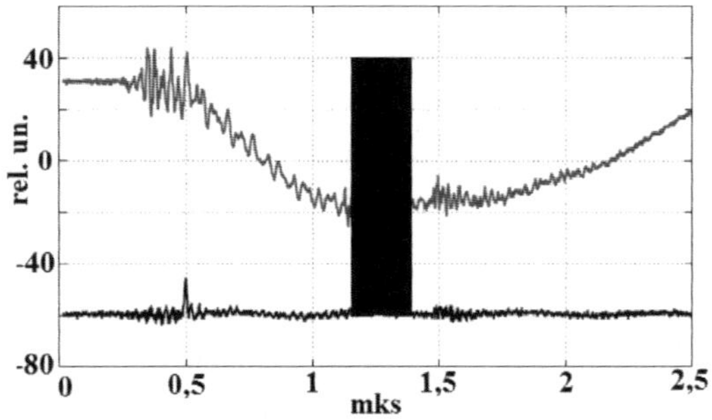


**Figure 3.** Temporal dispersion of X-ray emission in the first 200 discharges (average for two sequences).

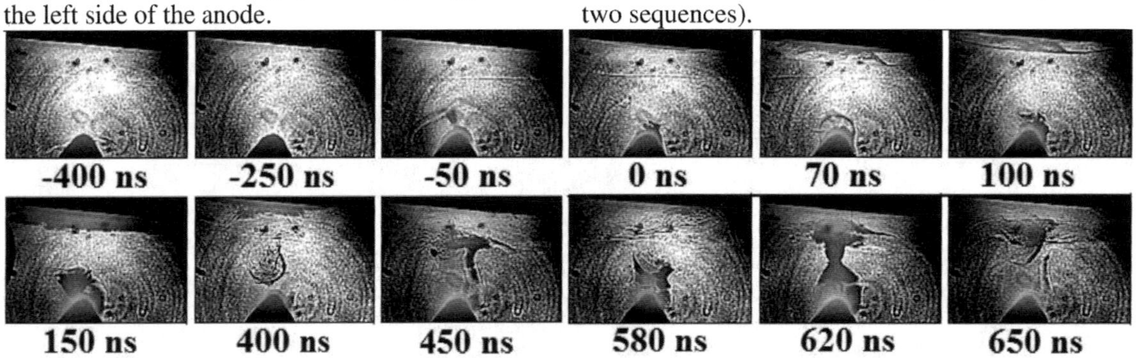


**Figure 4.** Sequence of shadowgramms for a fresh electrode system. Zero point corresponds to micropinch formation moment.

It was determined that when electrode system is still fresh (less than 200 discharges) the discharge is rather unstable – significant temporal (~200 ns) and spatial (~800 μm) dispersion is observed on pinhole images and pin-diode signals. As it appears from shadowgramms in this case HLVS discharge development is followed by rather high plasma density gradients, secondary constrictions are formed, which do not lead to X-ray emission, and primary constriction is shifted closer to the trigger. Distinctive pinhole images, temporal X-ray dispersion and shadowgramms sequence are presented on figures 2,3 and 4 respectively. On figure 3 width of the black rectangle shows temporal dispersion of X-ray pulse and height shows the disperse of pulse amplitude. In this case X-ray radiation was registered during ~ 75% of discharges.

As the electrode system becomes more exploited the discharge becomes more stable. Temporal dispersion decreases to ~80 ns (figure 6). Micropinch appears later in time, closer to discharge current maximum, which leads to optimization of energy input in plasma point. According to pinhole images (figure 5) and shadowgramms (figure 7) there is no such significant shift of constriction to the trigger side. X-ray radiation is more intense and is concentrated in anode area (see pinhole images on figure 5). It is seen from the sequence of shadowgramms on figure 7 that density gradients are not so defined in this case, and there are no secondary constrictions. Micropinch formation and development is more "laminar", and repeatability of discharge development is higher. In this case X-ray radiation was registered during ~ 90% of discharges.

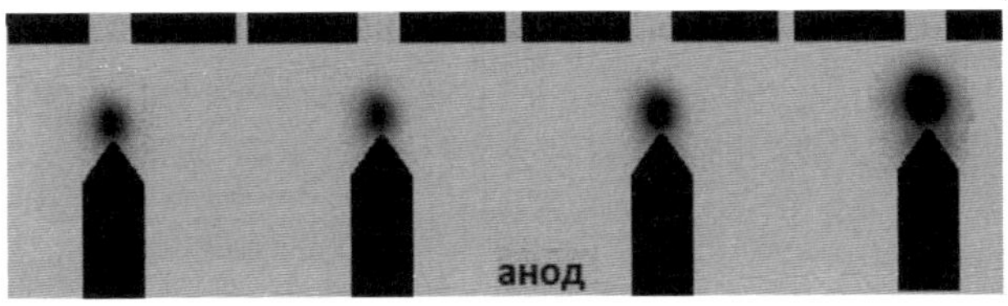


**Figure 5.** Pinhole images of the discharge for previously exploited electrode system. (The thinnest filter is on the right) Discharge trigger is on the left side of the anode.

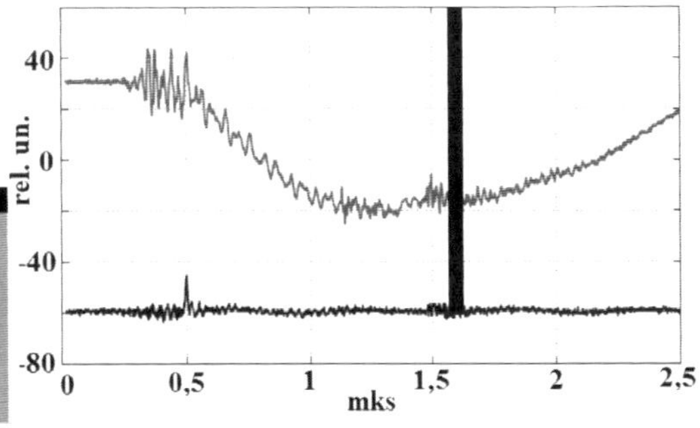


**Figure 6.** Temporal dispersion of X-ray emission from 300 to 600 discharge (average for two sequences).

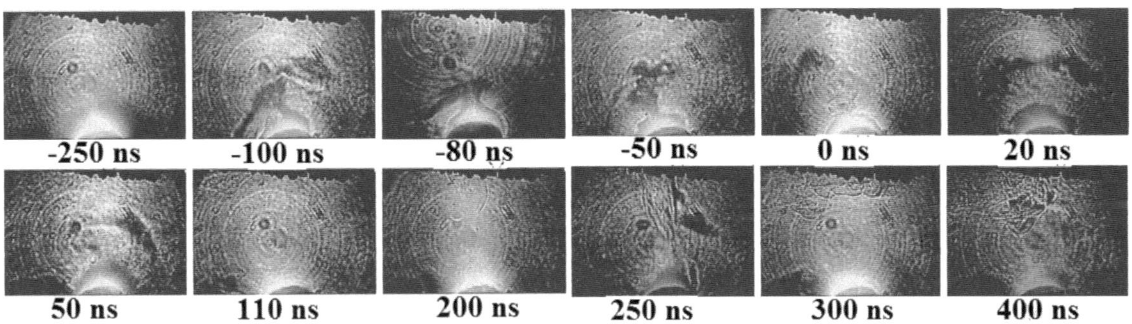


**Figure 7.** Sequence of shadowgramms for previously exploited electrode system. Zero point corresponds to micropinch formation moment.

## 3. Conclusions

Shadow photographing and pinhole imaging methods revealed number of patterns and features of high-current low-inductance vacuum spark development on "PION" installation during electrode system modification with increasing number of discharges. It was found that with increase of discharge numbers the behavior of HLVS is changed. There is a strong dependence of the constriction position from trigger position (it can be seen on pinhole pictures) during the development of HLVS in fresh electrode system (until 200 discharges), there are significant gradients of plasma density and secondary constrictions. The behavior of HLVS changes in previously exploited electrode system (more than 300 discharges), gradients of plasma density become less pronounced and secondary constrictions disappear. Pattern of plasma density gradients distribution in the electrode gap was slightly different from discharge to discharge in the electrode system with highly developed surface structure, constriction (plasma point) almost not moving toward the trigger, x-ray observed in ~90% discharge (versus ~ 75%) and intensity of x-ray was higher (emission area in pinhole pictures became higher), that indicates more stable discharge flowing.

**Acknowledgements**
This work was supported by the Competitiveness Program of NRNU MEPhI.

**References**
[1] Dodulad E I, Kostyushin V A and Bashutin O A 2016 *J. of Phys.: Conf. Series* **666** 012013
[2] Kirko D, Dodulad E and Savjolov A 2016 *J. of Phys.: Conf. Series* **666** 012019
[3] Pavez C, Pedreros J, Tarifeno-Saldivia A and Soto L 2015 *Physics of Plasmas* **22** 040705
[4] Dodulad E I, Kostyushin V A and Savjolov A S 2015 *Phys. Procedia* **71** 155-159
[5] Sarantsev S A, Dvoyeglazov Ya M and Raevskiy I F 2015 *Phys. Procedia* **71** 160-164
[6] Sarantsev S A 2015 *Phys. Procedia* **71** 165-170

# The upgraded heavy ion beam probe diagnostics on the T-10 tokamak

**M A Drabinskii[1,2,6], P O Khabanov[1,3], A V Melnikov[1,4], L I Krupnik[5], A S Kozachek[5], A D Komarov[5] and A I Zhezhera[5]**

[1] National Research Centre 'Kurchatov Institute', 1, Akademika Kurchatova pl., Moscow, 123182, Russian Federation
[2] Bauman Moscow State Technical University, 5, ul. Baumanskaya 2-ya, Moscow, 105005, Russian Federation
[3] Moscow Institute of Physics and Technology, 9, Institutskiy per., Dolgoprudny, Moscow Region, 141700, Russian Federation
[4] National Research Nuclear University MEPhI, 31, Kashirskoe highway, Moscow, 115409, Russian Federation
[5] National Science Center Kharkov Institute of Physics and Technology, 1, Akademicheskaya St., Kharkov, 61108, Ukraine

E-mail: drabinskii91@yandex.ru

**Abstract.** The upgraded Heavy Ion Beam Probe (HIBP) diagnostics on the T-10 tokamak (National Research Center 'Kurchatov Institute') is presented. HIBP is a powerful tool to study electric potential in the core and edge plasmas along with broadband turbulence and quasicoherent modes such as Geodesic Acoustic Mode (GAM) and Alfven Eigenmode (AE). To study broadband turbulence and AEs, which can be driven by fast electrons in regimes with auxiliary Electron Cyclotron Resonance Heating the frequency range of about several hundred kHz is needed. The upgrade is focused on the extension of the frequency range of HIBP signals up to 500 kHz, and on increasing of density operating limit up to $5 \cdot 10^{19}$ m$^{-3}$. It becomes possible due to a newly designed emitter-extractor unit of HIBP accelerator aiming to provide the primary beam with the current of 300 μA at the energy of 300 keV and diameter of 7-10 mm. The new in-vessel elements of a primary beamline – wire sensor and Faraday cup – were upgraded accordingly to be able to deliver the probing beam with advanced parameters to the plasma.

## 1. Introduction

Heavy Ion Beam Probe (HIBP) is instrumental to study magnetically confined plasmas. It is the only diagnostic to measure the electric potential in hot plasma of toroidal fusion devices, both tokamaks and stellarators [1, 2]. For the first time in tokamaks this technique was implemented by Jobes and Hickok in 1970 [3]. Unlike electric probes, HIBP does not disturb the plasma and the measurements are direct and local, taking place in both core and the edge plasmas. Moreover, along with electric potential, HIBP gives information on the electron density and the poloidal magnetic field or plasma current [4]. High time resolution provides the study of the fluctuations of all these quantities [5].

[6] To whom any correspondence should be addressed.

Remarkably, the measurements of all three HIBP quantities takes place simultaneously, which gives an important contribution to comprehensive analysis of the plasma turbulence.
HIBP has started to operate on T-10 tokamak with $Cs^+$ probing ions since late 1980-s [4, 6] in the regimes with low density ohmic plasmas and low magnetic field B=1.5 T. Then HIBP was upgraded several times to be able finally to operate in the ohmic and auxiliary Electron Cyclotron Resonance heated regimes with high magnetic field B≤2.5 T and densities up to $4\times10^{19}m^{-3}$ using $Tl^+$ probing ions [7]. The recent research with HIBP on T-10 is focused on studying of broadband plasma turbulence [8] and quasicoherent modes of plasma oscillations such as Geodesic Acoustic Mode (GAM) [9, 10, 11, 12] and Magneto Hydro Dynamic (MHD) tearing modes [13, 14]. GAM is a high-frequency type of zonal flows which is supposed to be a mechanism of turbulence self-regulation [15, 16]. It has been shown recently that GAM can interact with broadband turbulence at the ion drift wave frequencies (200-300 kHz) [17, 18], so the investigation of these processes in depth is important for understanding the mechanisms of the turbulent energy and particle transport. Interaction between fast ions, alpha-particles or fast electrons with a bulk plasmas can drive Alfven Eigenmodes (AE), which theoretically can have a considerable influence on fast particle transport [19]. Therefore, studies of AE can make a significant contribution to fusion reactor physics. Recently the Neutral Beam Injection induced AE were intensively studied with HIBP on the TJ-II stellarator [20, 21, 22, 23]. On top of that, the fast electron induced modes were found on TJ-II [24]. In T-10 it is expected to find AE driven by fast electrons at frequencies up to 500 kHz.
The modernization, described in the present paper aims to expand the observation area towards the plasma core (with densities up to $5\times10^{19}$ $m^{-3}$) and to broaden the frequency range of studied fluctuations to 500 kHz in order to provide the study of the broadband turbulence, GAMs, MHD modes and AEs.

## 2. Principles of operation and the measurement technique

HIBP is a test particle experiment using heavy ions as test particles. Commonly, singly charged ions of $Na^+$, $K^+$, $Cs^+$, $Tl^+$ or $Au^+$ are used as particles of a probing beam (the heavyweight $Tl^+$ is used now on T-10). The probing (primary) beam after being formed in the emitter-extractor unit and accelerated in the injector enters plasma and moves along Larmour circle in a toroidal magnetic field of a tokamak (see figure 1 (a)). As the primary beam passes through plasma, its particles are further ionized due to the collisions with plasma electrons. Doubly charged (secondary) ions move along the trajectories with twice smaller Larmour radius and form a fan of secondary ions. Proca-Green parallel plate energy analyzer is set on the way of these secondary ions. The aperture of the analyzer cuts out a secondary beam from the fan so only the particles from a given ionization point (sample volume - SV) reach a given detector location.
At the SV the primary ion loses an electron with an energy of -eΦ, where e is an electron charge and Φ is the value of plasma electric potential at the SV. Thus, Φ can be calculated as

$$\Phi = (W_1 - W_2)/e \,,$$

where $W_1$ is the energy of the primary beam and $W_2$ is the energy of a secondary beam.

As far as plasma potential is measured as a small difference between two large values, both beam injector and energy analyzer are the instruments of a high precision and stability, providing $\Delta W_{1,2}/W_{1,2}\sim10^{-5}$. On T-10 the value of plasma potential has a range from several dozen Volts at the periphery to several hundred Volts near the centre, as measured with primary beam energy up to 300 keV.

T-10 energy analyzer has five entrance slits, so five SV are observed simultaneously. The beam from each SV enters to split plate detector (see figure 1 (b)). The primary beam current is 10-100 μA, and the secondary beam current is about 0.1-10 nA so the current signal from split plates is pre-amplified and converted to voltage with the factor of $10^7$ V/A, then digitized and stored in the database.

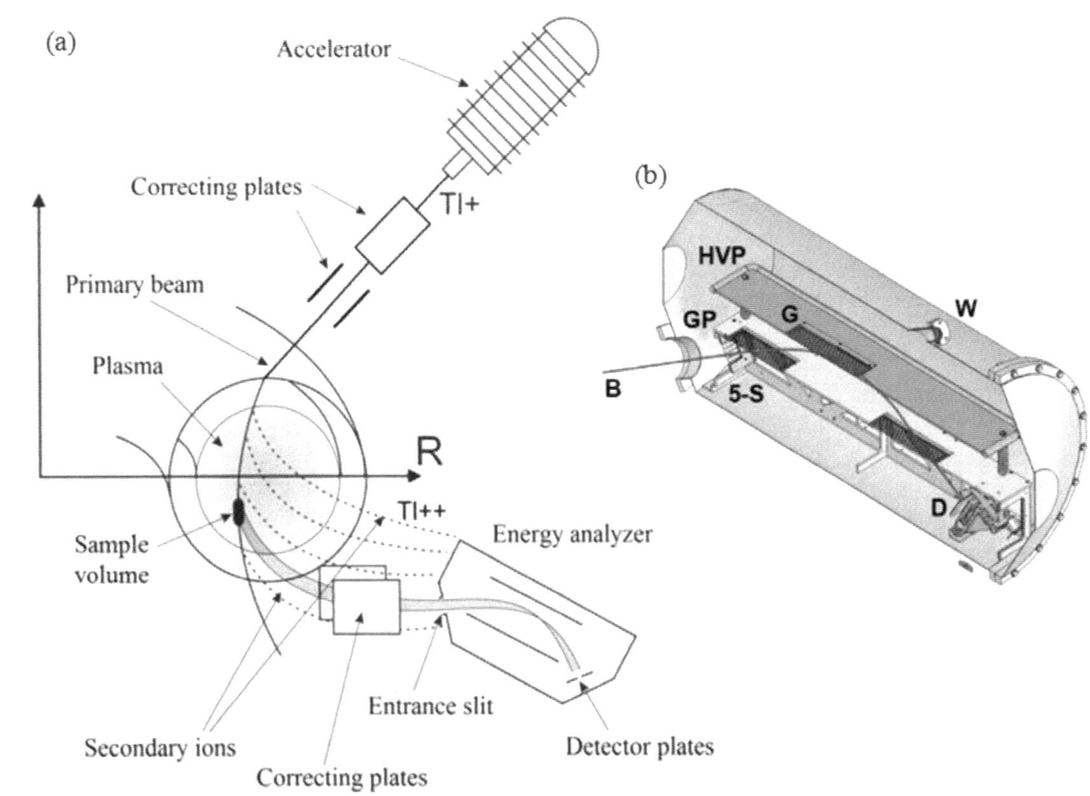


**Figure 1.** (a) basic principle of HIBP diagnostic; (b) Proca-Green parallel plate energy analyzer on T-10 (B – secondary beam, 5-S – five entrance slits, GP – ground plate, HVP – high voltage plate, G – wire grid, W – window, D – 5 channels split plate detector assembly).

Total secondary beam current I on the detector plates is linked with electron density $n_e$ at SV as

$$I \sim I_0 F_1 F_2 \sigma_{eff} n_e l$$

where $I_0$ is the primary beam current, $F_{1,2}$ are attenuation factors for primary and secondary beams, $\sigma_{eff}$ is an effective cross-section of ionization, $l$ is the length of the SV.

$$F_j = \exp(-\int \sigma_j n_e dl)\text{ ; j=1,2;}$$

$\sigma_j$ is the ionization cross-section from j-state, the integration is carried over the entire path of the primary/secondary beam. For low-to-moderate densities, when factors $F_{1,2}$ are not dominating, the density fluctuations are proportional to total current fluctuations:

$$\frac{\delta n(t)}{n} \sim \frac{\delta I(t)}{I}$$

## 3. Key features of the modernization

There are two reasons to increase the primary beam current. As soon as the attenuation factors $F_{1,2}$ increase with the electron density, at high densities ($n_e > 4\times10^{19}$ m$^{-3}$) the secondary beam current becomes comparable to the noise level. So, for measurements at higher densities a higher value of primary beam current density is required. In addition, the necessity to study plasma fluctuations claims the continuous particle flow, not the particle counting character of the signal. The frequency resolution is limited by the number of particles per oscillation period. So higher frequency resolution claims the higher value of primary beam current density. Finally, to expand the frequency range of potential and density fluctuations up to 500 kHz and to increase the density operating range up to $5\times10^{19}$ m$^{-3}$, it was targeted to increase primary beam current to 300 μA at the beam diameter at the SV of 7-10 mm and beam energy of 300 keV.

### *3.1. Emitter-extractor unit*

To solve this task a new emitter-extractor unit was designed and prepared for assembly. The 3D-model of the new unit is shown in figure 2 (a). A three-electrode lens configuration was chosen: the first one is the emitter tablet holder, the second one is the extraction electrode and the third one is the focusing electrode (the first ring of the accelerator tube) – see figure 2 (a) and (b). Calculations using Simion 3D code showed that this configuration will maintain the aimed values of primary beam current.
Figure 2 (b) shows the scheme and the circuit for the unit. It includes two batteries of high capacity (120 A×hour each) and two 10 kV voltage sources.

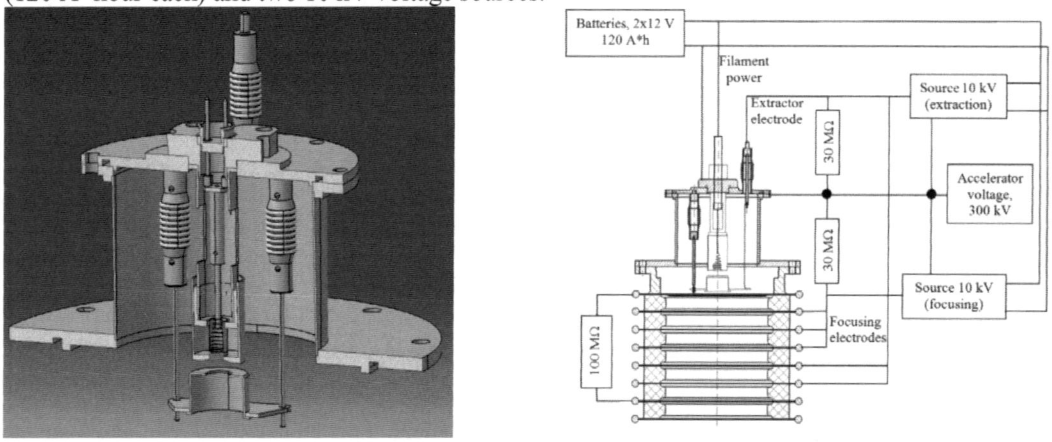


(a) (b)

**Figure 2.** (a) New design of the emitter-extractor unit; (b) emitter-extractor unit power system circuit; red marker shows thallium emitter covered by Pierce electrode with 6 mm hole.

### *3.2. Primary beamline*

Since the primary beam current grows from 100 μA to 300 μA, its diameter at the beamline may also increase. To avoid the beam contact with elements of the primary beamline, the in-vessel elements were also upgraded, as presented in figure 3. The distance between first correcting plates in a vertical direction (α-plates) was increased to 50 mm, and the Faraday cup diameter was increased to 56 mm. Faraday cup has movable bottom driven by traction relay. It is open during the steady state phase of the tokamak discharge to provide the beam pass into the plasma and then into the energy analyzer.

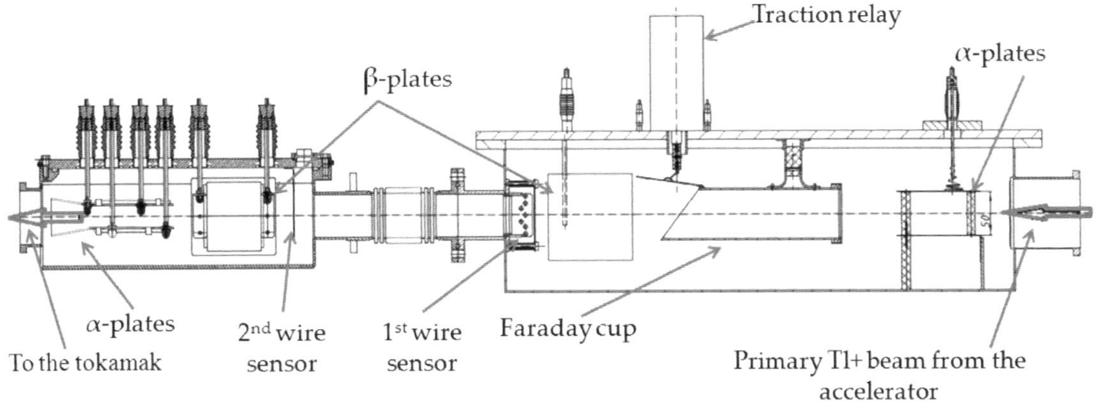


**Figure 3.** Primary beamline. Faraday cup bottom is open.

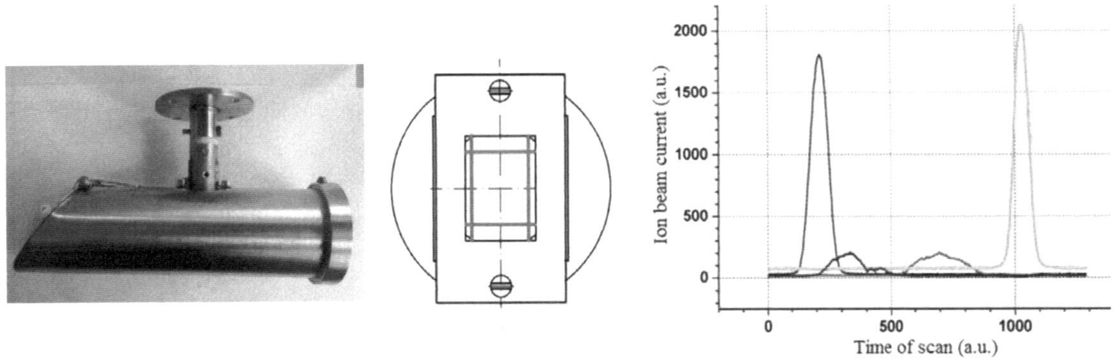


(a) (b) (c)

**Figure 4.** (a) Faraday cup (bottom closed); (b) the 1st wire sensor (front view, red lines show wires); (c) typical current signals from two wire sensors, black and green are signals from the 1st sensor, blue and red – from the 2nd one; the beam is focused on the 1st wire sensor, extraction voltage is 2.58 kV.

It is closed between tokamak discharges to provide primary beam current measurements and/or to prevent the beam interaction with chamber wall or residual working gas. The secondary electrons generated due to this interaction may be accelerated during the ramp up of the toroidal field and cause hard X-ray emission entering the chamber well. To avoid this harmful effect the movable bottom opens after the start of the plasma current. Photo of the new Faraday cup is shown in figure 4 (a). Wire sensors are aimed to measure the primary beam profile and position. Combination of two beam positions gives us the beam entrance angle to the plasma. Both upgraded 1st and 2nd wire sensors have similar design with 2 horizontal and 2 vertical wires. The minimal possible number of wires was chosen to minimize their influence on the beam. Front view of the 1st sensor is shown in figure 4 (b), wires are painted in red. By changing the voltage on α- or β-plates, orthogonal to α- we move the beam across the vertical or horizontal wires, thus measuring the beam profile and position by the wire currents. The typical currents from two wires from two sensors for α-scan are shown on figure 4 (c). In the present case the beam is focused on the 1st sensor with extraction voltage 2.58 kV.

## 4. Conclusion

The main physical aim of HIBP upgrade on the T-10 tokamak is to expand the diagnostic capabilities to study the Geodesic Acoustic Modes interaction with broadband turbulence to 500 kHz in the extended density operating range up to $5\times10^{19}$ m$^{-3}$. This requires the growth of the primary beam current to 300 μA. To achieve this value the new emitter-extractor unit was designed and the elements of the primary beamline were upgraded.

**Acknowledgements**
The work was carried out by Russian Science Foundation, project 14-22-00193.

**References**

[1] Fujisawa A, Ido T, Shimizu A *et al.* 2007 Experimental progress on zonal flow physics in toroidal plasmas *Nucl. Fusion* **47** S718

[2] Melnikov A V, Hidalgo C, Eliseev L G *et al.* 2011 Plasma potential and turbulence dynamics in toroidal devices (survey of T-10 and TJ-II experiments) *Nucl. Fusion* **51** 083043

[3] Jobes F C and Hickok R L 1970 A direct measurement of plasma space potential *Nucl. Fusion* **10** 195-197.

[4] Dnestrovskij Yu N, Melnikov A V, Krupnik L I and Nedzelskij I S 1994 Development of Heavy Ion Beam Probe Diagnostics *IEEE Trans. Plasma Sci.* **22 (4)** 310-331

[5] Melnikov A V, Eliseev L G, Jiménez-Gómez R *et al.* 2010 Internal measurements of Alfvén eigenmodes with heavy ion beam probing in toroidal plasmas *Nucl. Fusion* **50** 084023

[6] Melnikov A V, Bondarenko I S, Efremov S L *et al.* 1995 HIBP diagnostics on T-10 *Rev. Sci. Instrum* **66** 317

[7] Melnikov A V, Andreev V F, Grashin S A *et al.* 2013 Electric potential dynamics in OH and ECRH plasmas in the T-10 tokamak *Nucl. Fusion* **53** 093019

[8] Vershkov V A, Shelukhin D A, Subbotin G F *et al.* 2015 Density fluctuations as an intrinsic mechanism of pressure profile formation *Nucl. Fusion* **55** 063014

[9] Melnikov A V, Eliseev L G, Gudozhnik A V *et al.* 2005 Investigation of the plasma potential oscillations in the range of geodesic acoustic mode frequencies by heavy ion beam probing in tokamaks *Czech. J. Phys* **55** 349-360

[10] Melnikov A V, Vershkov V A, Eliseev L G *et al.* 2006 Investigation of geodesic acoustic mode oscillations in the T-10 tokamak *Plasma Phys. Control. Fusion* **48** S87-S110

[11] Zenin V N, Eliseev L G, Kozachek A S *et al.* 2014 Study of poloidal structure of geodesic acoustic modes in the T-10 tokamak with heavy ion beam probing, *Problems Atomic Sci. Techn. Series: Plasma Physics* **6 (94)** 269

[12] Melnikov A V, Eliseev L G, Perfilov S V *et al.* 2015 The features of the global GAM in OH and ECRH plasmas in the T-10 tokamak *Nucl. Fusion* **55** 063001

[13] Eliseev L G, Ivanov N V, Kakurin A M, Melnikov A V and Perfilov S V 2015 Magnetic island and plasma rotation under external resonant magnetic perturbation in the T-10 tokamak *Phys. Plasmas* **22** 052504

[14] Eliseev L G, Ivanov N V, Kakurin A M *et al.* 2015 Study of the large-scale MHD mode and its effect on GAM in the T-10 tokamak *42-nd EPS Conf. on Plasma Physics. Lisbon, Portugal, 22nd–26th June* Rep. P5.159 http://ocs.ciemat.es/EPS2015PAP/html/

[15] Winsor N, Johnson J L and Dawson J M 1968 Geodesic acoustic waves in hydromagnetic systems *Phys. Fluids* **11** 2448

[16] Diamond P H, Itoh S-I, Itoh K and Hahm T S 2005 Zonal flows in plasma - a review *Plasma Phys. Control. Fusion* **47** R35

[17] Nakashima Y, Hoshino K, Ejiri A *et al.* 2005 Observation of Nonlinear Coupling between Small-Poloidal Wave-Number Potential Fluctuations and Turbulent Potential Fluctuations in Ohmically Heated Plasmas in the JFT-2M Tokamak *Phys. Rev. Lett.* **95** 095002

[18] Ido T, Miura Y, Kamiya K *et al.* 2006 Geodesic–acoustic-mode in JFT-2M tokamak plasmas *Plasma Phys. Control. Fusion* **48** S41

[19] Heidbrink W W 2008 Basic physics of Alfvén instabilities driven by energetic particles in toroidally confined plasmas *Phys. Plasmas* **15** 055501

[20] Melnikov A V, Eliseev L G, Jiménez-Gómez R *et al.* 2010 Study of Alfvén Eigenmodes in the

TJ-II stellarator *Plasma and Fusion Research.* **5** S2019

[21] Jiménez-Gómez R, Könies A, Ascasíbar E *et al.* 2011 Alfvén eigenmodes measured in the TJ-II stellarator *Nucl. Fusion* **51** 033001

[22] Melnikov A V, Eliseev L G, Ascasibar E *et al.* 2012 Alfvén eigenmode properties and dynamics in the TJ-II stellarator *Nucl. Fusion* **52** 123004

[23] Melnikov A V, Ochando M, Ascasibar E *et al.* 2014 Effect of magnetic configuration on frequency of NBI-driven Alfvén modes in TJ-II *Nucl. Fusion* **54** 123002

[24] Melnikov A V, Hidalgo C, Ido T *et al.* 2012 Plasma Potential in Toroidal Devices: T-10, TJ-II, CHS and LHD *Plasma and Fusion Research* **7** 2402114

Journal of Physics: Conference Series **747** (2016) 012018 doi:10.1088/1742-6596/747/1/012018

# Diagnostics of a low-pressure ICP discharge using a magnetic sector mass analyser

**D V Kolodko, A V Kaziev[1], D G Ageychenkov and A A Pisarev**
National Research Nuclear University MEPhI (Moscow Engineering Physics Institute), 31 Kashirskoe highway, 115409 Moscow, Russian Federation

E-mail: kaziev@plasma.mephi.ru

**Abstract.** A simple magnetic sector mass analyser with an extraction system has been tested as a tool for monitoring ion contents in a low-pressure inductively coupled plasma in Ar, $N_2$, and their mixtures. The spectrometer demonstrated mass-resolution suitable for ion flux analyses in plasmas of various compositions. The capabilities of the mass analyser are discussed.

## 1. Introduction

Studies of surface modification processes in plasma facilities as well as improvement of their performance require knowledge and control of the plasma parameters. The results of plasma processing depend on plasma composition and degrees of dissociation and ionization. For example, plasma nitriding of metal alloys is very sensitive to the concentration of atomic particles and ions [1], and there are continuous debates about relative contribution of atoms and ions in the nitriding. Besides, the surface modification process can be strongly affected by small amounts of gas impurities and admixtures to the working gas [2, 3]. Therefore, investigation of plasma composition is crucial for technological applications.

Plasma composition is usually estimated using gas mass analysers. However, they are incapable of providing information about atoms and ions in plasma. Optical spectrometry is a useful tool, but it requires nontrivial knowledge how to convert the optical emission signals to real concentrations of atoms and ions [4]. Mass-spectrometry allows accurate determination of relative particle fluxes, but it is not often used since the equipment is usually expensive and complex. Nevertheless, the problem of ion contents measurements in plasma facilities for surface modification has been addressed in a number of papers [5–12], and various mass filters have been developed and applied.

We have recently investigated plasma parameters in an inductively-coupled plasma (ICP) set-up used for nitriding by means of Langmuir probes and optical emission spectroscopy (OES) [13, 14]. Here we describe a simple magnetic sector mass analyser and results of its experimental testing, with the aim of further investigation of plasma ion composition and comparison with OES and gas analyses data.

## 2. Experimental

The experiments were carried out in a plasma facility (figure 1) used for surface modification and described in [13]. The ICP discharge was excited by an internal flat circular antenna Beams&Plasmas

[1] To whom any correspondence should be addressed.

RPG-250 cooled by water and protected by quartz shield. External magnetic field was produced by a pair of DC electromagnetic coils. The magnetic field at the axis $z$ was kept at 2 mT. The forward RF power was 1 kW, and the reflected power did not exceed 10%.

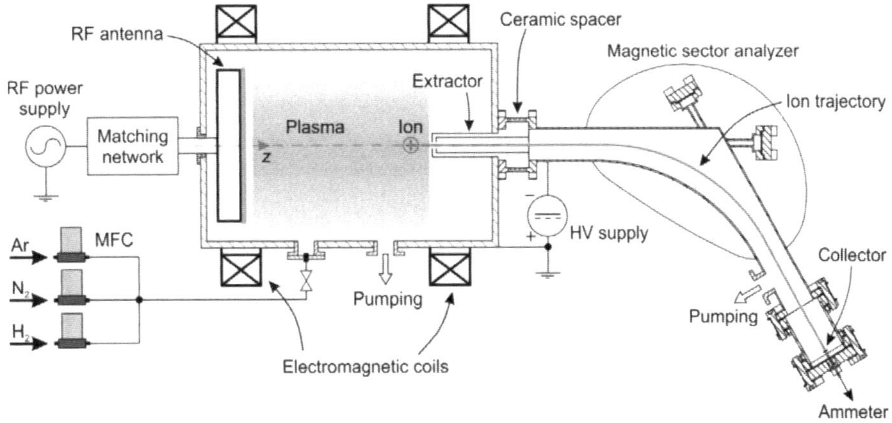


**Figure 1.** Experimental setup scheme.

The ICP vacuum chamber was evacuated to a base pressure of $1\times10^{-4}$ Pa with a turbomolecular pump backed by a dry scroll pump. Argon, nitrogen and their mixtures were used as working gases, and the total pressure was in the range of 0.1–1.5 Pa.

In order to mass-filter the plasma ions, a simple magnetic sector mass analyser (MA) with an electrostatic extractor was attached coaxially to the ICP chamber. The extractor, which was immersed into plasma, was comprised of two-electrode electrostatic lens with circular apertures. The entrance aperture and the extracting aperture were 2 mm and 3 mm in diameter, respectively. The entrance aperture was grounded, while the extracting one and the entire MA unit including the electronic equipment were under a high voltage (up to 5 kV). MA was individually evacuated with a turbomolecular pump backed by a diaphragm pump.

The ion beam was restricted by 2-mm wide slits located at both ends of the separator. The maximum magnetic field produced by the electromagnet was 0.265 T, while the main ion path had the radius of 26 cm. This filter design allowed detection of ions up to $E\times M = 210$, where $E$ is the ion energy in keV, and $M$ — its atomic mass. Therefore, argon ions with energies up to 5 keV could be detected.

The electromagnet was powered by a Matsusada RB5-6P supply. The collector current was recorded with a Keithley 6485 picoammeter. The measurement and control interface was based on NI LabVIEW. The mass analysing unit was fully automated, and the data were transmitted through a wireless protocol because of a high voltage applied.

Plasma density $n$ and electron temperature $T_e$ in the low-pressure ICP discharge used here are nearly uniform along the $z$-direction as our previous studies [13, 14] demonstrated. Therefore, regardless of the extractor immersion depth, the measured ion contents should satisfactorily represent those in the bulk plasma where the samples are usually processed.

One must note that the extractor in plasma becomes surrounded by a multicomponent double layer, and the extracted fluxes of various species, which have different masses and mobilities, would therefore be different even if their concentrations are equal. This makes the quantitative determination of ion concentrations in plasma a challenging task. Nevertheless, the extractor is in about the same surrounding as the sample embedded in plasma for plasma treatment. Therefore, the mass analyser gives reasonable information about ion fluxes incident on the sample during plasma processing.

The energy distribution of ions in plasma is characterized by their temperature $T_i$, which we assume to be $T_i < 0.1$ eV. Hence, the energies of ions in plasma are much less than the energy they receive when accelerating to about 1–5 keV. Therefore, one can consider monoenergetic energy distribution of ions entering the magnetic field to be a good approximation.

## 3. Results and discussion

The test measurements were performed with the ICP discharge operated in Ar, $N_2$ and their mixtures at RF power of 1 kW and total pressure of 1 Pa. The relative mass spectra are presented in figures 2 and 3. Note that the intensities in figures are normalized to different peak currents.

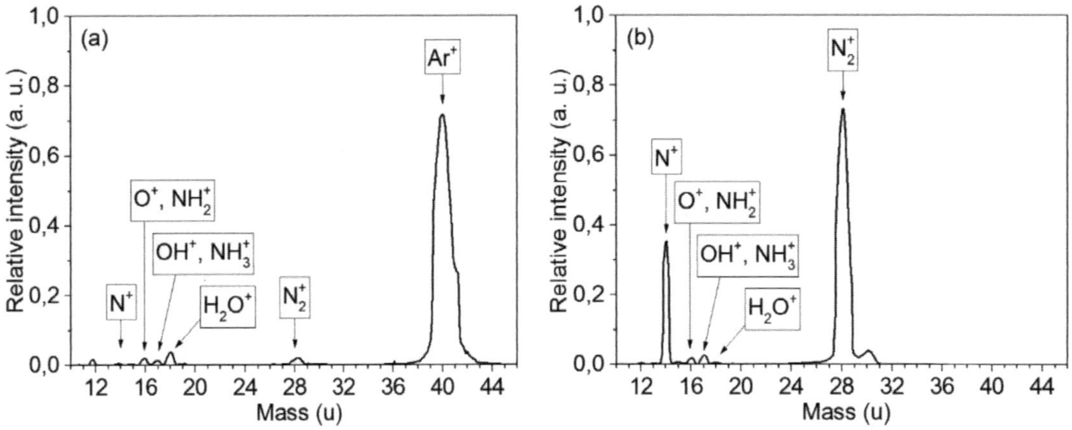


**Figure 2.** Mass spectra of ions extracted from the ICP discharge operated in (a) Ar and (b) $N_2$ at RF power 1 kW and total pressure 1 Pa.

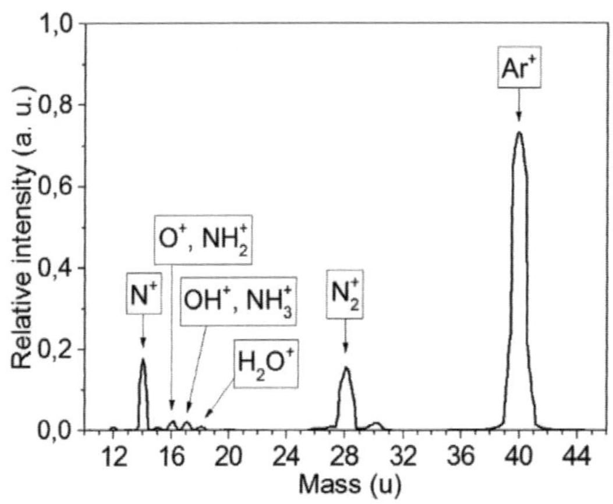


**Figure 3.** Mass spectrum of ions extracted from the ICP discharge operated in an Ar–$N_2$ mixture at RF power 1 kW and total pressure 1 Pa.

One can see that the dominant peaks in figure 2 are ions of the working gas. Besides, one can see masses $M = 16$–19 in all spectra. In the case of argon plasma, the relative intensities of these masses are similar to those observed by gas analysers. In the case of nitrogen plasma, however, the intensity of $M = 17$ exceeds that of $M = 18$, and this can be explained by formation of $NH_x$ molecules due to interaction of nitrogen with residual or impurity hydrogen. Formation of $NH_x$ molecules was discussed in many publications [15].

If the discharge takes place in the gas mixture Ar–$N_2$, relative intensities of $N_2^+$ and $N^+$ ions change: more atomic ions are produced, and this may be due to more intense dissociation of nitrogen molecules in presence of argon. The effect of correlation between the $N^+$ content and Ar flow will be examined in future.

## 4. Conclusion

A magnetic ion mass analyser with an extraction system was developed and tested in a low-pressure RF plasma in Ar, $N_2$, and Ar–$N_2$ mixtures. The mass resolution is sufficient to control ion spectra in technological plasma. The obtained mass spectra correspond to ion fluxes that can be expected on samples embedded in plasma. Main ion species are ions of working gas. Formation of $NH_x$ molecules in nitrogen discharge was suggested to explain the behaviour of ions with $M = 17$. Change of relative intensities of molecular and atomic nitrogen ions was observed if Ar was added to $N_2$ in the discharge.

Since the ions entering the magnetic sector are extracted from different equipotential surfaces, their energies must be mass-dependent. To take this effect into account, an energy analyser is planned to be developed for this mass spectrometer.

## Acknowledgments

The work was supported by contract no. 14.Y26.31.0008 with the Ministry of Education and Science of the Russian Federation and grant no. 16-32-00825\16 from the Russian Foundation for Basic Research.

## References

[1] Mittemeijer E J 2013 Fundamentals of Nitriding and Nitrocarburizing *Steel Heat Treating Fundamentals and Processes* (*ASM Handbook*, Volume 4A) eds J Dossett and G E Totten (Materials Park: ASM International) pp 619–646
[2] Koval N N, Schanin P M, Akhmadeev Yu Kh, Lopatin I V, Kolobov Yu R, Vershinin D S, and Smolyakova M Yu 2012 *J. Surf. Investig.* **6** 154
[3] Vershinin D S and Smolyakova M Yu 2012 *J. Surf. Investig.* **6** 159
[4] Debal F, Bretagne J, Jumet M, Wautelet M, Dauchot J P, and Hecq M 1998 *Plasma Sources Sci. Technol.* 7 219
[5] Kim G H, Rim G H, and Nikiforov S A 2001 *Surf. Coat. Technol.* **136** 255
[6] Maliska A M, Egert P, de Souza A R, Speller V, and Klein A N 1997 *J. Mater. Sci.* **32** 6375
[7] Tuszewski M 1996 *Rev. Sci. Instrum.* **67** 2215
[8] Tang B Y, Fetherston R P, Shamim M, Breun R A, Chen A, and Conrad J R 1993 *J. Appl. Phys.* **73** 4176
[9] Wang E Y, Schmitz L, Ra Y, LaBombard B, and Conn R W 1990 *Rev. Sci. Instrum.* **61** 2155
[10] Zolotukhin D B, Tyunkov A V, Yushkov Yu G, and Oks E M 2015 *Rev. Sci. Instrum.* **86** 123301
[11] Renaud D, Gerst D, Mazouffre S, and Aanesland A 2015 *Rev. Sci. Instrum.* **86** 123507
[12] Sorokin I A, Vizgalov I V, Gutorov K M, and Podolyako F S 2015 *Bull. Lebedev Phys. Inst.* **42** 350
[13] Meshcheryakova E, Zibrov M, Kaziev A, Khodachenko G, and Pisarev A 2015 *Phys. Procedia* **71** 121
[14] Meshcheryakova E A, Kaziev A V, Zibrov M S, Stepanova T V, Berdnikova M M, Kharkov M M, and Pisarev A A 2016 *Bull. Russian Acad. Sci. Phys.* **80** 175
[15] Priest J M, Baldwin M J, and Fewell M P 2001 *Surface and Coatings Technology* **145** 152

Journal of Physics: Conference Series **747** (2016) 012019 doi:10.1088/1742-6596/747/1/012019

# Pulse-periodic abnormal glow discharge in nitrogen, argon, hydrogen and their mixtures

**Yu V Borisyuk[1], N M Oreshnikova, D V Mozgrin, T V Stepanova and A A Pisarev**
MEPHI, 115409 Kashirskoe highway 31, Moscow, Russia

E-mail: yvb65@mail.ru

**Abstract.** Abnormal glow discharge in the pulse-periodic mode in Ar, $H_2$, $N_2$, Ar+$N_2$, and $N_2$+$H_2$ was investigated. Current-voltage characteristics, Langmuire probe and optical spectroscopy data were analysed. The electron temperature of 2-5 eV and concentration of 1-3×$10^{10}$ $cm^{-3}$ have been measured. Lines of $N_2$ $N_2^+$, N, $N^+$, $NH_x$, $Ar^+$, Ar, Ti, Fe, H, and OH have been detected by optical spectrometry. Dependences of intensities of spectral lines of $N_2$, $N_2^+$, N, and $N^+$ on Ar and $H_2$ content in gas mixtures were measured.

## 1. Introduction

Plasma nitriding and film deposition are widely used in industry for surface modification [1,2], and glow discharge is often used for this purpose. Therefore, investigations of the discharge are performed to choose optimum processing parameters [3,4]. Well-known drawback of the classical glow discharge at elevated pressures is its sudden transition in the arc mode. Nevertheless, if to operate the discharge at high frequency, the arc has no time to develop [5], which makes this regime attractive for technological applications.

Mixtures of gases are often used to optimise technological treatment; particularly during plasma nitriding, argon and hydrogen are often added to nitrogen to activate its penetration into the material. Though the pulsed-periodic abnormal glow discharge is widely used in industry, there are few works devoted to its properties. The discharge in $N_2$+$H_2$ mixture was investigated in [6], and a sharp increase of the intensity of the spectral lines was observed after adding up to 20% of hydrogen to nitrogen. The discharge in $N_2$+Ar mixture was investigated in [7], and it was found, that the intensity of $N_2^+$ decreased, while the intensity of $N_2$ increased with increasing argon fraction above 20%.

This work describes some results of emission-spectroscopic investigations, probe diagnostics, and the discharge current–voltage characteristic measurements in nitrogen, argon, hydrogen, and their mixtures in a wide range of compositions.

## 2. Experimental

The glow discharge installation is described in [8]. In brief, it consists of a vacuum chamber with pressure sensors, a dual-channel pumping system, a gas inlet system, and a power supplier. The cathode was a rod 100 mm in lengh coaxially installed inside a cylindrical anode 180 mm in diameter and 420 mm in length. The cathode was made either of russian stainless steel 12X18H10T (analogue

[1] To whom any correspondence should be addressed.

AISI 321) or russian titanium alloy BT-6 (analogue Ti-6Al-4V). Diameters were 12 mm and 10 mm, respectively. Coaxial thermal screen made of several layers of stainless steel sheets 0.5 mm thick was the anode.

The discharge was powered by a generator of rectangular pulses with the amplitude of 400-650V, pulse repetition frequency variable from 1 to 100 kHz, and a duty cycle variable from 10 to 80%. The discharge voltage and current were measured by a four-channel digital oscilloscope Tektronix TPC 2024B using a voltage divider and a shunt.

Discharges in Ar, $N_2$, $H_2$, Ar+$N_2$, and $N_2$+$H_2$ at a pressure of 1 torr and 1.5 torr were investigated.

Optical radiation from plasma was measured through a quartz glass window by a 3-channel spectrometer AvaSpec-ULS2048L-USB2-RM in the wavelength range from 200 to 810 nm with an optical resolution of 0.12 nm to 0.18 depending on the channel. The signal was analyzed by a standard software package Avantes and identified using data bases [9-11].

The Langmuir probe was made from a W wire 0.5 mm in diameter and had the working length of 8.5 mm (working surface 13.5 $mm^2$). The probe was placed about 5-8 mm from the cylindrical surface of the cathode. The decreasing voltage pulses with the amplitude of 50 V and duration of 35 ms were applied to the probe. The plasma concentration and the electron temperature were determined according classic theory in the collisionless regime [12-14].

## 3. Experiment results

Typical dependencies of the discharge voltage and current are shown in Figure 1. This example is given for the voltage of power supply of 500V, frequency of 3 kHz, and the duty cycle of 30%. One can see that the discharge reaches about the steady state regime at this frequency and duty cycle. The mean discharge power increased with the duty cycle.

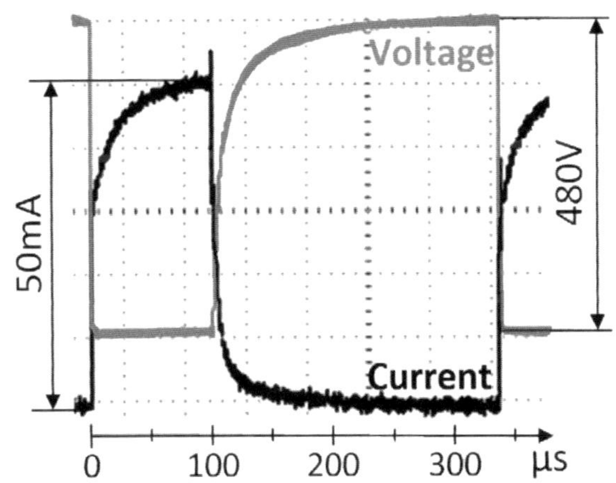


**Figure 1.** An example of the time dependencies of the discharge voltage and current for the discharge with SS cathode ignited in $N_2$ at 1 Torr, 500 V, 3 kHz, and the duty cycle of 30%.

Figure 2 and 3 shows the current-voltage characteristics for the cathode made of steel and titanium alloy. One can see that the current rises with the voltage, and this is the signature of the abnormal glow discharge [5, 13]. The average discharge power increases with gas pressure, and this was also observed in [15]. The discharge in the gas mixture Ar+$N_2$ is more powerful than discharges in pure argon and pure nitrogen, and the same result was found from comparison of discharges in Ar+$N_2$ and in pure $N_2$ in [16]. The discharge current depends on the cathode material used: it is higher for SS cathode. This difference can be connected with the difference of secondary electron emission efficiency from SS and Ti.

Figure 4 shows a typical current-voltage characteristics of the Langmuir probe. Typical values of the electron temperature and the plasma concentration were 2-5 eV and ~1-3×$10^{10}$ $cm^{-3}$, respectively, both for Ar and for $N_2$ in experiments with steel cathode at a pressure of 1 torr, discharge voltage of 600 V, frequency 3 kHz, and the duty cycle of 80%. The temperature and concentration increased with increase of voltage, and duty cycle.

Journal of Physics: Conference Series **747** (2016) 012019 doi:10.1088/1742-6596/747/1/012019

The temperature and the concentration obtained in other works are lower. For example, 0.1-0.3 eV and $10^9$ $cm^{-3}$ were obtained in [17] in $N_2$- $H_2$ mixture at 1 kHz and a duty cycle of 60%.

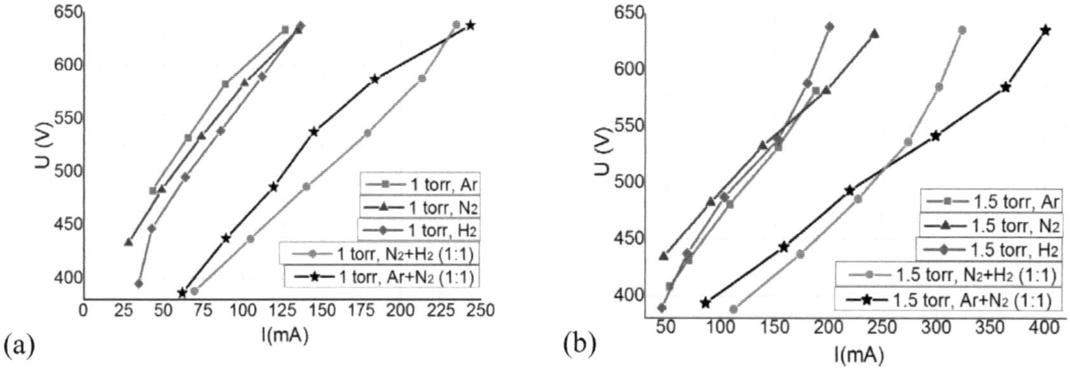


**Figure 2.** The current-voltage characteristics with the SS cathode at frequency 3 kHz; 30% duty cycle, and pressures (a) 1 torr and (b) 1,5 torr.

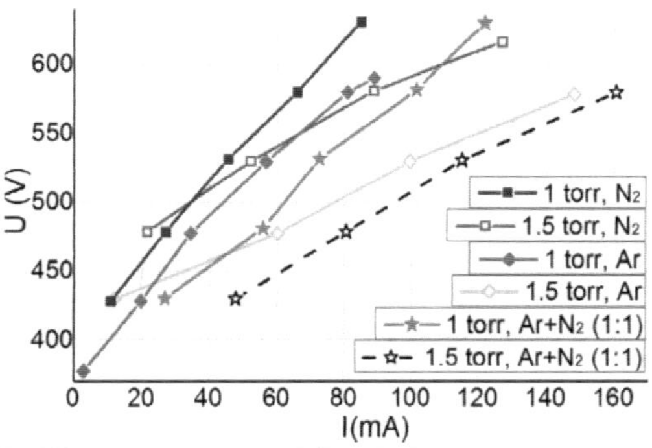


**Figure 3.** The current-voltage characteristics of the discharge. The cathode - titanium alloy. Frequency 3 kHz, duty cycle 30%.

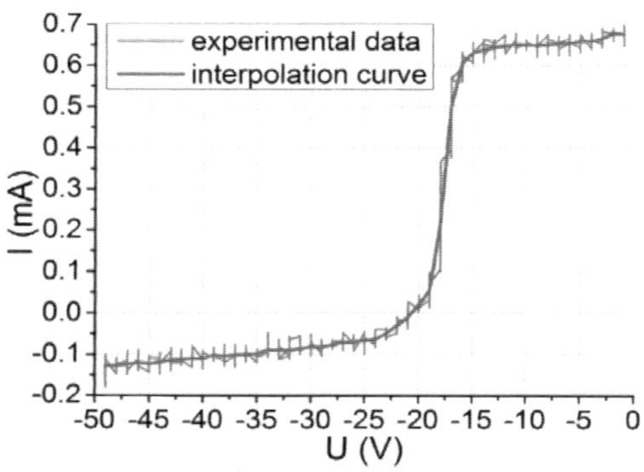


**Figure 4.** The current-voltage characteristic of the probe. Cathode - SS, gas - nitrogen, pressure 1 torr.

Figure 5 shows examples of optical spectra from the discharges in Ar, $N_2$, Ar+$N_2$ (1:1), and $N_2$+$H_2$ (1:1), obtained at a pressure of 1.0 torr, voltage of 480-650V, pulse frequency of 3 kHz, and a duty cycle of 60-80%.

A comparison of the experimental spectra with reference spectra indicates the presence of the following lines in the discharge in argon: $Ar^+$, Ar, Ti, $Ti^+$, Fe, H, OH. Lines in nitrogen were $N_2$, $N_2^+$, N, $N^+$, Fe, H, and OH. Similar results were observed in [17]. The intensity of radiation depends on the pressure and the applied voltage. The energy of electrons mainly depends on the applied potential and on the gas pressure. Optical emission increases with the gas pressure due to increase of the probability of inelastic collisions [18].

It is known that the nitriding process is activated by $N_2^+$, N, $N^+$, and $NH_x$, species in plasma as they easily penetrate through the surface. One can see that all these components exist in the optical spectra. Lines OH and H in nitrogen plasma appear due to water vapor, while lines Fe, $Fe^+$, Ti, $Ti^+$ appear due to sputtering of the cathode and structural elements. The absence of twice ionized atoms and molecules indicates that the electron temperature is relatively low.

One can roughly estimate the electron temperature from optical spectra with an error of about 50% [19]. This estimation gives 1.5 eV for a glow discharge in argon and nitrogen, which is in a good agreement with the data from probe measurements.

Journal of Physics: Conference Series **747** (2016) 012019 doi:10.1088/1742-6596/747/1/012019

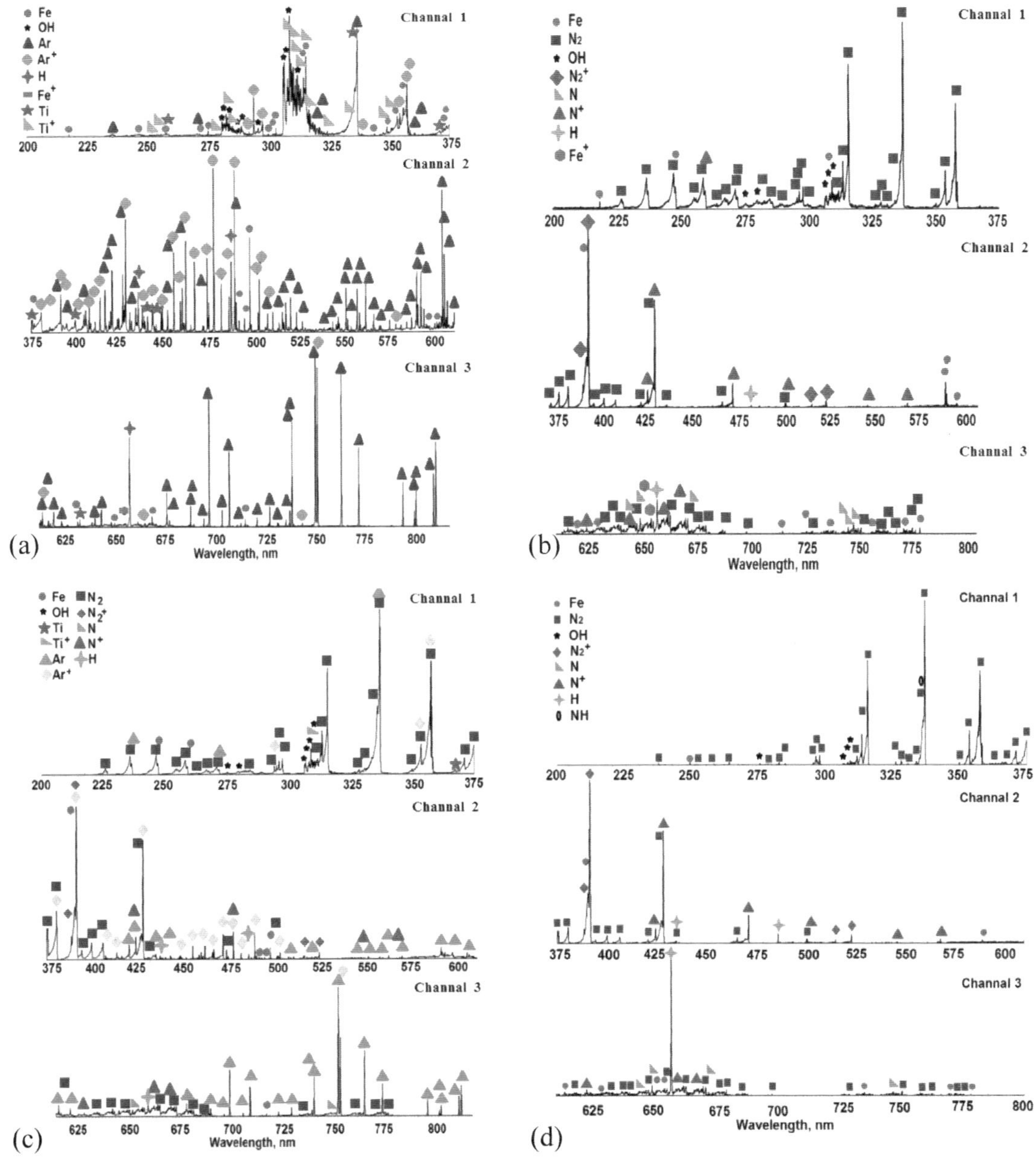


**Figure 5 .** Emission spectra of an abnormal glow discharge in: a) Ar, b) $N_2$, c) Ar+$N_2$, d) $N_2$+$H_2$.

Figure 6 shows dependencies of intensities of several optical lines on gas composition. The analyzed lines were $N_2(C^3\Pi_u \rightarrow B^3\Pi_g)$, $N_2^+(B^2\Sigma_u^+ \rightarrow X^2\Sigma_g^+)$, NH $(A^3\Pi \rightarrow X^3\Sigma^-)$, N (2p3s→2p3s), $N^+$(2p3s→2p3p), and $H_\alpha$.

The intensities of ion lines have a maximum if about 10-20% of argon is added to nitrogen, and the maximum intensity is roughly 1.5 times the intensity in pure nitrogen. Further increase of argon concentration leads to gradual decrease of nitrogen ion line intensities down to the background level at 100% Ar, mainly due to decrease of the nitrogen amount in mixture.

Intensities of lines of neutrals $N_2$ and N increase with increase of Ar concentration. At 90% Ar, a sharp decrease is observed, but the intensities do not drop to zero at 100% Ar. This effect may be explained by release of nitrogen from the cathode due to argon ion impact. The maximum intensity of lines of neutrals, which is observed at low concentration of nitrogen, is about 2.5 times their intensity in pure nitrogen. Qualitatively similar behavior of neutral and ion lines was observed in [7] in the classic glow discharge. The explanation was based on comparison of the energies for reactions in plasma. Increase of argon content leads to increase of the number of excited argon atoms in plasma. These argon atoms can transit to the metastable state, and this needs 11.55 eV for transition to $^3P_2$ and 11.72 eV to $^3P_0$. The alternative process is so-called Penning excitation where excited argon transfers its energy to nitrogen molecule in inelastic collision and excites it ($Ar^*+N_2=N_2^*+Ar$). The threshold excitation energy of $N_2$ is 11.1 eV, which is smaller than the energy necessary for excitation of metastable argon atoms. Therefore, increase of $N_2$ radiative states with increase of the argon concentration is observed due to Penning excitation. The Penning excitation is possible also with participation of nitrogen molecular ions $Ar^*+N_2^+=N_2^{+*}+Ar$, but this reaction needs a higher energy of about 18.7 eV, and experimentally the rate of this reaction even decreases with increase of argon concentration. There are two other reactions $Ar^*+N_2=N_2^++Ar$ (15,57 eV) and $Ar^*+N_2^+=N_2+Ar^+$ (15,75 eV), which have also a higher energy. Therefore excited neutral molecules are preferentially produced with adding argon to nitrogen. The intensities of ion lines become less at large argon content just because the content of nitrogen in the mixture decreases.

Though the energy of electrons increases with argon addition, and the ionization of $N_2$ is more sensitive to high-energy electrons than excitation of ($C^3P_u$) state, the concentration of exited ions decreases with argon content due to prevailed Penning excitation of $N_2$ molecules by metastable argon atoms.

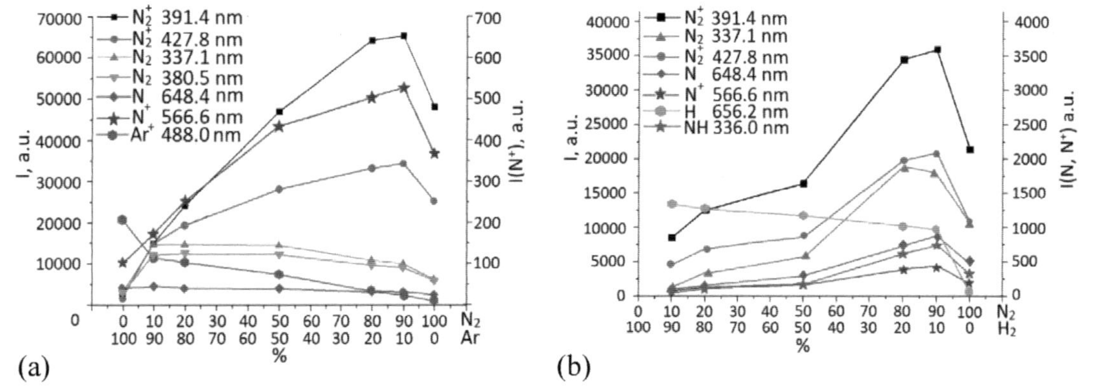


**Figure 6.** Spectral line intensities vs. (a) relative nitrogen - argon concentration (b) relative nitrogen - hydrogen concentration.

Figure 6 shows also variations of intensities of lines in mixture of nitrogen and hydrogen. Intensities of all ion and neutral lines have maxima if 10-20% of hydrogen is added to nitrogen. Increase of nitrogen excitation at about 20% of hydrogen was observed also in [6]. The increase of intensity of nitrogen lines with the addition of small quantities of hydrogen can be connected with increase of the secondary electron emission coefficient. The decrease at high hydrogen concentrations may be connected with decrease of the nitrogen content in plasma.

## 4. Conclusion

Glow discharge powered with the frequency of 3 kHz in Ar, $H_2$, $N_2$, $Ar+N_2$, and $N_2+H_2$ have been investigated at pressures of 1.0 - 1.5 torr. The discharges has increasing current-voltage characteristics typical for abnormal glow discharge. The discharge current in the gas mixtures was higher than that in pure gases. The current increased with the gas pressure. The discharge was almost in the steady state regime during the pulse, and the mean discharge power increased with the duty cycle.

The probe measurements gave the electron temperature of 2-5 eV , and the plasma density of $10^{10}$ $cm^{-3}$.

All main lines $N_2$, $N_2^+$, N, and $N^+$ are found in optical spectra, and $NH_x$ was additionally found in presence of hydrogen. These species are active in nitriding of steels and alloys. In the mixture of $N_2$+Ar, the maximum concentration of nitrogen ions was at about 80-90% of $N_2$, while the maximum concentration of excited neutrals is at about 10% of $N_2$. In the mixture of $N_2+H_2$, the maximum concentration of both nitrogen ions and neutrals is at about 80-90% of $N_2$.

**Acknowledgements**
This work was partially supported by Ministry of Education and Science of the Russian Federation (Contract № 14.Y26.31.0008) and RFBR (Contract № 26 16-32-00825/16).

**References**

[1] Kashaev N, Stock H-R and Mayr P 2005 *J. Surf. Coat. Technol.* **200** 502–506
[2] Ramazanov K N, Ishmukhametov D Z and Sadkova N S 2011 *Vestnik UGATU* **15 3 (43)** 67-71
[3] Kumar S and Ghosh P K 1993 *J. Phys. D: Appl. Phys* **26** 1419-1426
[4] Garcia-Cosio G, Martinez H, Calixto-Rodriguez M and Gomez A 2011 *J. Quant. Spectrosc. Radiat. Transfer* **112** 2787–2793
[5] Granovsky V D 1974 *Electric current in gases. The steady current* (Moscow: Nauka)
[6] Kim Y.M., Kim J.U., Han J.G. Investigation on the pulsed DC plasma nitriding with optical emission spectroscopy. 2002 *Surf. Coat. Technol.* **151-152** 227-232
[7] Qayyum A, Shaista Zeb, Naveed M A, Rehman N U, Ghauri S A and Zakaullah M 2007 *J. Quant. Spectrosc. Radiat. Transfer* **107** 361–371
[8] Borisyuk Yu V, Oreshnikova N M, Berdnikova M A, Tumarkin A V, Khodachenko G V and Pisarev A A 2015 *Phys. Procedia* **71** 105-109
[9] Zaidel A N, Prokofiev V K and Raisk S M 1962 *Tables of Spectral Lines* (Moscow).
[10] Pearse R W B and Gaydon A G 1975 *The indentification of molrecular spectra* (London, New York: Chapman and Hall)
[11] Ochkin V N 2009 *Spectroscopy of low temperature plasma* (Wiley-VCH)
[12] Lebedev U A *Electrical probe in low-pressure plasmas* (URL: http: // plasma. karelia.ru > pub / fntp / Lebedev)
[13] Raizer Yu P 1991 *Gas Discharge Physics* (Springer, Berlin, New York)
[14] Demidov V I, Kolokolov N B and Kudryavtsev A A 1996 *Probe Methods of Diagnostics of Low Temperature Plasma* (Moscow: Energoatomizdat)
[15] Sharma M K and Saikia B K 2008 *Indian J. Pure Appl. Phys.* **46** 463-470
[16] Sharma M K, Saikia B K and Bujarbarua S 2008 *Surf. Coat. Technol.* **203** 229–233
[17] Hannemann M, Hamann S, Burlacov I, Börner K, Spies H-J and Röpcke J 2013 *Surf. Coat. Technol.* **235** 561–569
[18] Saeed A, Khan A W, Jan F, Waqar M, Abrar M, Zaka-Ul-Islam M, Hussnain A and Zakaullah M 2014 Pulsed dc discharge in the presence of active screen for nitriding of high carbon steel *Mat. Res.* **17**
[19] Kolesnikov V N 2007 *Spectroscopic diagnostics of plasma* (Moscow: MEPhI)

Journal of Physics: Conference Series **747** (2016) 012020 doi:10.1088/1742-6596/747/1/012020

# Monitoring of the relationship between $H_\alpha$ and $D_\alpha$ emission as a detection method for water microleaks into ITER

**I.V. Vizgalov**[1], **I.A. Sorokin, V.A. Kurnaev**
National Research Nuclear University MEPhI, 115409, Russia, Moscow, Kashirskoe highway 31

E-mail: vizgalov@plasma.mephi.ru

**Abstract.** The validation of spectroscopic method for water microleakage detection in ITER plasma chamber by IO Vis/IR diagnostic system is presented. The method is based on evaluation of the intensity ratio between Hα and Dα spectroscopic lines. They have rather large intensity in low pressure discharges (glow, RF, microwave, for instance) and correspond to the middle of the range 615-700 nm with maximum of the IO Vis/IR sensitivity.

## 1. Introduction

Utilization of an emitted light from water molecules and their fragments exited in some plasma discharges is the most perspective method for operative detection and localization of water cooling system microleakages in ITER plasma vessel. The spectroscopic method usually is based on controlling intensity level of the spectral lines corresponded to exited fragments of the water molecule. Previous investigations [1, 2] have shown that $OH^*$ hydroxyl band (in vicinity of 310 nm wavelength) is a preferred candidate for stationary diagnostics. The limit of sensitivity is estimated as $3\cdot10^{-5}$ $Pa\,m^2s^{-1}$. However, the spectroscopic microleak detection using hydroxyl radiation is possible at the first stages of ITER operation, because of the fast degradation of the optic elements for UV range under neutron radiation. Also proposed wavelength is strongly attenuated as it passes through the IO Vis/IR diagnostic systems optics as they are designed presently [3].

Proposed method is based on evaluation of the intensity ratio between $H_\alpha$ and $D_\alpha$ spectroscopic lines. These lines are shifted by about 0.2 nm (656.28 and 656.10 nm). They have rather large intensity in low pressure discharges (glow, RF, microwave, for instance) and correspond to the middle range 600-700 nm (the maximum of Vis/IR sensitivity). Experiments on validation of the method were carried out at PR-2 facility. In normal conditions the discharge in He will stimulate desorption of deuterium and background protium (in ITER it must be at the level below 0.6 %) from the wall tiles. So, if the water vapors would ingress into rather hot and dense discharge He plasma, $H_2O$ molecules would excite and dissociate. Discharge parameters have been optimized on maximum of $H_\alpha$ radiation (as a fragment of $H_2O$ dissociation). Thus, the enhanced radiation of $H_\alpha$ relative to $D_\alpha$ would point on the leakage vicinity. This method can be considered as an alternative to previously developed method based on the detection of hydroxyl radiation in UV range.

## 2. Experimental setup

PR-2 beam plasma facility in MEPhI [4] can operate in different discharge modes in broad range of plasma parameter. Plasma with electron densities $10^{14}$-$10^{19}$ $m^{-3}$ is routinely obtained with density

[1] To whom any correspondence should be addressed.

gradient up to $10^{21}$ $m^{-4}$. PR-2 is equipped by a number of the plasma diagnostics. The local plasma parameters are controlled by single Langmuir probe, the ion mass-spectrum in plasma column – by the stationary magnetic mass-spectrometer (MS) [5] using own magnetic field of the plasma device, the residual gas composition – by the standard quadruple mass analyzer QMA EXTOR XT100M. Stationary optical emission spectroscopy system is based on MDR-206 spectrometer [1] mounted in front of the quartz window equipped by the collimator narrowing a light flux from the plasma and the cylindrical lens with focal length 15 cm used as light concentrator. Scheme of spectroscopic measurements is shown on figure 1.

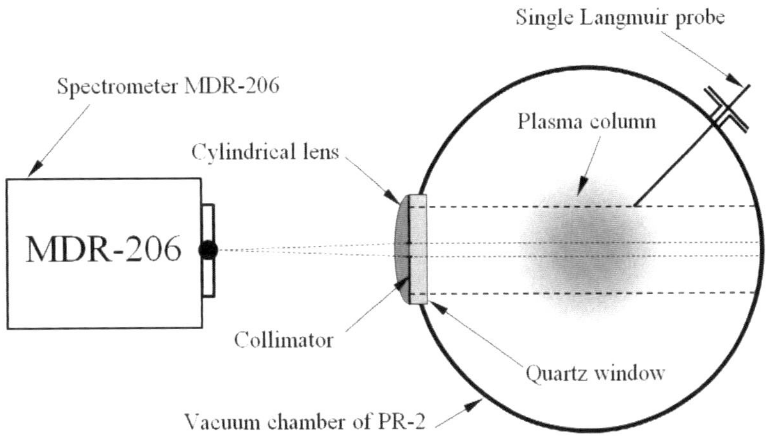


**Figure 1.** Scheme of spectroscopic measurement of $H_\alpha/D_\alpha$ radiation from a beam-plasma in PR-2.

The technical parameters of MDR-206 allow us to divide $H_\alpha$ and $D_\alpha$ lines with sufficiently high sensitivity. Figure 2 shows the example of spectra for electron beam driven discharge in deuterium with different admixtures of protium.

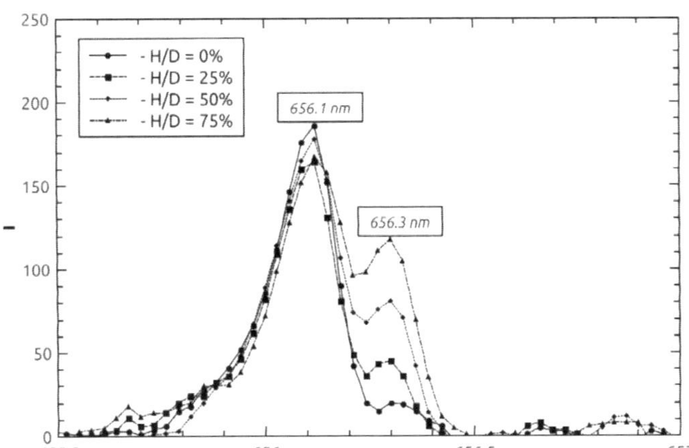


**Figure 2.** Enhancement of $H_\alpha$ line intensity with growing of $H_2$ admixture to $D_2$ (according to settings of mass flow controllers).

Water vapor injection into vacuum vessel was carried out by calibrated thermochemical water source based on $Ca(OH)_2$ thermal destruction [1, 2, 6]. It has fine-tuning of a water vapor flow in the range from $10^{12}$ molecules/s to $10^{22}$ molecules/s by a heater temperature. Low limit of the range is caused by pressure of a saturated vapor of a calcium hydroxide ($10^{-5}$ Pa),

## 3. Experimental results

First experiments were carried out without water vapor injection. The relative intensity $H_\alpha/D_\alpha$ in the emission spectrum of deuterium plasma was studied for different hydrogen gas flow. Relative gas puff was set in range from 1 % to 9 % by calibrated mass-flow controllers. Figure 3 shows the change between $H_\alpha$ and $D_\alpha$ with smooth change in $D_2$ and $H_2$ composition in the gas puff. It is seen that line intensity is approximately proportional to relation between $D_2$ and $H_2$ flows. Some deviations can be explained by a higher mobility of $H_2$ and HD molecules.

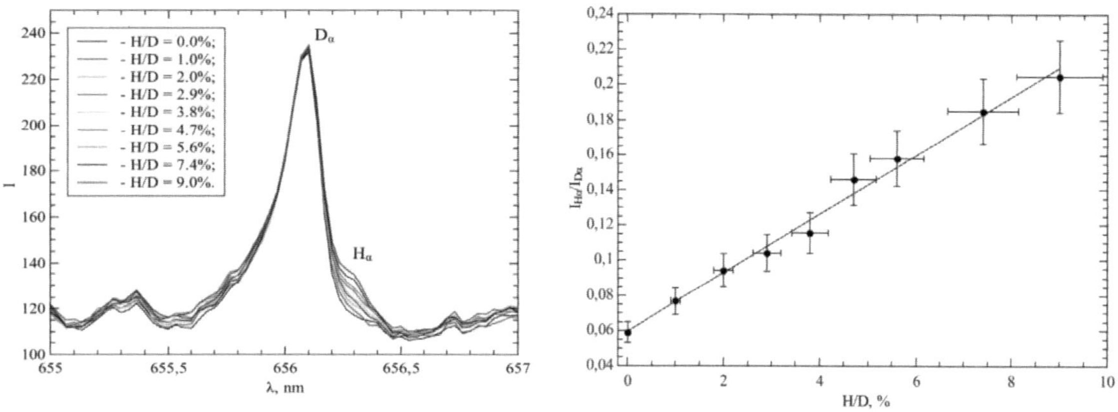


**Figure 3.** $H_\alpha/D_\alpha$ change with low $H_2$ admixtures to $D_2$ into the electron beam driven discharge at PR-2.

QMA shows an increase of $H_2$ and HD generated in plasma. The change of the H-components in residual gas was negligible compared with absolute value of the main molecular deuterium peak. Evaluation of the QMA spectrum is shown on Fig. 4. The molecular ions which are generated in the gage and have not neutral forms are shown in brackets.

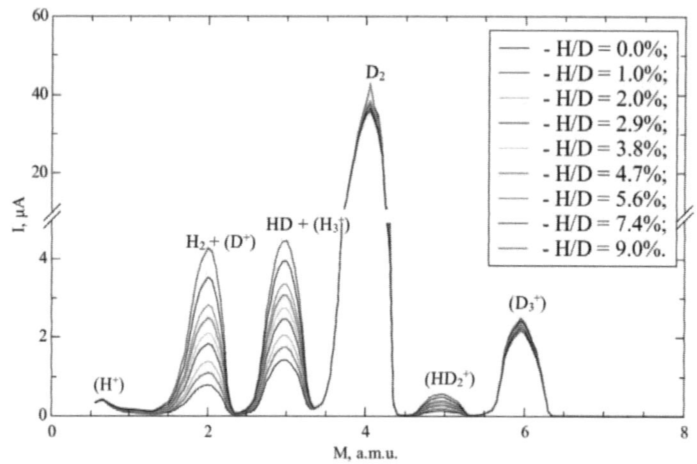


**Figure 4.** QMA signals with protium concentration change in $H_2/D_2$ mixture.

Small additive of hydrogen leads to increase of heteronuclear ions in ion mass-spectrum of the plasma column. Figure 5 shows the redistribution of the ion composition during the experiment.

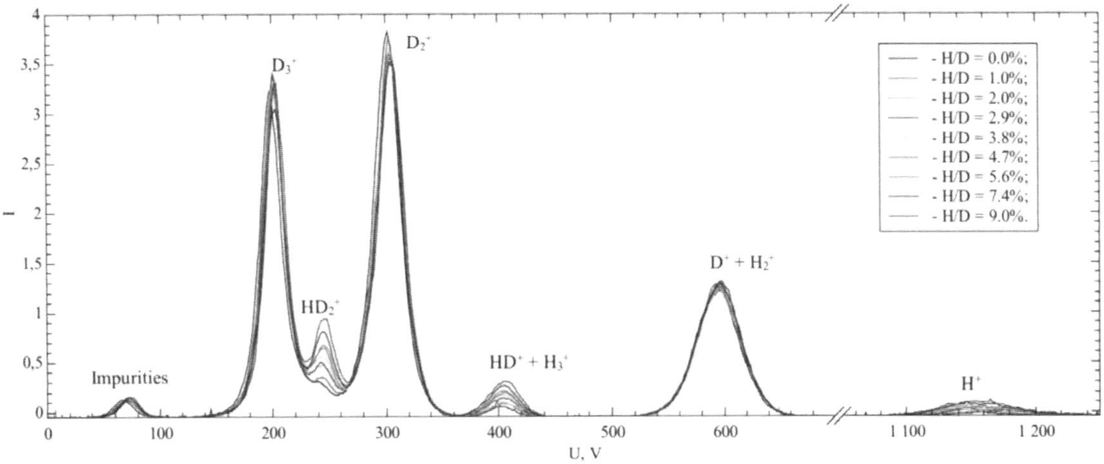


**Figure 5.** Signals from MS with protium concentration change in $H_2/D_2$ mixture.

The $H_\alpha/D_\alpha$ method can be easily applied for cleaning discharges. It can't be applied during ITER campaign with protium discharges. In powerful deuterium working discharges the broadening of the main $D_\alpha$ line would mask the appearance of $H_\alpha$. The experiment with water vapor injection simulates cleaning regimes of a discharge in He without strong magnetic field ($B < 0.1$ T) and as a consequence without a noticeable Zeeman effect. Zeeman broadening would split $H_\alpha/D_\alpha$ lines into triplets. As an example the calculated shape of the $D_\alpha$ line for divertor JET conditions (2.8 T) is presented [7]. It shows that the right satellite of $H_\alpha$ can be resolved from the $D_\alpha$ one.

PR-2 works predominantly with protium. So, in He discharges the protium desorption background prevails, the presence of protium in the $D_2$ balloon also limits its minimal quantity in the vacuum chamber. We used the deuterium addition into He puff for the $H_\alpha/D_\alpha$ intensity ratio change during water injection. The most close and brightest He line is the one with 667.8 nm wave length. The rise of $H_\alpha/D_\alpha$ ratio above the background level with water source ampoule heating is clearly seen (fig.6). The raise of $H_2O$ concentration in the vessel is calculated via vacuum gage signal readings with taking into account of component sensitivities. The reaching of equilibrium value for concentrations of water vapor in the chamber and of excited hydrogen atoms $H^*$ in plasma are determined by rather complicated kinetics. The gain of $H_\alpha$ intensity is more rapid in the beginning of water ingress.

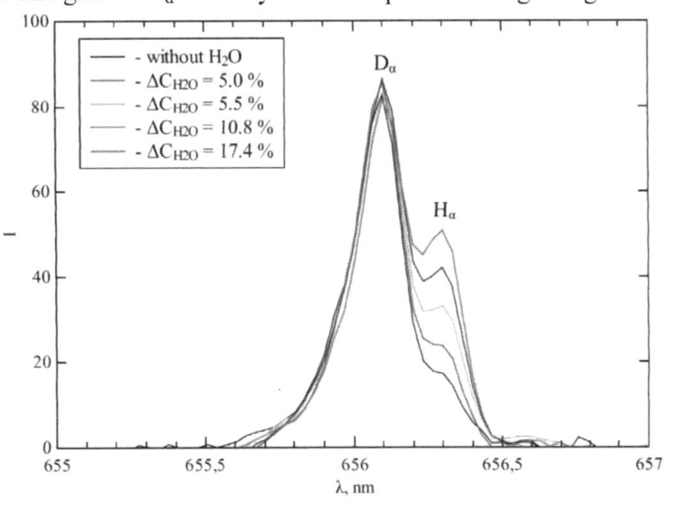


**Figure 6.** Enhancement of $H_\alpha/D_\alpha$ ratio with water ingress into He (with 7% $D_2$) discharge.

The change of the plasma ion composition with vapor ingress rise is shown in fig.7. The most noticeable growth for $H^+$ correlates with enhancement of $H_\alpha$ radiation. It is also seen that the water source of excited and ionized protium is less effective relative to gas $D_2$ admixture (7%) to He discharge. So, discharge parameters can be optimized on further increase of $H_\alpha$ radiation.

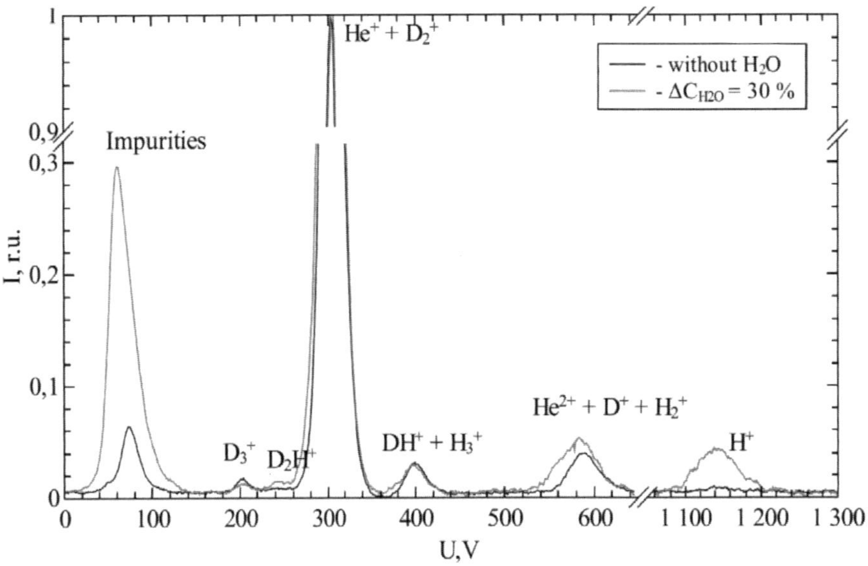


**Figure 7.** Plasma ion composition change with vapor ingress.

It must be mentioned that measurements have been made for a global water vapor pressure raise. Its raise above 1% can be registered by $H_\alpha/D_\alpha$ method for He discharge in PR-2. This level corresponds to the ingress of about $10^{16}$ water molecules per second.

## 4. Conclusion

The spectroscopic method for water microleakage detection in ITER plasma chamber by IO Vis/IR diagnostic system based on relationship between intensities of $H_\alpha$ and $D_\alpha$ lines was validated. The sensitivity of the method was about $10^{16}$ molecules/second of the water vapor inflow. It must be mentioned that measurements have been made for a global water vapor pressure raise. The method's sensitivity can be improved for localized water plume observation. So, with further enhancement of the monochromator resolution and detector photosensitivity, the detectable water ingress level can be decreased by at least an order to correspond the requirements for the sensitivity of the water leakage detection system in ITER. The $H_\alpha/D_\alpha$ method can be easily applied for He cleaning discharges, but not during ITER campaign with protium discharges.

**Acknowledgements**

This work was conducted in accordance to contract with Lebedev Physical Institute of the Russian Academy of Sciences (LPI RAS) within grant 14-12-00784 of the Russian Scientific Fund.

**References**

[1] G.S. Voronov *et al.* 2013 *Plasma Physics Reports* **39** 277
[2] V.A. Kurnaev *et al.* 2013 *Fus. Eng. and Design* **88** 1414
[3] M. Davi *et al.* 2008 *Rev. Sci. Instrum.* **79** 10F509
[4] K. M. Gutorov *et al.* 2007 *Proc. of 28th ICPIG* 378
[5] I.A. Sorokin *et al.* 2015 *Bulletin of the Lebedev Physics Institute* **42** 350
[6] I.A. Sorokin *et al.* 2015 *Physics Procedia* **71** 116
[7] A.B. Kukushkin *et al.* 2012 *Proc. 24th IAEA Fusion Energy Conf., San Diego, USA* ITR/P5-44

# Remote network control plasma diagnostic system for Tokamak T-10

**V I Troynov[1,3], A M Zimin[1], V A Krupin[2], G E Notkin[2] and M R Nurgaliev[2]**

[1] Plasma Power Engineering Department, Bauman Moscow State Technical University, 5, 2nd Baumanskaya Street, Moscow, 105005, Russia
[2] Tokamaks Department, National Research Centre Kurchatov Institute, Moscow, 1, Kurchatov Square, 123182, Russia

E-mail: 3off_vlad@mail.ru

**Abstract.** The parameters of molecular plasma in closed magnetic trap is studied in this paper. Using the system of molecular diagnostics, which was designed by the authors on the «Tokamak T-10» facility, the radiation of hydrogen isotopes at the plasma edge is investigated. The scheme of optical radiation registration within visible spectrum is described. For visualization, identification and processing of registered molecular spectra a new software is developed using MatLab environment. The software also includes electronic atlas of electronic-vibrational-rotational transitions for molecules of protium and deuterium. To register radiation from limiter cross-section a network control system is designed using the means of the Internet/Intranet. Remote control system diagram and methods are given. The examples of web-interfaces for working out equipment control scenarios and viewing of results are provided. After test run in Intranet, the remote diagnostic system will be accessible through Internet.

## 1. Introduction

Among the processes occurring in edge plasma, hydrogen isotope recycling is of special importance. In the course of this process atoms and molecules, formed as a result of ion-wall interaction, enter plasma. The ratio between the flow of particles (molecules and atoms) returning to plasma and the ion flow characterizes recycling efficiency. Since the particle flow from plasma can lead to the release of the particles previously captured by the wall, the value of this coefficient can be far greater than unity. Uncertainty of recycling characteristics, particularly of radioactive tritium, can further complicate the interpretation of tokamak's modes. Analysis of radiation arising as a result of optical transitions of hydrogen isotopes molecules facilitates the studies of edge-plasma processes which is proved by the experiments on tokamaks TEXTOR, DIII-D, JET, etc [1-3]. For instance, results in [1] show that the rotational temperature $T_{rot}$, obtained experimentally using molecular spectra, is a function of electron concentration $n_e$ and limiter surface temperature, which characterize plasma-wall interactions. Thus, the analysis of molecular spectra provides us with an opportunity to estimate these parameters.

[3] To whom any correspondence should be addressed.

## 2. Diagram and hardware resources of the molecular diagnostic system

To study molecular radiation of hydrogen-deuterium plasma within the visible range of wavelengths we developed and optimized several optical schemes for detecting radiation from two diagnostic cross-sections (A and D) of the tokamak T-10 [4]. One of molecular diagnostic subsystems is located within cross-section D (multi-channel subsystem) and allows us to register spectra at the tokamak edge plasma coming from 14 different observation lines (figure 1).

Another subsystem, with monochromator MDR-23 as its basis, was placed in limiter cross-section A within direct view of graphite circular limiter used for restricting the radius of discharge. The monochromator's characteristics are the following: diffraction grid – 1200 gratings/mm, wavelength range – 400-900 nm, inverse linear dispersion – 1.3 nm/mm, spectral resolution – 0.08 nm. In the exit slit plane of this spectral device there is a digital camera Proscan HS101H (CCD-matrix, 512x512 pixels, pixel size – 24x24 μm, resolution - 16 bits), which allows registering of spectrum areas up to 16 nm wide for one discharge of tokamak. The monochromator has a drive with step motor, which provides an opportunity to choose desirable spectrum areas using a computer.

For this diagnostic channel we installed a remote control system for monochromator MDR-23 and digital camera Proscan HS101H. This system allows adjusting of remote equipment before each tokamak discharge and can be used independently by several groups of scientists. Note that the like technology is planned to be used in the ITER project for remote diagnostics.

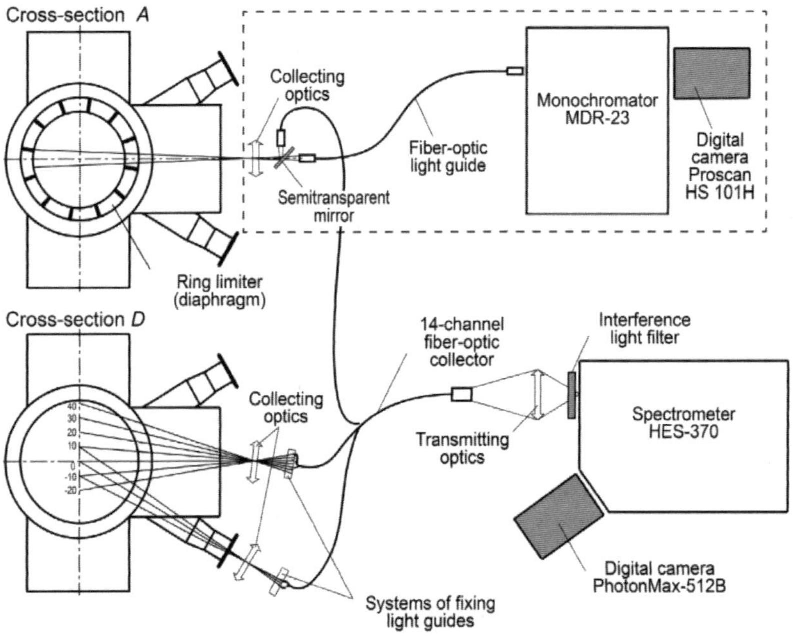


**Figure 1.** Scheme of registration of molecular spectra within diagnostic cross-sections of tokomak. Dotted lines show equipment used remotely.

The electronic atlases of radiative transitions of deuterium and protium has been created earlier for the quick identification and analysis of the registered molecular spectra [4]. The work with tables of the each atlas is performed using DBMS MySQL, and the access to the atlas can be performed using a Web-interface or a specialized software package of automated identification of atomic lines and molecular bands created in the MATLAB environment. The paper [5] served as basis for creation of electronic atlas of transitions in deuterium molecules. At present, this atlas includes information about more than 7000 deuterium rotational lines grouped according to the electronic transitions. Wavelengths, quantum characteristics of the corresponding transition, tabular intensity, and affiliation to any electronic system are given for each rotation line. The electronic atlas for protium molecules includes data on more than 13400 lines. Specialized software with MATLAB-based Graphical User Interface (GUI) was also used for further analysis and processing of molecular spectra.

## 3. System for remote control of diagnostic channel

To study the parameters of molecular plasma we developed the system for remote control of spectral equipment and analyzed different scenarios of its work. Remote network control plasma diagnostic includes automated dispatch and information system [6]. The software for the developed system on the one hand provides a dialogue between a remote experimenter and a web-server and, on the other it allows establishing regime for operation of equipment and translation of results to remote computer. Figure 2 shows a diagram illustrating the used method.

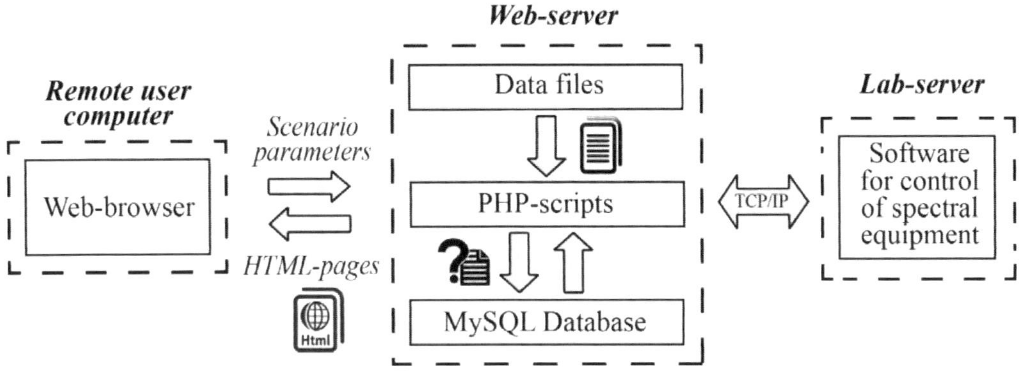


**Figure 2.** Diagram of remote control system for spectral equipment.

The control computer (Lab-server) is connected by local network with the Web-server accessible through the Intranet/Internet. The control interface for remote experimenter is generated by the Web-server using PHP-scripts and dynamic HTML-pages. User indicates the desired wavelength range and necessary camera parameters in the corresponding fields of an HTML-page. Next, these scenario characteristics are sent to Web-server, where they are put into integrated MySQL database.

The Lab-server software for control of spectral equipment repeatedly reads information on spectrum registration queries produced by remote experimenters from corresponding database tables. After correctness check scenarios are placed in the queue and special table field show the operation status. Remote experimenter is also informed about the course of experiment by means of special messages («scenario is placed in the queue», «spectrum is being registered», and «results can now be viewed», etc).

The Lab-server software determines the work regime of the spectral equipment in accordance with the scenario parameters indicated by remote experimenter and saves the spectrum registration results in the database. Experiment results can be accessed using special PHP-scripts, which provide remote experimenter with data on registered spectra in the form of graphs and tables.

Remote user can execute the following operations:

- log in the system and, if necessary, restore the lost data on registration;
- order remote diagnostics sessions through their personal cabinet;
- create scenarios for remote control of diagnostic equipment;
- view the spectra obtained at the end of scenarios.

Some Web-interfaces of remote experimenter are given below. Figure 3 shows the main page of the remote control system web-site. There are the following sections: «Remote experiment», «Equipment» and «Contacts». User can use the first section to register and authorize in the system. The «Equipment» section provides information on characteristics and regimes of used diagnostic equipment.

After filling out the registration form and entering the system, user gains access to their personal cabinet. Using a special Web-form user makes up and sends a request for conducting a remote control session at the convenient time. The administrator scans the submitted requests; if necessary, changes the time periods for sessions and assigns a status to the experiment (allowed/banned).

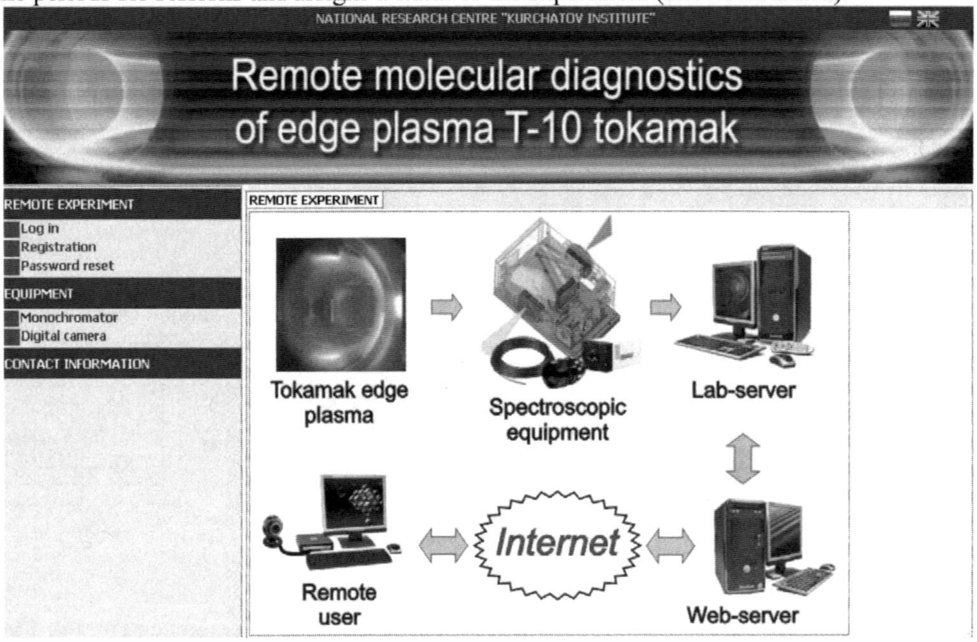


**Figure 3.** Main page of remote control web-site.

If the requested session is allowed, remote user gains access to Web-interface for equipment control (figure 4).

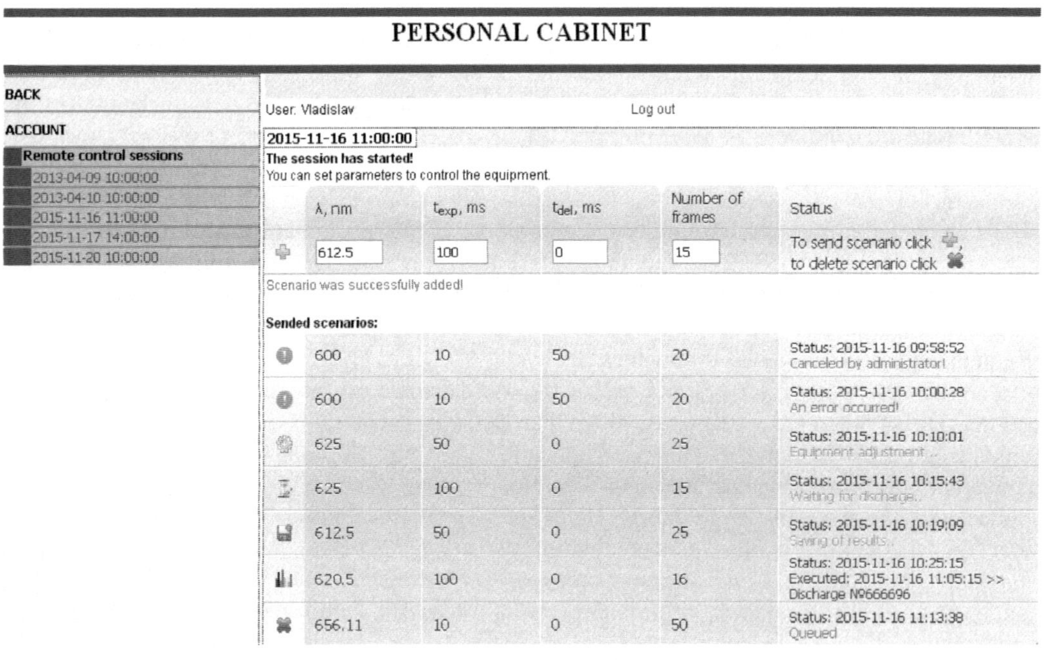


**Figure 4.** Web-interface for equipment control.

The developed system allows storing spectrum data, obtained by remote users. This data, as well as all the necessary information on spectrum registration parameters, is provided to users through their personal cabinet in the form of graphs (figure 5) or tables and can be viewed both in the course of session and on its completion.

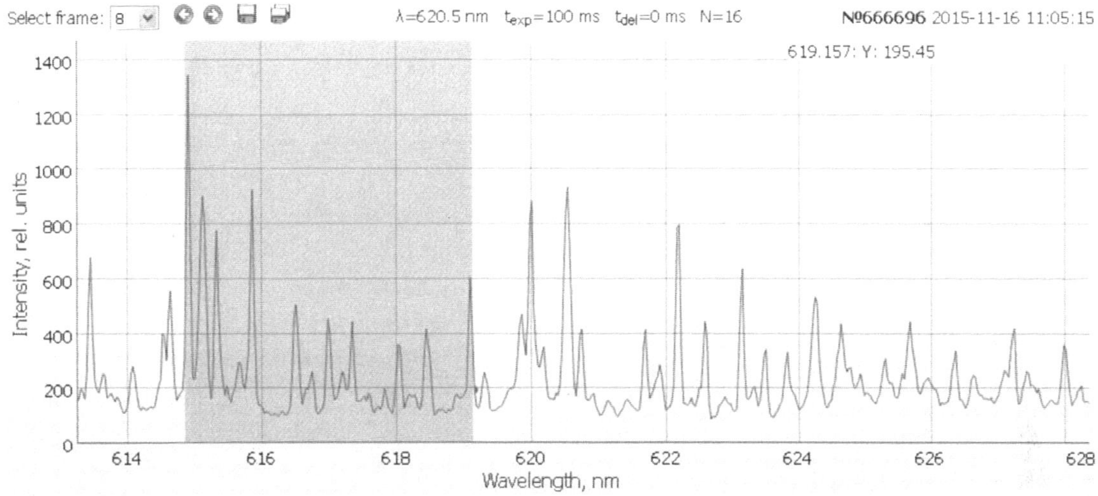


**Figure 5.** Web-interface for viewing and saving of results.

After experimental operations of the system in Intranet we plan to make it accessible through global network.

## 4. Conclusion

The diagnostic facility for registration of molecular spectra in two diagnostic cross-sections of tokamak T-10 is designed. The electronic atlases of radiative transitions of hydrogen isotopes (protium, deuterium) with Web-oriented user interface are developed. The software package for processing of molecular spectra and for identification of rotational lines is created. Vibrational and rotational structure of hydrogen isotopes molecular spectra is identified.

Rotational and vibrational temperatures for ground ($X^1\Sigma_g^+$) and excited ($d^3\Pi_u$) states are estimated using the spectra registered in the course of experiments. The dependencies of temperatures on electron density of edge plasma are obtained. Some experimental results are published in paper [4].

Method and software for remote control of spectral equipment of molecular diagnostics using Internet are developed.

## References

[1] Brezinsek S, Mertens Ph, Pospieszczyk A, Sergienko G and Greenland P T 2002 *Contrib. Plasma Phys.* **42** 668-74

[2] Hollmann E M, Pigarov A Yu, Rudakov D L, Brezinsek S, Brooks N H, Groth M and McLean A G 2006 *Plasma Phys. Control. Fusion* **48** 1165-80

[3] Meigs A, Stamp M, Pospieszczyk A, Brezinsek S, Sergienko G, Huber A, Mertens Ph, Samm U, Wiesen S and Greenland P T 2005 *Journal of Nuclear Materials* **337-339** 500-4

[4] Zimin A M, Krupin V A, Troynov V I and Klyuchnikov L A 2015 *Phys. Atom. Nucl.* **78** 1319-25

[5] Freund R S, Schiavone J A, Crosswhite H M 1985 *J. Phys. Chem. Ref. Data.* **14** 235-383

[6] Zimin A, Shumov A and Troynov V 2015 *Online experimentation: emerging technologies and IoT*, ed M T Restivo, A Cardoso and A M Lopes (Barcelona: IFSA Publishing) pp 275-94

Journal of Physics: Conference Series **747** (2016) 012022 doi:10.1088/1742-6596/747/1/012022

# Formation of dielectric silicon compounds by reactive magnetron sputtering

**D S Veselov[1] and Yu A Voronov**
National Research Nuclear University MEPhI (Moscow Engineering Physics Institute), 31 Kashirskoe highway, Moscow 115409, Russia

E-mail: dsveselov@mephi.ru

**Abstract.** The paper is devoted to the study of reactive magnetron sputtering of the silicon target in the ambient of inert argon gas with reactive gas, nitrogen or oxygen. The magnetron was powered by two mid-frequency generators of a rectangular pulse of opposite polarity. The negative polarity pulse provides the sputtering of the target. The positive polarity pulse provides removal of accumulated charge from the surface of the target. This method does not require any special devices of resistances matching and provides continuous sputtering of the target.

## 1. Introduction

Dielectric silicon compounds in thin films are widely used in electronic technology. One of their special applications is the formation of thin-film dielectric membrane structures, such as shown in figure 1. These structures are the heat-insulating constructive basis of the sensitive elements for semiconductor gas sensors [1-3]. Membrane films for these structures can be obtained by the method of chemical deposition from the gas phase or by various methods of the vacuum sputtering. Vacuum methods, unlike chemical deposition, do not require using of toxic and explosive gases (such as monosilane, ammonia, etc.). It makes them more interesting in the practical application regarding safety [4].

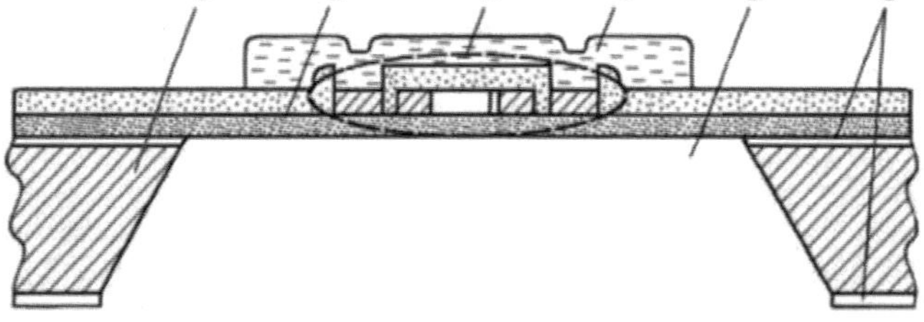


**Figure 1.** Sensitive element on dielectric membrane structure: 1 – silicon substrate; 2 – dielectric membrane; 3 – structural elements (resistive heater, resistive temperature sensor, contacts to sensitive layer); 4 – sensitive layer; 5 – cavity under the membrane; 6 – thermal silicon oxide layer.

---

[1] To whom any correspondence should be addressed.

Journal of Physics: Conference Series **747** (2016) 012022 doi:10.1088/1742-6596/747/1/012022

Reactive ion-plasma sputtering is the most widely used vacuum method for dielectric film formation. In comparison to other methods, it allows obtaining dielectric films with better uniformity of composition and thickness. But, at the same time, it has the lowest sputtering rate. Initially the diode sputterers were used. Then triode sputterers were developed to increase the sputtering rate. For such sputterers, the increase of the sputtering rate is due to the increase of the gas ionization degree in the reaction chamber. For this purpose, an additional source of electrons is introduced into the sputterer – a thermionic cathode. Moreover, the presence of additional ions stimulates sputtering at lower pressures. This ensures better purity of the formed films [5].

Further increase of the sputtering rate was provided by the systems with reactive magnetron sputtering. Herein the target is located in the area of the magnetic field action. Electrons emitted from the target surface began to move along spiral trajectories. Thus, the probability of collisions of electrons with gas atoms was increased. It allowed increasing the gas ionization degree in the reaction chamber. Moreover, it also contributed to the improvement of the purity of the formed films. At present, the reactive magnetron sputtering is the most popular method for formation of dielectric compounds in thin films [5].

## 2. Formation of dielectric compounds

Reactive magnetron sputtering of the silicon target in the ambient of argon with nitrogen or oxygen was conducted in the unit UVN-71P3. For this purpose, the unit was equipped with the three-channel gas injection system in the reaction chamber. The especially pure gases were used for the study. The gas flows into the reaction chamber were controlled by electronic regulators of the gas rate.

In the process of reactive magnetron sputtering, formation of dielectric compounds may take place on the substrate surface, on the target surface with subsequent sputtering and in the gaseous ambient. Formed dielectric compounds are deposited both on the substrate and the target. The dielectric film gradually covers the target surface. Subsequently, the positive ions partially remove this film and are partially deposited on it. Thus, the target surface accumulates the positive charge. As a result, the negative voltage on the target decreases, and the glow discharge is extinguished.

To maintain the glow discharge it is necessary to remove the positive charge periodically from the target surface. Usually, this is done by supplying the alternating voltage of the industrial frequency 13.56 MHz to the target. For effective use of the power of the high-frequency voltage source, it is necessary to match the resistance of the glow discharge with the output resistance of the power source. For this purpose, special matching devices are used, which significantly complicates application of high-frequency power sources for supplying the magnetrons.

At present, the mid-frequency sources of pulsed voltage with the frequency of 10-100 kHz are widely used for deposition of dielectric compounds. These power sources do not require any special devices for resistances matching. To provide the alternating power supply for the magnetron in this work two power sources with opposite signs of the voltage supply were used. During the negative polarity pulse the sputtering of the target material was conducted. During the positive polarity pulse the removal of accumulated positive charge from the target surface was performed. The duration of the negative polarity pulse was equal to 40 μs, while the duration of positive polarity pulse was equal to 6.5 μs, and the pulse frequency was 36 kHz.

The feature of the magnetron sputtering is that only about 5% of energy of bombarding ions is expended on the sputtering process. The remaining energy is spent on heating of the target. The heating of the target has a significant effect on the deposition rate of dielectric films. To reduce this effect, it is necessary to provide continuous cooling of the target during the sputtering process. For this purpose, the base of the target (cathode) was equipped with the flowing water cooling system. But, as it turned out, the presence of the cathode cooling system does not guarantee the required heat removal from the target. Also, it is necessary to provide the acceptable heat contact of the target with the cathode.

The base of the target is the part of the copper cylinder with the diameter of 150 mm and the length of 320 mm. The target represents the cut of the silicon ingot with the thickness of about 1.5 cm. The reverse side surface of the target is wavy, as it is shown in figure 2 (a). It does not provide the acceptable heat contact of the target with the cathode. To improve the heat contact, grinding of the reverse side surface of the target was conducted, as it is shown in figure 2 (b). Herein, to increase the magnetic field induction, the target thickness was reduced to 1 cm. It allowed reducing the pressure in the reaction chamber at sputtering. This allowed increasing the purity of the formed film.

After grinding, the target was fastened by screws to the cathode. Then, the target temperature at sputtering was studied. The results have shown insignificant improvement of the heat contact between the target and the cathode. To further improve the heat contact, the target was soldered to the cathode using the tin-lead solder. To provide the required adhesion level of the solder to the silicon, a layer of nickel was deposited on the reverse side of the target. Thus, the target temperature at sputtering was significantly reduced, which led to the increase of the sputtering rate.

a b

**Figure 2.** Silicon target (reverse side): (a) wavy surface; (b) grinding of the target surface.

**3. Study of properties of obtained dielectric compounds**

The deposition rate, element composition and physical properties of dielectric compounds were studied. Pre-oxidized monocrystalline silicon wafers of orientation (100) with the diameter of 76 mm and with the thickness of (380±20) μm were used as the substrates for deposition of dielectric compounds. Oxidation was conducted in the ambient of dry oxygen at the temperature of 1000 °C. The oxide layer thickness was about 0.15 μm. Initially, substrates were washed for 10 minutes in the peroxide-ammonia solution at the temperature of 70 °C. Then substrates were washed in flowing deionized water of grade A. Immediately before deposition of dielectric compounds, in single vacuum cycle, the ion cleaning of substrates in the ambient of argon was conducted. The ion cleaning mode was as follows: the accelerating voltage - 1500 V, the ion current - 120 mA, the pressure in the reaction chamber - 0.83 Pa, duration - 3 minutes.

Deposition of dielectric films was conducted in the ambient of argon with oxygen and in the ambient of argon with nitrogen. It was revealed that the deposition rate of dielectric films depends on the power, supplied to the magnetron, as it is shown in figure 3.

The element composition of the obtained films was studied in the unit XSAM-800 of the company Kratos. The method of XPS (X-ray photoelectron spectroscopy) was used. The element composition was studied on the surface of the obtained films and in some depth after ion etching in the ambient of argon for 1 and 4 minutes. On the surface of the films a lot of contaminants were present, which were formed during storage between the technological stages. Therefore, the results of the element composition study on the surface of the films were not taken into account.

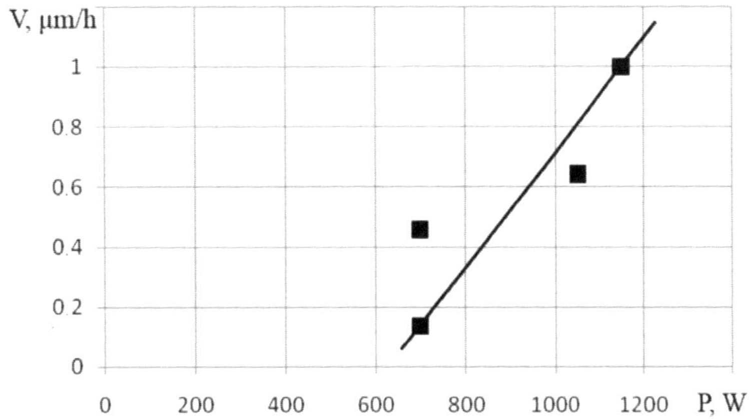


**Figure 3.** The deposition rate of dielectric films versus discharge power.

Figure 4 presents the composition of the dielectric films, which were formed in the ambient of argon with nitrogen. The column with the stoichiometric composition $Si_3N_4$ (right) is also shown for comparison. As can be seen in figure 4, these films contain the significant amount of oxygen and some amount of carbon. The presence of carbon in the films was caused by the oil of the forepump, which provides the preliminary pumping of the vacuum chamber. The presence of oxygen in the films was caused by its significantly higher chemical activity, compared to nitrogen. Oxygen gets into the vacuum chamber due to leakages from the external environment. Also, oxygen desorbs from the chamber walls and the equipment elements. Oxygen is reacting much faster with silicon than nitrogen, which is specially supplied into the reaction chamber. Thus, at the reactive magnetron sputtering, the oxygen absorption rate of the growing film significantly exceeds the nitrogen absorption rate. Moreover, the ratio of these rates strongly depends on the magnetron current. The increase of the magnetron current from 1 A to 5 A reduces the ratio of the oxygen absorption rate to the nitrogen absorption rate from 70 to 10. Consequently, to obtain the silicon nitride films with the minimum oxygen content, it is necessary to provide the sputtering of the target at the high power discharge. Also, the sputterer requires thorough hermitization of the vacuum system.

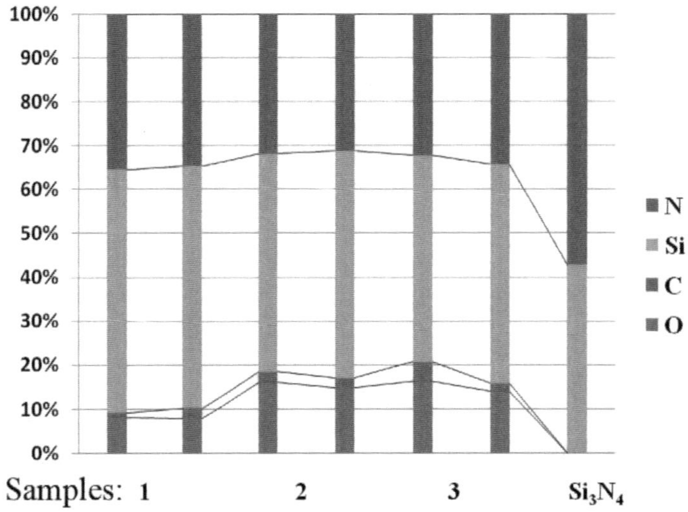


**Figure 4.** Element composition of obtained silicon nitride films: after ion etching during 1 minute (left), after ion etching during 4 minutes (right), for all samples.

The element composition study of the obtained films has shown that the oxygen content in the silicon oxide films and nitrogen content in the silicon nitride films depend on the magnetron current. With the increase of the magnetron current, the composition of the films approaches the stoichiometric one. For the silicon dioxide films, it corresponds to the ratio O/Si=2. For the silicon nitride films, it corresponds to the ratio N/Si=4/3. The silicon dioxide films, close to the stoichiometric composition, are formed when the magnetron current is about 1.2 A. The silicon nitride films, close to the stoichiometric composition, are formed when the magnetron current is about 4.5 A.

The influence of the magnetron supply voltage on the oxygen content in the silicon dioxide films was revealed. The increase of the supply voltage leads to decrease of the oxygen content in the film, as it is shown in figure 5. Herein, deviation from the stoichiometric composition of the film increases. In addition, some electrical parameters of the obtained dielectric films were measured. The measurements were made by the digital measuring device of type *L*, *C*, *R*. For this purpose, by using of magnetron sputtering through the mask, aluminum contacts with the size of $1\times1$ mm$^2$ were deposited on the surface of obtained films. The silicon substrates on the reverse side were completely covered with aluminum. Thus, the continuous contacts to the reverse side of the substrates were made. By the conductivity measurements of thus obtained MOS-structures, the resistance and the electrical strength of the films were estimated. The resistivity of the films is in the range of $5.2\times10^8 - 4.3\times10^9$ $\Omega\times$cm. The dielectric breakdown voltage of the films is in the range of $(3.1 - 7.4)\times10^6$ V/cm. This indicates good dielectric properties of the obtained films.

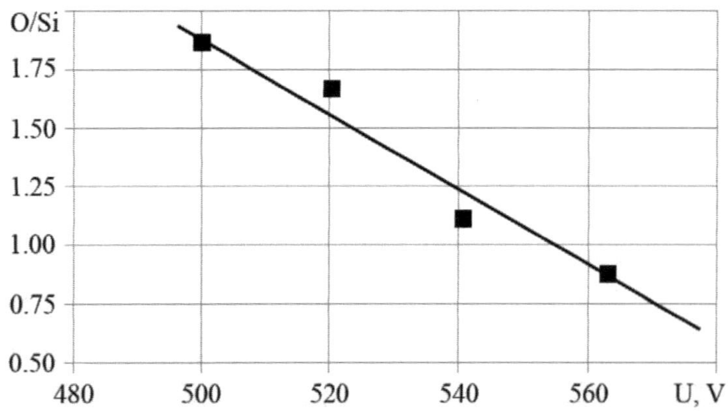

**Figure 5.** The ratio of the oxygen content to the silicon content in the silicon dioxide film versus magnetron supply voltage.

## 4. Conclusions

The formation processes of the dielectric silicon compounds by reactive magnetron sputtering of the silicon target in the ambient of argon with nitrogen or oxygen have been studied. At that, the magnetron was powered by the alternating pulse voltage with the frequency of 36 kHz. To improve the heat contact between the target and the cathode, the target was soldered to the cathode using the tin-lead solder. Thus, the target temperature at sputtering was reduced significantly, which led to increase of the sputtering rate.

It was revealed that the deposition rate of the dielectric films depends on the glow discharge power. It was shown, that the element composition of the obtained films primarily depends on the current and the voltage of the magnetron. To the lesser degree, it depends on the flow change of reactive gas into the vacuum chamber. To approach of the element composition of the films to the stoichiometric one, it is necessary to increase the current and reduce the voltage of the magnetron. When the current of about 1.2 A was supplied to the magnetron, the silicon dioxide films were formed, close to the stoichiometric composition. The silicon nitride films, close to the stoichiometric composition, were formed when the current of about 4.5 A was supplied to the magnetron. It was also revealed that during the formation processes of the silicon nitride films by reactive magnetron sputtering it is almost impossible to exclude the presence of oxygen completely in the film. This is due to the significantly higher chemical activity of oxygen, in comparison with nitrogen.

**Acknowledgements**

This work was supported by the Competitiveness Program of NRNU "MEPhI" (M.H.U.).

**References**

[1] Veselov D S 2015 *Physics Procedia* **72** 500-503
[2] Veselov D S and Voronov Yu A 2015 *Physics Procedia* **71** 423-427
[3] Veselov D S, Voronov Yu A and Vanyukhin K D 2015 *Physics Procedia* **72** 495-499
[4] Lieberman M A and Lichtenberg A J 2005 *Principles of Plasma Discharges and Materials Processing* (New York: John Wiley & Sons)
[5] Roth J R 1995 *Industrial Plasma Engineering (Principles* vol 1)ed T Francis (Bristol: IOP Publishing)

# Magnetic and transport properties of HTS $MgB_2$ wires

**D A Abin**[1,2,3]**, N A Mineev**[1,2]**, M A Osipov**[1,2]**, S V Pokrovskiy**[1,2] **and I A Rudnev**[1,2]

[1] National Research Nuclear University MEPhI (Moscow Engineering Physics Institute), Kashirskoe sh. 31, 115409 Moscow, Russia
[2] International Laboratory of High Magnetic Fields and Low Temperatures, Gajowicka 95, 53-421 Wroclaw, Poland

E-mail: Dima_abin@mail.ru

**Abstract.** Critical current $I_c$ and magnetization $M$ of industrial $MgB_2$ tape and wire were measured in order to estimate their suitability for production of liquid helium free cryomagnetic system. Samples were subjected to bending with different diameters $D_b$ and critical current was measured at temperature $T$ = 4.2 K in magnetic fields up to 10 T. The dependences of the critical current on the diameter of the bend $I_c$ ($D_b$) were found and used for optimization of liquid helium free cryomagnetic system. The magnetization of the wire was measured by a vibrating sample magnetometer at temperatures $T$ = 4.2,10, 15, 20, 25 K in magnetic fields up to 14 T. It was found that the magnetization curve $M$ ($H$) is influenced by ferromagnetic response of the metal matrix. Ferromagnetic contribution had been taken into account and magnetization loops caused by the diamagnetic contribution of the superconducting phase were extracted. Dependencies $I_c(T)$ and $I_c(H)$ were obtained from the data.

## 1. Introduction

Great attention is paid to $MgB_2$ superconductor because of its unique properties such as relatively high critical temperature, big coherence length and low anisotropy [1]. Low cost of magnesium and boron powders makes that material attractive for industrial application. Nowadays a lot of technological innovations in wire production allows to achieve high critical current values and good mechanical properties are developed. The most promising applications of $MgB_2$ wires are the cable industry and the development of MRI (magnetic resonance imaging) machines. By virtue of the application features, MRI coils should have large diameter (tens of centimeters) and provide uniform magnetic field. Using $MgB_2$ wire at liquid helium temperatures allows us to achieve the required parameters. However, there are problems concerned with creation of high-current leads and winding the coils bended with a small diameter (few centimeters). It is known, that the characteristics of wires and tapes based on $MgB_2$ are affected by bending deformations [2]. Starting with a certain critical diameter, the critical current starts to degrade. However, the literature provides little information about change of characteristics with decreasing diameter of the bending deformation. In our previous studies [3] for ex situ $MgB_2$ wires and tapes produced by Columbus Superconductor it was shown, that the critical current decreases with bending diameter decrease. Even for diameter of $D$ = 100 mm the bending deformation leads to drop of the critical current. Depending on amplitude of external magnetic field the critical current for $D$ = 30 mm is in the range 39 - 57% of that for $D$ = 100 mm. At the same time,

[3] To whom any correspondence should be addressed.

despite the significant degradation of the critical current, our sample did not break down and kept superconducting properties even at such small bending diameter.

In present work the $MgB_2$ wires are considered within the scope of the creation of liquid helium free cryomagnetic system with operating temperatures of 4.2 - 30 K and a small diameter warm bore (4-10 cm). To create $MgB_2$ coil with small diameter the "wind and react" technology is usually used [4]. That avoids bending deformation, and hence degradation of the wire, since the formation of the superconducting phase takes place under annealing of already formed coil. However, this technology requires a complex process of homogeneous heat treatment in the furnace and does not allow pre-study the characteristics of the wire. Thus, we decided to explore the possibility of creating a coil from an industrial superconducting wire by "react and wind" method. For this purpose, at different temperatures the magnetization curves of for $MgB_2$ wire and tape were measured along with isothermal ($T = 4.2$ K) dependences of critical current on the mechanical bending of various diameters. The magnetization was measured using a vibrating sample magnetometer and allowed us to calculate $I_c$ ($T$) dependence. The data obtained can be used to optimize the cryomagnetic system.

## 2. Samples

In this paper we investigated the flat tape with the size of 3.5 x 0.6 mm and round wire with diameter $d$ = 1 mm, commercially manufactured by Columbus Superconductor by PIT ex-situ technology. Number of filaments in the metal matrix is equal to 12 and the filling factor of $MgB_2$ iss 10.7% for the wire and 12.2% and for the tape. The filling factors were calculated as the ratio of $MgB_2$ filaments area to the total cross-section area on photographs presented in Figure 1. There is a copper filament in the center of the wire, encased in an iron shell. Filaments of $MgB_2$ are enclosed in a matrix of Ni.

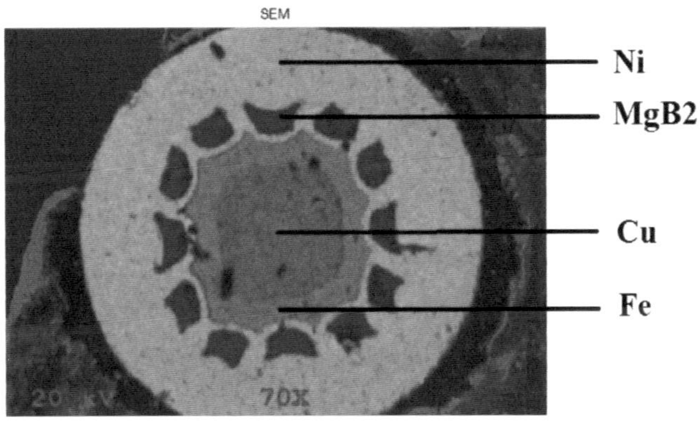


**Figure 1.** Photograph of cross-sections of ex situ $MgB_2$ wire

## 3. Mechanical bending

In order to determine the dependence of the critical current of tapes and wires on the diameter of the mechanical bending, $V$-$I$ characteristics were measured at temperature of liquid helium. Measurements were carried out for bending diameters $D$ = 30, 40, 60, 70, 80, 90, 100 mm in magnetic field applied perpendicular to the direction of the current. A detailed description of the measurement methods and analysis of experimental data is given in [3]. It is shown that with increasing external magnetic field, the critical current density decreases exponentially. However, the superconductivity of the tape does not disappear even for minimal bend diameter of $D$ = 30 mm, but $I_c$ is only 44% of that for $D$ = 100 mm. For comparison, significant degradation of the critical current of the 2G HTS tapes begins only when the bending diameter is less than 15mm for both compressing and tensile bending strain [5].

Journal of Physics: Conference Series **747** (2016) 012023 doi:10.1088/1742-6596/747/1/012023

## 4. Magnetic measurement

To study magnetic properties of wires and tapes the vibrating sample magnetometer was used [6]. Measurements were carried out in magnetic field applied perpendicular to the wire axis and the plane of the tape (force configuration). Heat-exchange helium gas was used to stabilize temperature of the sample.

To increase amount of superconducting material and amplitude of the measured signal we used two parallel pieces of the wire with equal lengths. Measurements were carried out at temperatures $T = 4.2$; 10; 15; 20; 25; 45 K. The necessity of measurements above the critical temperature ($T_c \approx 39$ K) was caused by presence of strongly pronounced reversal ferromagnetic contribution to magnetization from Fe and Ni contained in shell. As seen in Figure 2, the magnetization curves saturate near $H = 3$ T, so expand field to greater values is not advisable. Because of the loops symmetry, measurements of the curves in negative fields were restricted by -2 T.

The superconductive diamagnetic contribution to the magnetization in the temperature range 4.2 – 25 K was obtained by subtracting the curve measured at $T = 45$ K. We expect that variation of the ferromagnetic contribution at $T < 45$ K is not essential since these temperatures are small in comparison with the Curie temperatures for Fe and Ni. Magnetization of $MgB_2$ changes drastically in low field, see Figure 3. Thermomagnetic instability of $MgB_2$ and magnetic field deployment provide significant measurement error in low fields, so part of magnetization curve in vicinity of zero field was obtained as a result of polynomial fitting.

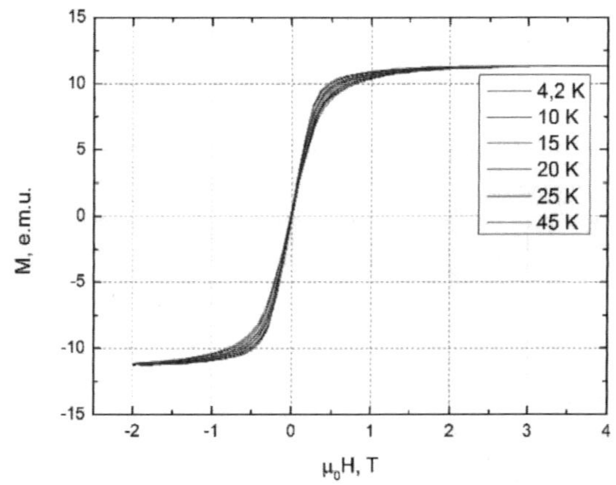


**Figure 2.** Magnetization loop of the wire. External field was applied perpendicular to the wire axis.

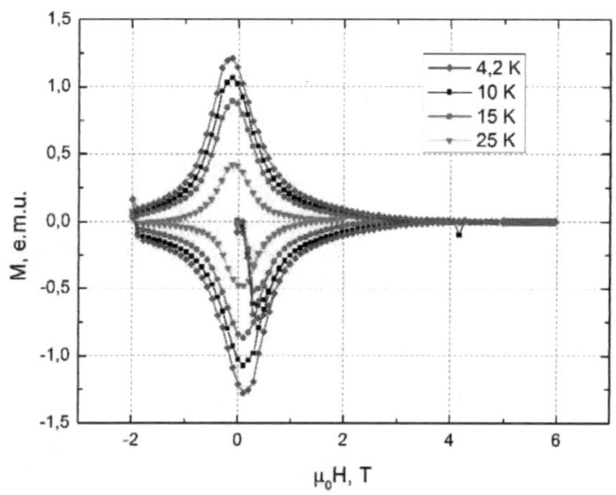


**Figure 3.** The superconductive hysteretic contribution to the magnetization loop for temperatures $T = 4.2 – 25$ K. External field was applied perpendicular to the wire axis.

The dependence of critical current on external magnetic field and temperature (Fig.4) was calculated by the formula

$$Ic(T,H) = Ic(4{,}2) * \frac{\Delta M(T,H)}{\Delta M(4.2,H)} \quad , \tag{1}$$

where $Ic(4{,}2)$ is the critical current in zero field at $T = 4.2$ K and $\Delta M(T,H)$ is width of the magnetization curve hysteresis.

The field and temperature dependences of the critical current are shown in Figures 4 and 5.

As seen from Figure 4, for all temperatures the zero-field points deviate from $Ic(H)$ dependences at higher fields. So the polynomial fitting represents only lower estimate of low-field magnetization and the critical currents can be significantly increased if the thermomagnetic instability is suppressed.

Critical current drastically decreases with increasing external magnetic field (5 times drop with field increasing from 0 T to 1 T) for all temperatures. A similar behavior of critical current occurs with

Journal of Physics: Conference Series **747** (2016) 012023 doi:10.1088/1742-6596/747/1/012023

increasing temperature at persistent external magnetic field (drop up to 25% for temperature rise from 4.2 K to 15 K).
The data obtained allow one to calculate load curves and maximal magnetic field for a solenoid of given sizes, to estimate its operating temperature and to choose optimal construction of a cryomagnetic system.

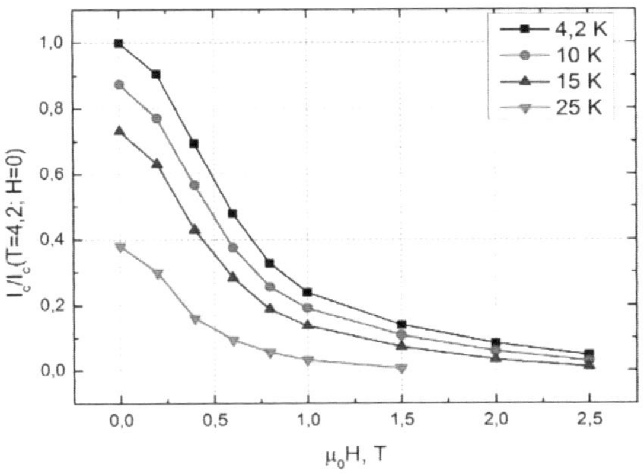


**Figure 4.** Dependences of critical current on external magnetic field obtained at different temperatures. The data are normalized on $I_c$ value measured at $T$ = 4.2 K and $H$ = 0 T

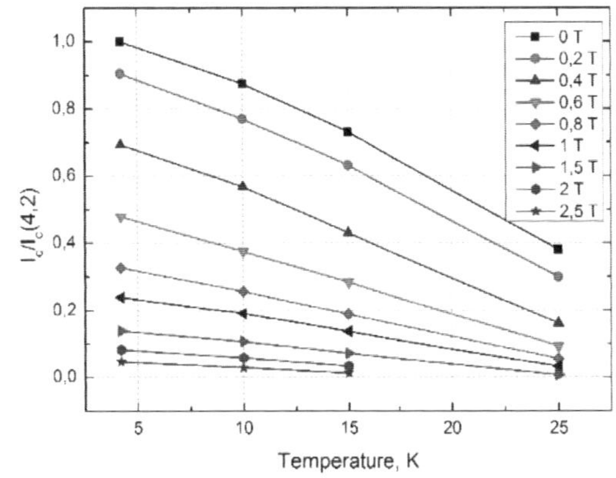


**Figure 5.** Temperature dependences of critical current obtained for different magnetic field. The data are normalized on $I_c$ value measured at $T$ = 4.2 K and $H$ = 0 T

In present work we consider two solenoids with inner diameters of $D_{in}$ = 60 and 100 mm, height $h$ = 140 mm, length of the winding wire $l$ = 400 m and expected fill factor of winding $k$ = 0.8. Load lines, calculated for field constants of the solenoids of 13.94 mT/A and 8.28 mT/A respectively, are shown in Figure 6. For $D_{in}$ = 60 mm the maximal value 1.45 T of the magnetic field $H_m$ at the solenoid center is achieved at temperature of 4.2 K for current $I$ = 104 A. Temperature increase up to 10 K or 25 K leads to decrease of $H_m$ by 9%, or 51%. For $D_{in}$ = 100 mm the $H_m$ value 1.35 T is achieved at 4.2 K for $I$ = 163 A. Temperature increase up to 10 K or 25 K leads to decrease of $H_m$ by 10%, or 52%.

Thus, solenoid of $D_{in}$ = 60 mm produces greater field for the same length of wire. In addition, essentially lower current is required for field producing. Increase of operating temperature to 10 K leads to small (within 10%) operating field decrease, so a cryocooler with low power can be used for liquid helium free cryomagnetic system to reduce its cost.

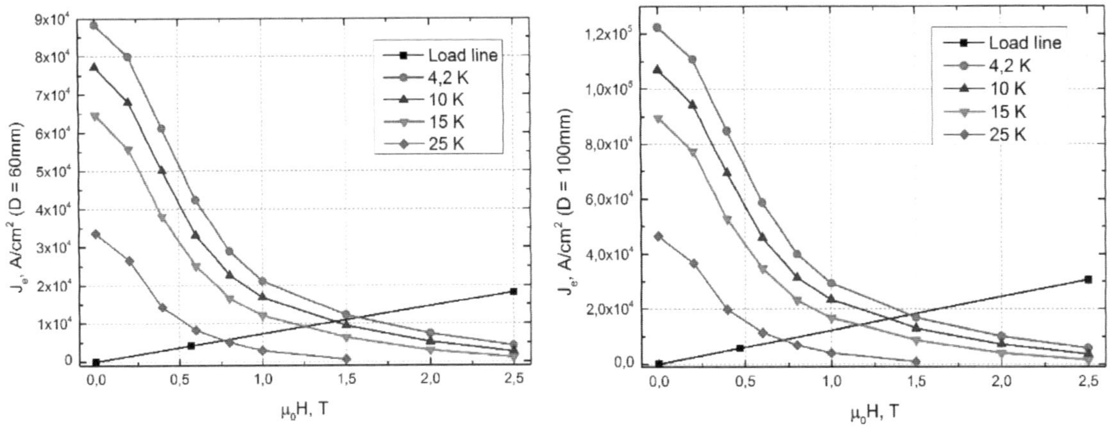


**Figure 6.** Field dependences of the critical current density obtained for $MgB_2$ wire under bending deformation with $D_{in}$ = 60 mm (left) and 100 mm (right). The straight lines are load lines for solenoids

with the above-mentioned bore diameters.

## 5. Conclusions

Despite the fact that the current-carrying capacity of bended $MgB_2$ wires significantly degrades, it remains suitable for the production of solenoids with small enough internal diameters. Maximal magnetic field calculated for solenoids of diameters 60 mm $\leq$ D $\leq$ 100 mm exceeds 1 Tesla. The solenoid of smaller internal diameter produces greater field and its operating current is lower. An increase of the operating temperature up to 10 K leads to field reduction only by 10%. In such conditions, a low-power cryocooler with operating temperature of about 10 K can be used for cooling the cryomagnetic system.

**Acknowledgments**
The work was done with the financial support of the Ministry of Education and Science of the Russian Federation (the Agreement №14.575.21.0050 (RFMEFI57514X0050)).

## References

[1] Buzea C, Yamashita T 2001 *Supercond. Sci. Technol.* **14** R115-R146
[2] Pavol K, Lubomir K, Tibor M, Gustavo S, Santiago S C, Silva B, Davide N, Matteo T 2015 *IEEE Trans. Appl. Supercond.* **25** 620067
[3] Abin D A, Mineev N A, Osipov M A, Pokrovsky S V, Rudnev I A 2015 *Phys. Procedia* **71** 412-416
[4] Sumption M D, Bhatia M, Buta F, Bohnenstiel S, Tomsic M, Rindfleisch M, Yue J, Phillips J, Kawabata S, Collings E W 2007 *Phys. C Supercond. Its Appl.* **458** 12-20
[5] Rudnev I A, Mareeva A, Mineev N A, Pokroskiy S V, Sotnikova A P 2014 *Journal of Physics: Conference Series* **507** 022029
[6] Nizhankovskii V I, Lugansky L B 2007 *Meas. Sci. Technol.* **18** 1533-1537

# Close order in the molecular system near hard surface

**Yu Agrafonov[1] and I Petrushin**

Irkutsk State University, Karl Marx str. 1, 664003 Irkutsk, Russia

agrafonov@physdep.isu.ru

**Abstract**. Surface forces in boundary layers or thin films of classic molecular systems should be considered when describing different phenomena in near-surface layers. In this case a molecular system has axial symmetry. We obtained the solution for two-particle distribution function at low densities. It is shown that the solution describes the transition from axial to spherical symmetry while each particle goes to infinity

## 1. Introduction

We should take into account surface forces boundary layers and thin films of classic molecular systems when describing phenomena (adsorption, wetting) in near-surface layers. In such cases a molecular system has axial symmetry, so we may apply a model of a liquid near a hard surface.

Far from the surface contact a molecular system is heterogeneous and isotropic so we should take into account the boundary condition for transition from axial to spherical symmetry. Moreover, properties of the system are described by pair molecular correlations, decreasing with the distance between particles as interaction potential *U(r)*.

In our work [1] the asymptotic behavior of pair distribution function was described for classic molecular system of liquid near hard surface. In this paper we get the particle distribution functions for arbitrary distance from surface. Interaction between particles is described by hard sphere potential.

## 2. Common equations

Statistics considering molecular system is based on Born-Green-Yvon-equations system (also known as BGY hierarchy) for l-particle functions $G_{l,\dots,i} = (\mathbf{r}_l,\dots,\mathbf{r}_i)$ for ensemble of identical particles. These particles interact with each other through potential $\Phi_{ij}\left(r_{ij}\right)$, where $r_{ij} = \left|\mathbf{r_i} - \mathbf{r_j}\right|$ is the distance between centers of particles *i* and *j*. The diameter of each particle is σ.

The BGY-equations system may be transformed to equations for one- and two-particle distribution functions, which may be written as Ornstein-Zernike [2, 4, 5]:

$$\omega_1 = n\int G_2 C_{12}^{(1)} d(2) + \ln a \;;\quad h_{12} = C_{12}^{(2)} + n\int C_{13}^{(2)} h_{23} d(3) \tag{1}$$

Here we integrating on coordinates of *i*-th particle: $d(i) \equiv dr_i$, $n$ – density, $G_1 = \exp(-\Phi_i (kT)^{-1} + \omega_1)$ one-particle distribution function, which describes particle position; $\Phi_i$

[1] To whom any correspondence should be addressed.

- potential energy in external field; $\omega_1$ - one-particle thermal potential; $a$ – activity coefficient, which is defined by condition of passing to isotropic system. $h_{ij} = [\exp(-\Phi_{ij}(kT)^{-1} + \Omega_{ij}) - 1]$ is the pair correlation function, which is connected with two-particle distribution function by expression: $G_{ij} = G_j G_i (1 + h_{ij})$; $\Omega_{ij}$ is the two-particle thermal potential, which takes into account indirect pair interaction; $C_{ij}^{(k)}$ - direct correlation functions:

$$C_{ij}^{(1)} = h_{ij} - \omega_{ij} - \tfrac{1}{2} h_{ij}\left(\omega_{ij} + M_{ij}^{(1)}\right); \qquad C_{ij}^{(2)} = h_{ij} - \omega_{ij} + M_{ij}^{(2)} \tag{2}$$

The functions $G_1(\mathbf{r}_1)$ and $G_{12}(\mathbf{r}_1,\mathbf{r}_2)$ are most critical since they describe internal structure. Let us obtain thermodynamic parameters of the system. To use the equations (1) and (2) in practice one should approximate these series by simple expressions (closures). Most known of them are hyperchain, the Percus-Yevick and Martynov-Sarkisov approximations.

Isotropic systems, such as spatial liquids when far from surface, are of special interest. In these systems $G_1(\mathbf{r}) \equiv 1$ and $\omega_1(\mathbf{r}) \equiv 0$. As a result, the first equation of the system (1-2) simplifies to

$$\mu = \ln a = -n \int C_{12}^{(1)}(r_{12})\, d\mathbf{r}_{12} \tag{3}$$

Second equation may be formed as

$$h_{12} = C_{12}^{(2)} + n \int C_{13}^{(2)} h_{23}\, d(3) \tag{4}$$

It defines the unknown function $h_{12} = G_{12}^{(0)}(r_{12}) - 1$. The structure and thermodynamic parameters are evaluated through two-particle distribution function $G_{12}^{(0)}(r_{12})$ depending on density $n = \sigma^3 N V^{-1}$.

Space-heterogeneous systems, such as a liquid near a hard surface, are described by one- and two-particle distribution functions: $G_1(\mathbf{r}_1)$ and $G_{12}(\mathbf{r}_1,\mathbf{r}_2)$. Boundary condition for these equations is a transition from a hard surface to a liquid.

### 3. Singlet approximation

Equations (1-2) are very hard for direct solution. In order to simplify this problem we may substitute function $G_{12}(\mathbf{r}_1,\mathbf{r}_2)$ by its boundary value $G_{12}^{(0)}(r_{12})$ for spatial liquid. We call such substitution as singlet approximation. As a result we obtain the equation for one-particle distribution function $G_1(\mathbf{r}_1)$, which describes local density profile $n(\mathbf{r}_1) = nG_1(\mathbf{r}_1)$ near a hard surface. Numerical solution for this equation is presented in [3].

We cannot directly estimate the precision of singlet approximation because of neglecting the close order change near surface but for indirect estimation we can compare the results of our calculation and numerical modeling (numeric experiment). Taking into account the close order change near surface for functions $G_1(\mathbf{r}_1)$ and $G_{12}(\mathbf{r}_1,\mathbf{r}_2)$ we increase the precision of a singlet approximation.

When solving equation (1-2) the origin is located in the center of the particle contacting with hard surface. The $z$ axis is perpendicular to the surface, thus the whole liquid is placed in upper half-space ( $z \geq 0$) and bottom one ( $z < 0$) is unavailable for the particles (Fig. 1). Such a system has axial symmetry,

$$G_1(\mathbf{r}_1) = G_1(z_1) = \exp(\omega_1(z_1)) \quad G_{12}(\mathbf{r}_1,\mathbf{r}_2) = G_{12}(z_1, z_2, r_{12}) \tag{5}$$

where $r_{12}$ is measured in particle diameter units, $z_i \geq 0$ – particle distance from the surface. Boundary conditions for $G_1(\mathbf{r}_1)$ and $G_{12}(\mathbf{r}_1,\mathbf{r}_2)$ are defined as follows

$$\omega_1(z_1) \to 0, \quad G_1(z_1) \to 1$$

$$z_1 \to \infty \qquad z_1 \to \infty \qquad G_{12}^{(0)}(r_{12}) = \lim_{\substack{z_1\to\infty \\ z_2\to\infty \\ r_{12}=|\mathbf{r}_1-\mathbf{r}_2|=const}} G_{12}(z_1, z_2, r_{12}) \tag{6}$$

System (1-2) with a singlet approximation takes the form

$$\omega_1 = n\int G_2 C_{12}^{(1,0)} d(2) + \ln a \tag{7}$$

$$h_{12}^{(0)} = C_{12}^{(2,0)} + n\int C_{13}^{(2,0)} h_{23}^{(0)} d(3) \tag{8}$$

One-particle potential may be obtained from equation (7), where direct correlation function $C_{12}^{(1,0)}(r_{12})$ and pair correlation function $h_{12}^{(0)}$ are considered to be known.

## 4. Beyond singlet approximation

Close order in a molecular system changes even at low densities. Let's take rarefied gas as an example. Thermal potential is obtained through series expansion by density powers. At the first order we get:

$$\omega_1(z_1) = n\omega_1^{(1)}(z_1), \quad \omega_{12}(\mathbf{r}_1,\mathbf{r}_2) = n\omega_{12}^{(1)}(\mathbf{r}_1,\mathbf{r}_2) \tag{9}$$

Substituting (9) into (7, 8) we obtain expression for factoring coefficients

$$\omega_1^{(1)} = \int f_{12} d(2) + \mu^{(1)} \tag{10}$$

$$\omega_{12}^{(1)} = \int f_{13} f_{23} d(2) \tag{11}$$

Here $f_{ij}$ – Mayer's function. Integration is carried out through the whole upper half-space. Constants $\mu^{(i)}$ are derived from border condition at infinity. When evaluating the integral (10) we place the origin at the hard surface (Fig. 1). The surface normal goes through the center of the particle. All upper half-space is reachable for the second particle. By integrating in cylindrical system we get:

$$\omega_1^{(1)}(z_1) = \frac{\pi}{3}\left(z_1^3 - 3z_1 + 2\right)\theta(1 - z_1) \tag{12}$$

When integrating (11) the origin is placed at mass center. Through the centers of first and second particles we pass a plane normal to hard surface (Fig. 1). Intersection field of Mayer's functions (11) is limited by two hemispheres with radii $\Delta_1$ and $\Delta_2$. Integrating at each quadrant, due to symmetry of particles exchange we obtain

$$f_{13} f_{23} = \theta\left(1 - r_{23}^2\right) = \theta\left(1 - \left(x_3^2 + z_3^2 + \left(R_{12} + y_3\right)^2\right)\right), \tag{13}$$

where $\theta(\mathrm{x})$ – Heavyside function. In particular, the integration range specified in Figure 1 is

$$2\left[\begin{array}{l}\int_{-\infty}^{+\infty} dx_3 \int_0^{\Delta_1} dy_3 \int_0^{\infty} dz_3 \theta\left(1-\left(x_3^2+z_3^2+\left(R_{12}+y_3\right)^2\right)\right)+ \\ \int_{-\infty}^{+\infty} dx_3 \int_0^{\Delta_2} dy_3 \int_0^{\infty} dz_3 \theta\left(1-\left(x_3^2+z_3^2+\left(R_{12}+y_3\right)^2\right)\right)\end{array}\right] \tag{14}$$

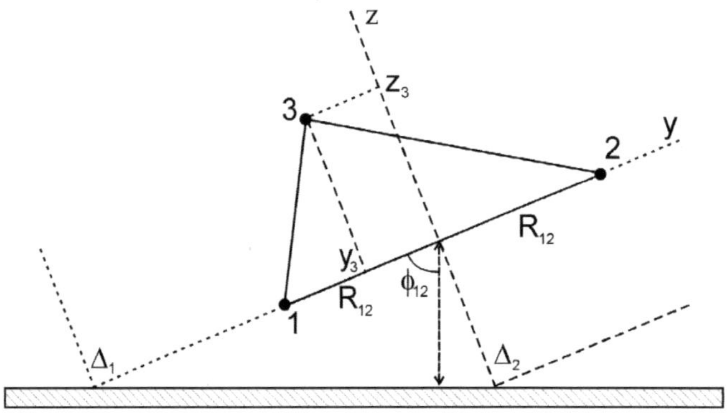


**Figure 1.** Axial symmetry of two-particle distribution function near hard surface

Integrating for all $\Delta_1$ and $\Delta_2$ values, we get

$$\omega_{12}^{(1)}=\frac{\pi}{3}\theta\left(1-R_{12}\right)\left[\begin{array}{l}\left(R_{12}^3-3R_{12}+2\right)\theta\left(\Delta_1+R_{12}-1\right)+\left(3\Delta_1\left(1-R_{12}\right)-3\Delta_1^2-\Delta_1^3\right)\times \\ \theta\left(1-\left(\Delta_1+R_{12}\right)\right)+\left(R_{12}^3-3R_{12}+2\right)\theta\left(\Delta_2+R_{12}-1\right)+ \\ \left(3\Delta_2\left(1-R_{12}^2\right)-3\Delta_2^2-\Delta_2^3\right)\theta\left(1-\left(\Delta_2+R_{12}\right)\right)\end{array}\right] \tag{15}$$

$$\Delta_1=\frac{z_0}{\cos\varphi_{12}},\quad \Delta_2=\frac{z_0}{\sin\varphi_{12}},\quad z_0=\frac{z_1+z_2}{2},\quad R_{12}=\frac{r_{12}}{2} \tag{16}$$

Where $z_1$, $z_2$ are the distances from the particles to the surface, $\varphi_{12}$ – radius-vector angle to the z-axis, $\theta(x)$ – the Heavyside function. When $z_1, z_2$ go to infinity, $\omega_1^{(1)}(z_1)$ converges to zero, and $\omega_{12}^{(1)}$ to volume value

$$\theta\left(1-R_{12}\right)\frac{2\pi}{3}\left(R_{12}^3-3R_{12}+2\right) \tag{17}$$

Thus, as in [2] the solution obtained describes transition from axial to spherical symmetry while each particle goes to infinity. For all other $z_1, z_2$ values the equation (15) describes the close order change near surface.

**Acknowledgements**
We are grateful to prof. A. P. Menushenkov and Dr. A. V. Kuznetsov for their interest to our work. Following work is supported by Russian Basic Research Fund (grant No. 15-02-08204a).

**References**
[1] Agrafonov Yu V, Petrushin I S 2015 *Physics Procedia* **71** 364
[2] Martynov G A *J. Chem. Phys.* **129** 244

[3] Tikhonov D A, Kiselyov O E, Martynov G A, Sarkisov G N, *J. Mol. Liquid.* **82** 3
[4] Martynov G A, 1992 *Fundamental Theory of Liquids; Method of Distribution Functions* (New-York: Adam Hilger)
[5] Apfelbaum E M, Vorobev V S, Martynov G A, *J. Chem. Phys.* **127** 064507

# Diamond single crystal-surface modification under high-fluence ion irradiation

**V A Anikin[1], A M Borisov[1,6], V A Kazakov[1,2], E S Mashkova[3], Yu N Palyanov[4], V P Popov[5], E A Shmytkova[2] and S K Sigalaev[2]**

[1] Moscow Aviation Institute (National Research University), Volokolamskoe sh. 4, 125993 Moscow, Russia
[2] Keldysh Research Center, Onezhskaya str. 8, 125438 Moscow, Russia
[3] Skobel'tsyn Institute of Nuclear Physics, Moscow State University, Leninskie gory, GSP-1, 119991 Moscow, Russia
[4] Sobolev Institute of Geology and Mineralogy, Ac. Koptyuga ave. 3, 630090 Novosibirsk, Russia
[5] A V Rzhanov Institute of Semiconductor Physics, pr. Lavrentieva 13, 630090 Novosibirsk, Russia

E-mail: anatoly_borisov@mail.ru

**Abstract**. The modification of (111) face of synthetic diamond has been studied experimentally for high-fluence 30 keV argon bombardment. It has been found that ion irradiation leads to the electrically conductive layer formation the sheet resistance of which decreases more than 100 times while changing the temperature of the irradiated diamond from 70 to 400 °C. This effect, as well as significant changes of optical transmittance after ion irradiation are associated with ion-induced structural changes of irradiated diamond obtained by the methods of Raman spectroscopy.

## 1. Introduction

Ion irradiation of diamond and diamond-like carbon (DLC) films is one way of producing nanostructures with unique properties [1, 2]. High-fluence focused ion beams can also be used to form buried amorphous layers. The graphitization of amorphous layers is of substantial interest in fabricating devices with diamond-graphite structures [3]. The conductivity of DLC films is raised by several orders of magnitudes after ion irradiation. This increase in conductivity is related not only to the transition of $sp^3$- to the $sp^2$- carbon, but to the ordering of $sp^2$- carbon substantially [2]. That causes great variation of the conductivity with the target temperature change from room temperature (RT) to 300 °C. As for diamond single crystal it is known that the high-fluence irradiation of the (111) face of synthetic diamond with 30 keV $Ar^+$ ions at a temperature of 400 °C produces a graphite-like conductive surface layer [4, 5]. In this paper we continue the study of ion-induced conductivity of single-crystal diamond. Similarly [2] the temperature of irradiated target has been varied from room temperature to 400 °C. Along with the conductivity optical transparency and the structural and morphological changes of diamond modified surface layer are studied.

[6] To whom any correspondence should be addressed.

Journal of Physics: Conference Series **747** (2016) 012025 doi:10.1088/1742-6596/747/1/012025

## 2. Experimental

1b synthetic diamond crystal was used as the target. Irradiation with 30 keV $Ar^+$ ions was conducted normal to the surface of the (111) face using a mass-monochromator of the Skobel'tsyn Research Institute of Nuclear Physics, Moscow State University [6], as described in [4, 5]. The ion current density was about 0.3-0.4 $mA/cm^2$ at a beam cross section of 0.3 $cm^2$ and the irradiation fluences of $10^{18}$ - $10^{19}$ $ion/cm^2$. The modified target was examined by AFM Veeco Dimension V, a Horiba Jobin Yvon T64000 Raman spectrometer and spectrophotometer-fluorimeter SFF-2 "FLUORAN". Electrical measurements were performed using the IUS-3 probe station.

## 3. Results and discussion

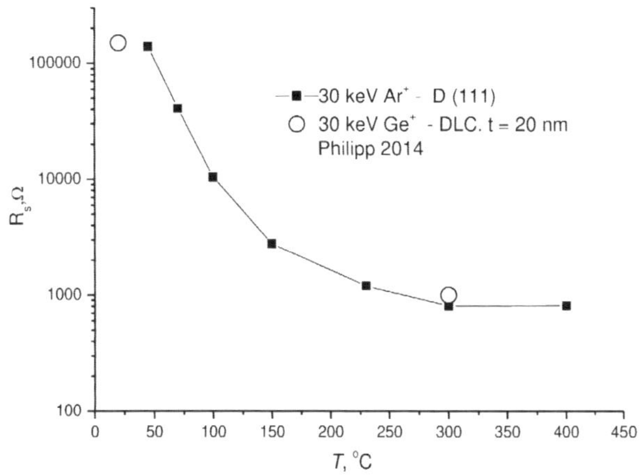


**Figure 1.** Sheet resistance $R_s$ dependence of modified diamond surface layer on temperature of irradiated diamond under 30 keV $Ar^+$ ion at high fluence ($\geq 10^{18}$ $ion/cm^2$). Data for DLC film irradiated by 30 keV $Ge^+$ions at $10^{17}$ $ion/cm^2$ were calculated from [2].

It has been found that high-fluence ion irradiation of the (111) face of 1b diamond leads to the formation of temperature-dependent conductive surface layer. The data presented in Figure1 show that the sheet resistance $R_s$ of ion-modified surface layer decreases more than 100 times when the temperature of irradiated diamond changes from 70 to 400 °C. The obtained data correlate with changes in electrical properties of the diamond and amorphous diamond-like materials [1-3, 7]. Namely, the results of the structural, electrical and optical properties of diamond-like carbon films irradiated at RT and elevated (till 300 °C) target temperatures by heavy ions (Si, Ge, Ga, Au and Bi) in both small and large fluence with the same (30 keV) energy [2] are comparable in terms of ion irradiation. One of the results was a reduction of resistivity ρ with fluence of irradiation at RT to stationary value at high fluencies (~$10^{17}$$ion/cm^2$), as well as a significant influence of the atomic masses of ions and temperature on the ρ stationary value. However, there is not irradiation by $Ar^+$ ions per atomic mass between Si and Ge in the work of [2]. The establishment of the stationary state of the modified layer allows making the quantitative comparison of data for $Ar^+$ high-dose irradiation with data [2] obtained by the most of large fluencies. Figure1 shows that the value of sheet resistance $R_s$, corresponding to the given in [2] the resistivity of DLC films irradiated by Ge ions with a fluence of ~$10^{17}$ $ion/cm^2$ at RT and 300 °C is very close to measured data for argon ion irradiation. The resistivity ρ of modified layer of diamond has been evaluated by assuming that the thickness of the modified layer when irradiated by 30 keV $Ar^+$ ions is 40-50 nm [4, 5]. The comparison shows that at a temperature of 70 °C ρ = 0.2 - 0.25 Ω·cm close to the value 0.3 Ω cm for Ge at RT, see Figure 8 in [2]. The value of ρ = 2 - 2.5 mΩ·cm obtained in present experiment after irradiation at 400 °C practically coincides with the value of 2 mΩ·cm for DLC film irradiated by Ge ions at 300 °C. The last value is close to the resistivity of graphite materials. One can see that graphite-like properties are observed only at elevated irradiation temperatures greater than 150 – 200 °C. With decreasing temperature ($T <$ 150 °C), the resistance increases exponentially. Note that the temperature interval of 150 – 200 °C corresponds to ion-induced annealing temperatures of the radiation damage in the graphites and glassy carbons [**8**].

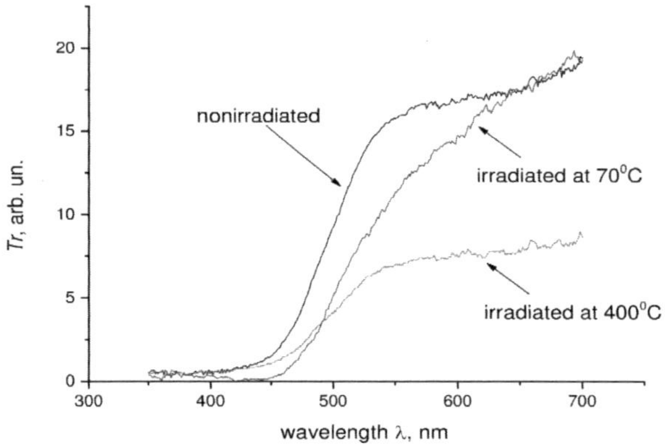


**Figure 2.** Spectral dependences of diamond transmittance *Tr* of synthetic diamond before and after ion irradiation at 70 and 400 °C.

Ion-irradiated formation of conductive surface layer leads to temperature-dependent spectral dependences of diamond transmittance *Tr*. The values of transmittance presented in Figures 2 and 3 are given in relative units due to the small surface area of the diamond crystal as compared with a cross section of the light beam in the spectrophotometer. One can see that transmittance spectrum of non-irradiated diamond is characterized by a threshold at λ = 450 nm caused apparently by the admixture of nitrogen, which gives the crystal a characteristic yellowish color [9]. Ion-irradiation at 70 °C as for non- irradiated diamond leads to a strong spectral dependence of light transmittance and at λ > 650 nm the conductive layer almost does not change the transparency of a diamond. At λ < 650 nm a modified layer introduces additional absorption, monotonically increasing with decreasing of λ. As a result, the transparency threshold is increased from 450 for non-irradiated diamond to 470 nm after irradiation.

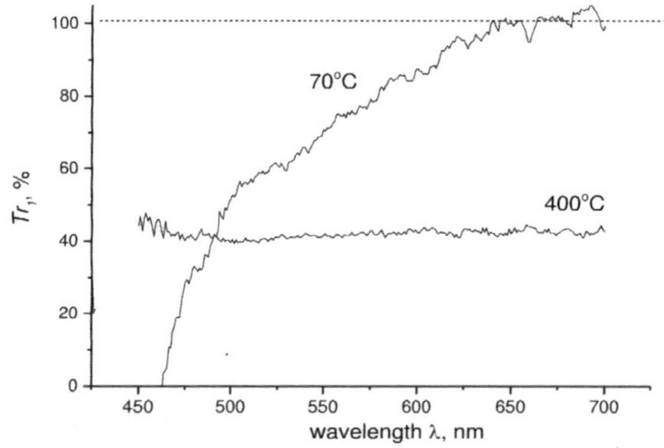


**Figure 3.** Relative transmittance $Tr_1$ of conductive layer formed at 70 and 400 °C.

After ion-irradiation at 400 °C the conductive layer reduces the transparency of the diamond more than twice in the absence of explicit spectral dependence in the measured spectral region of diamond transmission from 450 to 700 nm. A more detailed picture of the transmission spectra of modified layers at different temperatures gives Figure 3, where the transmittance $Tr_1$ is defined by dividing the transmission spectra of the irradiated samples on the transmission spectrum of non-irradiated diamond.

Large differences of $Tr(\lambda)$ dependencies shown in Figure 3 may not be associated with the influence the ion irradiation on the surface topography. AFM analysis shows that ion irradiation does not increase the surface roughness after machining polishing, see Figure 4. Indeed there is some effect of ion polishing. It is more noticeable for the ion irradiation at a temperature of 400 °C when the ion-induced nanostructure is finer than at RT-irradiation. One can see that on a polished surface of the original diamond (Figure 4a) after irradiation at a temperature of 70 °C a system of globules with sizes of 200 nm is formed (Figure 4b). After ion irradiation at elevated temperatures the globules with sizes less than 100 nm are observed (Figure 4c).

a b c

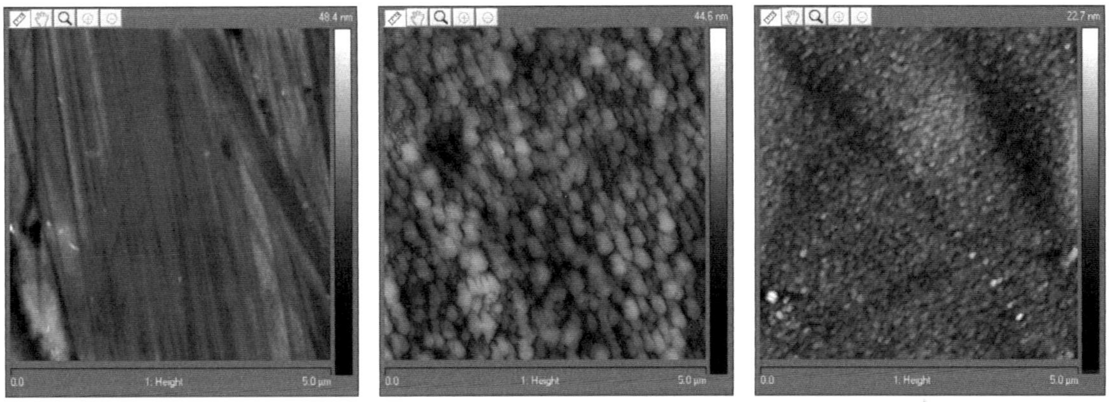


**Figure 4.** AFM images of (111) face of synthetic diamond crystal: (a) - original polished sample, (b) - irradiated at 70 °C, (c) – irradiated at 400 °C.

Qualitative difference of the $Tr_1$ ($\lambda$) dependencies may be due to significantly different structure of the modified layer at different irradiation temperatures, namely the structure of amorphous $sp^2$ carbon layer at $T$ = 70 °C and structure of the graphitized layer at 400°C, cf. [2]. For such modified layers both the conductivity mechanisms and optical properties must actually be different. Conductive surface diamond layer formed under ion irradiation manifests itself in Raman spectra as a broad band with gentle maximum in the region of the Raman shift 1000 – 1700 cm$^{-1}$. The maximum position is approximately 1500 and 1600 cm$^{-1}$ after ion irradiation at the temperatures $T$ = 70 and 400 °C respectively, see Figure 5. Modified layer transparency leads to the presence in the spectra of diamond peak at 1332 cm$^{-1}$. The peak intensity for a more transparent layer after irradiation at $T$ = 70 °C significantly higher than after irradiation at $T$ = 400 °C. For comparison the Raman spectra for glassy carbon irradiated by 30 keV $Ar^+$ ions both at RT and $T$ = 350 °C [10] are also presented in Figure 5. One can see that Raman spectrum at elevated temperature shows characteristic D- and G- bands of graphite. Under ion irradiation at RT an amorphization of carbon materials takes place, D- and G- bands are not separated and a broad band close to that described above for the irradiated diamond is observed. Ion irradiation of diamond at $T$ = 400 °C leads to the graphitized surface layer with a resistance two orders of magnitude less than the resistance of the layer after irradiation at RT. However, one can see from Figure 5 that the separation of the D- and G- bands in the Raman spectrum after irradiation at 400 °C is not happening, cf. [2]. Graphite-like diamond modified layer is revealed in the increasing of G-band and the shift of Raman spectrum maximum to the position of G-peak.

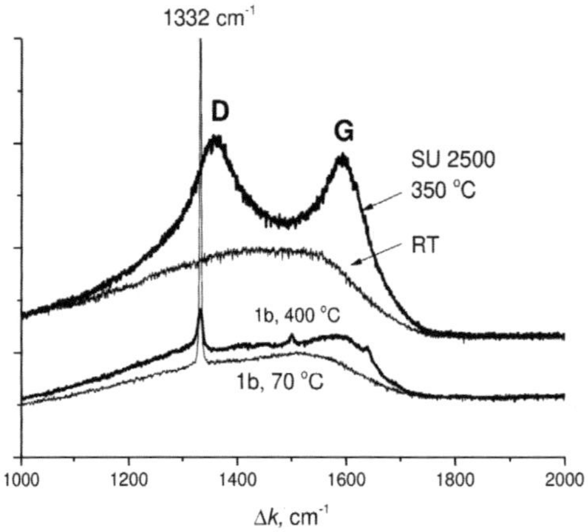


**Figure 5.** Raman spectra of (111) face of synthetic diamond measured after 30 keV $Ar^+$ ion irradiation at 70 and 400 °C for 488 nm excitation. Data for glassy carbon SU-2500 were taken from [10].

## 4. Conclusions

Surface modification of the (111) face of a synthetic diamond crystal by high-fluence ($\geq 10^{18}$ ion/cm$^2$) 30 keV $Ar^+$ions has been experimentally studied. Sheet resistance of created conductive surface layer depends on irradiation temperature and decreases more than 100 times while it changes from 70 to 400 °C. At irradiation temperatures more than 150 – 200 °C the resistivity is close to data for the graphite-like materials. At $T < 150$ °C exponential increase of resistivity is observed.

The conductive layer after ion-irradiation at 70 °C does not change the transparency of a diamond at $\lambda > 650$ nm. After ion-irradiation at 400 °C the conductive layer reduces the transparency of the diamond more than twice in the absence of explicit spectral dependence in the measured spectral region of diamond transmission from 450 to 700 nm.

Conductive surface diamond layer formed under ion irradiation manifests itself in Raman spectra as a broad band with gentle maximum in the region of the Raman shift 1000 – 1700 cm$^{-1}$. The maximum position is approximately 1500 and 1600 cm$^{-1}$ after ion irradiation at the temperatures $T =$ 70 °C and 400 °C respectively.

## References

[1] Popov V P *et al* 2011 *Advanced Materials Research* **276** 27
[2] Philipp P *et al* 2014 *Carbon* **80** 677
[3] Popov V P *et al* 2012 *Nucl. Instrum. Methods. Phys. Res.* B **282** 100
[4] Andrianova N N *et al* 2015 *J.Surface Investigation. X-ray, Synchrotron and Neutron Techniques* **9** 346
[5] Andrianova N N *et al* 2016 *Bull. Russian Academy of Science. Physics.***80** 156
[6] Mashkova E S and Molchanov V A 1985 *Medium-energy ion reflection from solids* (North-Holland, Amsterdam)
[7] Gippius A A *et al* 2001 *Physica* B **306-310** 573
[8] Andrianova N N *et al* 2013 *Horizons in World Physics* **280** 171 (Nova Science Publishing, New-York)
[9] Dobrinets I A, Vins V G and Zaitsev A M 2013 *HPHT-treated Diamonds* (Springer)
[10] Andrianova N N *et al* 2015 *J.Surface Investigation. X-ray, Synchrotron and Neutron Techniques* **9** 237

# Kinetic processes in heavily doped semiconductor heterojunctions

**A V Baskakova[1], K V Bukhenskyy[1], A B Dubois[1,3], S I Kucheryavyy[2], S N Mashnina[1], A S Safoshkin[1]**

[1] Ryazan State Radioengineering University, Gagarin str. 59/1, 390005 Ryazan, Russia
[2] National Research Nuclear University MEPhI (Moscow Engineering Physics Institute), Kashirskoe sh. 31, 115409 Moscow, Russia

E-mail: dubois.a.b@rsreu.ru

**Abstract.** We report results of study of electron - electron relaxation processes in the 2D electrons system with the fine structure of the energy spectrum and the spatial distribution of the electron density. For heavily doped heterojunction, when two subbands are filled, expressions for the time of intra- and inter-subband electron-electron interactions, matrix elements of a complete screening potential and the dielectric function were obtained beyond the long-wavelength limit. It is shown that the oscillations of temperature and concentration dependence of the electron - electron interactions are associated with the excitation of plasma oscillations components in 2D electron system.

## 1. Introduction

A number of experimental and theoretical studies of kinetic processes in low-dimensional semiconductor structures $Al_xGa_{1-x}As/GaAs$ [1-5] has discovered the small-angle oscillating dependence of the relaxation time $\tau_q$ on temperature $T$. These oscillations were observed for the heavily doped heterostructures ($n_s > 8.5\cdot 10^{11} cm^{-2}$), which are filled with main and second excited subband. The unique relationship of these anomalies with strong electron - electron interaction were established. An attempt to quantify the experimental curves $\tau_q(T)$ in the framework of existing theoretical concepts [6] was not successful. A qualitative explanation of the observed effects requires an examination of electronic interactions channels in complex 2D system of degenerate electrons. Using the method of self-consistent solution of the Poisson and Schrödinger equations [7], we found energy structure of the conduction band of the nanostructure and the spatial distribution of the electron density in the quantum-size levels. Figure 1 shows a typical nanostructure diagram for one of the samples with the filling of the two subbands, and the designation of the size quantization parameters used below.

[3] To whom any correspondence should be addressed.

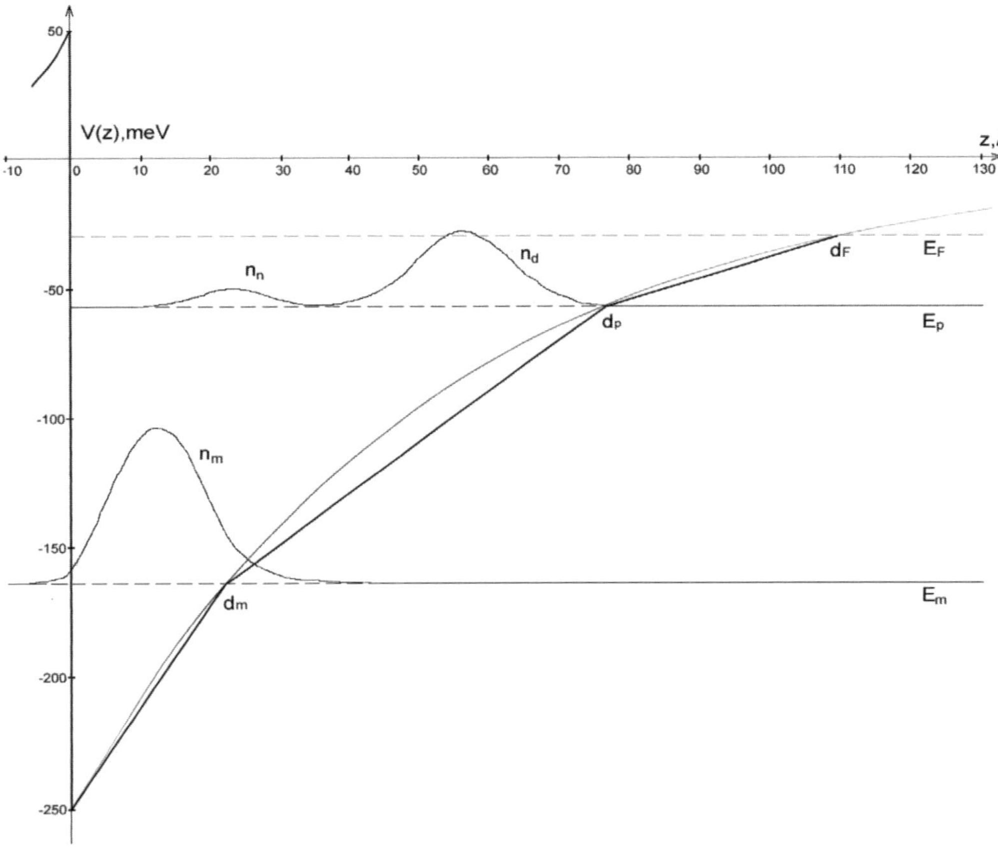


**Figure 1**. Diagram of the conduction band $V(z)$ of heterojunction with two occupied size-quantization subbands $E_m$ and $E_p$; $n_m$ and $n_{n,d}$ — surface concentrations of 2D electrons, $E_F$ — Fermi energy.

## 2. The matrix elements of the screening potential.

For a complete interaction Hamiltonian: $H = H_0 + H_{\text{int}}^{ijkl}$ with the field operators of creation and annihilation: $a_p^+(t) = \exp\left(-\frac{i}{\hbar}E(p)t\right)a_p^+$, $a_p(t) = \exp\left(-\frac{i}{\hbar}E(p)t\right)a_p$ we introduce one-particle fermion Green's function [7]:

$$G(p,t-t`) = \frac{\left\langle T\left(a_p(t)a_p^+(t)S_\alpha(-\infty,\infty|g)\right)\right\rangle}{\left\langle S_\alpha(-\infty,\infty|g)\right\rangle} \tag{1}$$

with $S$ - matrix:

$$S_\alpha(-\infty,\infty|g) = \sum_{n=0}^{\infty}\left(-\frac{i}{\hbar}\right)^n \frac{g^n}{n!}\int_{-\infty}^{+\infty}dt_1 \ldots \int_{-\infty}^{+\infty}dt_n \times e^{\alpha(|t_1|+\ldots+|t_n|)} T\left(H_{\text{int}}(t_1)\times\ldots\times H_{\text{int}}(t_n)\right) \tag{2}$$

where $H_{\text{int}}(t) = \exp\left(-\frac{i}{\hbar}H_0 t\right)H_{\text{int}}$, $H_0$ - non - interaction Hamiltonian operator.

Then the probability of particle transition from pulses $\mathbf{k}$ and $\mathbf{p}$ in state $\mathbf{k}+\mathbf{q}$ and $\mathbf{p}-\mathbf{q}$ will be proportional to the integral of collisions [8]

$$\sum_{\mathbf{k},\mathbf{p}} \delta\left(E_j(\mathbf{k}+\mathbf{q}) + E_l(\mathbf{p}-\mathbf{q}) - E_i(\mathbf{k}) - E_k(\mathbf{p})\right) f_{\mathbf{k}} f_{\mathbf{p}} \left(1 - f_{\mathbf{k}+\mathbf{q}}\right)\left(1 - f_{\mathbf{p}-\mathbf{q}}\right) \quad (3)$$

where $f$ — Fermi-Dirac distribution function. The indices $i$, $j$, $k$, $l$ are as follows: electron in a state $i$ interacts with an electron in the state $k$ whereby the transitions occur according to the states $j$ and $l$.

Then the interaction Hamiltonian can be written as::

$$H_{\text{int}}^{ijkl}(t) = \sum_{\mathbf{q}} g(\mathbf{q}) V_{tot}^{ijkl}(\mathbf{q},\omega) a_{\mathbf{p}+\mathbf{q}}^{+}(t) a_{\mathbf{p}}(t) + \sum_{\mathbf{p_i},\mathbf{p_j},\mathbf{q_i},\mathbf{q_j}} V_{ijkl}(\mathbf{q},\omega)\delta\left(\mathbf{p_i} + \mathbf{q_i} - \mathbf{p_j} - \mathbf{q_j}\right) a_{\mathbf{p}}(t) a_{\mathbf{p}+\mathbf{q}}^{+}(t) \quad (4)$$

In [2,6] an expression for the time of "e-e" interaction can be represented as:

$$\frac{1}{\tau_{ij}^{ee}} = \int_{-\infty}^{\infty} d\omega \sum_{k,m} \sum_{\mathbf{q}} \left|V_{tot}(\mathbf{q},\omega)\right|^2 \sum_{\mathbf{k},\mathbf{p}} \delta\left(E_j(\mathbf{k}+\mathbf{q}) + E_l(\mathbf{p}-\mathbf{q}) - E_i(\mathbf{k}) - E_k(\mathbf{p})\right) f_{\mathbf{k}} f_{\mathbf{p}} \left(1 - f_{\mathbf{k}+\mathbf{q}}\right)\left(1 - f_{\mathbf{p}-\mathbf{q}}\right) \quad (5)$$

where $V_{tot}(\mathbf{q},\omega)$ — the matrix element of the full potential of shielding, which is the Fourier transform of the external potential screening $V_{ext}(\mathbf{r})$:

$$V_{tot}(\mathbf{q},\omega) = \frac{1}{(2\pi)^2} \int_{-\infty}^{\infty} V_{ext}(\mathbf{r}) \exp\left[-(\mathbf{qr} - \omega t)\right] d^2\mathbf{r}. \quad (6)$$

In a Cartesian coordinate system, the expression for the external potential, double approximated triangular profile, will have such formula:

$$V_{ext}(z) = \begin{cases} \dfrac{E_1 - E_0}{d_m} z + E_0, \; z < d_m, \\ \dfrac{E_2 - E_1}{d_p - d_m}(z - d_m) + E_1, \; d_m < z < d_p, \\ \dfrac{E_F - E_2}{d_F - d_p}(z - d_p) + E_2, \; d_p < z < d_F, \\ E_0 \exp\left[-z\dfrac{\ln(E_0 / E_F)}{d_F}\right], \; z > d_F. \end{cases} \quad (7)$$

where $E_0 = V(0)$, $E_1 = V(d_m)$, $E_2 = V(d_p)$.

Thus, the solution to the problem of obtaining the analytical expressions for the relaxation time is reduced to calculation of the matrix elements of the screening for these types of interactions:

1) Intrasubband interaction, leading to transitions within the subbands (matrix elements $V_{tot}^{mmmm}$, $V_{tot}^{pppp}$, relaxation times $\tau_{mm}^{\text{intra}}$, $\tau_{pp}^{\text{intra}}$).

2) Intrasubband interaction, leading to transitions between subbands (matrix elements $V_{tot}^{mpmp}$, $V_{tot}^{pmpm}$, relaxation times $\tau_{mm}^{\text{intra}}$, $\tau_{pp}^{\text{intra}}$).

3) Intersubband interactions, leading to transitions within the subband (matrix elements $V_{tot}^{mmpp}$, $V_{tot}^{ppmm}$, relaxation times $\tau_{mm}^{\text{inter}} = \tau_{pp}^{\text{inter}}$).

4) Intersubband interactions, leading to transitions between subbands (matrix elements $V_{tot}^{mppm} = V_{tot}^{pmmp}$, relaxation times $\tau_{mp}^{\text{inter}} = \tau_{pm}^{\text{inter}}$).

The intra and inter superscripts correspond to intra- and inter-subband interactions and identical or different subscripts denote transitions within or between subbands. Then we transform (6) to a more convenient form:

$$\frac{1}{\tau_{ij}^{ee}} = \int_{-\infty}^{\infty} \frac{d\omega}{\cosh^2(\hbar\omega/2k_BT)} \sum_{k,l} \sum_{\mathbf{q}} \left| V_{tot}^{ijkl}(\mathbf{q},\omega) \right|^2 \chi_{ik}(\mathbf{q},\omega) \chi_{jl}^*(\mathbf{q},\omega) \quad (8)$$

where $k_B$ - Boltzmann constant, polarizing function in the random phase approximation [9] is described by

$$\chi_{ik}(\mathbf{q},\omega) = \frac{e^2}{q^2 n^2} \lim_{\alpha\to 0} \sum_{\mathbf{k}} \frac{f(\mathbf{k}) - f(\mathbf{k}+\mathbf{q})}{E(\mathbf{k}+\mathbf{q}) - E(\mathbf{k}) - \hbar\omega - i\hbar\alpha}. \quad (9)$$

Figure 2 illustrates the contributions of various mechanisms in the destruction of the Landau quantization, it shows the frequency dependence of the real and imaginary parts of the polarization function for different filling dimensional quantization subbands. It also illustrates the influence of the second subband filling degree on the spectrum of collective oscillations in 2D electron system. With increasing concentration $n_m$, the resonant frequency shifts to lower values in the temperature range 4-8 K.

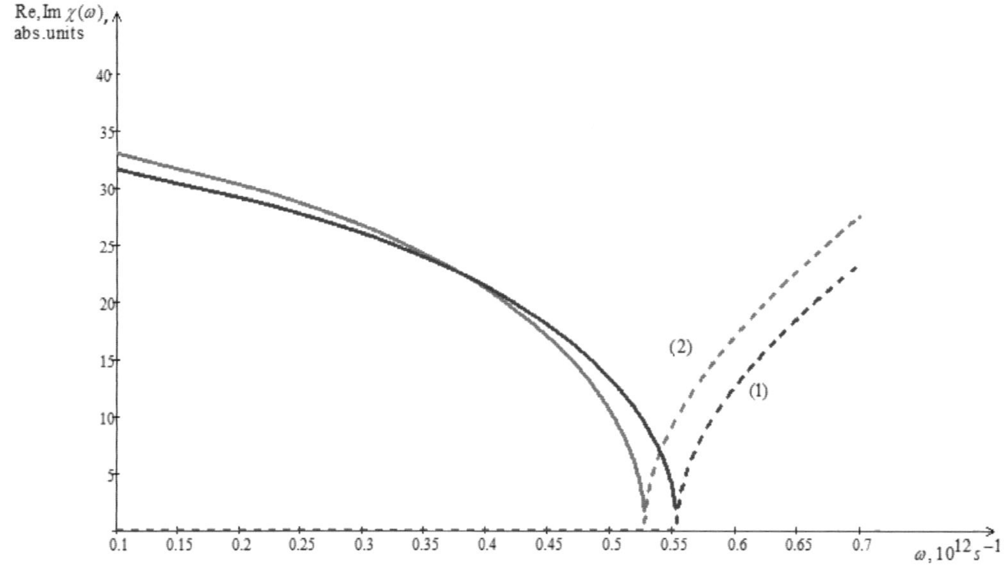


**Figure 2**. The frequency dependence of the polarization function for the main channel of interaction with the subband *n*-satellite: $n_m$: (1) – $9.1\cdot 10^{11} cm^{-2}$, (2) – $10^{12} cm^{-2}$, $d_p/d_m = 3.5$. Solid curves – $\mathrm{Re}\,\chi(\omega)$, dashed curves – $\mathrm{Im}\,\chi(\omega)$.

It should be noted that each resonant frequency can be associated with the temperature at which the screening mechanism changes and the temperature dependence of the electron-electron interaction becomes non-monotonic. It is shown that the resonant frequency and temperature anomalies can be related as $\hbar\omega_{res} = kT$.

**References**

[1] Kadushkin V I , Dubois A B, 2003 *Phys. Low-Dim. Struct.* **7/8** 7.
[2] Kadushkin V I, Dubois A B, 2002 *Journal of Radio Electronics* **5**
[3] Kadushkin V I 2000 P*hys. Low – Dim. Structures* **1/2** 93
[4] Dubois A B, 2008 *Journal of Physics: Conference series* **129** 012025
[5] Ambartsumyan V A, Andryushchenko E A, Bukhenskyy K V, Dubois A B, Dvoretskova E A, Gordova T V, Kucheryavyy S I, Mashnina S N, Safoshkin A S, 2014 *Nanosystems: physics, chemistry, mathematics* **5(3)** 343-353

[6] Slutzky M, Entin-Wohlman O, Berk Y, and Palevsky A, 1996 *Phys. Rev.* B **53**(**7**) 4065.
[7] Dubois A B, Zilotova M A, Kucheryavyy S I, Safoshkin A S, 2013 *Vestnik of RSREU* **3**(**45**) 88-92.
[8] Pines D, Nozieres P 1966 *The theory of quantum liquids* W A Benjamin, Inc. New York, Amsterdam, 383
[9] Bukhenskyy K V, Dubois A B, Gordova T V, Kucheryavyy S I, Mashnina S N, Safoshkin A S, 2015 *Physics Procedia* **71** 359 – 363

# A theoretical basis of the approach for the magnetic field penetration measurement

**P I Bezotosnyi[1], S Yu Gavrilkin, O M Ivanenko, K V Mitsen and A Yu Tsvetkov**

Lebedev Physical Institute, Russian Academy of Sciences, Leninskii pr. 53, 119991 Moscow, Russia

E-mail: bezpi@sci.lebedev.ru

**Abstract.** An approach for the assessment of London penetration depth of superconducting films is proposed. This approach is based on the analysis of linear response of the sample to a local low-frequency alternating magnetic field generated by the measuring coil disposed near the film surface. A visual "electrical engineering" model of induced currents distribution in the superconductor taking into account the kinetic inductance was developed for a description of this response. The possibility of determining of the penetration depth from changing the inductance of the system "coil-sample" is shown in the framework of this model. The sensitivity of the proposed method for the films with different thicknesses is considered.

## 1. Introduction

Currently the problem of determining the temperature dependence of the magnetic field penetration depth $\lambda$ is important, because it allows analyzing the microscopic properties of the superconducting condensate. Most of the existing methods for determining $\lambda$ require complex expensive equipment (precise magnetometers, microwave technique, etc.) and use rather complicated theoretical models, which not always exactly correspond to the real geometry of the experiment.

The low-frequency linear induction technique examines the impact of superconducting sample on the self-inductance of the measuring coil and is the most attractive from the point of simplicity of technical realization. For thin films and plates a coil is located close to a sample surface and its axis is perpendicular to the sample plane. This type of apparatus enables measurements both in linear and in nonlinear induction techniques [1].

Thus, to establish a relation between the measured change in inductance $\Delta L$ and the magnetic field penetration depth $\lambda$, a model of the distribution of induced currents in superconducting films taking into account the geometry of the experiment is needed.

## 2. Model

A coil of circular cross section, placed near the surface of a thin superconductive film, is considered. Circular currents are induced in the sample when a sinusoidal alternating current $I_0(t)$ is passed through the coil. Two-dimensional model was developed to find a circular currents distribution. In framework of this model to describe an inhomogeneous current distribution in the sample, we consider the system of isolated thin coaxial turns (circuits), inductively coupled with each other and the excitation coil. The sample is divided notionally into $N=N_r{\cdot}N_z$ coaxial circuits with the centres located

[1] To whom any correspondence should be addressed.

on the axis of the excitation coil. Here $N_r$ is the number of turns in the radial direction, and $N_z$ is the number of turns in the film thickness direction. For sufficiently thin films, with thickness not exceeding $\lambda$, we can put $N_z$=1 ("one-dimensional" model, 1D). "Two-dimensional" (2D) model with $N_z$ >1, allowing to obtain the distribution of the current density in depth, is used in general case. Hereinafter, we assume a homogeneous current distribution in each turn. That is fulfilled when coil thickness is significantly less than the London penetration depth $\lambda$.

We introduce the following notation: $r_k$, $dr$ and $dz$ are radius, width and thickness of $k$–th circuit, $I_k(t)$ is current induced in it, $L_0$ is self-inductance of the coil, $L_k$ is the full inductance of the $k$–th circuit (sum magnetic $L_k^{mag}$ and kinetic $L_k^{kin}$ inductances), $M_k$ is the mutual inductance of the $k$–th circuit and the excitation coil, $M_{ik}$ – mutual inductance of the $i$–th and $k$–th circuits coinciding with a magnetic part of $L_k$ when $i = k$.

An equivalent electrical diagram corresponding to the problem is shown in Figure 1.

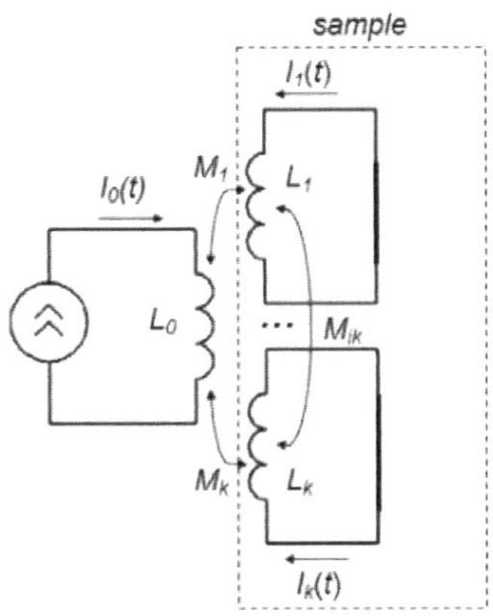


**Figure 1.** An equivalent electrical diagram corresponding to the problem.

Full magnetic fluxes through the excitation coil $\Phi_0$ and model turns $\Phi_k$ can be written as:

$$\Phi_0(t) = L_0 I_0(t) + \sum_{i=1}^{N} M_i I_i(t),$$

$$\dots$$

$$\Phi_k(t) = M_k I_0(t) + \sum_{i=1}^{N} M_{ik} I_i(t).$$

When the sample is fully in superconducting state (linear mode) the current distribution in the sample is determined by the system of equations:

$$\frac{dI_k(t)}{dt} = -\frac{1}{L_k^{mag} + L_k^{kin}} \left\{ M_k \frac{dI_0(t)}{dt} + \sum_{i=1, i\neq k}^{N} M_{ik} \frac{dI_i(t)}{dt} \right\}$$

For the voltage across the coil terminals we write:

$$E(t) = -\frac{d\Phi_0(t)}{dt} = -L_0 \frac{dI_0(t)}{dt} - \sum_{i=1}^{N} M_i \frac{dI_i(t)}{dt},$$

where the second term represents the response of the film $\Delta E(t)$.

Consider,

$$\left[\frac{dI_k(t)}{dt}\right] = -\mathbf{W} \cdot \left[M_k \frac{dI_0(t)}{dt}\right],$$

$$\mathbf{W} = \mathbf{M}^{-1},$$

where the expression in square brackets meets the $k$-th elements of the corresponding vector-columns, $\mathbf{M} = [M_{ik} + \delta_{ik}L_i^{kin}]$ - the matrix of mutual inductances of the turns and $\delta_{ik}$ - the Kronecker symbol . The main diagonal of this matrix contains full inductance $L_i$. In this case, we get the formula:

$$I_k(t) = -I_0(t)\sum_{i=1}^{N} W_{ik}M_i,$$

and the voltage response of the sample is expressed as:

$$\Delta E(t) = \frac{dI_0(t)}{dt}\sum_{k=1}^{N} M_k \sum_{i=1}^{N} W_{ik}M_i$$

The value

$$\Delta L = -\sum_{k=1}^{N} M_k \sum_{i=1}^{N} W_{ik}M_i$$

represents the change in self-inductance of the coil under the film influence and can be determined from the experiment. On the other hand, in the framework of the presented model it can be calculated as a function of the penetration depth $\lambda$. This gives the possibility of experimental estimation of $\lambda$ value. Let us consider results of numerical calculation of $\Delta L(\lambda)$ dependence.

## 3. Results of the numerical calculation

Dependence of the coil inductance change $\Delta L$ on the magnetic field penetration depth $\lambda$ calculated for different films thicknesses $d$ is presented in Figure 2. As seen, the thinner the film the stronger the change of $\Delta L(\lambda)$ for small $\lambda$ values. As low-temperature values of the London penetration depth $\lambda$ lie in submicron lengths for both type-I and type-II superconductors, so it is reasonable to use thin films in measurements.

An important condition for the model is smallness of the circuits thickness in comparison with the London penetration depth, $dz \ll \lambda$. But reduction of the turns thickness is limited by the capabilities of the computational techniques used. Excessive fragmentation of the thickness may lead to significant increase of the computation time.

Figure 3 shows results of the calculations made for several film thicknesses and different numbers of circuits on the plate thickness $N_z$. It is clearly seen for the thick plates ($d$=1000 nm) that calculated $\Delta L(\lambda)$ values are less for $N_z$ = 40 than for $N_z$ = 1 and the difference is more pronounced for small $\lambda$ values.

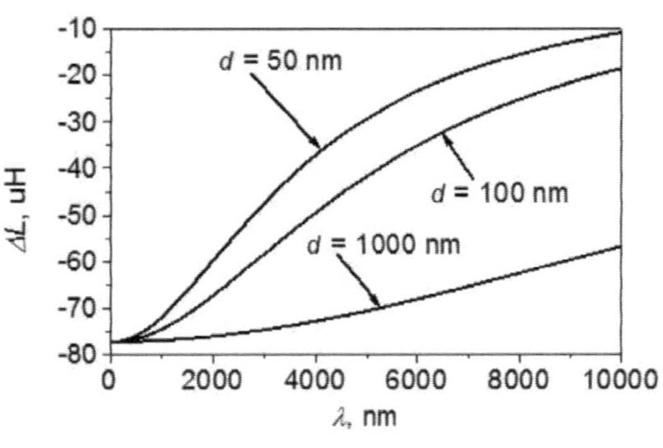


**Figure 2.** Dependences $\Delta L(\lambda)$ calculated for different film thicknesses. The calculations were made for $N_r$ = 200, and $N_z$ = 1. $\Delta L$ is in μH units.

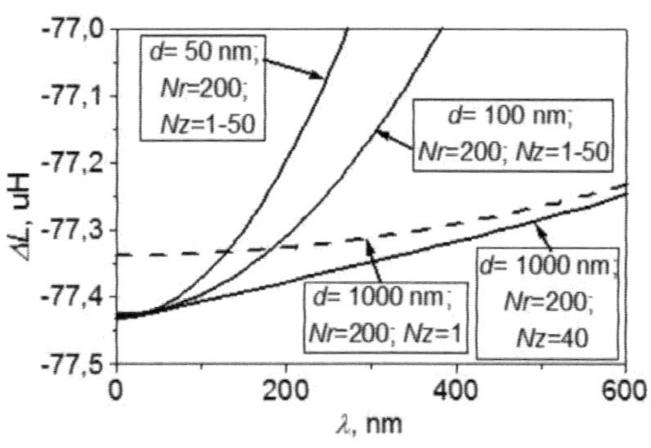


**Figure 3.** Dependences $\Delta L(\lambda)$ calculated for different film thicknesses. The calculations were made for $N_r$ = 200 and different $N_z$. $\Delta L$ is in μH units.

Let us estimate the calculation error for different thickness of the circuits. Assuming an exponential decrease of the magnetic field through the sample thickness the dependence of the kinetic inductance of the turn with rectangular cross section has the form:

$$L^{kin}{}_{kin\,hom} = \frac{\pi\mu_0\lambda r_k \mathrm{cth}(dz/2\lambda)}{dr}.$$

For a homogeneous distribution of magnetic field in the turn cross section, we can use more simple formula:

$$L^{kin}{}_{k\,hom} = \frac{\mu_0\lambda^2 2\pi r_k}{dzdr}.$$

In the described model this case corresponds to $dz \ll \lambda$. Ratio of the kinetic inductances calculated from these formula for different thicknesses of turns is shown in Table 1.

**Table 1.** Ratio of the kinetic inductances calculated for homogeneous and inhomogeneous distribution of magnetic field in circuits with different thicknesses. The thickness is in λ units.

| $dz/\lambda$ | $1 - L^{kin}{}_{hom}/L^{kin}{}_{inhom}$ |
|---|---|
| 0,01 | 0,00% |
| 0,1 | 0,08% |
| 0,2 | 0,33% |
| 0,3 | 0,75% |
| 0,4 | 1,33% |
| 0,5 | 2,07% |

It should be noted that accuracy of the model depends on the thickness $dz$ of the circuits and $dz$ should be small for precise estimation of the penetration depth $\lambda$.

## 4. Conclusions

The impact of superconducting film on low-frequency self-inductance of the measuring coil was analysed and the model for estimation of the London penetration depth $\lambda$ from the experimental data was developed. We found that precision of $\lambda$ estimation increases with decrease of both film thickness and thickness of the model turns $dz$. We obtained that for $dz = 0.3\lambda$ a contribution of the kinetic inductance value uncertainty to total error of $\lambda$ estimation is less than 1%. There are no severe restrictions for the model turns width $dr$.

## Acknowledgments

This work is supported by Russian Foundation for Basic Research (grant № 16-32-00586).

## References

[1] Gavrilkin S Yu, Ivanenko O M, Mitsen K V and Tsvetkov A Yu, 2014 *Bulletin of the Lebedev Physics Institute* **41** 47-52

Journal of Physics: Conference Series **747** (2016) 012028 doi:10.1088/1742-6596/747/1/012028

# The peculiarities of local structure of $YbNi_2$ and $YbCo_2$ intermetallics synthesized at high pressure.

**O V Chernysheva[1,3], A P Menushenkov[1], A V Tsvyashchenko[2], L N Fomicheva[2], A A Ivanov[1], A V Kuznetsov[1]**

[1] National Research Nuclear University MEPhI (Moscow Engineering Physics Institute), Kashirskoe sh. 31, 115409, Moscow, Russia
[2]Vereshchagin Institute for High Pressure Physics, Russian Academy of Sciences, Kaluzhskoe sh. 14, Troitsk, 142190 Moscow, Russia

E-mail: olgagrish@gmail.com

**Abstract**. Local structure of $YbCo_2$ and $YbNi_2$ was investigated by EXAFS and XANES spectroscopy. It was found that the bond Yb-Co(Ni) has the highest static disorder and its length increases with temperature decrease, while all other bonds remain almost unchanged. This phenomenon may be caused by short-range magnetic ordering at temperatures above the phase transition. XANES measurements above $L_{III}$-Yb absorption edge in $YbCo_2$ and $YbNi_2$ revealed that both compounds have almost the same density of free 5*d*-states

## 1. Introduction

Unusual magnetic and transport properties of Laves phases $YbM_2$, where M – transition metal, are determined in general by hybridization of 4*f* electron with *s*- and *d*- conduction band electrons. In the case of magnetic compounds there is also a contribution of rare earth 5*d* and metal 3*d* band hybridization [1-3]. Despite intensive study of intermetallic Yb compounds some questions are yet to answer, including influence of temperature and pressure (hydrostatic and chemical), which may change in the local interatomic distances, and hence the magnitude of the exchange interaction [4-7]. Investigation of the local structure of $YbCo_2$ and $YbNi_2$ by X-ray absorption spectroscopy may remove contradictions and gaps in the data on properties of these compounds.

## 2. Experiment

Samples were prepared by melting of the constituent materials at a constant pressure of 8 GPa in a chamber of the "Toroid" type [8]. Then samples were rapidly quenched to room temperature, and pressure was released. The resulting material was a homogeneous polycrystal.

The X-ray absorption measurements were carried out at I811 beamline of MAX IV facility (Lund, Sweden) in the transmission mode at *K*-Co and $L_{III}$-Yb absorption edges in 80-300 K temperature range with normalized energy resolution of $1\times10^{-4}$. EXAFS spectra analysis was performed using the VIPER [9] software pack with standard procedure of EXAFS-function $\chi(k)$ extraction and Fourier-analysis. Fitting of EXAFS-functions was performed in standard harmonic approximation. Back-scattering amplitude and phase shift calculated by FEFF [10] within the framework of structure model of cubic Laves phase C15.

[3] To whom any correspondence should be addressed.

### 3. Results and discussion

EXAFS-functions $\chi(k)k^2$, measured above *K*-Co and $L_{III}$-Yb absorption edges at different temperatures are given in Figures 1 and 2.

(a) (b)

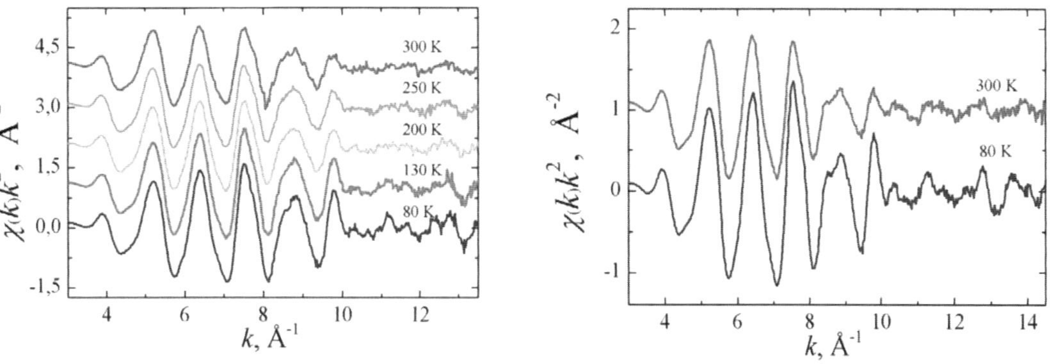


**Figure 1.** EXAFS-functions $\chi(k)k^2$, measured above $L_{III}$-Yb absorption edge at temperatures 80-300 K for $YbCo_2$ (a) and $YbNi_2$ (b).

The radii of coordination spheres Yb-M slightly change with temperature. At the same time the other distances stay almost unchanged. Bond lengths are shorter than the values extracted from XRD data for $YbNi_2$ and $YbCo_2$ intermetallics [11,12], except the bond length Yb-Yb. It is equal to published data for $YbCo_2$, and is 0.03 Å greater for $YbNi_2$. The value of the distance between Yb and Co(Ni) atoms is the most significant for the magnetic properties of these compounds (Fig. 3). P The radius of the first coordination shell, corresponding to the shortest distance between atoms Yb and Co(Ni), increases with decreasing temperature. Weak Co moments survive far above the phase transition temperature [13], forming the clusters with short-range magnetic order. Therefore, the increase in the unit cell volume can occur when approaching the phase transition temperature [14] leading to a corresponding increase in local distances.

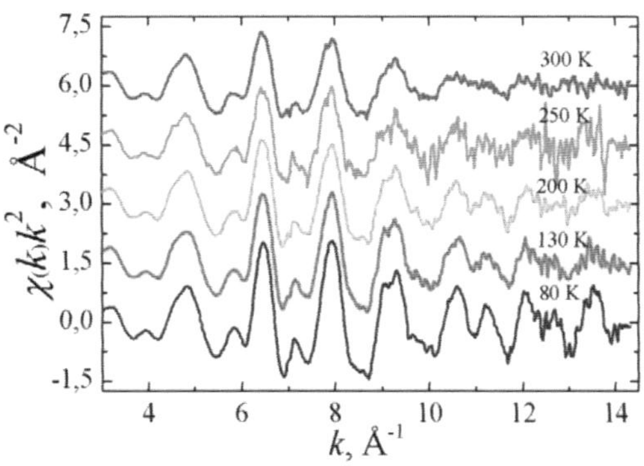


**Figure 2.** EXAFS-functions $\chi(k)k^2$, measured above *K*-Co absorption edge at temperatures 80-300 K

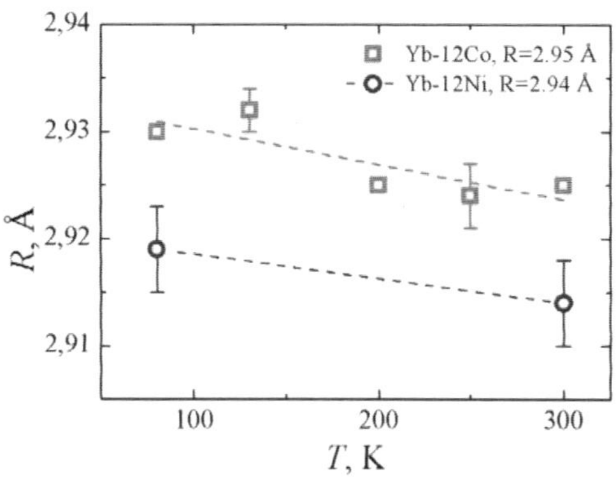


**Figure 3.** Temperature dependence of the distances between Yb and *d*-metal atoms in the first coordination sphere. The values in legend were calculated from data [3,23]

The temperature dependences of Debye–Waller factors $\sigma^2(T)$ for coordination shells around ytterbium and cobalt atoms were also obtained. Here Debye-Waller factor is defined as a standard deviation of the interatomic distance from the mean value $\sigma^2 = \sigma_s^2 + \sigma_d^2$ and consists of the temperature independent contribution $\sigma_s^2$ due to the statistical disordering of atoms and dynamic contribution $\sigma_d^2$ due to thermal vibrations. The temperature dependence of the dynamic contribution can be calculated using Einstein model: $\sigma_d^2 = (\hbar/2\omega\mu)$ cth $(\Theta_E/2T)$

Here $\omega$ – the frequency of longitudinal oscillations, $\mu$ – the reduced mass of the atomic pair, $\Theta_E = \hbar\omega/k_B$ - Einstein temperature. Experimental results are shown in Figure 4 and Table 1.

One can see that Einstein temperature and hence bond rigidity is the highest for Co-Co bond (Table 1) that correlates with the fact that this bond is the shortest. The greatest static disorder corresponds to Yb-Co bond. Magnetic ordering is determined by the exchange interaction between sublattices of Co and Yb. The spread of distances between the atoms and the corresponding change in the overlap of the wave functions affect the density of states in 3*d*-band of Co and magnetic moment of Co atoms.

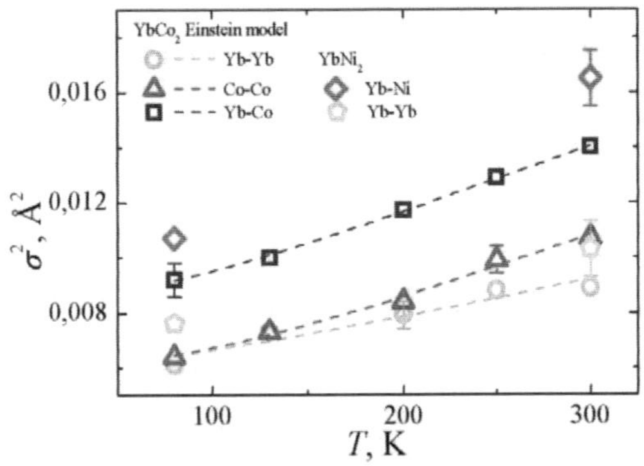


**Figure 4**. Temperature dependence of Debye–Waller factors for Yb-Co, Yb-Yb, Yb-Ni bonds

**Table 1.** Einstein temperatures and Debye-Waller factors for $YbCo_2$ interatomic bonds.

| Bond | $\Theta_E$, K | $\sigma_s^2$(Å$^2$) |
|---|---|---|
| Yb-Yb | 190 | 0,005 |
| Co-Co | 260 | 0,003 |
| Yb-Co | 207 | 0,006 |

For $YbNi_2$ it was not possible to distinguish the static and dynamic contribution since only two temperature points were available. However, we can say that in $YbNi_2$ the atoms have a higher degree of disorder than in $YbCo_2$ and Debye-Waller factor of Yb-Ni bond is also the largest.

The normalized experimental $L_{III}$-Yb XANES spectra for $YbCo_2$ and $YbNi_2$ measured at the temperature of 8 K are shown in Figure 5. Energy position of the absorption edge indicates that Yb atom is in $Yb^{3+}$ valence state. The shape and intensity of the white line are almost identical for both samples and do not vary with temperature. XANES spectroscopy above $L_{III}$-absorption edge is based on the excitation of an electron from an internal $2p_{3/2}$ level on the *d*-valence orbitals and carries information about the density of unoccupied electron states, therefore, the density of free 5*d*-states of these two compounds is almost the same.

## 4. Conclusion

The distance between Yb and Co (Ni) atoms increases with decreasing temperature. Perhaps this is due to the presence of clusters with short-range magnetic ordering at temperatures above the phase transition. The same bond has the greatest static disorder. The shape and intensity of the $L_{III}$-Yb absorption edge white line is practically identical for $YbCo_2$ and $YbNi_2$ samples, which indicates almost the same value of the density of free 5*d*-states in these two compounds.

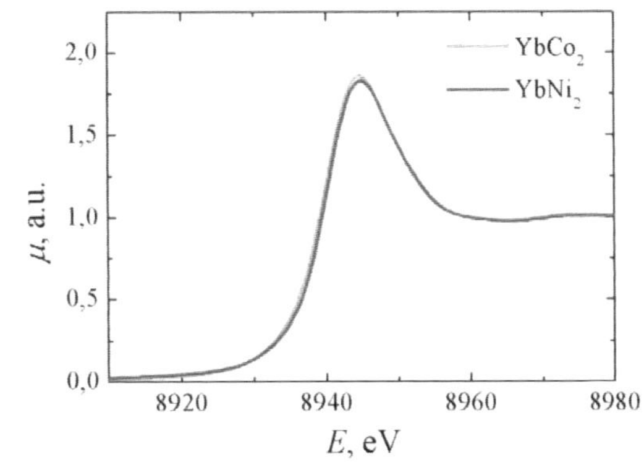


**Figure 5.** XANES spectra of $YbCo_2$ and $YbNi_2$ compounds measured above $L_{III}$- Yb absorption edge.

**Acknowledgments**
The work was supported by the Russian Foundation for Basic Research (grants No. 14-02-00001 and 14-22-01002-ofi_m) by the Competitiveness Program of NRNU MEPhI.

**References**

[1] Harada I, Asakura K, Fujiwara A, Kotani A 2004 *Journal of Electron Spectroscopy and Related Phenomena* **136** 125–133

[2] Matsunami M, Chainani A, Taguchi M, Eguchi R, Takata Y, Oura M, Yabashi M, Tamasaku K, Nishino Y, Ishikawa T, Kosaka M, and Shin S 2012 *Journal of the Physical Society of Japan* **81** 073702

[3] Okada H, Koyama K, Yamada M, Goto T, Makihara Y, Fujii H, Watanabe K 2006 *Journal of Alloys and Compounds* **408–412** 144–146

[4] Sahu P CH, Chandra Shekar N V 2006 *J. Mater. Sci.* **41** 3207–3228

[5] Dallera C, Anneseb E, Rueff J-P, Palenzona A, Vankó G 2004 *Journal of Electron Spectroscopy and Related Phenomena* **137–140** 651–655

[6] Kumar R S, Svane A, Vaitheeswaran G, Kanchana V, Bauer E D, Hu M, Nicol M F and Cornelius A L 2008 *Phys Rev B* **78** 075117

[7] Görlach T, Pfleiderer C, Grube K and v Löhneysen H 2005 *Phys Rev B* 2005 **71** 033101

[8] Khvostantsev L G, Vereshchagin L F and Novikov A.P 1977 *High Temp.-High Press.* **9** 637

[9] Klementev K V 2001 *J. Phys. D: Appl. Phys.* **34** 209

[10] Rehr J J and Albers R C 2000 *Rev. Mod. Phys.* **72** 621

[11] Rojas D P, Fernandez Barquın L, Echevarria-Bonet C, Rodrıguez Fernandez J 2012 $YbNi_2$: A heavy fermion ferromagnet *Solid State Communications* **152** 1834–1837

[12] Buschow K H J 1972 *Journal of the Less Common Metals* **26** 329-333

[13] Prchal J, Sebesta J, Valenta J, Misek M, Turcinkova D, Lapcak L, Prokleska J, Kratochvilova M and Sechovsky V 2014 *Acta Physica Polonica* **126** 288-289

[14] Syshchenko O, Fujita T, Sechovsky V, Divis M and Fujii H, 2001 Phys Rev B 63, 054433

# Quasi-two-dimensional metallic hydrogen inside di-phosphide at high pressure

**N N Degtyarenko and E A Mazur**[1]

National Research Nuclear University MEPhI (Moscow Engineering Physics Institute), Kashirskoe sh. 31, 115409 Moscow, Russia

eugen_mazur@mail.ru

**Abstract**. The method of mathematical modelling was used for the calculation of the structural, electronic, phononic, and other characteristics of various normal phases of phosphorus hydrides with stoichiometry $PH_k$. It was shown that the di-phosphine may form 2D lattice of the metallic hydrogen in it, stabilized by phosphorus atoms under high hydrostatic pressure. The resulting structure with the elements of H-P-H has a locally stable (or metastable) phonon spectrum. The properties of di-phosphine were compared with the properties of similar structures such as the sulphur hydrides.

## 1. Introduction

One of the search areas for the new superconducting materials with high critical temperatures is the study of compounds with high hydrogen content. In [1] it was discovered, that the transition to the superconducting state of the metallic phase of the hydrogen sulfide under pressure 170 GPa takes place at the temperature $T = 203$ K. The idea of the experiment was based on the theoretical works on the study of the metallic hydrogen properties, in which high critical temperature $T_c \sim 200 \div 400$ K [2] were predicted. In the works [3, 4] it was shown that such a hydrogen phase may be metastable. After publication of the results [1] the interest in searching for compounds that have higher $T_c$ temperatures has increased significantly. In particular, the preliminary experimental results on the compression of phosphide $PH_3$ and the effect of its superconductivity at 100 K [5] should be mentioned. The following fact wass taken into account in [1]. At the higher pressure values the hydrogen forms a hydride $SH_n$ ($n \leq 3$) [6] through dissociation of the sulfur compounds. The specific properties of the $SH_3$ can lead to the possible explanation of the high-temperature superconductivity existence. In [7-9] the $SH_2$ phase with the symmetry *I4/MMM* was found along with the study of the properties of the $SH_3$ compound, for which the electron and phonon spectra and the density of electron and phonon states in the pressure range $100 \div 225$ GPa were calculated.

In this paper we present the calculations of the electronic and phonon properties of the di-phosphine H2P-PH2 system compressed with up to the 250 GPA hydrostatic pressures. Di-phosphine under normal conditions is a colorless liquid with the structure of molecules presented in Figure 1a [10].

[1] To whom any correspondence should be addressed.

## 2. Method

The calculations of the structural, electronic and phonon characteristics of the molecules and the periodic crystal structure of $PH_k$ in the pressure range 0 ÷ 250 GPa were performed using ab-initio method. The DFT approximation with the plane wave basis with the correlation functional GGA – PBE pseudopotential which preserves the norm was used. The unit cell of the structure used to calculate the electronic properties contains an element such as H2P-PH2. All the calculations were performed in the spin-polarized approximation in order to compare the calculated energy values of different phases in an adequate manner.

## 3. Results

At the 25 GPa pressure the molecular di-phosphine P2H4 phase becomes metallic because of overlap of electron shells of the molecules. The compression of the combination of such molecules in the pressure range 25 ÷ 250 GPa leads to the formation of metallic crystalline phase. The main structural element of this phase H2P-PH2 acquires straight H-P-H bond directed at an angle of 180° (Fig. 1c). In calculations of the first cycle it was assumed that the pressure applying leads to the symmetry P1 of the initial cell containing the $P_2H_4$ molecule. The resulting crystal structure at the pressure of 25 ÷ 125 GPA acquires the *I4/MMM* symmetry of the unit cell. The structure formed in this pressure range with a final *I4/MMM* crystal symmetry (Fig. 1e) results from the structural modification of the unit cell in the studied phase with the stoichiometry $P_2H_4$. At the next stage of the calculations the cell with the symmetry *I4/MMM* was used as a starting one (second cycle). Crystal structure with this unit cell symmetry was studied in the pressure range 25 ÷ 250 GPa. In the entire pressure range 25÷250 GPa the structural modification of the $P_2H_4$ unit cell with the symmetry *I4/MMM* leads to formation of set of parallel planes containing either phosphorus or hydrogen atoms (Fig. 1e). The hydrogen atoms form a square lattice with a period depending on the pressure. As a result, the electronic properties of the system acquire quasi-two-dimensional character. The phosphorus atoms form their own planes. The first coordination sphere of phosphorus atoms consists of four P atoms located in above mentioned plane and two pairs of hydrogen atoms located above below it. At the pressure $P$ = 125 GPa the calculated unit cell volume amounts to $V = 13.54$ A$^3$, which gives the total electron density value $n = 10^{24}$ cm$^{-3}$ (14 electrons per cell, or about 1 elect./A$^3$). The observed effect of the redistribution of the electron density between atoms with increasing pressure, determined by the conditional Mulliken charges of phosphorus $Z_M(P)$ and hydrogen $Z_M(H)$, should be considered as an interesting one. The Mulliken charge is positive for phosphorus and negative for hydrogen, its magnitude increases with pressure increase.

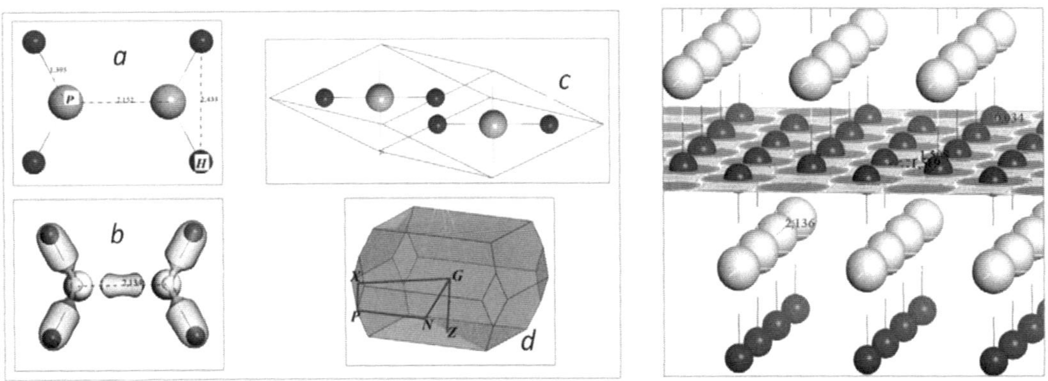


**Figure1**. *a* – structure of di-phosphine under normal conditions; *b* – isosurface of the electron density for molecules of di-phosphine at the level of 0.6, indicating the covalent bonding in the molecule; *c*– a primitive cell with two elements of H-P-H at the pressure $P$ = 150 GPa; *d* – the Brillouin zone corresponding to the primitive cell at the pressure $P$ = 150 GPa; *e* – the structure formed from molecules of di-phosphine at the pressure $P$ = 150 GPa; the two-dimensional distribution of electron density in the plane containing the hydrogen atoms is shown.

As follows from the results of the calculations presented here, the pressure *P* "pushes" the electron density away from the phosphorus atoms to the hydrogen ones. This effect can be interpreted as an actual "doping" of the hydrogen planes with the additional electrons. The results show a reduction of the distance between two hydrogen planes with pressure increase. The hydrogen atoms form a quasi-two-dimensional square lattice as result of this redistribution. The maximal concentration of hydrogen atoms within the planes occurs at the pressure $P$ = 150 GPa.

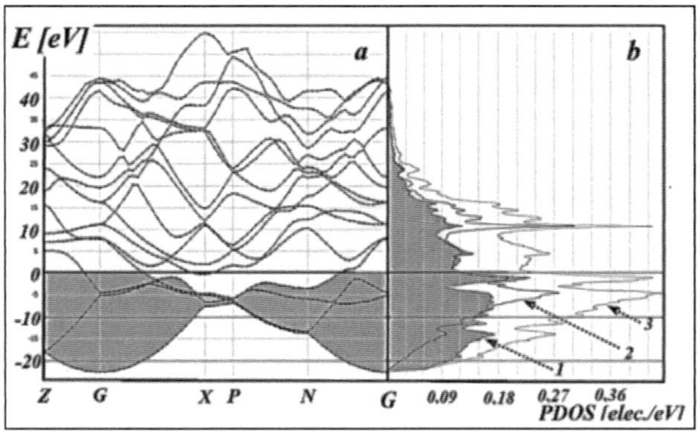


**Figure 2.** *a* – band structure of P2H4 electrons at the pressure $P$ = 175 GPA; *b* –the partial density of of the states number: 1 – for *s*–electrons, 2 – for *p*-electrons, 3 – the total DOS.

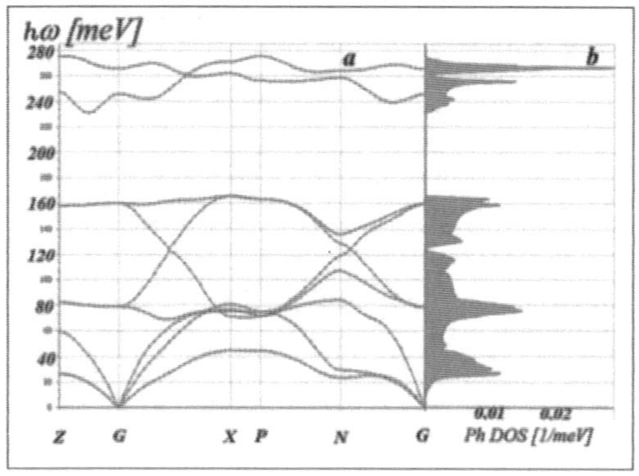


**Figure 3.** *a* – phonon dispersion curves P2H4 at the pressure $P$ = 175 GPA; *b* – PhDOS – phonon P2H4 density of states, the maximum energy of the phonons at 175 GPa – $h\omega \approx 270$ meV.

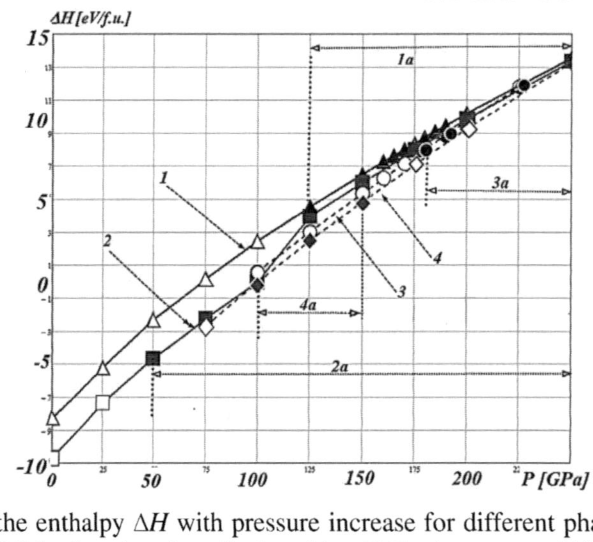


**Figure 4.** Change of the enthalpy $\Delta H$ with pressure increase for different phases: 1– (Δ ▲) – sulphide $SH_2$ (symmetry *I4/MMM*); 2 – (□ ■) – di-phosphine $P_2H_4$ (symmetry *I4/MMM*); 3 – (○ ●) – $SH_3$ (symmetry *IM-3M*); 4 – (◊ ♦) – $PH_3$ (symmetry *IMMM*). Open (Δ,□,○,◊) and closed (▲,■,●,♦) symbols correspond to presence or absence of imaginary frequencies. 1a, 2a, 3a and 4a are the pressure intervals, in which the structures have only real frequencies.

The Figure 2 shows the properties of the electron subsystem for the pressure $P$ = 175 GPa. As seen, the Fermi level is located at the electronic DOS peak position and at least one of the zones has "dropped" below the Fermi level at this pressure. The estimation of effective masses of carriers gives near the Fermi level a value of about 0.1 meV. At the pressure $P$ = 175 GPa the Fermi level crosses three subzones (Fig. 2).

There are no imaginary frequencies in the pressure range 50 ÷ 250 GPa in the phonon spectrum of the crystal $P_2H_4$ – *I4/MMM* phase (Fig. 3). Therefore the phase $P_2H_4$ – *I4/MMM* is stable in this pressure range. Maximum frequencies lie in the energy region of the order of 265 meV. Note that the optical modes corresponding to oscillation of hydrogen atoms (Fig. 3) exist at both at small $q$ (point $G$) and segments connecting the symmetry points of the Brillouin zone (Fig. 1c). The phonon spectrum becomes more rigid with increasing pressure.

Pressure dependence of the enthalpy $\Delta H$ is presented in Figure 4 for the most stable crystalline phases: $SH_2$ and $SH_3$ sulfide, di-phosphide $P_2H_4$ and phosphine $PH_3$. As follows from the Figure, the XH2 compounds should be metastable, since their enthalpy is more than that for XH3 phase. Laser-heating of $SH_2$ compressed in the diamond cell was used in experiments [1] to stimulate the XH2 to XH3 phase transition. The metastable XH2 phase with the *I4/MMM* symmetry can be formed under compression of di-phosphine P2H4 without heating since in the range 50 ÷ 250 GPa its spectrum does not contain imaginary frequencies. In addition, for the phase having the $PH_3$ stoichiometry and the *IMMM* symmetry we obtained relatively small range of the pressure (100 ÷ 150 GPa), in which imaginary frequencies are absent.

## 4. Conclusions

The $P_2H_4$ di-phosphine molecular phase becomes metallic at pressures more than 25 GPa because of overlap of the electron shells of molecules. In the range of pressure 50 ÷ 250 GPa the structural modification of the $P_2H_4$ unit cell causes formation of the *I4/MMM* symmetry (Fig. 1e). As a result, parallel planes of P and H atoms are formed and a square lattice is observed in hydrogen planes. The electronic properties of the system acquire quasi-two-dimensional metallic character. There are no imaginary frequencies in the phonon spectrum of the di-phosphine $P_2H_4$ – *I4/MMM* crystal phase in the pressure range 50 ÷ 250 GPa, so it is stable (or metastable) this pressure range. The maximal frequency lies in the 265 meV energy region. In the narrow pressure range 100 ÷ 150 GPa the competing $PH_3$ phosphine phase has lowest enthalpy.

Hydrostatic pressure forms the hydrogen and phosphorus (or sulfur) planes and "pushes away" the electron density from the many-electron atoms (P or S) in hydrogen planes. Conventionally, this effect can be described as a "doping" of the planes.

Basing on the calculations we conclude that the optimal pressure for superconductivity in di-phosphine is about 150 GPa and suppose that its critical temperature is more than $T_c$ for $PH_3$ [5] because in a wide range of pressures it has the properties of two-dimensional metallic hydrogen.

**Acknowledgments**
The work was supported by the Competitiveness Program of NRNU MEPhI.

**References**

[1] Drozdov A P, Eremets M I, Troyan I A, Ksenofontov V, Shylin S I, 2015 *Nature* **525** 73.
[2] Ashcroft N W, 1968 *Phys. Rev. Lett.* **21** 1748.
[3] Kagan Y, BrovmanE G, 1971 *Uspekhi Fizicheskikh Nauk* **105** 777.
[4] Brovman E G, KaganY, KholasA, PushkarevV, 1973 *JETP letters* **18-4** 269.
[5] DrozdovA P, Eremets M I, Troyan I A, *arXiv*:1508.06224.
[6] DuanD, LiuY, TianF et al, *Sci. Rep.* **4 (04)** 6968.
[7] Degtyarenko N N, Mazur E A, 2015 *Journal of Exper. and Theoretical Physics* **148-2(8)** 275.
[8] Degtyarenko N N, Mazur E A, 2015 *Physics Procedia* **71** 374.
[9] Degtyarenko NN, Mazur E A, 2015 *Journal of Experimental and Theoretical Physics* **148-6(12)** 1215.
[10] Nekrasov B, 1973 *Fundamentals of General Chemistry*. - Vol. 1. - M.: Chemistry 444.

# Influence of fast proton irradiation with energies of 12.4 and 12.8 MeV on magnetic characteristics and microstructure changes of superconducting intermetallic compound $Nb_3Sn$

**P N Degtyarenko[1,5,6], A Ballarino[2], L Bottura[2], S Y Gavrilkin[3], R Flükiger[2], I A Karateev[1], V S Kruglov[1,4], S T Latushkin[1], C Scheuerlein[2], A I Ryazanov[1,4], E V Semenov[1], S V Shavkin[1], T Spina[2], V N Unezhev[1], A L Vasiliev[1]**

[1] National Research Centre "Kurchatov Institute", Akademika Kurchatova pl. 1, 123182 Moscow, Russia
[2] CERN, CH-1211 Geneva 23, Switzerland
[3] P.N. Lebedev Physical Institute of the Russian Academy of Sciences, Leninskij Prospekt 53, 119991 Moscow, Russia
[4] National Research Nuclear University MEPhI (Moscow Engineering Physics Institute), Kashirskoe sh. 31, 115409 Moscow, Russia
[5] Joint Institute for High Temperatures of Russian Academy of Sciences, Izhorskaya st. 13 Bd.2, 125412 Moscow, Russia

E-mail: degtyarenkopn@gmail.com

**Abstract.** Data on magnetization, magnetic susceptibility and microstructure modification of irradiated $Nb_3Sn$ platelets are presented. The irradiation was produced at room temperature by fast protons with the energies of 12.4 and 12.8 MeV with fluencies of $5x10^{17}$ and $1x10^{18}$ $cm^{-2}$. Variation of the superconducting transition temperature *versus* irradiation dose was determined. Temperature dependence of magnetic susceptibility of $Nb_3Sn$ platelet with 160 μm thickness demonstrates several steps corresponding to different superconducting transition temperatures. We supposed that there are layers inside the sample with significantly different radiation damage levels caused by particles movement termination (Bragg peak). It was found that after the irradiation a lot of randomly oriented platelet-like Nb-enriched particles of 0.1-0.5 μm size appear in volumes with maximal damages.

## 1. Introduction

Multifilament wires of superconducting intermetallic compound $Nb_3Sn$ are the main candidates for utilising in superconducting magnetic system of future large projects of accelerators at CERN and in future fusion reactors. So study of influence of fast particle irradiation on critical parameters of these materials is very important at least for estimation of magnetic system lifetime. It is well known, that irradiation of $Nb_3Sn$ leads to strong changes of superconducting parameters [1-5] and the changes depend on both the initial state of material and irradiation parameters. Previous investigations of $Nb_3Sn$ platelets irradiated by protons with energies of 65 MeV and 24 GeV up to fluence of $3x10^{17}$ $cm^{-2}$ revealed almost 100% increase of the critical current density $J_c$ and ~ 5% increase of the upper critical field $B_{c2}$ while at higher fluencies both $J_c$ and $B_{c2}$ decreased [3, 6]. It was found also that

[6] To whom any correspondence should be addressed.

the critical temperature $T_c$ decreases monotonically with the fluence increase [3, 6, 7]. Irradiation of $Nb_3Sn$ multi-filamentary wires by 32 MeV protons demonstrated similar behavior [8]. However, the investigation of changes of $Nb_3Sn$ microstructure in wires is quite difficult because of complicated design of modern commercial wires. In contrast, irradiation of bulk $Nb_3Sn$ material is very helpful to understand the main physical mechanisms of radiation defects formation in these materials. Therefore, in this paper the results of study of $T_c$ and microstructure changes in $Nb_3Sn$ thin platelets after irradiation by fast protons are presented.

## 2. Platelets and experimental technique

Uniform and pure $Nb_3Sn$ platelets were cut from a cylindrical ingot of $Nb_3Sn$ produced by a reaction under HIP conditions, described in details in Refs. 3 and 6. The platelets parameters are presented in the Table 1. Typical area of platelets was 2x2 mm$^2$. Thickness of the unirradiated platelet was 0.19 mm. In order to observe at the same fluence both the effects of steady loss and Bragg peak (see Figure 1), pairs platelets in a stack were subjected to the irradiation simultaneously. The first pair with platelets of 0.17 and 0.14 mm thicknesses was irradiated by 5x10$^{17}$ cm$^{-2}$ fluence, while the second one with 0.22 and 0.16 mm thicknesses was irradiated by 10$^{18}$ cm$^{-2}$ fluence.

The irradiations were done on the isochronous cyclotron in NRC "Kurchatov institute" at room temperature using water-cooled targets separated by an aluminum foil from vacuum. The energies and fluencies of protons were 12.4 and 12.8 MeV, 5x10$^{17}$ and 1x10$^{18}$ cm$^{-2}$, correspondingly. The scheme of irradiation is presented on Figure 1a. The proton energies were adjusted for different fluencies to make equal the penetration depth of protons and total thickness of two platelets. Numerical calculations of damage profiles were made with SRIM-2013 code [9] for $Nb_3Sn$ density 8.9 g/cm$^3$ and displacement energy 25 eV for both Nb and Sn atoms. According to calculations, the mean path for 12.8 MeV protons passing the front platelet exceeds 300 μm (see Fig. 1b). Thus, it was assumed that the stack of #1a and #2a platelets should be completely penetrated by protons and protons should be completely stopped in #1b and #2b stack, forming the peak of radiation damages in the last platelet.

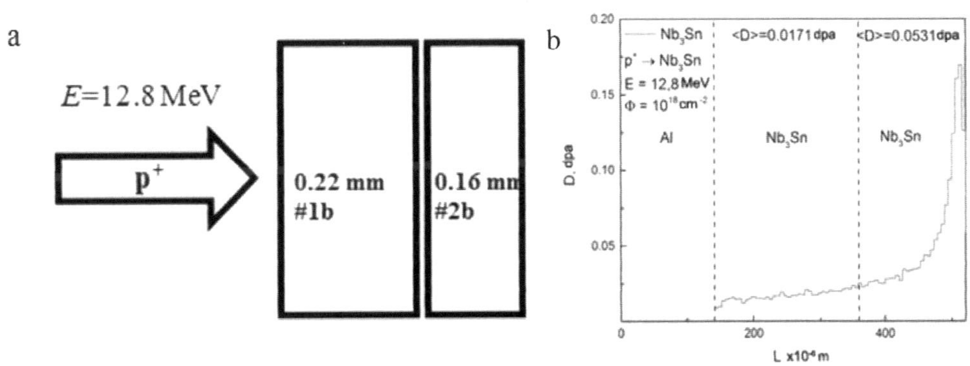


**Figure 1.** (a) The scheme of irradiation; (b) calculated radiation damage profile for $Nb_3Sn$.

Magnetic measurements were performed on the PPMS station in Lebedev Institute in magnetic fields up to 8 T and at temperatures up to 20 K. To estimate the critical current, the magnetization curves at $T$ = 7 K in fields up to 6 T and temperature dependence of remanent magnetization were measured on Vibrating Sample Magnetometer (VSM). For detailed $T_c$ characterization AC magnetic susceptibility was measured in field of 5 Oe at frequency of 37 Hz in the temperature range from 2 up to 20 K.

Detailed microstructural analysis was performed using scanning/transmission electron microscope TITAN 80-300 TEM/STEM (FEI, US) with Probe-Cs corrector and EDX spectrometer (EDAX, US) operated at 300 kV in NRC "Kurchatov Institute". The high angle annular dark field (HAADF) detector (Fischione, US) was used in STEM mode. Platelets were prepared by standard FIB procedure in a Helios (FEI, US) SEM/FIB dual beam microscope.

**Table 1.** The characteristics of $Nb_3Sn$ platelets.

| Type of platelets | Fluence, $cm^{-2}$ | Thickness, mm | Size, $mm^2$ | Mass, g |
|---|---|---|---|---|
| Reference platelet | unirradiated | 0,19 | 2,53x2,47 | 0,00822 |
| Platelet #1a | $5x10^{17}$ | 0,17 | 2,5x1,9 | 0,00620 |
| Platelet #2a | $5x10^{17}$ | 0,14 | 2,5x2,1 | 0,00540 |
| Platelet #1b | $10^{18}$ | 0,22 | 1,1x2,3 | 0,00340 |
| Platelet #2b | $10^{18}$ | 0,16 | 1,5x1,6 | 0,00286 |

## 3. Results and discussion

The results of remanent magnetization measurements for unirradiated and irradiated $Nb_3Sn$ platelets are presented on Figure 3. The curves normalized to magnetization at 7 K demonstrate the essentially different behavior for unirradiated and irradiated samples. For the unirradiated platelets the temperature dependence decreases abruptly and the magnetic moment completely vanishes at the critical temperature 17.9 K of $Nb_3Sn$ [9]. The curves for irradiated platelets #1a, #2a and #1b are similar to the unirradiated one but they drop to zero at lower temperatures. On the contrary, the curve for irradiated sample #2b passes below other curves and drops to zero at 16.2 K. As we assumed, this is the manifestation of non-uniformity of radiation defect on depth in the sample #2b.

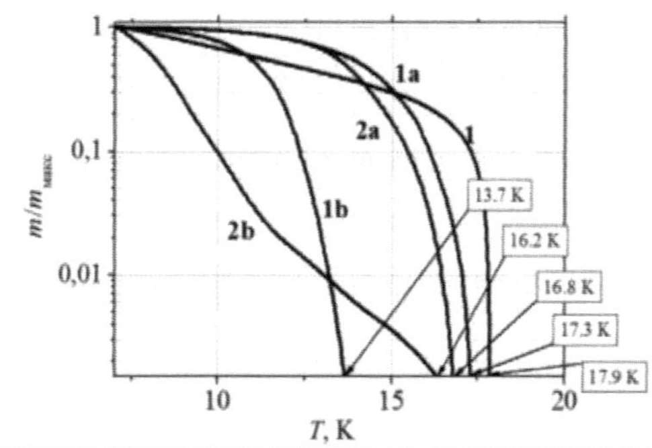


**Figure 2.** Temperature dependence of remanent magnetization of platelets: 1 – 0.19 mm, unirradiated; 1a and 2a – 0,17 and 0,14 mm, irradiated by 12.4 MeV protons with $5x10^{17}\,cm^{-2}$ fluence; 1b and 2b – 0.22 and 0.16 mm, irradiated by 12.8 MeV protons with $1x10^{18}\,cm^{-2}$ fluence.

The HAADF STEM image of the platelet with the marks, pointed to the areas of the EDX microanalysis, is presented in Figure 3b. An uniform image contrast indicated the good homogeneous composition of the platelet. Some contrast variations appeared due to the difference in specimen thickness aroused during FIB platelet preparation. Additional confirmation of the homogeneity was obtained from the EDX microanalysis. The Nb:Sn ratio is very close to the $Nb_3Sn$ stoichiometry.

The measured magnetic susceptibility is very sensitive to $T_c$ variations. The superconducting transition for unirradiated platelets corresponds to the critical temperature of pure $Nb_3Sn$ (see Fig. 3). The critical temperature for platelets #1a and #2a (at the fluence $5x10^{17}\,cm^{-2}$) is close to 15.3 K (Fig. 4). For platelet #1b (at the fluence $1x10^{18}\,cm^{-2}$) it decreases to 12.4 K (Fig. 5). For platelet #2b the $\chi_{AC}$ curve has several steps at temperatures – 16.2, 13.7, 9.9, 8.7 and 2.7 K, what also confirms our assumption.

Journal of Physics: Conference Series **747** (2016) 012030 doi:10.1088/1742-6596/747/1/012030

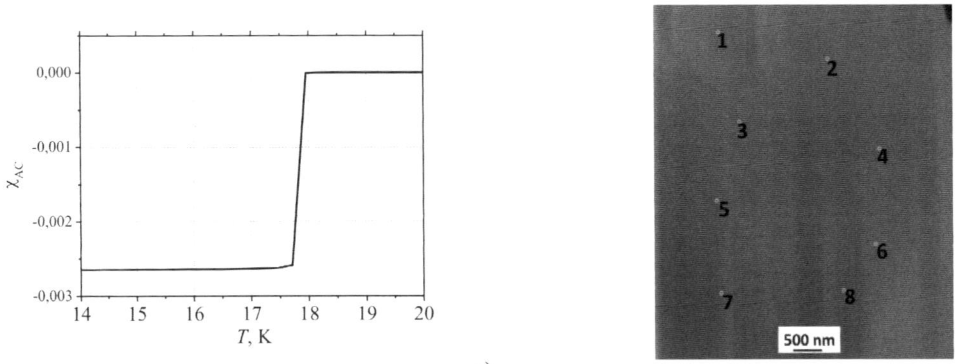


a) b)

**Figure 3.** Temperature dependence of $\chi_{AC}$ magnetic susceptibility (left) and HAADF STEM image (right) for unirradiated platelet #1. The small circles point to the areas of EDX microanalysis (1 – 73 at.% Nb and 27 at.% Sn; 2 – 74 at.% Nb and 26 at.% Sn; 3 – 75 at.% Nb and 25 at.% Sn; 4 – 76 at.% Nb and 24 at.% Sn; 5 – 74 at.% Nb and 26 at.% Sn; 6 – 76 at.% Nb and 24 at.% Sn; 7 – 75 at.% Nb and 25 at.% Sn; 8 – 74 at.% Nb and 26 at.% Sn).

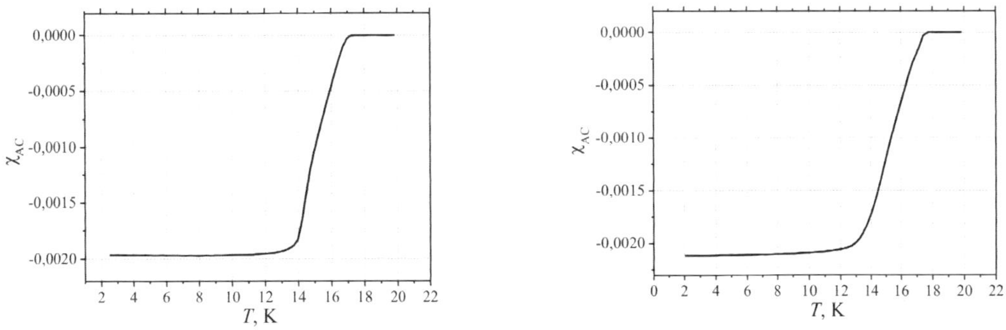


a) b)

**Figure 4.** Temperature dependence of $\chi_{AC}$ magnetic susceptibility for platelets #1a (left) and #2a (right) after proton irradiation with the energy 12.4 MeV and fluence $5x10^{17}$ $cm^{-2}$.

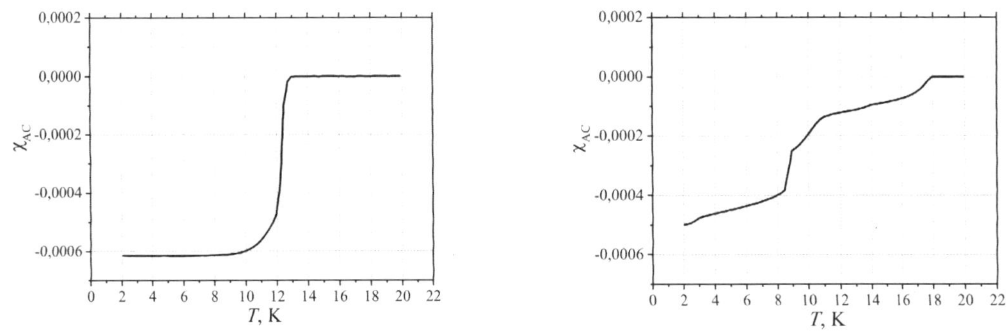


a) b)

**Figure 5.** Temperature dependence of $\chi_{AC}$ magnetic susceptibility for platelets #1b (left) and #2b (right) after proton irradiation with the energy 12.8 MeV and fluence $1x10^{18}$ $cm^{-2}$.

The bright field (BF) TEM image of platelet #2b presented in Figure 6a demonstrates appearance of randomly oriented particles of 0.1-0.5 μm size after irradiation of the sample. The contrast, similar to Moiré fringes, is clearly visible in vicinity of these particles. The selected area diffraction pattern

(SADP) from the area with several similar defects is shown in Figure 6b. The SADP consists of both set of spot reflexes corresponding to A15 structure of $Nb_3Sn$ in [120] zone axis and diffraction rings. Estimation of inter-planar distances, corresponding to these rings, unambiguously demonstrated that they originated from metallic Nb plane indexes (see Fig. 6b). The dark field (DF) TEM images of Nb particle was obtained by selecting part of the $110_{Nb}$ diffraction ring shown in Figure 6b and corresponding image is presented in Figure 6c. The particle exhibits bright contrast in these DF TEM images with dislocations, which are supposed to originate at the Nb particle – $Nb_3Sn$ interface and arise from Nb-$Nb_3Sn$ lattice mismatch. Surprisingly, the EDX microanalysis did not show noticeable presence of extra Nb (see Fig. 7). That means that Nb particle are very thin and do not impact to the EDXS signal. We suppose that these Nb particles adopted platelet morphology with the habit plane parallel to the specimen surface and perpendicular to $e^-$-beam.

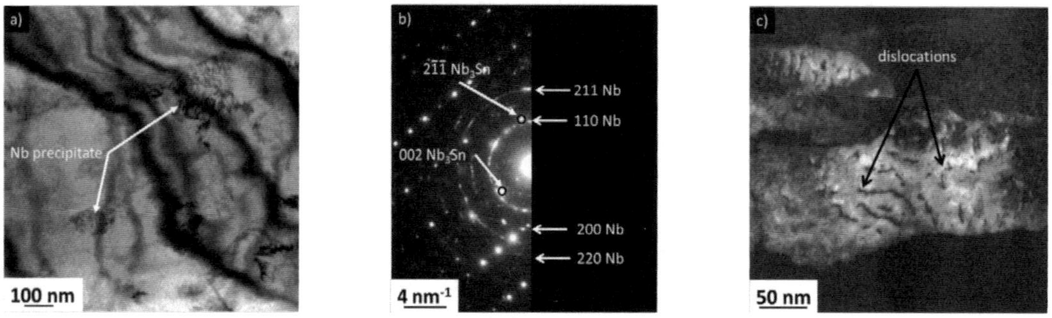


a) b) c)

**Figure 6.** BF TEM image of platelets #2b, the areas with typical radiation defects are arrowed (a); SADP from the sample with radiation defects (b) and DF TEM image of Nb precipitate (Note, that dislocations look like dark lines) (c).

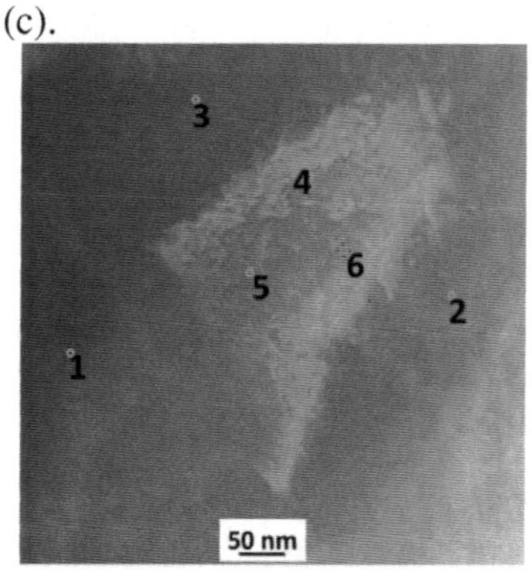


**Figure 7.** HAADF STEM image of the platelet #2b. The small circles point to the areas of EDX microanalysis (1 – 68 at.% Nb and 32 at.% Sn; 2 – 73 at.% Nb and 27 at.% Sn; 3 – 75 at.% Nb and 25 at.% Sn; 4 – 75 at.% Nb and 25 at.% Sn; 5 – 73 at.% Nb and 27 at.% Sn; 6 – 74 at.% Nb and 26 at.% Sn).

## 4. Summary

The experimental investigations of the magnetic characteristics and microstructure analyses of thin $Nb_3Sn$ platelets with different thickness after irradiation by fast protons with the fluencies $5x10^{17}$ (energy 12.4 MeV) and $10^{18}$ cm$^{-2}$ (energy 12.8 MeV) were performed. The curve of temperature dependence of remanent magnetization for irradiated platelet #2b at fluence of $10^{18}$ cm$^{-2}$ passes below the curve for unirradiated platelet. We supposed that an inhomogeneous structure was formed in this platelet under irradiation. This assumption is confirmed by data on magnetic susceptibility. The $\chi_{AC}$ transition curve demonstrated several steps at temperatures 16.2, 13.7, 9.8, 8.7 K (approximately to temperature of critical temperature of pure Nb) and 2.7 K. It was found that thin randomly oriented

Nb-enriched particles 0.1-0.5 μm in size are formed in platelet-like shape under irradiation. The observed structural transformations of sample lead to decrease of critical current density. The results will be extended in further investigations of irradiated $Nb_3Sn$ samples.

**Acknowledgements**
We would like to thank our colleagues P. Bezotosny and A. Tsvetkov from P.N. Lebedev Physical Institute of the Russian Academy of Sciences, and colleagues N. Degtyarenko from NRNU MEPhI and N. Balashov from Joint institute for high temperature of the Russian Academy of Sciences for profitable discussions, and management of NRC "Kurchatov institute" for support. The magnetic measurements were done at LPI Shared Facility Centre for Studies of HTS and other Strongly Correlated Materials in Lebedev Institute. The TEM measurements were performed on equipment of Resort Center for electronic microscopy nanozond in NRC "Kurchatov Institute".

**References**
[1] Ischenko G, Mayer H, Voit H, Besslein B and Haindl E 1972 *Z. Physik* **256** 176-184
[2] Baumgartner T, Eisterer M, Weber H W, Flükiger R, Sheuerlein C, Bottura L 2014 *Supercond. Sci. Technol.* **27** #1 015005
[3] Spina T July 2015 *PhD thesis* Nr. 2015-086 University of Geneva
[4] Voronova I V, Mihailov N N, Sotnikov G V and Zaikin V J, 1978 *J. Nucl. Mater.* **72**(1/2) 129–141
[5] Bode H J and Wohlleben K 1967 *Phys. Lett.* A, **24**(1) 25–27
[6] Spina T, Scheuerlein C, Richter D, Bordini B, Bottura L, Ballarino A and Flukiger R, 2015 *IEEE Trans. Appl. Supercond.* **25** 6000505
[7] Spina T, Scheuerlein C, Richter D, Ballarino A, Cerutti F, Esposito L S, Lechner A, Bottura L, Flükiger R, 2016 *IEEE Trans. Appl. Supercond.* **26**(3) 6001405
[8] Degtyarenko P N, Ballarino A, Bottura L, Flukiger R L, Gavrilkin S, Latushkin S, Primenko A E, Ryazanov A, Scheuerlein C, Shavkin S, Shikov A, Spina T, Unezhev V 2014 In Abstract book of ASC-2014 Charlotte North Carolina
[9] Dew Hughes D 1975 *Cryogenics* **15**(8) 435-454

# Mössbauer spectroscopy possibilities in investigation of the phase changes in alloy near surface layers under plasma, optic and ion irradiation influence

**V P Filippov**[1] **and V A Salomasov**

National Research Nuclear University MEPhI (Moscow Engineering Physics Institute), Kashirskoe shosse 31, 115409 Moscow, Russia

E-mail: vpfilippov@mephi.ru

**Abstract**. In this paper Mössbauer spectroscopy data on optic, plasma and ion irradiation influence on iron state are analyzed. It was shown the formation of new iron phases under plasma interaction with surface of steel samples. The appearance of paramagnetic austenite phases up to 20 microns from surface was found. The laser irradiation of thin Fe layers on V and Al layers on iron samples results into formation of solid solutions of these elements. The influence of ion irradiation with Ar+ on phase changes on surface of alloys is shown.

## 1. Introduction

One of the priorities of modern materials is the development of new methods to increase the performance of metallic materials. Reliability, durability and efficiency of parts and components of systems and mechanisms are largely determined by their surface properties. To obtain the needed characteristics of used construction details, sometimes it is enough to change the surface properties. Therefore, the aim of many investigations is focused on obtaining the necessary structural and phase states and properties of the surface layers of the material. To do this, it is necessary to study the changes in the structural and phase state occur in the surface layers after laser irradiation, plasma exposure and ionizing radiation. Mössbauer spectroscopy allows us to study the thin surface layers, so it can be helpful to solve these problems. The aim of the article is to review and show the Mössbauer spectroscopy possibilities in study of Nano structural and phase state changes of iron in the alloys which have been exposed to the influence of the laser, ion and plasma irradiation.

## 2. Experiment

The used Mossbauer spectrometer operated in the mode of constant acceleration. $^{57}$Co (Cr) was used as a source of resonant radiation. Isomer shifts are given relatively of metallic iron. Conversion Electron Mössbauer Spectra (CEMS) were registered with proportional counter. The fitting of spectra is carried out using Univem MS. Laser irradiation has been used for modification of near surface regions of the samples. The films of iron-57 and Al with thickness 200 nm on, respectively, vanadium and iron substrates were used as samples for investigations. The layers have been obtained by thermal evaporation of iron and aluminum at vacuum ~0.0013 Pa. The laser with wavelength of =1.06 microns

[1] To whom any correspondence should be addressed.

and pulse duration of 30 nano seconds was used. The details of experimental technique are described in paper [1].

The $Ar^+$ ions irradiation has been made using the installation, forming $Ar^+$ beam with Gaussian distribution of the ions energy in the range 3.5 - 15 keV, with an average energy of 10 keV and the ion flux density $j$ = 15 - 20 μA*cm$^{-2}$. The doze of irradiation was in the range of (0.1-5) *10$^{18}$ ion * cm$^{-2}$ [2]. The cold worked zirconium alloy zircaloy-2 (Zry-2) was used for irradiation.

For studying of plasma influence the steel samples were prepared in the form of discs with a diameter of ~20 mm and thickness of 3–5 mm. The high-temperature pulsed plasma flux of $q$ = 0.85 mW/cm$^2$ at a pulse duration τ = 20 ms was used. Working installation and surge chamber were previously evacuated to a pressure ~ 10$^{-1}$ Pa, and then the chamber was filled with plasma gas (nitrogen p = 36 Pa) [3]. The energy content of the stream varies from 10 to 50 kJ [3].

## 3. Results and discussion.

### *3.1 Laser irradiation*

Results of the Mössbauer investigations are shown in Figures 1, 2. Before laser irradiation the spectrum consists of the lines of hyperfine splitting of alpha-iron (Figure 1a) (only 1 and 6 lines are shown) with hyperfine magnetic field $H_{eff}$ =33.0±0.3 T. After laser irradiation the shape of CEMS spectrum is changed. The first and six lines show the superposition lines of hyperfine magnetic splitting of the magnetic phases (Figure 1b) with 3 hyperfine magnetic fields $H_{eff1}$=33.0±0.3 T, $H_{eff2}$=29.8±0.3 T and $H_{eff3}$=27.8±0.3 T. According to results in paper [4], the solid solutions of Al in Fe are formed under laser irradiation. The intensities of lines with $He_{ff1}$=33.0±0.3 T correspond to the share of iron without Al atoms in nearest neighborhood, lines with $H_{eff2}$=29.8±0.3 T and $H_{eff3}$=27.8±0.3 T correspond to the coordination spheres containing 1 and 2 Al, respectively. It means that the laser irradiation mixes Al atoms with iron and initiates formation of solid solutions (SS).

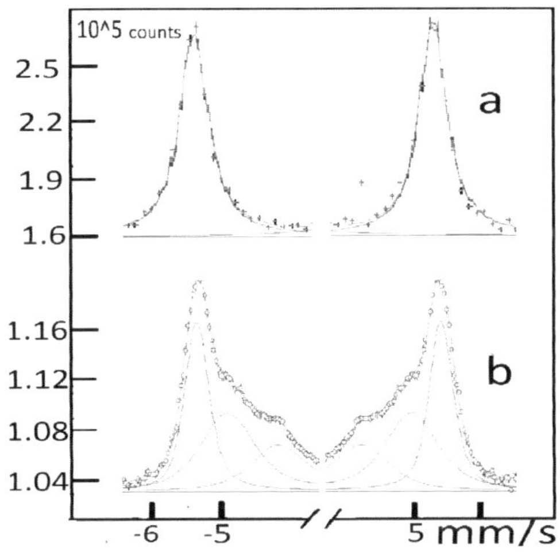


**Figure 1.** CEMS spectra of iron layer and iron layer under Al after irradiation by laser beam.

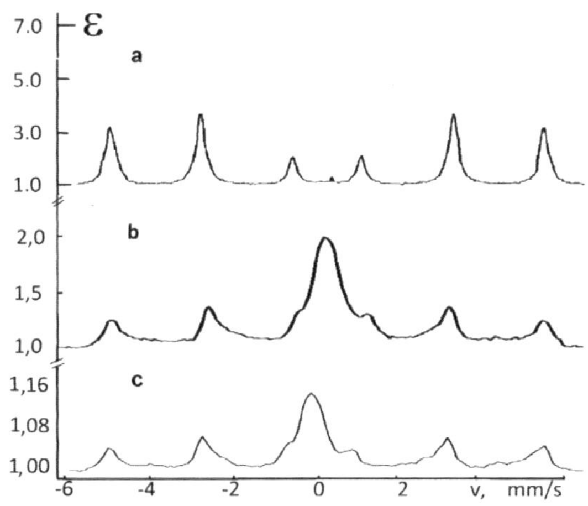


**Figure 2.** Spectra of iron layer on vanadium substrate before and after laser irradiation and after annealing.

The same laser irradiation were exposed on iron film on vanadium substrate. The CEMS spectrum is shown in Figure 2. The shape of initial spectrum (Figure 2a) presents the shape of alpha iron spectrum with $H_{eff}$=33.0±0.3 T. The intensities of second and fifth lines are greater than that of first and sixth lines. It means that majority of magnetic moments on the surface are oriented parallel to surface. After the laser influence, the spectrum shape is changed. In its central part the large line of paramagnetic phase is appeared (Figure 2b). In addition to lines of hyperfine magnetic splitting of alpha-iron the lines of hyperfine splitting of another magnetic phases are detected in this spectrum too. Fitting gives parameters of these lines such as $H_{eff2}$=29.8±0.3 T and $H_{eff3}$=27.8±0.3 T ($IS_1$=0.00±0.01 mm/s, $IS_2$=0.00±0.01 mm/s). Approximately the same results are obtained after annealing of the same type specimen (Figure 2c). Comparing these results allows us to make the conclusion that the laser irradiation results in formation of SS of V in Fe. The spectrum parameters of the central part are as follows IS=0.00±0.01 mm/s, QS=0.00±0.01 mm/s. Comparing them with parameters known for V-Fe alloys [5] we conclude that the central line of paramagnetic phase belongs to intermetallic compound Fe-V, and the lines of hyperfine magnetic splitting belong to SS of V in Fe. So the laser irradiation and annealing provides the same changes, i.e. formation of the intermetallic compound Fe-V and SS of V in Fe.

### *3.2 Ar+ irradiation*

Influence of ion irradiation by ions $Ar^+$ is shown in the Figure 3. Much changes are visible in CEMS spectra (Figure 3d). In transmission geometry the intensity of lines of intermetallic compound $Zr(Fe, Cr)_2$ in the form (C14) decreases relatively to intensity of lines of intermetallic compound $Zr(Fe,Cr)_2$ in the form (C15). Comparison of lines intensities in the CEMS spectra (Figure 3a, 3b) shows that part of $Zr(Fe, Cr)_2$ in the form (C14) was transformed to the compound $Zr_3Fe$.

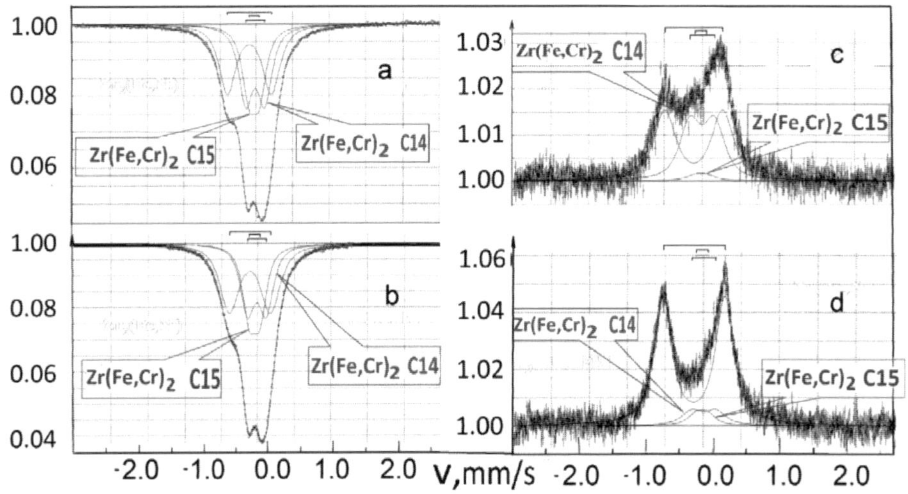


**Figure 3.** Spectra of $^{57}Fe$ in the zirconium alloys (Zry-2) in transmission mode: a) before irradiation, b) after irradiation, in CEMS mode: c) before irradiation, d) after irradiation.

### *3.3 Plasma irradiation*

Interaction of plasma with surface of steel results in the phase changes in the nearest neighbor layers. This follows from analysis of CEMS spectra of initial specimen of steel and those after plasma influence (Figure 4). The spectrum of the initial steel specimen consists of lines of hyperfine magnetic splitting (HMS) of 3 magnetic phases (Figure 4a).

The values of IS = 0.00±0.02 mm/s, the values of magnetic fields are $H_{eff1}$ = 33.0±0.1 T, $H_{eff2}$ = 30.4±0.2 T, $H_{eff3}$ = 27.5±0.5 T. After the interaction they are equal to $H_{eff1}$ = 33.6±0.1 T, $H_{eff2}$ = 30.4±0.2 T, $H_{eff3}$= 28.1±0.5 T. Before the interaction the line of paramagnetic phase (IS = 0.28±0.03 mm/s, A = 61%) is detected. After the interaction the lines of great intensity (IS = - 0.07±0.05 mm/s, A = 14.1%) and low intensity (IS =0.28±0.03 mm/s, A =41%) are visible in the central part of the spectrum. Existence of three sextets can be explained by formation solid solutions of doping atoms, for example, Cr or another atom.

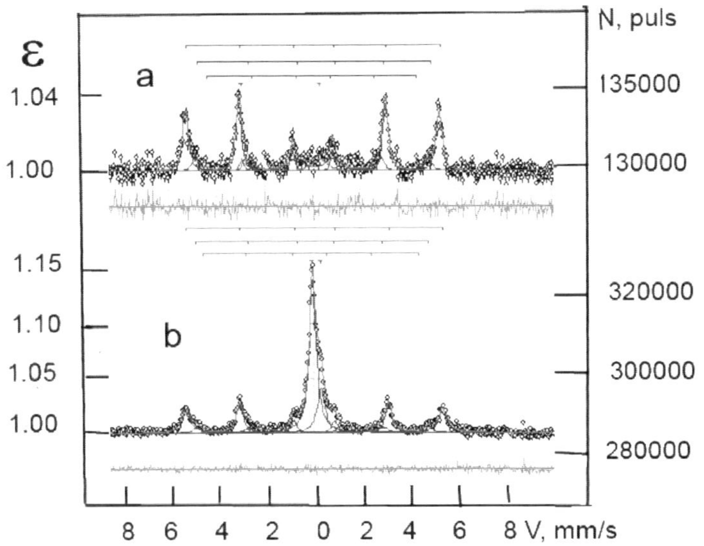


**Figure 4**. Spectra of $^{57}Fe$ in the steel at $T = 295$ K a) before (a) and after (b) interaction with plasma.

These doping nonmagnetic atoms such as Cr or Ni decrease the magnitude of $H_{eff}$. Sextet with $H_{eff1}$=33.0±0.1 T is explained by existence of iron without another atom in nearest neighborhood, the other two ($H_{eff2}$, $H_{eff3}$) –by existence of iron states with coordination spheres containing one or two nonmagnetic atoms. So Mössbauer spectroscopy allows us to detect changes in structure of near surface layers and appearance of paramagnetic (possibly austenite) phase and to estimate the relative concentration.

## 4. Conclusion

Mössbauer spectroscopy is applied to investigations of the influence on surface layers of laser irradiation, irradiation by $Ar^+$ ions and interaction with plasma. It was shown:

a) Laser irradiation of Al films on Fe and Fe films on V results in formation of solid solutions of Al and V atoms in iron and formation of intermetallic compound Fe-V;

b) Irradiation by Ar+ ions of zirconium alloys (E635) results in transformation of intermetallic compound $Zr(Fe, Cr)_2$ in the form (C14 ) to compound $Zr_3Fe$;

c) Plasma influence on surface of steel results in formation of paramagnetic austenite phase.

**Acknowledgments**

The work was supported by the RF Ministry of Education and Science, agreement no. 14.575.21.0086, dated 20 October 2014,Unique project identifier RFMEFI57514X008 and by the Competitiveness Program of NRNU MEPhI.

## References

[1] Ali-Zade et. all 1987 *J Voprosi pricladnoy yadernoy fiziki* (Moscow, Energoatomizdat) 75-81 (Rus).

[2] Filippov V P, Volkov N V, Kalin B A, Petrov V I and Lauer D E 2010 *Bulletin of the Russian Academy of Sciences: Physics* **74**(3) 374–378.

[3] Filippov V P and Bokov A V 2015 *Bulletin of the Russian Academy of Sciences. Physics* **79**(8) 1031–1035.

[4] Virender K Sharma 2014 *Mossbauer Spectroscopy: Applications in Chemistry, Biology, and Nanotechnology* ed. Gostar Klingelhofer and Tetsuaki Nishida 42-72

[5] Ovchinikov V V 2006 *Mössbauer Analysis of the Atomic and Magnetic Structure of Alloys*, CISB 247.

Journal of Physics: Conference Series **747** (2016) 012032 doi:10.1088/1742-6596/747/1/012032

# Phase changes of iron near surface layers under argon ion irradiation

**V P Filippov[1], B A Kalin, V I Petrov and N V Volkov**

National Research Nuclear University MEPhI (Moscow Engineering Physics Institute), Kashirskoe shosse 31, 115409 Moscow, Russia

E-mail: vpfilippov@mephi.ru

**Abstract.** Mössbauer spectroscopy data on $Ar^+$ ion irradiation influence on iron state are analyzed. It was shown the formation of new iron phases under $Ar^+$ ion interaction with surface of zirconium alloy samples with iron and without iron layers. The appearance of paramagnetic phase in the surface layers was detected. Transformations of intermetallic compounds in zirconium alloys was established. Much changes are found in near surface layers.

## 1. Introduction

One of the main problems of nuclear power production is at present the safe maintenance of nuclear reactors with increasing reactor service time and fuel burnout rate. Raising the corrosion stability of zirconium alloys is important for the solution of this problem. This deals with creation of modern materials. The main priorities of modern materials is the development of new methods to increase the performance of metallic materials. Reliability, durability and efficiency of parts and components of systems and mechanisms are largely determined by their surface properties. In particular, to improve the corrosion resistance and to improve the tribological characteristics of products it is often sufficient and economically advantageous the modification of only the surface layers. Therefore, the challenge now is focused to obtain the necessary structural and phase states and properties of the surface layers of the material. To do this, we must study the changes in the structural and phase state in the surface layers after $Ar^+$ ion radiation. Mössbauer spectroscopy allows us to study the thin surface layers, so it can be used to solve these problems. The aim of the article is to show the changes in phase compositions of zirconium alloy under $Ar^+$ ion irradiation by Mössbauer spectroscopy.

## 2. Experimental

The used Mossbauer spectrometer operated in the mode of constant acceleration. $^{57}Co$ (Cr) was used as a source of resonant radiation. Isomer shifts are given relatively of $\alpha$–iron. Conversion electron Mössbauer spectra (CEMS) were registered with proportional counter [1, 2]. The fitting of spectra is carried out using Univem MS. The scheme unit for production of iron films and irradiation with argon ions is shown in Figure 1.

[1] To whom any correspondence should be addressed.

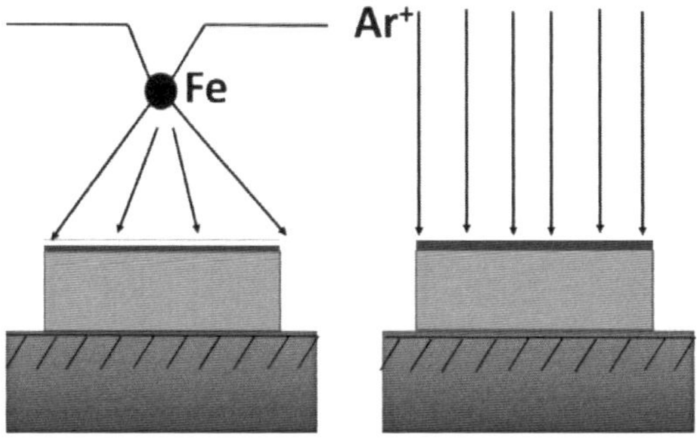


**Figure** 1. Schema of iron layer preparation and irradiation by Ar+ atoms.

Irradiation by Ar+ ions have been made using the installation, forming a beam of $Ar^+$ ions with Gaussian distribution of the ions energy in the range of 3.5 - 15 keV, with an average energy of 10 keV and the ion flux density $j = 15 - 20$ μA*sm$^{-2}$. Doze of irradiation was in the range of (0.1-5)*$10^{18}$ ion*cm$^{-2}$ [1].

The thickness of iron layer was determined by gravimetric and photometric methods. The experimental error is 20 nm.

## 3. Results and discussion

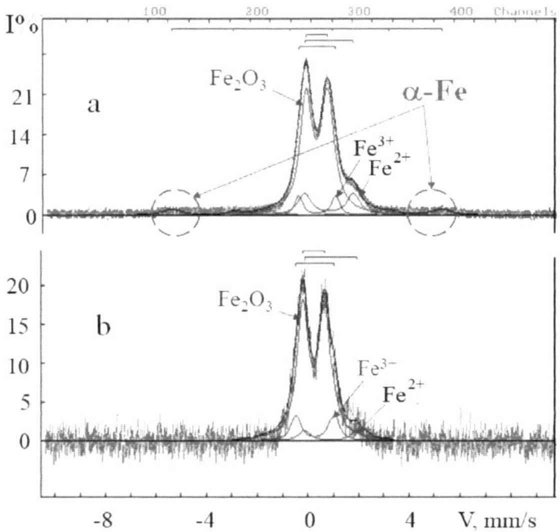


**Figure 2.** CEMS spectra of 50 nm $^{57}Fe$ layer on surface of E635 before (a) and after (b) irradiation by $Ar^+$ ions. $E$ = 10 keV, dose 5*$10^{17}$ cm$^{-2}$.

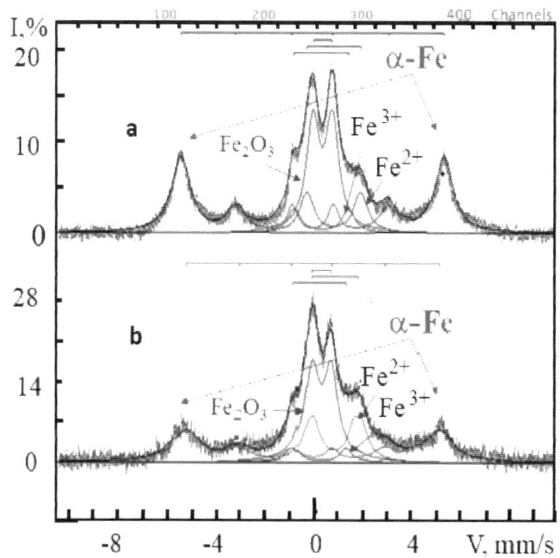


**Figure 3**. CEMS spectra of 500 nm $^{57}Fe$ layer on surface of E635 before (a) and after (b) irradiation by $Ar^+$ ions. $E$ = 10 keV, dose 5*$10^{17}$ cm$^{-2}$.

Results of the Mössbauer investigations are shown in Figures 2, 3 and Tables 1, 2. CEMS spectrum of $^{57}Fe$ layer (thickness 50 nm) can be fitted by superposition of lines of quadrupole splitting of paramagnetic phases (Figure 2 a, b) such as $Fe_2O_3$, $Fe^{3+}$, $Fe^{2+}$. During obtaining iron layer on any substrate by iron evaporation the lines of $Fe_2O_3$ in paramagnetic state appeared in the beginning of this process. Before irradiation the low intensity lines of α-Fe are visible in the spectra (Figure 2a). After irradiation these lines are not visible. The spectra parameters are given in Table 2. The irradiation results in disappearance of α-Fe and increasing of $Fe_2O_3$ quantity in paramagnetic state.

Existence of lines of quadrupole splitting of $Fe_2O_3$ paramagnetic phase can be explained by formation of iron nanoparticles on the surface and their interaction with air components and thin oxide layer on surface of zirconium alloy. Existence of iron paramagnetic phases with valence state $Fe^{3+}$, $Fe^{2+}$ has not good explication at this stage of research. One can suppose an interaction with thin oxide layer $ZrO_2$ which is always formed on surface of zirconium alloys. Irradiation by $Ar^+$ ions changes the phase composition in the surface layers both for thin (50nm) and thick (500nm) iron films (Tables 1 and 2).

**Table 1**. Mössbauer spectra parameters of $^{57}Fe$ in 50 nm layer.

| Phase | S, % | IS, mm/s | QS, mm/s | W, mm/s | H, T |
|---|---|---|---|---|---|
| | Before irradiation | | | | |
| $Fe_2O_3$ | 69.3 | 0.38±0.02 | 0.83±0.02 | 0.58±0.03 | |
| $Fe^{3+}$ | 14.3 | 0.38±0.02 | 1.42±0.02 | 0.64±0.03 | |
| $Fe^{2+}$ | 7.70 | 0.85±0.02 | 1.90±0.02 | 0.45±0.02 | |
| α–Fe | 10.6 | 0.03±0.02 | 0.03±0.02 | 0.30±.02 | 32.80.2 |
| | After irradiation | | | | |
| $Fe_2O_3$ | 81.2 | 0.24±0.02 | 0.87±0.02 | 0.54±0.03 | |
| $Fe^{3+}$ | 13.2 | 0.28±0.02 | 1.48±0.02 | 0.45±0.02 | |
| $Fe^{2+}$ | 6.2 | 0.90±0.02 | 2.06±0.02 | 0.55±0.02 | |
| α–Fe | - | - | - | - | |

**Table 2.** Mössbauer spectra parameters of $^{57}Fe$ in 500 nm layer.

| Phase | S, % | IS, mm/s | QS, mm/s | W, mm/s | H, T |
|---|---|---|---|---|---|
| | Before irradiation | | | | |
| $Fe_2O_3$ | 33.2 | 0.43±0.02 | 0.81±0.02 | 0.65±0.03 | |
| $Fe^{3+}$ | 13.3 | 0.38±0.02 | 2.25±0.02 | 0.69±0.03 | |
| $Fe^{2+}$ | 7.2 | 0.90±0.03 | 2.25±0.02 | 0.67±0.02 | |
| α–Fe | 47.3 | 0.02±0.02 | 0.00±0.02 | 0.90±0.02 | 33.1±0.2 |
| | After irradiation (information to depth 300 nm) | | | | |
| $Fe_2O_3$ | 37.2 | 0.30±0.02 | 0.82±0.02 | 0.75±0.03 | |
| $Fe^{3+}$ | 5.2 | 0.26±0.02 | 1.97±0.02 | 0.75±0.02 | |
| $Fe^{2+}$ | 20.2 | 0.85±0.02 | 2.25±0.02 | 0.65±0.02 | |
| α–Fe | 38.1 | 0.02±0.02 | 0.00±0.02 | 0.90±0.02 | 33.1±0.2 |
| | After irradiation from surface (to depth ~20 nm) | | | | |
| $Fe_2O_3$ | 43.2 | 0.31±0.02 | 0.83±0.02 | 0.74±0.03 | |
| $Fe^{3+}$ | 10.1 | 0.39±0.02 | 2.21±0.03 | 0.75±0.04 | |
| $Fe^{2+}$ | 14.2 | 0.92±0.02 | 2.06±0.02 | 0.69±0.02 | |
| α–Fe | 33.3 | 0.02±0.02 | 0.00±0.02 | 0.90±0.02 | 32.7±0.2 |

There is decreasing of α–Fe content for thick layer (500 nm) and its disappearing for thin layer (50nm). The nearer the surface the lower α–Fe content (Table 2 column 2). The increasing of paramagnetic phase $Fe_2O_3$ from 33% to 43 % is detected (Table 2). The increasing of the share of $Fe^{2+}$ ions is accompanied with simultaneous decreasing of the $Fe^{3+}$ share. The changes of the share are approximately equal. This can be explained by influence of zirconium. Comparing the parameters of CEMS spectra for different iron layer thicknesses (Tables 1 and 2) we can see some regularity in the behavior of the isomer shift and large changes of the values of quadrupole splitting under irradiation influence. The irradiation results in the decreasing of isomer shifts magnitudes from 0.38 to 0.24 mm/s for 50 nm layer and from 0.43 to 0.31 mm/s for 500 nm layer. This can be explained by large changes in electron density on the iron nuclei. The decreasing of isomer shift of $Fe_2O_3$ paramagnetic phase indicates an increase in the density of *s*-electrons at the nucleus. As there are no significant changes in the magnitudes of QS for $Fe_2O_3$ paramagnetic phase, so there are no significant changes in surrounding of iron atoms in this phase. Part of iron ions is in $Fe^{2+}$ and $Fe^{3+}$ states. The magnitudes of

QS of these phases containing iron in paramagnetic states such as $Fe^{2+}$ and $Fe^{3+}$ are abnormally high. There is no any regularity in the behavior of QS (Tables 1, 2). As follows from high magnitudes of QS, great distortions of crystal structure presenting in these phases containing iron in paramagnetic $Fe^{2+}$ and $Fe^{3+}$states results in increasing of electric field gradient and consequently in the increase of the QS magnitudes. As to the alpha-iron spectra parameters, there are significant changes in their contents under irradiation and non-significant changes in magnitudes of IS, QS, $H_{eff}$. The latter can be explained by fluctuation in atoms surrounding caused by irradiation in this magnetic phase.

We can state that the second and fifth lines of alpha-iron spectra do not correspond to the law of intensities 3:2:1 for chaotic distribution of the magnetic moments in the sample or layer (Figures 3). When the vectors of magnetic moments are perpendicular to surface the intensities of 2-d and 5-th lines decrees or disappear. So in iron layers on zirconium alloys the vectors of magnetic moments are preferentially orientated perpendicular to surface, while in pure iron their preferential orientation is usually parallel to surface [3]. The significant magnitudes of line width (W=0.45-0.90 mm/s ) (Tables 1, 2) for all phases are due to different surroundings of iron atoms in the layers.

## 4. Conclusion

Thus, Mössbauer spectroscopy allows us to obtain data on the iron atoms state in the surface layers of different thicknesses. It was found that the iron layers have a heterogeneous structure as different phases or compounds. Iron layers have different phase composition depending of iron layer thickness. There are paramagnetic phases such as $Fe_2O_3$, $Fe^{3+}$, $Fe^{2+}$ and magnetic phase as α-Fe in the iron layers. The phase composition is changed under irradiation, in particular the quantity of α-Fe decreases, while the quantity of paramagnetic phase $Fe_2O_3$ increases. These compounds are no uniformly distributed through the thickness of the layers.

## Acknowledgements

The work was supported by the RF Ministry of Education and Science, agreement no. 14.575.21.0086, dated 20 October 2014, Unic project identifier RFMEFI57514X0086 and by the Competitiveness Program of NRNU MEPhI.

## References

[1] Filippov V P, Volkov N V, Kalin B A, Petrov V I and Lauer D E. 2005 *Bulletin of the Russian Academy of Sciences: Physics* **74**(3) 374

[2] Filippov V P and Bokov A V. 2015 *Bulletin of the Russian Academy of Sciences. Physics* **79**(8) 1031

[3] Milan Seber′ıni and Jozef Sitek 2008. *Journal of Electrical Engineering* **59**(5) 234

# Mössbauer spectroscopy research of interaction of alumosilicic reagent and iron dissolved in water

**D Y Feklistov[1], V P Filippov[1,4], I M Kurchatov[1], N I Laguntsov[1], V A Salomasov[2] and Yu V Permyakov[3]**

[1] National Research Nuclear University MEPhI (Moscow Engineering Physics Institute), Kashirskoe shosse 31, 115409 Moscow, Russia
[2] JSC Aquaservice, Kashirskoe shosse 31, 115409 Moscow, Russia
[3] VNIIFTRI, 141570, Mendeleevo, Solnechnogorskiy district, Moscow region, Russia

E-mail: vpfilippov@mephi.ru

**Abstract.** The aim of this work is to reveal the results of alumosilicic reagent interaction with iron compounds contained in the water. This reagent is simultaneously coagulant-flocculant and adsorbent. The iron atoms state is studied in the reagent and in reacted sediment. The valence state of iron atoms are determined in the reagents and sediments. The existence of iron containing superparamagnetic particles in the sediment is shown.

## 1. Introduction

Artesian waters contain complexes of iron compounds. The surface waters are vulnerable to anthropogenic pollution. People's health depends on water purity, for example, consumption of water with the high iron content could develop the blood disease. Therefore the process of water treatment from dissolved iron is necessary. In modern technological scheme of natural and wastewater treatment coagulants and flocculants are used. To purify water from all contaminants the alumosilicic reagent has been used [1]. This reagent is a coagulant-flocculant and simultaneously an adsorbent[2,3]. Water was purified from iron and the precipitates or sediment were obtained with coloured rust. The main form of iron in the artesian waters taken from Moscow region is complexes of ferrous iron in divalent state with dissolved inorganic and organic compounds. The most effective and direct method to determine the iron chemical state and to detect iron nanoparticles is the Mössbauer spectroscopy which supplies an information about the isomer shift and quadrupole splitting.

## 2. Experimental

### *2.1. Method of preparation of sorption-coagulation-flocculation reagent*

The sorption-coagulation-floccilation reagent is produced from cheap aluminosilicate raw which is the nepheline [3]. The model formula of the reaction is:

$$(Na\,,K)_2O \cdot Al_2O_3 \cdot 2SiO_2 + 4H_2SO_4 + aq \rightarrow (Na,K)_2SO_4Al_2(SO)_3 \cdot mH_2O + 2SiO_2 \cdot nH_2O \qquad (1),$$

[4] To whom any correspondence should be addressed.

Journal of Physics: Conference Series **747** (2016) 012033 doi:10.1088/1742-6596/747/1/012033

which includes hydrated silicic acid (nanoparticles).

The active reagent such as (Na, K)$_2$ $SO_4$ Al2$(SO)_3$· m$H_2O$ with different concentrations of the main components is introduced into the aqueous solution. Its chemical interaction in water leads to an association and formation of silico-supramolecular structures with high rate adsorption. Then the self-organization of these structures forms macrocomplexes, such as zeolite matrix. Dynamics of synergetic effect in the alumosilicic reagent is not well studied. Size of the reagent particles in the range 40-200 nm was determined, in our case, using the laser particle size analyzer Nanotrac Ultra (ISO 13321).

The resulting matrix molecules adsorb pollutants such as impurities in contaminated water and as a result increase in size and weight. During this process an intense flocculation occurs and nascent flocs drop to the bottom of the container. The flocs were picked out and dried. Thus, obtained little crystals of sediment were used for investigations.

### 2.2. *Sample preparation for Mössbauer spectroscopy.*

The Mössbauer absorbers were prepared from the reagent before interaction with water and from sediments. The crystals of the sediment and initial reagent were pulverized using a porcelain mortar to obtain a powder of same-dimension crystals. The crystal density was 2.0 g cm$^{-3}$. The resulting powder filled in the molten paraffin and then mixed until a homogeneous mass. The thickness of absorbers were 40-100 mg cm$^{-2}$. The content of the Mössbauer isotope $^{57}$Fe was approximately 0.012 mg cm$^{-2}$. The used Mossbauer spectrometer operated in the mode of constant acceleration. Mössbauer radiation detector with a thin NaJ (Tl) crystal (200 microns) was used. The spectrometer was calibrated using the standard thin absorber of alpha-iron. $^{57}$Co (Cr) was used as a source of resonant radiation. The isomer shifts are given relatively a-Fe. Mössbauer spectra were obtained at room temperature and 77 K. The fitting of spectra is carried out using Univem MS.

## 3. Results and discussion

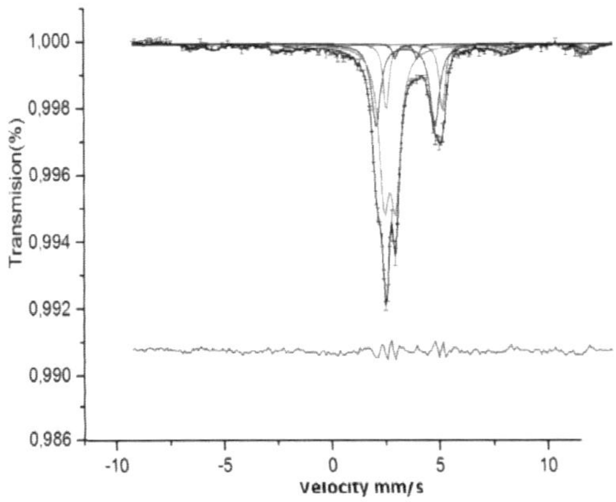


**Figure 1.** Room temperature Mössbauer spectra of nepheline minerals.

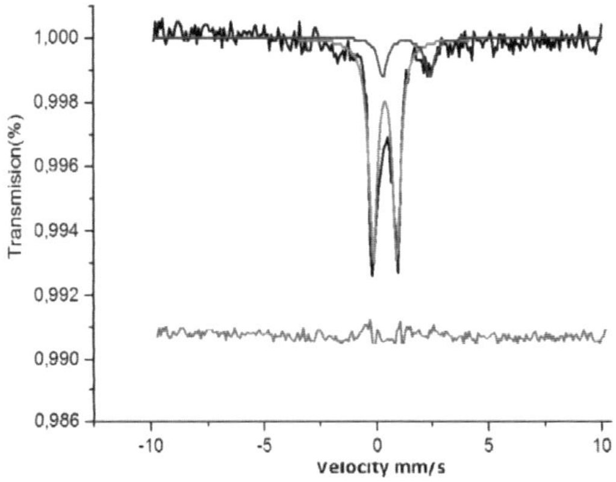


**Figure 2.** Room temperature Mössbauer spectra of dried reagent

In Figure 1 the room temperature Mössbauer spectra and fitting for nepheline minerals before reaction with sulfuric acid are shown. The spectra were fitted with two magnetic components and two or three paramagnetic components. The main subspectra of the iron (~60%) is in trivalent paramagnetic state (IS = 0.36 ± 0.02 mm s$^{-1}$, QS = 0.46 ± 0.02 mm s$^{-1}$). All parameters of this spectra are presented in Table 1. Results of the fitting and analysis of the experimental spectra of nepheline shown that γ-$Fe_2O_3$ nanoparticles with sizes approximately of 30 – 70 nm are contained in nepheline [6]. The room temperature Mössbauer spectra of the dried reagent are shown in Figure 2 as an addition. Main part (~90%) of these spectra is doublet with IS = 0.37 ± 0.02 mm $s^{-1}$, QS = 1.01 ± 0.02 mm $s^{-1}$.

**Table 1.** Mössbauer parameters of the nepheline minerals used for synthesis alumosilicic reagent $T$ = 293 K.

| Phase | IS, mm/s | QS, mm/s | $H_{eff}$, T | G, mm/s |
|---|---|---|---|---|
| Sextet 1 | 0.43 ± 0.05 | 0.00 ± 0.05 | 49.1 | 0.4242 |
| Sextet 2 | 0.88 ±0.05 | -0.21 ± 0.05 | 46.3 | 0.5826 |
| Doublet 1 | 0.36 ± 0.05 | 0.46 ± 0.05 | | 0.4884 |
| Doublet 2 | 1.34 ± 0.05 | 2.24 ± 0.05 | | 0.2609 |
| Doublet 3 | 1.02 ± 0.05 | 0.87 ± 0.05 | | 0.2328 |

The result spectra of the reaction of reagent with dissolved iron in the water are shown in Figure 3. The spectra were obtained at 293 K (a) and 77 K (b). The shape of the spectra corresponds to lines of hyperfine quadrupole splitting of paramagnetic phases. The main share of 90% of the iron is in the paramagnetic phase in the trivalent state (isomer shift IS = 0.35 ± 0.02 mm/s, quadrupole splitting QS = 0.82 ± 0.02 mm/s). The fitting of the spectra at 77 K gives doublet with paramagnetic phase (IS=0.45 ± 0.02 mm/s, QS =0.87 ± 0.02 mm/s). The measurements at $T$ = 77 K showed that, along with the existence of the paramagnetic phase, a small amount of magnetic phases are contained in the sample. In this phase the iron is in trivalent and divalent states. This indicates there are the iron superparamagnetic particles in precipitates. Calculated size of these particles is $d$ ~ 3.7 nm [5]. The distribution of hyperfine magnetic fields of the sediment at 77 K was obtained (Figure 3c). This distribution allows distinguishing four magnetic phases. Fitting of the spectrum obtained at 77 K by one doublet and four sextets gives precise results for IS, QS, $H_{eff}$ in these magnetic phases (Table 2). As the lines of these phases are not visible in spectra measured at room temperature, we conclude that these phases are in superparamagnetic state. So, the process of water treatment by reagent gives sediment in which the iron is in the form of nanoparticles. Using known data [6], the phase with lines of quadrupole splitting (doublet 1, Table2) can be identified as paramagnetic $\gamma$-$Fe_2O_3$ compound. The sextets 3 and 4 may be caused by presence of $\gamma$-$Fe_2O_3$ $\alpha$-$Fe_2O_3$ nanoparticles in the superparamagnetic state. The magnetic phases with $H_{eff}$ = 24.9 ± 0.2 T and $H_{eff}$ = 34.5 ± 0.2 T (sextets 1 and 2 Table2) were not identified at this step of investigations.

**Table 2.** Mössbauer parameters of the sediment at liquid $N_2$ temperature.

| Phase | IS, mm/s | QS, mm/s | $H_{eff}$, T | G, mm/s |
|---|---|---|---|---|
| Doublet 1 | 0.45 ± 0.05 | 0.86 ± 0.05 | | 0.55 ± 0.05 |
| Sextet 1 | 0.16 ± 0.05 | 0.02 ± 0.05 | 24.9 ± 0.2 | 0.77 ± 0.05 |
| Sextet 2 | 0.18 ± 0.05 | -0.21 ± 0.05 | 34.5 ± 0.2 | 0.26 ± 0.05 |
| Sextet 3 | 0.60 ± 0.05 | -0.48 ± 0.05 | 49.9 ± 0.2 | 0.76 ± 0.05 |
| Sextet 4 | 0.28 ± 0.05 | 0.19 ± 0.05 | 54.9 ± 0.2 | 0.39 ± 0.052 |

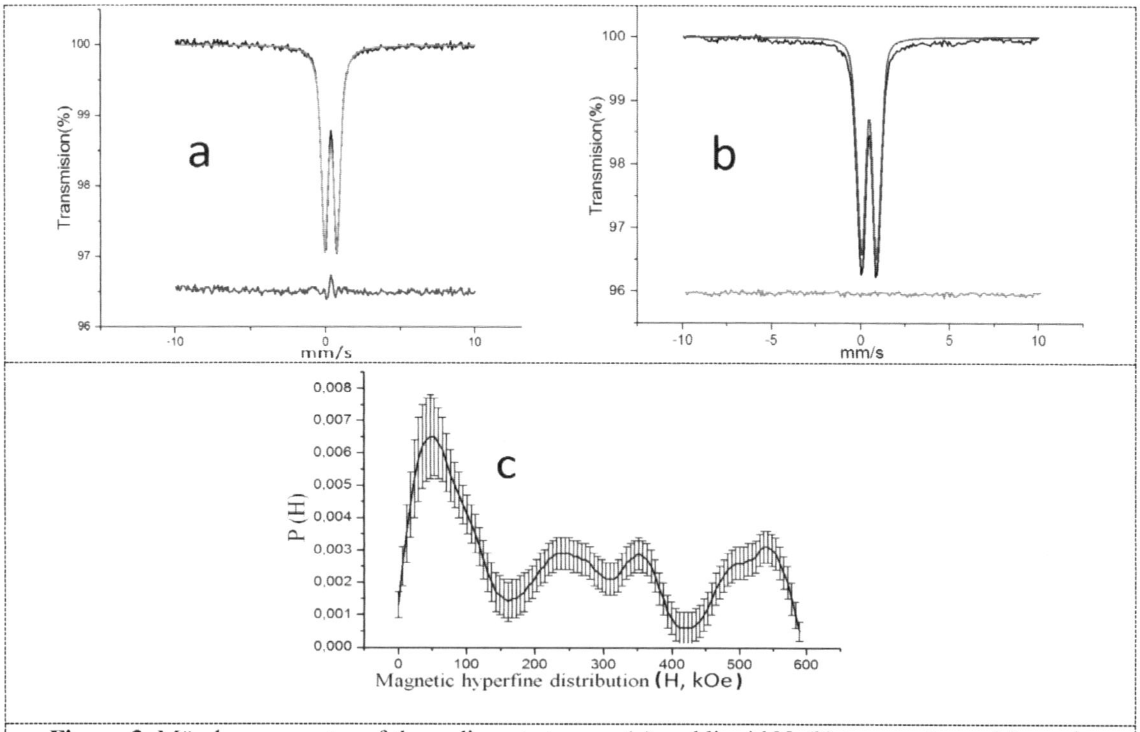


**Figure 3.** Mössbauer spectra of the sediment at room (a) and liquid $N_2$ (b) temperatures. Magnetic hyperfine distribution of the sediment at 77 K.

## 4. Conclusion

Mössbauer spectroscopy was applied to investigate nepheline minerals, initial component for reagent preparation, reagent itself and product of its reaction with water, sediment.

The following was shown.

a) In the nepheline minerals the iron forms several complex paramagnetic and magnetic phases with $Fe^{2+}$ and $Fe^{3+}$ ions. Size of nanoparticles in magnetic state was estimated as 30 –70 nm.

b) The sediment contains one paramagnetic and four superparamagnetic phases of iron in $Fe^{2+}$ or $Fe^{3+}$ valance state. One of these phases was identified as $\gamma$-$Fe_2O_3$ in magnetic and paramagnetic states. One of the magnetic phases was identified as $\alpha$-$Fe_2O_3$ nanopartcles.

**Acknowledgements**

The work was financially supported by the Ministry of Education and Science of the Russian Federation, agreement № 14.575.21.0086, dated 20 October 2014, Unique project identifier RFMEFI57514X0086 and by the Competitiveness Program of NRNU MEPhI.

## References

[1] Laguntsov N I, Neschimenko Y P, Feklistov D Y 2008 Nanotechnology in water treatment processes *Proceedings of Nanotechnology International Forum «Rusnanotech 08»* (Moscow) 609-611

[2] Kim V E, Laguntsov N I , Karpukhin V F, Lisujk B S 1998 The method of water purification. *Patent RU* № 2114787 B.I. № 19

[3] Kurchatov I M, Laguntsov N I, Neschimenko Y P, Feklistov D Y 2015 The reagent-sorption technology of water treatment *Physics Procedia* **72** 89-92

[4] Brent Fultz 2011 Mössbauer Spectrometry *in Characterization of Materials* ed. Elton Kaufmann (John Wiley, New York)

[5] Morup S, Frandsen C, Hansen M F, 2010 Uniform excitations in magnetic nanoparticles *Belstein J. Nanotechnol. 1, 48-54*

[6] Skryarova A, 2010 Mössbauer spectra of iron oxides in the bulk and nanocrystalline states, *Lapeenranta university of technology Faculty of Technology Master's Degree Programme in Technomathematics and Technical Physic*

# Apparent anisotropy effects of upper critical field in high-textured superconducting Nb-Ti tapes

**V V Guryev[1,2,3], S V Shavkin[1], V S Kruglov[1,2], P V Volkov[1], A L Vasiliev[1], A V Ovcharov[1], I A Likhachev[1], E M Pashaev[1], R D Svetogorov[1], Y V Zubavichus[1]**

[1] National Research Centre "Kurchatov Institute", Kurchatov sq. 1, 123182 Moscow, Russia
[2] National Research Nuclear University MEPhI (Moscow Engineering Physics Institute), Kashirskoe sh. 31, 115409 Moscow, Russia

E-mail: GuryevVV@mail.ru

**Abstract**. Analysis of the voltage-field characteristics of superconducting niobium-titanium tapes is the objective of this work. We used both the original cold-rolled and heat-treated tapes, whose structure and texture we had studied in detail. We observed anisotropy in the upper critical field and the reduction of the transition width with increasing angle between the plane of tape and the direction of the magnetic field. In addition, we have registered a significant difference of the upper critical field values, obtained from the voltage-field characteristics and field dependence of the pinning force. Considering our results, we suggest an explanation of the observed features within the macro-inhomogeneity model.

## 1. Introduction

One of the key problems of applied superconductivity is electrodynamics of high pinning superconductors (HPSC) [1]. Usually the nano-sized defects (the so-called "pinning centers") are formed in HPSC during a manufacturing process. Such defects cause high current-carrying capacity and, on the other hand, they often lead to the broadening of transitions and to critical current anisotropy against magnetic field and major directions (e.g. as rolling direction, direction of broaching etc.), which is noticeable even in body-centered-cubic Nb-Ti material [2]. Obviously these features should be considered in device designing process.

It is well known that Nb-Ti is suitable and reproducible material for application and study of HPSC electrodynamics [1-3]. A complex structure of thin ribbon-like α-Ti precipitates is formed in commercial multi-filamentary Nb-Ti wires [4]. In contrast, the cold-rolled Nb-Ti tape is simpler object and thus allows investigating of native effects related to the variations of the pinning centers system. Therefore, this paper analyses changes in material microstructure and voltage-field characteristics (VFC) of cold-rolled and heat-treated Nb-Ti tapes.

## 2. Methods and samples

Two kinds of samples were investigated. First series of samples was fabricated from long-length 80-mm width and 10-μm thick cold-rolled tape of the Nb-50wt.%Ti, manufactured at the metallurgic factory in 1978 [2]. The second one was made from the same tape, and additionally annealed in high

[3] To whom any correspondence should be addressed.

vacuum at 385° C during 25 hours to precipitate non-superconducting α-Ti nano-particles. We expected that such heat treatment should change pinning system significantly.

Flat samples with bridge geometry along the rolling direction were made by laser cutting (the bridge was 1.0 mm in width and 6.7 mm in length). The VFCs of the samples were obtained by direct transport method. The sample was fixed on a rotating probe, which allowed them to rotate in an external field and change the angle $\theta$ between the sample plane and the field direction. The current direction was perpendicular to the field at all times. Measurements were carried out in liquid helium in the superconducting magnet with field up to 13 Tesla.

The criterion for the upper critical magnetic field value ($H_{c2}$) was half of the resistance in the normal state ($\rho_n$) and the width of the transition was determined by the difference in field values at 0.9 and 0.1 fracture of the $\rho_n$. VFCs were studied at various angles $\theta$ to investigate the $H_{c2}$ anisotropy.

Titan S-TWIN 80-300 (FEI) microscope was used to carry out SEM investigation and to determine the average size of the grains. The detailed texture analysis was performed using Bruker D8 Discover diffractometer with Cu-Kα radiation and was verified on Kurchatov Synchrotron Radiation Source. The set of pole figures were used to construct the orientation distribution function (ODF) [11].

## 3. Results and discussion

### *3.1. Micro-structural and textural studies*

Typical SEM images of cross-sections of cold-rolled and heat-treated tapes are shown in Figure 1. The Nb-Ti grains resemble stacks of thin long ribbons extended along the rolling direction (RD). The main grains sizes in all directions were determined by averaging of statistical data from several dozen of images. Calculated main grains sizes were: 65 nm in the direction of normal to the plane of the tape; 0.4 μm in the transverse direction; and more than 1 μm in the RD. Pronounced grain boundaries were observed only in parallel to the plane (Fig. 1). Distributions of Nb-Ti grain sizes were similar in cold-rolled and heat-treated tapes.

Figure 2 shows the pole figures of Nb-Ti phase for the reflections (110), (200) and (222) in the cold-rolled tape. Analysis of ODF showed almost uniform distribution of the orientations from {100} <110> to {112} <110>, which is typical for the rolled β-Ti alloys [5]. Heat treatment does not change the texture of the main Nb-Ti phase. Additional α-Ti reflections arise on X-ray patterns.

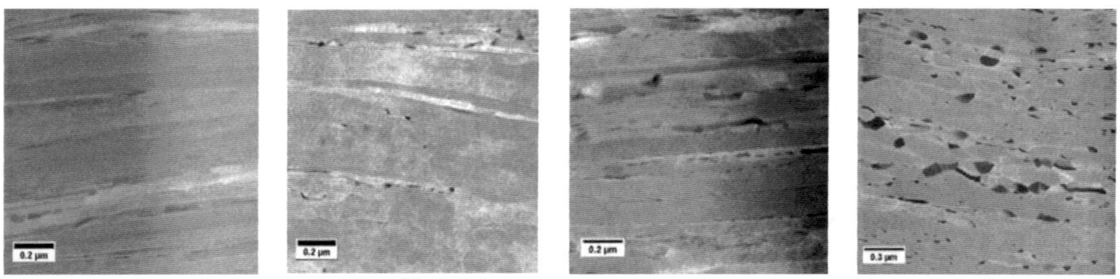

a) b) c) d)

**Figure 1**. SEM cross-section images of samples: a) cold-rolled, rolling direction is horizontal; b) cold-rolled, rolling direction is perpendicular to the image plane; c) heat-treated, rolling direction is horizontal; d) heat-treated, rolling direction is perpendicular to the image plane. Black spots are clearly visible on images of heat-treated samples and were interpreted as α-Ti precipitates.

### *3.2. Voltage – field characteristics*

Figure 3 shows the VFCs for cold-rolled and heat-treated samples at the applied field orientated perpendicular to the tape ($\theta$ = 90°) and in plane of the tape ($\theta$ = 0°). The differences in the $H_{c2}$ values for such orientations are about 0.3 T and 0.2 T for the cold-rolled and heat-treated samples. The anisotropy of $H_{c2}$ for Nb-Ti tapes (processed with complicated thermo-mechanical treatment) was also

Journal of Physics: Conference Series **747** (2016) 012034 doi:10.1088/1742-6596/747/1/012034

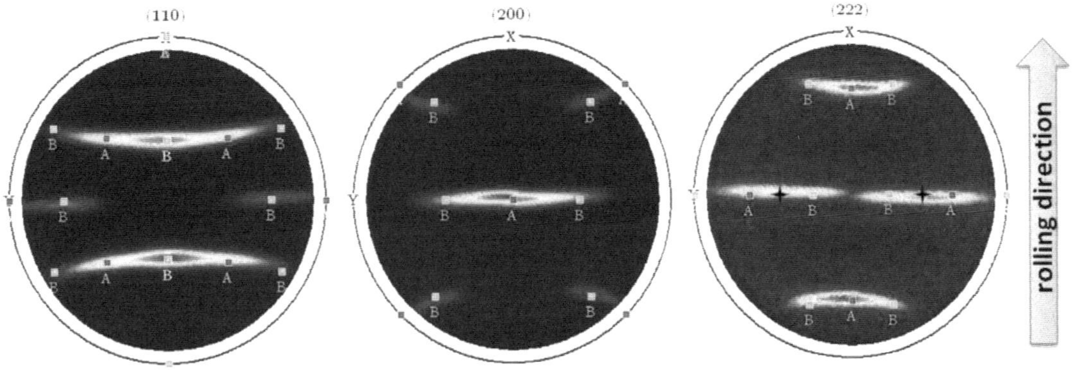


**Figure 2**. The direct pole figure for the (200), (110), (222) reflections on Nb-Ti phase. The rolling direction is vertical. One can see almost uniform distribution of the orientations from {100} <110> (corresponding to the direction marked as "A") to {112} <110> (marked as "B"). Crosses signs in the (222) pole figure correspond to maximum of intensity (is about 43° to the plane of the tape)

observed in [3] where authors suggested an explanation of the effect: the α-Ti precipitations form the weak links between the Nb-Ti grains due to their elongated shape and a small thickness comparable to the coherence length ξ. Theory [6] predicts that anisotropy arises due to weak links. This assumption is not supported by our results, as we have observed the most pronounced anisotropy in cold-rolled samples, which had very small amount of α-Ti precipitations.

Another possible source of anisotropy is high degree of texture. Indeed, in single crystals with a cubic symmetry one can expect variations in physical properties in directions [100], [110] and [111]. Such anisotropy of $H_{c2}$ was observed in the niobium single crystal [7]. According to [7], the $H_{c2}$ increases in following sequence of directions: [100] → [110] → [111]. From the pole figure (222) (Fig. 2) it is clear that at $\theta = 43^{\circ}$ the majority of the grains are oriented in the [111] direction, and one can expect the maximum value of $H_{c2}$ at this angle. However, $H_{c2}$ reaches the maximum value when the field is in the plane of the tape ($\theta = 0^{\circ}$) (see Table 1). This means that apparent $H_{c2}$ anisotropy does not depend on crystalline orientation of grain in Nb-Ti tapes.

**Table 1**. $H_{c2}$ and transition width at different angles $\theta$ for two cold-rolled samples.

| Sample# | Measuring current, mA | $H_{c2}$, T (Transition width, T) | | |
|---|---|---|---|---|
| | | $\theta = 0^{\circ}$ | $\theta = 43^{\circ}$ | $\theta = 90^{\circ}$ |
| 1 | 5 | 11.89 (0.434) | - | 11.58 (0.279) |
| | 5 | 12.06 (0.436) | - | - |
| | 1 | 12.10 (0.387) | 11.75 (0.225) | - |
| 2 | 0.1 | 12.11 (0.354) | 11.75 (0.195) | - |
| | 0.01 | 12.11 (0.347) | 11.74 (0.189) | - |
| | 0.001 | 12.12 (0.334) | 11.74 (0.183) | - |

Finally, to analyse the influence of the measuring current on $H_{c2}$, we carried out series of measurements using different currents in the range between 1 μA to 5 mA (Table 1). Both the $H_{c2}$ and transition width values do not depend significantly on the current, so we suggest that the observed anisotropy is directly linked to anisotropy of $H_{c2}$ rather than any other kind of anisotropy (e.g. critical current anisotropy).

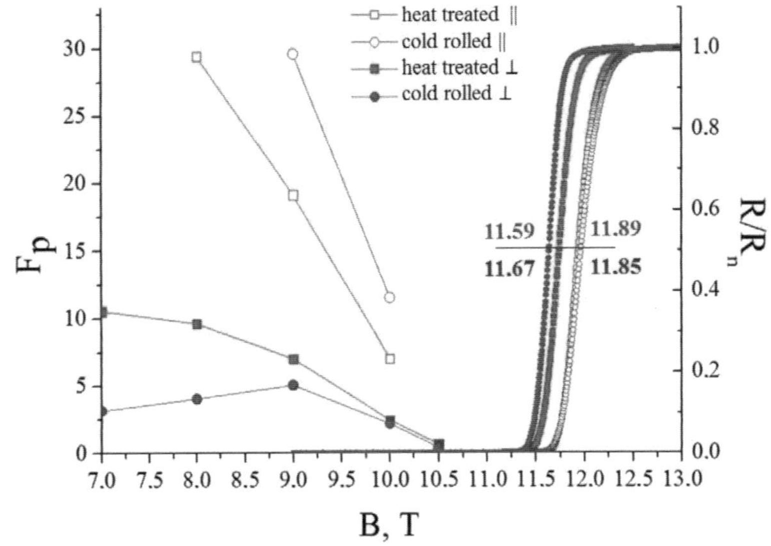


**Figure 3**. On the left: The magnetic field dependence of pinning force $F_p$. On the right: The magnetic field dependence of the normalized resistance (VFCs at measuring current 5 mA). The legend shows symbols indicating the orientation of the field and the type of samples.

Figure 3 also shows the magnetic field dependences of the pinning force $F_p$ ($F_p$ calculated as $I_c*B$, where $I_c$ was determined by a standard criterion of 1 μV/cm). Extrapolation of these curves into a higher field shows that in the absence of inflections the pinning force should equate to zero at the field above 10.7 T.

All these features can be explained by the macro-inhomogeneity model [8] where the length of an electron's free path is locally reduced at the Nb-Ti grain boundaries, which leads to reduction of the coherence length $\xi$ and increase of $H_{c2}$. The similar microscopic mechanism determines the pinning in the cold-rolled tape at the field below ~10.7 T [9]. Above this field only the Nb-Ti grain boundaries form percolating superconducting paths; while significant part of the Nb-Ti grain matrix is not superconducting. This explains the lack of dependence of $H_{c2}$ anisotropy on crystallographic orientation of grains.

From this point of view Nb-Ti tape in the high field can be roughly represented as a stack of superconducting films (grain boundaries) with thicknesses of about ξ. The distance between them is equal to the mean grain thickness $d$ (~ 65nm) which is significantly larger than ξ (~5 nm). Therefore the films are not coupled, which is in contrast with the model [3]. So observed anisotropy $H_{c2}$ is similar in nature to the well-known $H_{c2}$ anisotropy observed in thin superconducting films. As seen from Figure 1 the α-Ti precipitates are concentrated basically at the grain boundaries, violating their flatness, and reducing the anisotropy. As a result, the anisotropy is more pronounced in cold-rolled rather than heat-treated tape. This is also consistent with the experimental results of [3], where authors observed the growth of anisotropy with the increasing intensity and degree of rolling.

The macro-inhomogeneity model implies that there are at least two superconducting components in Nb-Ti tape: the grain boundaries and the grain bodies, which have different average values of $H_{c2}$. It should be noted that the method used to determine $H_{c2}$ does not give the "true" $H_{c2}$ value for grain boundaries because they are shunted by grain bodies. We cannot calculate correctly the shunting effect, since a magnetic field dependence of grains resistance as well as statistical dispersion of $H_{c2}$ in the grains are unknown. Quantitative estimation of the relationships between $H_{c2}$ values of grain bodies and grain boundaries is complicated task which is beyond the frame of this work.

The macro-inhomogeneity model also explains the augmentation in the transition width with $\theta$ reduction (see Table 1). Indeed, there is dispersion in the grain boundaries orientation in our samples (see Figure 1). In accordance with the angle dependence of $H_{c2}$ for the non-connected stack of thin films [10] the derivative $dH_{c2}/d\theta$ increases quickly with the angle reduction. So for the fixed dispersion $\delta\theta$ (i.e. fixed misorientation degree of grain surfaces) corresponding dispersion of the upper critical field $\delta H_{c2}$ increases with decreasing of angle $\theta$.

## 4. Conclusion

In this study the upper critical field anisotropy in Nb-Ti cold-rolled and heat-treated tapes was observed by the direct transport method. The apparent anisotropy is not related to the crystallographic texture and has been observed in both kinds of samples. It was shown that this anisotropy is also not directly related to α-Ti precipitations. The extrapolated upper critical field obtained from the field dependence of pinning strength is of about 1 Tesla lower than the one derived from voltage - field characteristics. The transition width decreases with the increasing angle between the tape and the direction of the magnetic field. All these features can be explained by the macro-inhomogeneity model. It assumes that in the high magnetic field near $H_{c2}$ only boundaries of Nb-Ti grains are in superconducting state. So Nb-Ti tape can be roughly represented as a stack of thin non-coupled superconducting films with thicknesses of about the coherence length.

## Acknowledgments

The authors are very grateful to Prof. E. Yu. Klimenko for contributing to the discussions of the results obtained in the work. The authors would also like to thank the management of NRC "Kurchatov institute" for the support. The work was also partly supported by RFBR grant #16-32-00224. The SEM and X-ray measurements were performed on equipment of the Resource Center of Probe and Electron Microscopy, Resource Center of X-ray methods (Kurchatov Complex of NBICS-Technologies, NRC "Kurchatov institute") and on Kurchatov Synchrotron Radiation Source.

## References

[1] Klimenko E Yu 2011 Electrodynamics of High Pinning Superconductors *Superconductivity Theory and Applications* ed A M Luiz (Rijeka: InTech) chapter 15 pp. 329-346
[2] Klimenko E Yu, Shavkin S V and Volkov P V 1997 *Journal of Experimental and Theoretical Physics* **112**(3) 1055-1081
[3] Cooley L and Patel A 2001 *IEEE Trans. On App. Supercon.* **11** 3792-3795
[4] Cooley L D and Motowidlo L R 1999 *Supercond. Sci. Technol.* **12** 135–151
[5] Gepreel M 2013 Texturing Tendency in β-Type Ti-Alloys *Recent developments in the study of* ed P Wilson (Rijeka: InTech) chapter 4
[6] Takahasi S and Tachiki M 1986 *Physical review* B **7** 4620-4631
[7] Williamson S J 1970 *Physical review* B **2** 3545-3556
[8] Klimenko E Yu, Shavkin S V and Volkov P V 2001 *Phys. Met. Metallogr.* **92** 552-556
[9] Yetter W E, Thomas D A and Kramer E J 1982 *Phil. Mag. B* **46** 523-537
[10] Takezawa N, Koyama T and Tachiki M 1993 *Physica* C **207** 231-238
[11] Hielscher R, Schaeben H 2008 *J. of Appl. Cryst.* **41** 1024-1037

Journal of Physics: Conference Series **747** (2016) 012035 doi:10.1088/1742-6596/747/1/012035

# Restoration of density of states for FeAs-based superconductors

**V A Kashurnikov, A V Krasavin[1] and Ya V Zhumagulov**

National Research Nuclear University MEPhI (Moscow Engineering Physics Institute), Kashirskoye sh. 31, 115409 Moscow, Russia

E-mail: avkrasavin@gmail.com

**Abstract**. The spectral and total electron density of states were calculated for two-dimensional FeAs-clusters modeling iron-based superconductors, with use of the generalized quantum Monte Carlo algorithm within the limits of the two-orbital model. The spectra have been obtained by means of the stochastic procedure, which was modified to restore the kernel of the integral equation relating the Matsubara Green's function and the spectral density. The calculations were made for clusters with sizes up to 10×10 FeAs-cells. The data on the total density of states near the Fermi level are obtained. There are two branches of the quasiparticle spectrum on both sides of the Fermi level. The effect of the values of the interaction parameter, the cluster size and temperature on the excitation spectra is discussed.

## 1. Introduction

One of the most interesting objects of research in the physics of superconductivity nowadays is iron-based high-temperature superconductors (HTSC) [1–3]. Similar to the copper-based HTSC, these compounds have a layered structure. However, due to the features of the multi-gap band structure their phase diagram is more complicated and includes antiferromagnetic, structural, and superconducting phase transitions [4–8].

The pronounced anisotropy of iron-based HTSC allows applying two-dimensional tight-binding models for their description. One of the most appropriate model for these compounds is the two-orbital model [9, 10], which is a typical multi-band generalized Hubbard model. The lack of expansion parameters for analytical approximations, and the presence of strong Coulomb correlations require the use of exact numerical quantum methods for the study of iron-based HTSC.

The present work focuses on the calculation of the spectrum of elementary excitations which determines the Fermi surface and the density of states. The quantum continuous time world line Monte Carlo (MC) algorithm (CTWL-algorithm) was used to simulate the full two-orbital model [11], which allows calculation of the Matsubara Green's function and obtaining the quasiparticle spectrum and its dependence on the system size and interaction parameters.

## 2. Model

The Hamiltonian of the two-orbital model is as follows [9, 10]:

$$H = H_{in} + H_{kin} \tag{1}$$

[1] To whom any correspondence should be addressed.

$$H_{in} = U\sum_{i,\alpha} n_{i,\alpha,\uparrow} n_{i,\alpha,\downarrow} + V\sum_{i} n_{i,x} n_{i,y} - \mu\sum_{i} n_{i} - J\sum_{i}\left(n_{i,x,\uparrow} n_{i,y,\uparrow} + n_{i,x,\downarrow} n_{i,y,\downarrow}\right)$$

$$-J\sum_{i}\left(a^{+}_{i,x,\downarrow} a_{i,x,\uparrow} a^{+}_{i,y,\uparrow} a_{i,y,\downarrow} + a^{+}_{i,x,\uparrow} a_{i,x,\downarrow} a^{+}_{i,y,\downarrow} a_{i,y,\uparrow} + a^{+}_{i,x,\uparrow} a_{i,y,\downarrow} a^{+}_{i,x,\downarrow} a_{i,y,\uparrow} + a^{+}_{i,y,\uparrow} a_{i,x,\downarrow} a^{+}_{i,y,\downarrow} a_{i,x,\uparrow}\right)$$

$$H_{kin} = -t_1\sum_{i,\sigma}\left(a^{+}_{i,x,\sigma} a_{i+\hat{x},x,\sigma} + a^{+}_{i,y,\sigma} a_{i+\hat{y},y,\sigma}\right) - t_2\sum_{i,\sigma}\left(a^{+}_{i,y,\sigma} a_{i+\hat{x},y,\sigma} + a^{+}_{i,x,\sigma} a_{i+\hat{y},x,\sigma}\right)$$

$$-t_3\sum_{i,\sigma}\left(a^{+}_{i,x,\sigma} a_{i+\hat{x}+\hat{y},x,\sigma} + a^{+}_{i,x,\sigma} a_{i+\hat{x}-\hat{y},x,\sigma} + a^{+}_{i,y,\sigma} a_{i+\hat{x}+\hat{y},y,\sigma} + a^{+}_{i,y,\sigma} a_{i+\hat{x}-\hat{y},y,\sigma}\right)$$

$$+t_4\sum_{i,\sigma}\left(a^{+}_{i,x,\sigma} a_{i+\hat{x}-\hat{y},y,\sigma} + a^{+}_{i,y,\sigma} a_{i+\hat{x}-\hat{y},x,\sigma} - a^{+}_{i,x,\sigma} a_{i+\hat{x}+\hat{y},y,\sigma} - a^{+}_{i,y,\sigma} a_{i+\hat{x}+\hat{y},x,\sigma}\right) + h.c.$$

Here $a^{+}_{i,x(y),\sigma}\left(a_{i,x(y),\sigma}\right)$ creates (annihilates) an electron with spin $\sigma$ on site $i$ and orbital $x(y)$; $t_i, i = 1,\ldots,4$ are the hopping amplitudes between $d_{xz}$ and $d_{yz}$ orbitals defined as *x* and *y*, respectively; *U* and *V* are the parameters of the Coulomb interaction within and between orbitals; *J* is the exchange integral; μ is the chemical potential.

The parameters of $H_{kin}$ in (1) were chosen similarly to [10]:

$$t_1 = 0.058;\ t_2 = 0.22;\ t_3 = -0.0208;\ t_4 = -0.079; \tag{2}$$

the relationship between the parameters of $H_{in}$ was defined as following:

$$V = 0.5U;\ J = 0.25U. \tag{3}$$

Hereafter, all the energy parameters are normalized to the value of 0.2 eV.

## 3. Results

The Matsubara Green's function

$$\left\langle T_{\tau} a_{i\alpha\sigma}(\tau) a^{+}_{j\beta\sigma'}(0)\right\rangle, \tag{4}$$

where $i, j$ are coordinates of iron atoms; $\alpha, \beta$ are orbitals; $\sigma, \sigma'$ are spin projections, was calculated for clusters of sizes up to 10×10 FeAs cells. To properly restore the spectral density $A_{\sigma}(k,\omega)$ from data on the Matsubara Green's function, and to calculate the total density of states $N(\omega) = \sum_{k,\sigma} A_{\sigma}(k,\omega)$ we must determine the kernel of the integral equation

$$G_{\sigma}(k,\omega) = -\int \frac{A_{\sigma}(k,\omega) e^{-\tau\omega}}{1+e^{-\beta\omega}} d\omega, \tag{5}$$

$\beta = 1/T$, $T$ is the temperature. The problem (5) is a typical ill-posed problem; its solution is a non-trivial mathematical task, and a method of direct iterations usually does not give the correct result if no information about a kernel is provided.

The density of states $N(\omega)$ was restored for the whole Brillouin zone using the combined method "Monte Carlo + gradient descent", first introduced in [12] and adapted by us for the Green's function of the two-orbital model. The input parameter for (5) were the data for the Matsubara Green's function

(4), obtained with the use of the CTWL-algorithm [11]. It was found that for clusters with sizes greater than 4×4 the results are no longer dependent on the size of the system, which suggests the applicability of the data presented for the analysis of the properties of a macroscopic system.

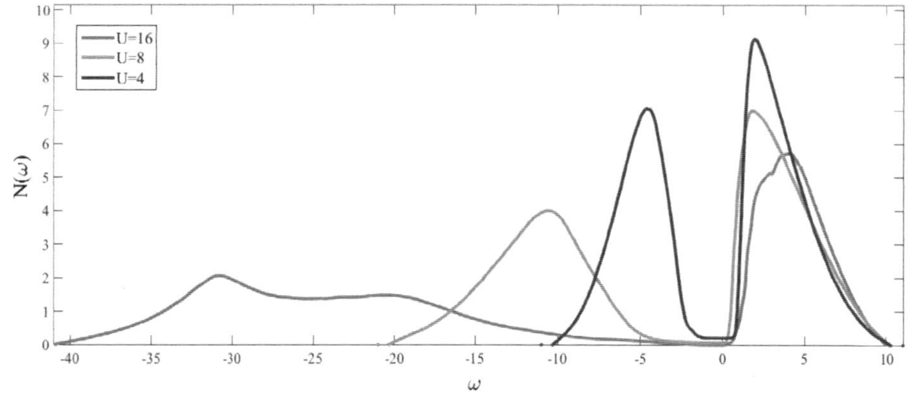


**Figure 1.** The dependence of the electron states density on the value of the interaction parameter $U$ calculated for cluster 8×8 and temperature $\beta = 20$.

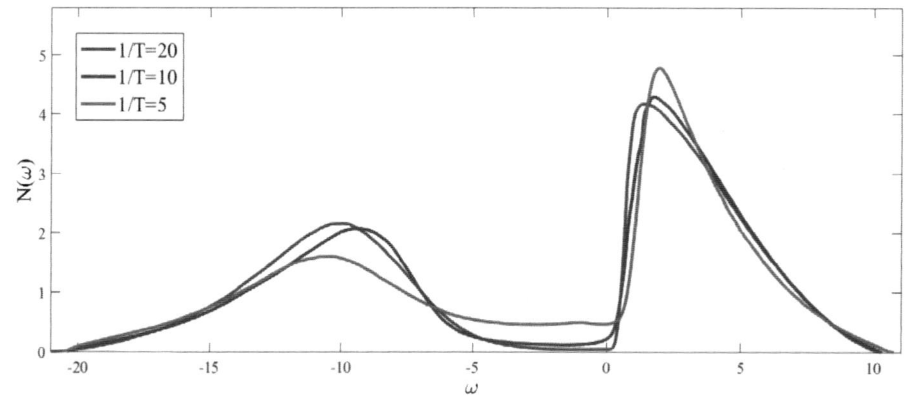


**Figure 2.** The dependence of the electron states density on temperature calculated for cluster 6×6 and $U = 8$.

Figure 1 shows the dependence of the density of states on the interaction parameter $U$ of the model (3). The Fermi energy corresponds to zero on the energy scale. There are two zones above and below the Fermi level. The density of states is weakly dependent on the interaction parameter near the Fermi level. The increase of the interaction parameter almost does not change the electron zone above the Fermi level, only slightly smoothing the sharp peak at the maximum. For the hole zone located substantially lower on the energy scale, the changes are more significant. The maximum is smoothed and shifted by the value of $\sim U$ down the energy scale, the zone itself is broadened, and the gap in the spectrum increases. Similar behavior is observed for the upper and lower zones in the standard Hubbard model, which is also mentioned in LDA + DMFT calculations [13]. From this perspective, it can be assumed that FeAs compounds belong to the strongly correlated systems.

Figure 2 shows the density of states as a function of temperature. The zones are separated by the "gap", a quite pronounced decrease in the value of the total density. In the gap, however, the total density differs from zero, indicating that undoped FeAs are metals; this fact is confirmed experimentally [4, 14]. The dependence of the total density of states on temperature at $\beta \leq 10$ is weak. When the temperature rises above this value, the "gap" begins to close.

## 4. Conclusions

Calculations of FeAs clusters of size up to 10×10 within the limits of the full two-orbital model were performed using quantum Monte Carlo algorithm at half-filling and different values of temperature and the interaction parameter. The spectral density was restored from Matsubara Green's function with the use of the combined method "Monte Carlo + gradient descent". The obtained data on the excitation spectra and the density of states indicate the possibility of a strong correlation between characteristics of the spectrum and the strength of interaction, in qualitative agreement with the available experimental data and results of other calculations.

It is shown that the density of states weakly depends on the parameters of the model near the Fermi level, but it is strongly dependent on the interaction far from it for the case of hole filling. There is a "gap" in the spectrum; at the Fermi level between the zones the density of electron states decreases sharply, remaining, however, finite, which indicates the metallic state of undoped FeAs-systems.

**Acknowledgements**

The work was supported by the Russian Science Foundation (project # 14-22-00098).

**References**

[1] Kamihara Y, Watanabe T, Hirano M, and Hosono H 2008 *J. Am. Chem. Soc.* **130** 3296
[2] Isyumov Yu, Kurmaev E 2010 *High-$T_c$ Superconductors Based on FeAs Compounds* Springer
[3] Dagotto E 2013 *Rev. Mod. Phys.* **85** 849
[4] Sadovskii M V 2008 *Physics-Uspekhi* **51** 1201
[5] Ishida K, Nakai Y, and Hosono H 2009 *J. Phys. Soc. Jpn.* **78** 062001
[6] Sadovskii M V, Kuchinskii E Z, Nekrasov I A 2012 *Journal of Magnetism and Magnetic Materials* **324** 3481
[7] Medvedev M V, Nekrasov I A, Sadovskii M V 2012 *JETP Lett.* **95** 37
[8] Nekrasov I A, Sadovskii M V2014 *JETP Lett.* **99** 598
[9] Raghu S, Qi Xiao-Liang, Liu Chao-Xing *et al.* 2008 *Phys. Rev.* B **77** 220503(R)
[10] Moreo A, Daghofer M, Riera J A, and Dagotto E 2009 *Phys. Rev.* B **79** 134502
[11] Kashurnikov V A, Krasavin A V 2013 *JETP Lett.* **97** 333
[12] Mishchenko A S 2005 *Physics-Uspekhi* **48** 887
[13] Nekrasov I A, Sadovskii M V 2014 *JETP Lett.* **99** 598
[14] Liang S, Alvarez G, Sen C, Moreo A, and Dagotto E 2012 *Phys. Rev. Lett.* **109** 047001

Journal of Physics: Conference Series **747** (2016) 012036 doi:10.1088/1742-6596/747/1/012036

# Interactive graphical system for small-angle scattering analysis of polydisperse systems

**P V Konarev[1,3], V V Volkov[1] and D I Svergun[2]**

[1] Laboratory of Reflectometry and Small-angle Scattering, A.V. Shubnikov Institute of Crystallography, Russian Academy of Sciences, Leninsky pr. 59, 119333 Moscow, Russia
[2] European Molecular Biology Laboratory, Hamburg Outstation, Notkestrasse 85, 22607 Hamburg, Germany

E-mail: konarev@ns.crys.ras.ru

**Abstract**. A program suite for one-dimensional small-angle scattering analysis of polydisperse systems and multiple data sets is presented. The main program, *POLYSAS*, has a menu-driven graphical user interface calling computational modules from *ATSAS* package to perform data treatment and analysis. The graphical menu interface allows one to process multiple (time, concentration or temperature-dependent) data sets and interactively change the parameters for the data modelling using sliders. The graphical representation of the data is done via the Winteracter-based program *SASPLOT*. The package is designed for the analysis of polydisperse systems and mixtures, and permits one to obtain size distributions and evaluate the volume fractions of the components using linear and non-linear fitting algorithms as well as model-independent singular value decomposition. The use of the *POLYSAS* package is illustrated by the recent examples of its application to study concentration-dependent oligomeric states of proteins and time kinetics of polymer micelles for anticancer drug delivery.

## 1. Introduction

Small-angle scattering (SAS) is an effective technique for structural characterization of compounds of different nature: gels, films, microemulsions, solutions of biological macromolecules, polymers, nanoparticles. Especially SAS method has advantages for the studies of biological macromolecules as it does not require special sample preparation, allows one to measure the samples at native physiological conditions and is applicable in the wide range of molecular weights (from 5 kDa to 50 MDa). The progress in instrumentation and software development during two last decades permitted solution SAS technique to solve important questions about the structure of macromolecules and their biological functions. At the same time it enormously increased the amount of the measured data and thus the efforts for the data analysis.

There are a number of software packages developed for the analysis of SAS data, such as ATSAS [1,2], GIFT [3], IRENA [4], IMP [5], etc. They permit to obtain low resolution shapes of the macromolecules, refine the quaternary structure of macromolecular complexes, restore missing portions and characterize protein flexibility, estimate size distributions of the particles. However, most

[3] To whom any correspondence should be addressed.

of the programs deals with single experimental data curves and for multiple data sets (e.g. time, concentration or temperature scans) the data processing becomes tedious.

Here we present a program suite for one-dimensional small-angle scattering analysis of polydisperse systems. It includes the support for processing of multiple data sets as well as the possibility for interactive modeling.

## 2. The graphical interactive system *POLYSAS* and its functionality

The main program, *POLYSAS*, has a menu-driven graphical user interface calling computational modules from *ATSAS* package to perform data treatment and analysis. The graphical menu interface allows one to process multiple (time, concentration or temperature-dependent) data sets and interactively change the parameters for the data modeling using sliders. The graphical representation of the data is done via the Winteracter-based program *SASPLOT* [6].

For multiple data sets POLYSAS permits to automatically estimate overall structural parameters such as the radius of gyration ($R_g$), the maximum size of the particle ($D_{max}$), the molecular weight (*MW*) and the excluded hydrated volume of the particle ($V_{Porod}$). For these purposes POLYSAS calls the computational modules DATRG, DATGNOM and DATPOROD [7] from ATSAS package. Estimated parameters are plotted using SASPLOT in real time and can be easily saved in an ASCII format or as a picture file. The list of the loaded experimental curves can be modified and reprocessed using the buttons in the graphical menu.

In the case of polydisperse systems POLYSAS can provide the size distributions of the particles. Depending on the type of the particles it estimates the following size distributions: 1) distance distributions functions $p(r)$ for the monodisperse particles of arbitrary shape; 2) volume size distributions of spherical particles; 3) cross-section size distributions for the rod-like particles; 4) length size distributions for cylindrical particles. It is performed by calling GNOM module [8] from ATSAS package. The smearing effects for different geometrical setups can be taken into account and the desmeared curves can be obtained. The graphical menu of POLYSAS allows one to interactively change the structural parameters of the particle and/or the geometrical parameters of the experimental setup and monitor how it influences the quality of the fitted curve and the calculated size distribution function.

There are several options for the analysis of mixtures in the POLYSAS. In the case when the high resolution models of the mixture components are available (e.g. in PDB format), POLYSAS permits to fit experimental data sets with the linear combinations of the scattering curves calculated from the selected PDB models. The volume fractions of the components are calculated using OLIGOMER module [6] from ATSAS package and the results are displayed via SASPLOT.

POLYSAS has the option for the non-linear analysis of the multi-component systems using the simple geometrical bodies of three types (spheres, cylinders and dumb-bells) and for maximum of five components. It is possible to take into account the polydispersity effects and interparticle interactions (the latter can be applied for the spherical particles within sticky hard sphere potential approximation). The refined parameters are calculated by MIXTURE module [6] from ATSAS package. The parameters of the models can be interactively changed using the sliders and spin buttons and the fit from the model to the experimental data will be updated. The upper and lower limits of the parameters for each component can be set before the minimization process. If the upper and lower limits coincide with each other, the corresponding parameter will be fixed.

The model independent analysis using the singular value decomposition (SVD) method is also available in POLYSAS. It can be applied for the systems measured at different conditions (different pH, temperature, buffer type, sample and salt concentration, ligand addition etc.). The eigenvalues and the eigenvectors of the SVD method are calculated and the number of independent components in the system is estimated using the statistical test for the non-random oscillations of the eigenvectors. The SVDPLOT module [6] from ATSAS package is called by POLYSAS. The graphical menu of the POLYSAS provides options for interactive selection of the full data sets and partial data subsets of the experimental curves and allows one to check the stability of the estimated number of components in the system.

## 3. Applications of POLYSAS to experimental data

The usage of POLYSAS can be illustrated on several experimental *X*-ray data sets that were recorded in collaborative user projects on the X33 beamline of the EMBL [9] at the storage ring DORIS-III (DESY, Hamburg).

The first example concerns the studies of the complex equilibrium mixture of the AXH domain of the Ataxin-1 protein [10]. Ataxin-1 is a human protein responsible for ataxia type 1, a hereditary disease associated with protein aggregation and misfolding. The AXH domain of Ataxin-1 displays a dimer of dimers arrangement in the crystal asymmetric unit. In solution, the domain is present as a complex equilibrium mixture of monomeric, dimeric, and higher molecular weight species. This behavior, together with the tendency of the AXH fold to be trapped in local conformations, makes the AXH domain an unusual example of a chameleon protein whose properties bear potential relevance for the aggregation properties of ataxin-1 and thus for disease.

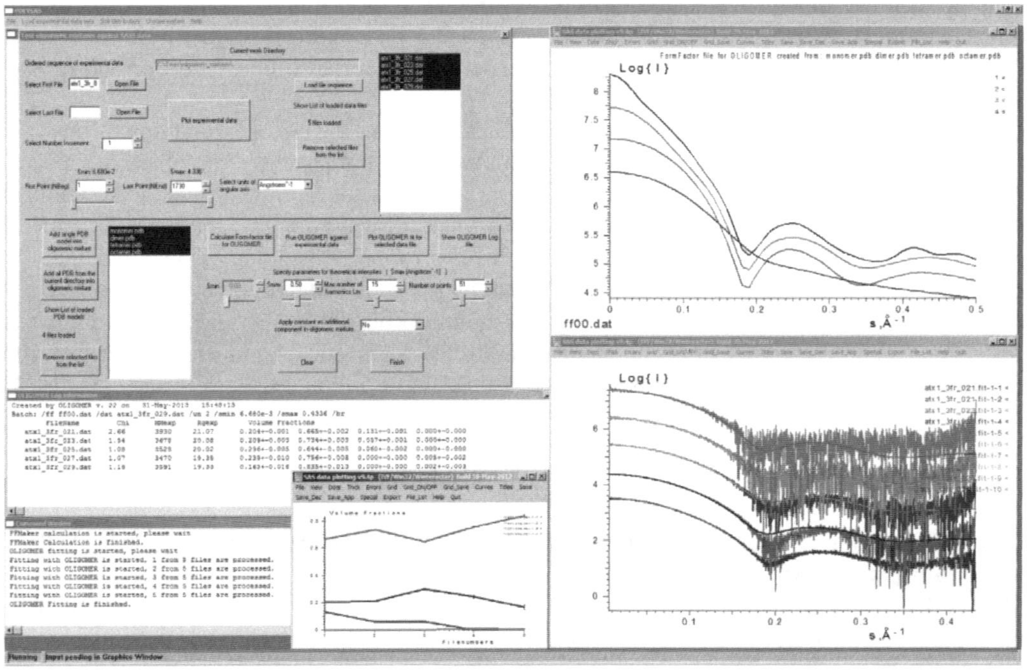


**Figure 1.** POLYSAS GUI for SAXS data modelling of Ataxin-1 using an oligomeric mixture

In Figure 1 the graphical user interface of POLYSAS is presented for modelling of Ataxin-1 in solution as an oligomeric mixture. The graph with the theoretical curves calculated from PDB models (monomers, dimers, tetramers and octamers) is displayed in the top-right corner of the screen, the graph with the fits to the experimental data set is shown in the bottom-right corner of the screen and the graph with the estimated volume fractions of the components is displayed in the bottom-left corner. The usage of POLYSAS enables to quantitatively characterize the system and provides the opportunity for interactive monitoring of different hypotheses during the data modelling.

The second example is devoted to time-dependent studies of hydrolytically degradable polymer micelles intended for controlled drug delivery [11]. Highly hydrophobic cholesterol moieties as well as the anticancer drug doxorubicin were attached to the polymer backbone by a pH-sensitive hydrazone bond. Time-dependent SAXS measurements were performed after changing pH from a typical blood value (pH=7.2) to that of tumor cells (pH=5.0) to characterize the drug release and changes in particle size and shape. It was found that nanoparticles composed of the conjugates containing Dox were generally bigger than the drug-free ones. For most conjugates, nanoparticle growth or decay was observed in the time range of several hours. The growth/decay rate and the steady-state size of nanoparticles depended on the spacer structure.

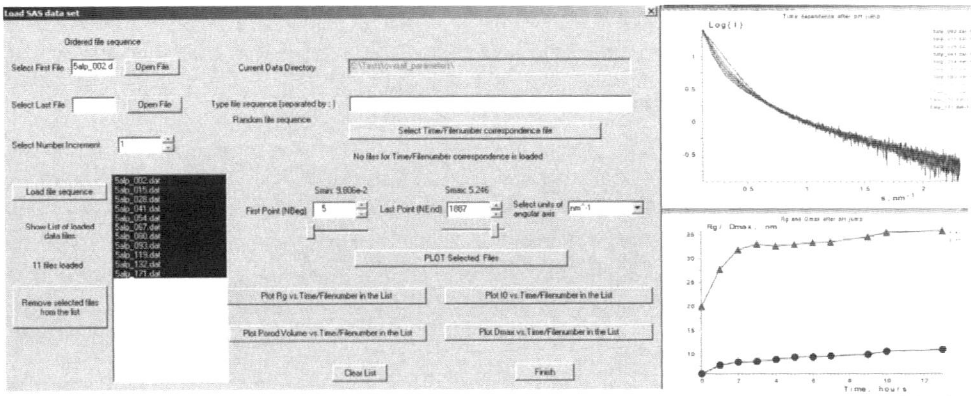


**Figure 2.** POLYSAS GUI for time-dependent SAXS data studies of hydrolytically degradable polymer micelles bearing the anticancer drug

In Figure 2 the graphical user interface of POLYSAS is shown that was applied for modelling of time dependence of the overall parameters (the radius of gyration $R_g$ and the maximum size of the particles $D_{max}$) of polymer micelles with the anticancer drug. The graph with the experimental time dependent data set is displayed in the top-right corner of the screen, the time dependence of $R_g$ and $D_{max}$ is shown in the bottom-right corner of the screen. This dependence was automatically obtained by clicking on a single button in POLYSAS GUI.

Thus, we have demonstrated that POLYSAS provides enhanced and convenient options for structural studies of macromolecular solutions using small-angle scattering data. The program POLYSAS is included in the ATSAS package (http://www.embl-hamburg. de/biosaxs/software.html), is freely available to academic users, together with other ATSAS programs as from the 2.6 release.

**Acknowledgements**
The authors acknowledge the support of the Bundesministerium fur Bildung und Forschung (BMBF), project BIOSCAT, grant No. 05K2091.

**References**

[1] Konarev P V, Petoukhov M V, Volkov V V and Svergun D I 2006. *J. Appl. Cryst.* **39** 277
[2] Petoukhov M V, Franke D, Shkumatov A V, Tria G, Kikhney A G, Gajda M, Gorba C, Mertens H D T, Konarev P V and Svergun D I 2012 *J. Appl. Cryst.* **45** 342
[3] Weyerich B, Brunner-Popela J and Glatter O 1999 *J. Appl. Cryst.* **32** 197
[4] Ilavsky J and Jemian P R 2009. *J. Appl. Cryst.* **42** 347
[5] Russel D, Lasker K, Webb B, Schneidman D, Velázquez-Muriel J and Sali A 2012 *PLoS Biol.* **10** e1001244
[6] Konarev P V, Volkov V V, Sokolova A V, Koch M H J and Svergun D I 2003 *J. Appl. Cryst.* **36** 1277
[7] Svergun D I 1992 *J. Appl. Cryst.* **25** 495
[8] Petoukhov M V, Konarev P V, Kikhney A G and Svergun D I 2007 *J. Appl. Cryst.* **40** s223
[9] Blanchet C E, Zozulya A V, Kikhney A G, Franke D, Konarev P V, Shang W, Klaering R, Robrahn B, Hermes C, Cipriani F, Svergun D I and Roessle M 2012 *J. Appl. Cryst.* **45** 489
[10] de Chiara C, Rees M, Menon R P, Pauwels K, Lawrence C, Konarev P V, Svergun D I, Martin S R, Chen Y W and Pastore A. 2013 *Biophys. J.* **104** 1304
[11] Filippov S K, Franklin J M, Konarev P V, Chytil P, Etrych T, Bogomolova A, Dyakonova M,

Papadakis C M, Radulescu A, Ulbrich K, Stepanek P and Svergun D I, 2013 *Biomacromol.* **14** 4061

# Regions of the nanoparticle confinement by the electrodynamic linear Paul trap

**D S Lapitsky**[1]
Joint Institute for High Temperatures of the Russian Academy of Sciences, Izhorskaya st. 13 Bd.2, 125412 Moscow, Russia

E-mail:

**Abstract.** The possibility of the nanoparticles confinement by the electrodynamic Paul trap is shown. The areas of a nanoparticle confinement in gas flow are found. The electric potential of a nanoparticle for its capturing in the Paul trap should be of order 0.5 V to 150 V for particle sizes from 0.03 nm to 1 μm.

## 1. Introduction

Nanosized particles are present in atmospheric aerosols and astrophysical dusty plasmas [1, 2, 3]. Recent investigations have demonstrated a strong correlation between the presence of aerosols in the atmosphere and their effect on climate parameters and the quality of life [4, 5]. Therefore, a removal of charged particles is a very important issue. One way of filtering is a particle charging by the corona discharge and their deposition on the precipitation electrodes. For example, in electrostatic filters particles gain charge in the corona discharge and then are accumulated by the electrodes [6]. However, electrostatic filters are not efficient in capturing particles with sizes of 0.6–1.6 microns. Unfortunately, the problems of a selective particles removal cannot be solved by the corona discharge precipitators.

The other possibility of charged particles capturing is their confinement in alternating electric fields generated by electrodynamic traps [7]. The observation of dynamic traps for the charged particles confinement is presented in [8]. The analysis of charged particle motion in alternating electric fields is presented in [9].

In previous works [10, 11, 12, 13] the confinement of micron sized particles by the alternating electric fields has been studied in static gas media and in gas flows.

The goal of this work is a theoretical study of charged nanoparticles confinement by the electrodynamic Paul trap [7] in gas flows. The regions of the nanoparticle and trap parameters necessary for the particles confinement are investigated at normal conditions.

## 2. Mathematical simulation of charged nanoparticle dynamics in Paul trap in gas flows

The sketch of a linear Paul trap is presented in Figure 1*a*. The trap consists of four cylindrical electrodes with radius $R_1 = 1.5$ mm and length $L_m = 6$ cm. The alternating voltage is applied to electrodes: $U_\omega \sin(\omega t + \pi)$ to pair electrodes with number 1 and $U_\omega \sin(\omega t)$ to those with number 2, where $U_\omega$ is the alternating voltage magnitude, $\omega = 2\pi f$ and $f$ is the alternating voltage frequency. The distance between the axes of the neighboring electrodes varies from $L = 8$ mm to 28 mm. To simulate the charged nanoparticle dynamics in the trap and to find the regions of nanoparticle confinement the Brownian dynamics has been used. The simulations took into account stochastic forces of random collisions with

[1] To whom any correspondence should be addressed.

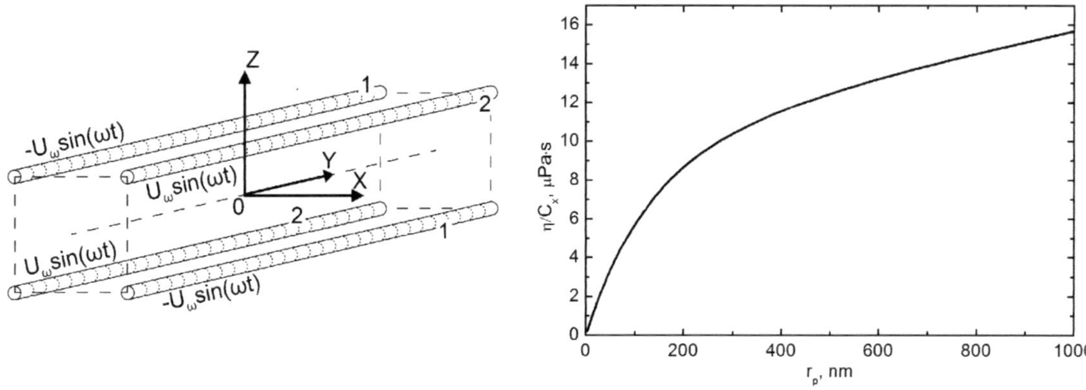


a) b)

**Figure 1.** (a) scheme of Paul trap; (b) the dynamic viscosity with the relevant Cunningham correction factor $\eta/C_x$.

neutral particles, viscosity of the gas medium, regular forces of the trap electrodes and the gravitational force. Thus, the particle dynamics was described by the following Langevin equation [14]:

$$m_p \frac{d^2 r}{dt^2} = F_t(r) - 6\pi \frac{\eta}{C_x} r_p \left( \frac{dr}{dt} - \upsilon_f \right) + F_b + F_g \tag{1}$$

where $r$ is the radius-vector of particle, $r_p$ is the particle radius, $m_p$ is the particle mass in the assumption of spherical particles ($m_p = \frac{4}{3}\pi r_p^3 \rho_p$), in simulations $\rho_p = 3700$ kg/m$^3$, $\eta$ is the dynamic viscosity of gas medium (18.2 μPa·s [15]), $C_x$ is the Cunningham factor [16], $\upsilon_f$ is the gas flow velocity, $F_t(r)$ is the force of trap electrodes, $F_b$ are stochastic delta-correlated forces accounting for stochastic collisions with neutral particles, $F_g$ is the gravitational force. To solve the stochastic differential equation (1) the numerical method developed in [17] was used.

To simulate nanoparticle dynamics in the air at the normal conditions the Cunningham correction factor $C_x$ was taken into account as the dependency on particle size. In Figure 1*b* the dependence of $\eta/C_x$ on the particle size is presented obtained from the data in [16]. For the micron sized particles the Cunningham factor is close to 1.

To simulate the interaction of the electric field of the trap with a charged nanoparticle the model of point charges distributed along each electrode [14] was used.

Figure 2 presents the lower bounds of the regions of nanoparticle confinement for the charged nanoparticle for gas flow velocities 1, 5 and 10 cm/s at different interelectrode distances $L$. As the $F_t(r)$ depends on the $q_p U/L$ ($U = 2U_\omega$) [14], the lower bounds of confinement regions are presented as the dependencies of the expression $q_p U/L$ on $f$, where $U = 2U_\omega$. From Figure 2 one can see that at increasing of the gas flow velocity the minimum frequency of alternating voltage for particle catching increases. To catch nanoparticles in the gas flow with the velocity of 1 cm/s the frequency of the alternating voltage should be higher than 150 Hz (Figure 2*a*). In gas flows with the velocity of 5 cm/s the lowest frequency should be 400 Hz and for the velocity of 10 cm/s the frequency should be higher than 700 Hz. To catch micronsized particles the frequency $f$ should be one and a half times lower. Also from Figures 2*a–c* one can see that at increasing the interelectrode distance $L$ the frequency $f$ for particle cathing becomes lower.

From Figure 2 it is possible to find the potentials of nanoparticle needed for its confinement in the Paul trap for a given magnitude of the electric field strength ($2U_\omega/L$). For example, for the electric field strength of 20 kV/cm in Figure 3 the lower bounds of particle potentials $U_p$ are presented.

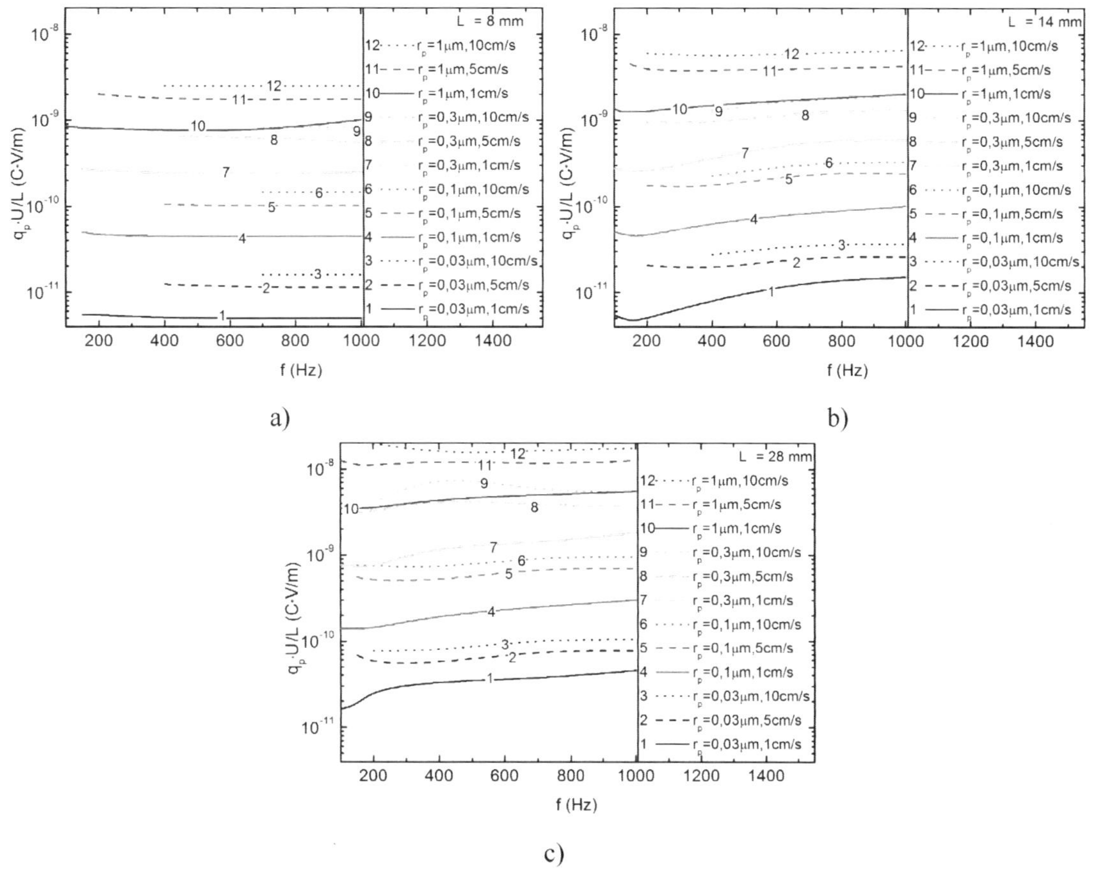


**Figure 2.** The lower bounds of the nanoparticle confinement for different particle sizes and gas flow velocities in trap with (*a*) $L$ = 8 mm, (*b*) $L$ = 14 mm and (*c*) $L$ = 28 mm.

Figure 4 shows that at increasing the interelectrode distance the particle electric potential for its confinement rises despite of the same value of electric field strength in trap $2U_{\omega}/L = 20$ kV/cm.

## 3. Conclusion

The possibility of the nanoparticle confinement by the electrodynamic Paul trap was shown. The areas of the nanoparticle confinement in gas flow were found. The electric potential of the nanoparticle for its capturing in the Paul trap should be of order 0.5 V to 150 V for particle sizes from 0.03 nm to 1 μm and electric field potential up to 20 kV/cm. As the interelectrode distance increases the particle electric potential for its confinement rises. Also as the gas flow velocity increases, the frequency of alternating voltage needed for particle capturing grows as well. To catch particles in gas flow with 1 cm/s the frequency of alternating voltage should be higher than 100 Hz, gas flows 2 cm/s the lowest frequency should be 200 Hz and for 10 cm/s – 400 Hz.

**Acknowledgments**
The work was done by the financial support of the Russian Science Foundation via grant 14-50-00124.

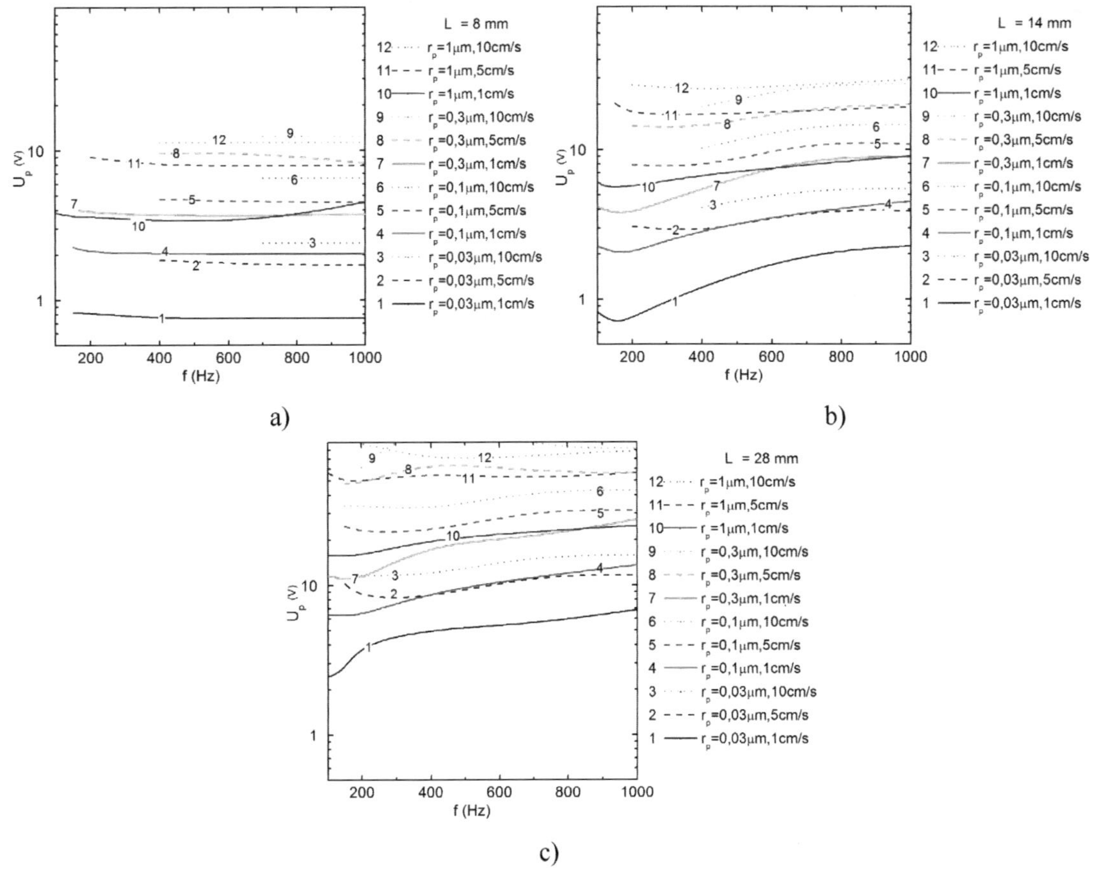


**Figure 3.** The lower bounds of the electric potentials of the nanoparticle for different particle sizes and gas flow velocities in trap with (*a*) $L = 8$ mm, (*b*) $L = 14$ mm and (*c*) $L = 28$ mm.

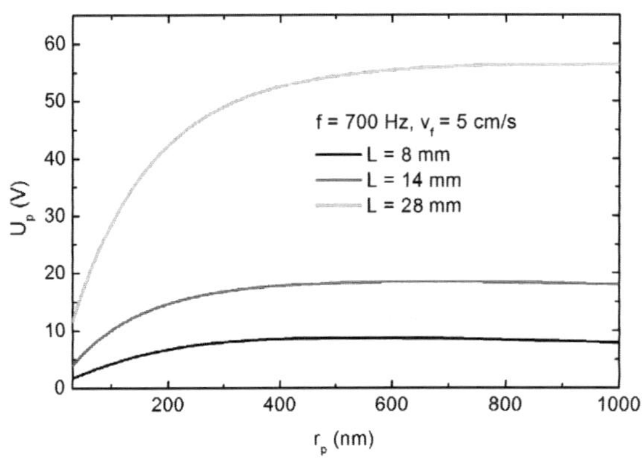


**Figure 4.** The lower bounds of the electric potentials of the nanoparticle for different interelectrode distances.

## References

[1] Pandis S N, Wexler A S an Seinfeld J H 1995 *J. Phys. Chem.* **99** 9646–9659
[2] Seinfeld J H, Pandis S N 2006. *Atmospheric Chemistry and Physics: From Air Pollution to Climate Change* (Wiley)

Journal of Physics: Conference Series **747** (2016) 012037 doi:10.1088/1742-6596/747/1/012037

[3] Draine B T 2003 *Annual Review of Astronomy and Astrophysics* **41** 241–289
[4] Trevitt J A 2006 *Ion Trap Studies of Single Microparticles: Optical Resonances and Mass Spectroscopy* (The University of Melbourne)
[5] Kulkarni P, Baron P A and Willeke K 2011 *Aerosol Measurement: Principles, Techniques and Applications* (Wiley)
[6] Peukert W , Wadenpohl C 2001 *Powder Technology* **118** 136–148
[7] Paul W 1990 *Usp. Fiz. Nauk* **160** 109
[8] Molhave Kristian 2000 *Construction of and Eperiments with a Linear Paul Trap* (Institut for Fysik of Astronomi, Aarhus Universitet).
[9] Lapitsky D S 2015 *J. Phys.: Conf. Ser.* **653** 012130
[10] Lapitsky D S, Filinov V S, Deputatova L V, Vasilyak L M, Vladimirov V I and Pecherkin V Ya 2015*EPL* **110** 15001
[11] Deputatova L V, Filinov V S, Lapitsky D S, Pecherkin V Ya, Syrovatka R A, Vasilyak L M and Vladimirov V I 2015 *JPCS* **653** 012131
[12] Vasilyak L M, Vladimirov V I, Deputatova L V, Lapitsky D S, Molotkov V I, Pecherkin V Ya, Filinov V S and Fortov V E 2013 *New Journal of Physics* **15** 043047
[13] Hart M B, Sivaprakasam V, Eversole J D, Johnson L J and Czege J 2015 *Applied Optics* **31** 54
[14] Lapitsky D S, Filinov V S, Deputatova L V, Vasilyak L M, Vladimirov V I and Pecherkin V Ya 2015 *High Temperature* **5** (1) 1–8
[15] Tsilingiris P 2008 *Energy Conversion and Management* **49** 1098
[16] Gussman R A 1969 *Journal of applied meteorology* **8** 999–1001
[17] Skeel R and Izaguirre J 2002 *Molecular Physics* **100** (24) 3885

# $Pr_2M_2O_7$ (M=Zr, Hf) local and crystal structure changes in a fluorite-pyrochlore phase transition.

**A P Menushenkov[1], V V Popov[1,2], B R Gaynanov[1,3] and Ya V Zubavichus[2]**

[1] National Research Nuclear University MEPhI (Moscow Engineering Physics Institute), Kashirskoe sh. 31, 115409 Moscow, Russia
[2]NRC "Kurchatov Institute", pl. Akademika Kurchatova 1, 123182, Moscow, Russia

E-mail: BRGaynanov@gmail.com

**Abstract**. The formation of nanocrystal powders of $Pr_2M_2O_7$ (M=Zr, Hf) compounds upon calcinations up to 1400°C has been investigated by using of X-ray absorption fine structure (XAFS) spectroscopy combined with X-ray powder diffraction (XRD). The structure ordering upon calcination was estimated independently from XRD and EXAFS data. It was found that the phase transition fluorite-pyrochlore occurs at calcination temperature in the range of 900-1000 °C. The appearance of the ordered pyrochlore structure at calcination temperatures above 1000 °C was manifested as splitting of the first oxygen coordination shell of the local praseodymium environment and supported by a sharp reduction of the Debye-Waller factor values of interatomic bonds. Besides we observed the $L_3$-Hf X-ray absorption near edge line shape splitting with increasing of annealing temperature of amorphous precursor due to crystal field influence in high ordered pyrochlore-type structure.

## 1. Introduction

In the "$Ln_2O_3$-$MO_2$" (*Ln* = lanthanide, M = Zr, Hf) systems, the formation of several intermediate phases were found: phases of stoichiometry formula $Ln_2M_2O_7$ with pyrochlore (*Fd-3m*) and fluorite (*Fm-3m*) types of crystalline structure and a phase with hexagonal structure (*R-3*), formulated as $Ln_4M_3O_{12}$ [1]. The thermodynamic stability of the pyrochlore type compounds can be determined using empirical rule, which identifies the ratio between ionic radii as $1.46 < (r_{Ln3+}/r_{M4+}) < 1.78$ [2]. Earlier we investigated the fluorite-pyrochlore phase transition in a "boundary" zirconate $Gd_2Zr_2O_7$ and hafnates $Ln_2Hf_2O_7$ (*Ln* = Sm – Dy) with the cation ratio $r_{Ln^{3+}}/r_{M^{4+}}$~1.46 [3-6]. Among the complex oxides based on Zr, Hf and rare earth ions, $Pr_2Me_2O_7$ (M=Zr, Hf) has the pronounced inclination to the formation of pyrochlore structure. In the "$HfO_2$ - $Pr_2O_3$» phase diagram there exists a considerable pyrochlore phase area and $Pr_2Hf_2O_7$ ($r_{Pr3+}/r_{Hf4+} = 1.586$) melts congruently at 2420°C (i.e. the liquid has the same composition as the solid compound) [7]. The "$ZrO_2$ - $Pr_2O_3$" phase diagram similarly exhibits significant pyrochlore phase area, but $Pr_2Zr_2O_7$ ($r_{Pr3+}/r_{Zr4+} = 1.564$) melts incongruently at 2220°C [1]. $Pr_2Zr_2O_7$ and $Pr_2Hf_2O_7$ reveal the magnetic frustration structure properties [8-9] and may be used as the thermal barrier coatings [10].

The aim of the present work is to study the crystallization processes of an amorphous precursor upon isothermal calcination and to make clear the peculiarities of fluorite - pyrochlore phase transition in $Pr_2M_2O_7$ (M = Zr, Hf), using combined X-ray diffraction (XRD) and X-ray absorption spectroscopy (XAFS) at synchrotron radiation.

[3] To whom any correspondence should be addressed.

## 2. Experimental section

The starting materials were $Pr(NO_3)_3 \cdot 6H_2O$ (99.8% purity), $ZrOCl_2 \cdot 8H_2O$ (99% purity) and $HfOCl_2 \cdot 8H_2O$ (99.7% purity). Precursor (mixed Pr-Zr or Pr-Hf hydroxide) was prepared by co-precipitation of metal salts solution with ammonia $NH_3 \cdot H_2O$ (analytical grade) [11]. The $Pr_2Zr_2O_7$ or $Pr_2Hf_2O_7$ powders were prepared by calcination of an amorphous precursor at 600-1400°C for 3 h in air. A more detailed experimental procedure was described in Refs [3-4, 6].

X-ray diffraction investigations of the structure were carried out at the "Structural Materials" beamline of the Kurchatov synchrotron radiation source. Standard mode in the Debye-Scherrer geometry (transmissive mode), was implemented in an environment optimized for high-quality experimental data: the distance the sample - detector is 200 mm; beam size 200x200 mm; the exposure time of 15-30 minutes, the wavelength $\lambda = 0.68886$ Å. Two series of $Pr_2M_2O_7$ compounds with (M = Zr, Hf) were examined. Each series included 7-9 samples obtained by calcination of initial precursor at different temperatures in a temperature range 600 -1400 C °. The Rietveld full-profile analysis of X-ray diffraction patterns was performed with the Jana2006 software [12].

X-ray absorption spectra (EXAFS) of $Pr_2Me_2O_7$ (Me=Zr, Hf) above the $L_3$-Pr (5964эB), $L_3$-Hf (9561 эB) and *K*-Zr (17998 эB) absorption edges, were measured at i811 beamline of MAXLab (Lund, Sweden) in a transmission mode at room temperature. Processing and modeling of XAFS-spectra were carried out using the VIPER [13]. The amplitudes and phases of photoelectron back scattering in the local environment of zirconium, praseodymium and hafnium atoms were calculated by program FEFF-8.20 [14] using the crystal structure parameters of pyrochlore and fluorite, taken from the work of Blanchard et al. (2013)[15] and from the diffraction data for $Pr_2Me_2O_7$ (where Me = Zr, Hf).

## 3. Results and discussion

The XRD study showed that precursors calcinated at 600 °C have the X-ray amorphous structure. However, their structure can't be considered as completely disordered, since one to several broad peaks were observed in the region of double Bragg angle. Calcination of $Pr_2Zr_2O_7$ at 700 °C leads to the appearance of crystalline domains in the amorphous structure, which is manifested as the presence of broad diffraction peaks. Upon calcination at 800 °C the beginning of nanocrystallization observed in both $Pr_2Zr_2O_7$ and $Pr_2Hf_2O_7$ systems. The increasing of the calcination temperature above 900 °C leads to appearance of about 10 x-ray reflections corresponding to a fluorite structure for all the samples. Besides above 900 °C we observed the gradual narrowing of the diffraction reflexes and the appearance of superlattice peaks (Figure 1), which point to increase of the structure ordering. This is also manifested in increase of crystallite size and decrease of values of microstrain obtained from the analysis of the diffraction patterns (Figure 2). These superstructure reflections are substantially wider than the main peaks of the fluorite-type structure. It suggests that cationic ordering with formation of the pyrochlore phase upon calcination occurs inside the separate regions (nanodomains) distributed over crystals matrix with the fluorite structure.

Analysis of EXAFS-data showed that spectra for calcination temperature up to 800-900 °C are well reproduced by the model of fluorite structure, and bond lengths Pr-O and Me-O slightly increase (Figure 3). However, at higher calcination temperatures the fluorite structure can no longer provide an adequate simulation of EXAFS-function and the analysis was conducted using the pyrochlore structure model. Thus above 1000 °C there is clearly observed splitting of the Pr-O bond length with a weak reducing of M-O distances (Figure 3). This directly points to the restructuring of the local oxygen environment of rare-earth ions as a result of the observed fluorite-pyrochlore phase transition. The values of the Debye-Waller factor (Figure 4) of interatomic bonds decrease with calcination temperature, further pointing to the growth of the degree of local structure ordering in the transition from a disordered fluorite structure to an ordered structure of pyrochlore.

Journal of Physics: Conference Series **747** (2016) 012038 doi:10.1088/1742-6596/747/1/012038

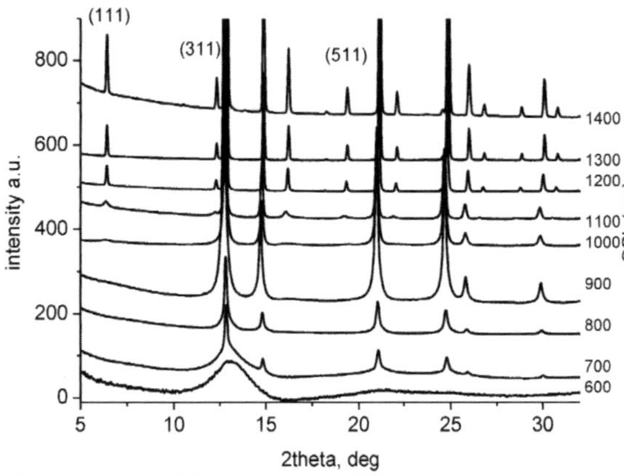


**Figure 1.** XRD patterns of $Pr_2Zr_2O_7$ powders obtained by calcination of the precursor at different temperatures at 3 h in the air.

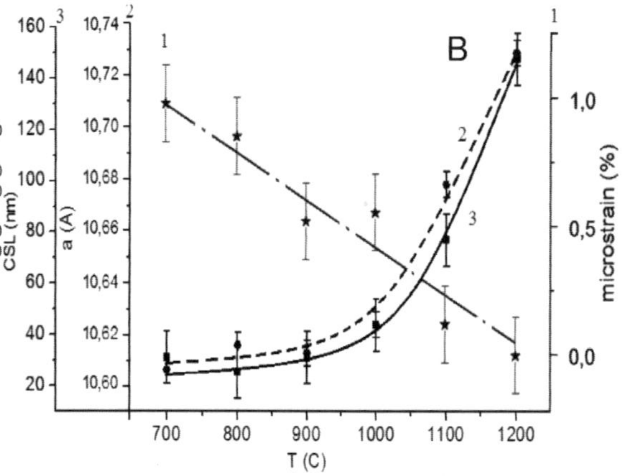


**Figure 2.** Microstrains (1), unit cell parameters (2) and coherent scattering lengths (3) of the series $Pr_2Zr_2O_7$ established by Rietveld method. Similar results were obtained for the $Pr_2Hf_2O_7$ compounds.

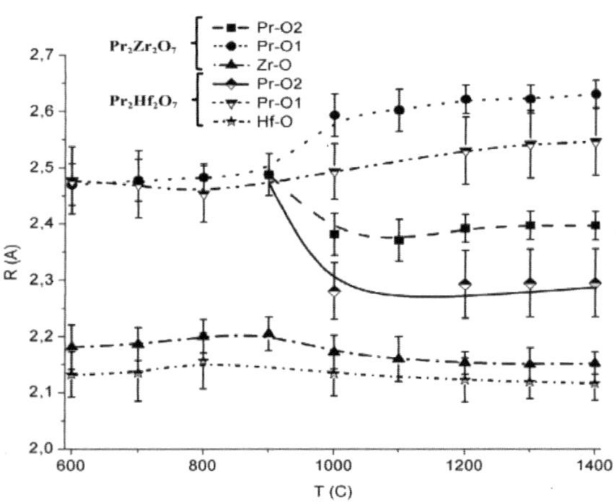


**Figure 3.** Bond lengths evolution of $Pr_2Zr_2O_7$ and $Pr_2Hf_2O_7$ with a calcination temperature increase.

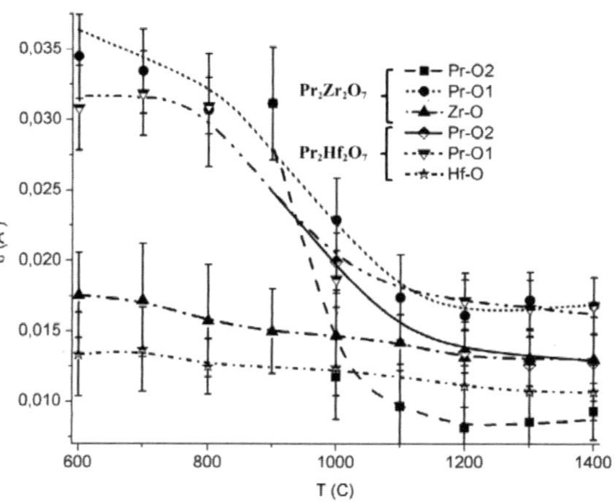


**Figure 4.** Debye-Waller factors evolution of $Pr_2Zr_2O_7$ and $Pr_2Hf_2O_7$ with a calcination temperature increase.

Additionally, the analysis of $L_3$-Hf of the absorption edge (XANES) in $Pr_2Hf_2O_7$ showed that increase of calcination temperature of the initial precursor up to 800 °C leads to the essential widening of the so-called "white line", corresponding to the main peak in the absorption spectrum (Figure 5). Further increase of temperature above 900 °C leads to splitting of the "white line", which is clearly manifested in the spectral dependence of the second derivative (Figure 6). As the XANES for $L_3$-Hf of the absorption edge arises from the dipole transition $2p_{3/2} \rightarrow 5d_{5/2}$, we believe that the observed effect is connected to the removal of degeneracy 5d level of hafnium and its splitting into $e_g$ and $t_{2g}$ parths as a result of the crystal field influence in the ordered pyrochlore structure. This observation is an additional confirmation of the phase transition fluorite-pyrochlore in $Pr_2Hf_2O7$ with increase of calcination temperature above 1000 °C.

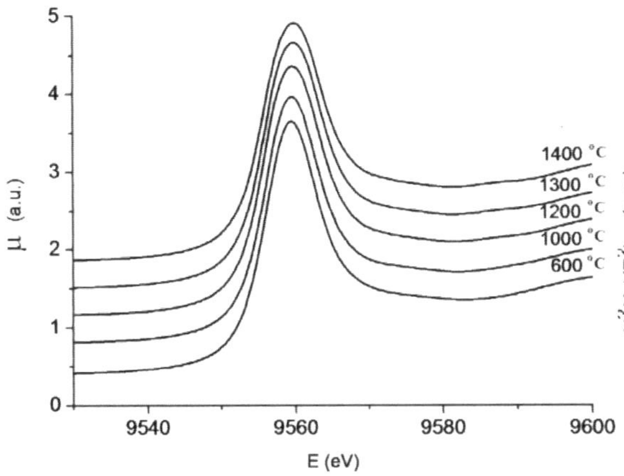


**Figure 5.** $L_3$-Hf XANES spectra of $Pr_2Hf_2O_7$ normalized to the white line for different calcination temperature.

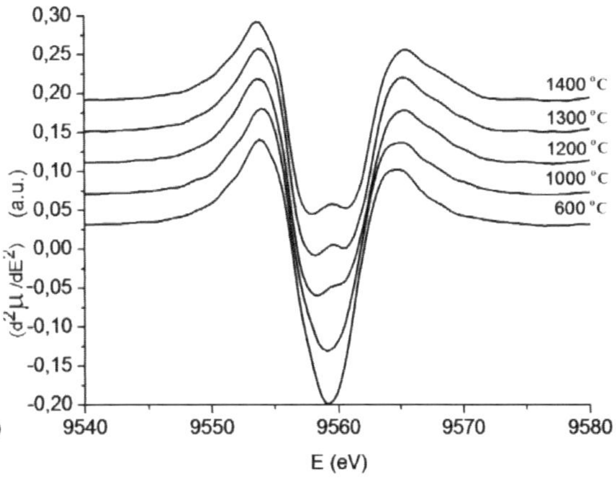


**Figure 6.** Second derivative XANES spectra of $Pr_2Hf_2O_7$ for different calcination temperature.

## 4. Conclusion

Local and crystal structure evolution of $Pr_2Zr_2O_7$ and $Pr_2Hf_2O_7$ upon calcinations of the initial X-ray amorphous precursors in the temperature range 600 - 1400 °C has been studied by a combination of X-ray powder diffraction (XRD) and XAFS-spectroscopy using synchrotron radiation. Splitting of the first oxygen coordination shell of the local praseodymium environment and "white line" $L_3$-Hf of the absorption edge into two components was observed in pyrochlore structure at calcination temperature higher than 900-1000 °C. The splitting of $L_3$-Hf edge caused by crystal field influence in high ordered pyrochlore-type structure.

**Acknowledgements**

The authors thank Drs. S. Carlson, K. Klementiev (MAX IV lab.), A. Yaroslavtsev (European XFEL), R. Chernikov (DESY Photon Science) and A. Ivanov (NRNU "MEPhI") for their help in experiments and the Russian Science Foundation (grant 14-22-00098) for financial support.

**References**

[1] Andrievskaya E R 2008 Phase equilibria in the refractory oxide systems of zirconia, hafnia and yttria with rare-earth oxides *J. Eur. Ceram. Soc.* **28** pp 2363-2388.

[2] Subramanian M A, Aravamudan G, Subba Rao G V 1983 Oxide pyrochlores—a review. *Prog. Solid State Chem.* **15** pp 55-143.

[3] Popov V V, Petrunin V F, Korovin S A et al 2011 Formation of nanocrystalline structures in the $Ln_2O_3$-$MO_2$ systems (Ln = Gd, Dy; M = Zr, Hf) *Russ. J. Inorg. Chem.* **56** pp 1538-1544.

[4] Popov V V, Menushenkov A P, Zubavichus Ya V *et al* 2013 Characteristic Features of the Nanocrystalline Structure Formation in $Ln_2Hf_2O_7$ (Ln = Gd, Dy) compounds *Russ. J. Inorg. Chem.* **58** pp 1400-1407.

[5] Popov V V, Zubavichus Ya V, Menushenkov A P *et al* 2014 Short- and Long-Range Order Balance in Nanocrystalline $Gd_2Zr_2O_7$ Powders with a Fluorite–Pyrochlore Structure *Russ. J. Inorg. Chem.* **59** pp 279-285.

[6] Popov V V, Zubavichus Ya V, Menushenkov A P *et al* 2015 Lanthanide Effect on the Formation and Evolution of Nanocrystalline Structures in $Ln_2Hf_2O_7$ Compounds (Ln = Sm–Dy) *Russ. J. Inorg. Chem.* **60** pp 16-22.

[7] Stanek C R 2003 Atomic scale disorder in fluorite and fluorite related oxides (Ph. D. Thesis Imperial college London UK) p 195.

Journal of Physics: Conference Series **747** (2016) 012038 doi:10.1088/1742-6596/747/1/012038


[8] Craig H A 2009 Magnetic Frustration and Phase Purity in Pyrochlore Materials: $Pr_2Zr_2O_7$, $Pr_2Hf_2O_7$ and $Tb_2Hf_2O_7$ *Explorations: The UC Davis Undergraduate Research J.* **12** pp 23-41.

[9] Matsuhira K, Sekine C, Paulsen C *et al* 2009 Spin freezing in the pyrochlore antiferromagnet. $Pr_2Zr_2O_7$ *J. Phys.: Conf. Ser.* **145** p 012031(5).

[10] Pan W, Phillpot S R, Wan C *et al* 2012 Low thermal conductivity oxides. *MRS Bull* **32** pp 917-922.

[11] Popov V V 2015 Formation Regularities of Dispersed Hydrated Oxide Systems *Russ. J. Inorg. Chem.* **60** pp 420-427.

[12] Petricek V, Dusek M, Palatinus L 2014 Crystallographic computing system JANA2006: General features *Z. Kristallogr.* **229** pp 345-352.

[13] Klementiev K V 2001 VIPER for Windows *J. Phys. D: Appl. Phys.* **34** p 209.

[14] Newville M 2001 IFEFFIT: interactive XAFS analysis and FEFF fitting *J. Synchr. Rad.* **8** p 322.

[15] Blanchard P E R, Liu S and Kennedy B J *et al* 2013 Investigating the Local Structure of Lanthanoid Hafnates $Ln_2Hf_2O_7$ via Diffraction and spectroscopy *J. Phys. Chem.* **117** pp 2266-2273.

Journal of Physics: Conference Series **747** (2016) 012039 doi:10.1088/1742-6596/747/1/012039

# Effect of zirconium addition and microstructure refinement on the local magnetic properties of the $Nd_2Fe_{14}B$-based alloys

**A P Menushenkov[1,4], V G Ivanov[1], I V Shchetinin[2], D G Zhukov[2], V P Menushenkov[2], I A Rudnev[1], F Wilhelm[3], A Rogalev[3] and A G Savchenko[2]**

[1] National Research Nuclear University "MEPhI" (Moscow Engineering Physics Institute), Kashirskoe sh. 31, 115409 Moscow, Russia
[2] National University of Science and Technology "MISIS", Leninskiy pr. 4, 119049 Moscow, Russia
[3] European Synchrotron Radiation Facility (ESRF), CS40220, F-38043 GrenobleCedex 9, France

E-mail: APMenushenkov@mephi.ru

**Abstract**. The effect of Zr addition and microstructure refinement on the local magnetic properties of intermetallic compound $Nd_2Fe_{14}B$ in micro- and nano-crystalline alloys was studied by X-ray Magnetic Circular Dichroism. XMCD spectra were measured at the beamline ID-12 of ESRF (Grenoble, France) at the Fe-*K* and Nd-$L_{3,2}$ absorption edges in a magnetic field of 10 T at room temperature. The noticeable differences of XMCD amplitude signals, measured on the samples with different content of Zr and various parameters of the microstructure were found, which clearly indicates the influence of the both factors on the magnetic state of Nd and Fe ions in the $Nd_2Fe_{14}B$-based compounds.

## 1. Introduction

$Nd_2Fe_{14}B$-based alloys belong to the most popular and commercially used materials for highest performance permanent magnets [1]. Ability to change the properties of these alloys by the addition of different doping elements, as well as the use of various types of heat treatment for directed changes of their microstructure can significantly expand the scope of their practical use. For example, the Zr addition into $Nd_2Fe_{14}B$ increases the coercivity and remanence of the respective alloys [2]. Similar effect is also observed with grain size decreasing to the nano-scale range at melt spinning.

In the present work the comparative study of the effect of nano-structuring and Zr addition on local magnetic properties of $Nd_2Fe_{14}B$ alloys was performed by means of X-ray Magnetic Circular Dichroism (XMCD) at the *K*-edge of Fe and $L_{2,3}$-edges of Nd.

## 2. Experiment

Samples of hard magnetic alloys with a nominal compositions $Nd_2Fe_{14}B$ and $Nd_{10,4}Zr_{4,0}Fe_{79,2}B_{6,4}$ were prepared by copper mold casting and quenching from the liquid state on the rapidly rotating steel disc. X-ray diffraction (XRD) studies were carried out using the diffractometer Rigaku Ultima-IV focusing a Bragg-Brentano, with $Co_{K\alpha}$-radiation and a graphite monochromator in the diffracted beam.

[4] To whom any correspondence should be addressed.

The grain size in the micro- and nano-structure phases was detected by a JEM-1400 transmission electron microscope (TEM). Measurements of magnetic properties were performed using physical property measurement system PPMS (EverCoolII, Quantum Design) in fields up to 9 T.

To get a deeper insight into microscopic magnetic properties of these systems, we have used an element-selective technique, such as X-ray Magnetic Circular Dichrtoism (XMCD) [3-5]. XMCD spectra at the $L_{2,3}$ edges of Nd and at the *K*-edge of Fe have been measured at the ID-12 beamline of the European Synchrotron Radiation Facility, Grenoble, France [5]. The source was the helical undulator APPLE-II, which emits either right or left circularly polarized X-rays with a polarization rate in excess of 97%. The X-ray absorption spectra at the $L_{2,3}$-edges of Nd and at the *K*-edge of Fe were recorded using the total fluorescence yield detection mode in backscattering geometry. XMCD spectra were obtained as direct difference of absorption spectra recorded with right and left-circularly polarized photons and a magnetic field (10 Tesla) generated by a superconducting solenoid was applied along the X-ray beam direction. To make sure that the obtained XMCD spectra are free of any experimental artefact, the same measurements were repeated for opposite direction of applied magnetic field. The samples were kept at room temperature. Since the penetration depth of the X-rays is smaller than the sample thickness, the recorded XAS and XMCD spectra were corrected for so-called self-absorption effects. The correction took into account the chemical composition and thickness of the sample, incidence angle of the X-ray beam, and the solid angle of the detector [6].

## 3. Results and discussion

According to XRD studies the microcrystalline as-cast $Nd_2Fe_{14}B$ and $Nd_{10,4}Zr_{4,0}Fe_{79,2}B_{6,4}$ samples consist of single $Nd_2Fe_{14}B$ phase ($P4_2/mnm$, tP68). The measurement of magnetic properties of as-cast samples exhibited nearly zero coercivity and remanence. Quenching from the liquid state led to the formation of an amorphous phase in $Nd_{10,4}Zr_{4,0}Fe_{79,2}B_{6,4}$ sample together with small amount of soft magnetic phases, such as α-Fe and $Nd_2Fe_{17}$. Melt spinning resulted in an increase of the coercive force and residual magnetization, however, the magnetic hysteresis loop exhibits an inflection that could be due to the presence of soft magnetic amorphous phase (Fig. 1).

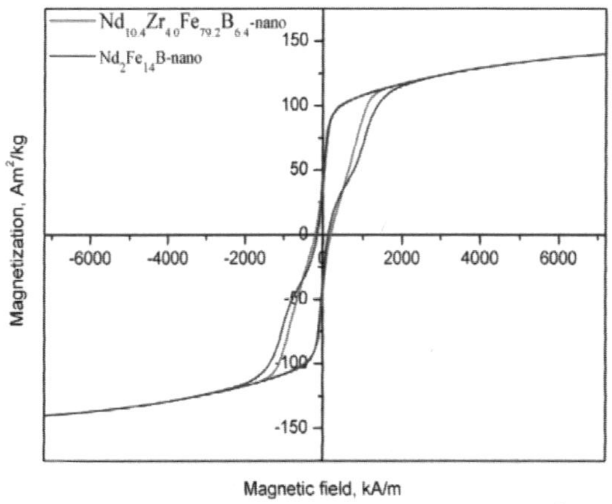


**Figure 1**. Magnetic hysteresis loops of melt spun $Nd_2Fe_{14}B$ and $Nd_{10,4}Zr_{4,0}Fe_{79,2}B_{6,4}$ samples

The results of a comparative analysis of XMCD spectra of Zr-doped and undoped $Nd_2Fe_{14}B$ samples are presented below. The effect of melt spinning of $Nd_{10,4}Zr_{4,0}Fe_{79,2}B_{6,4}$ alloy on element selective magnetic properties is also analyzed.

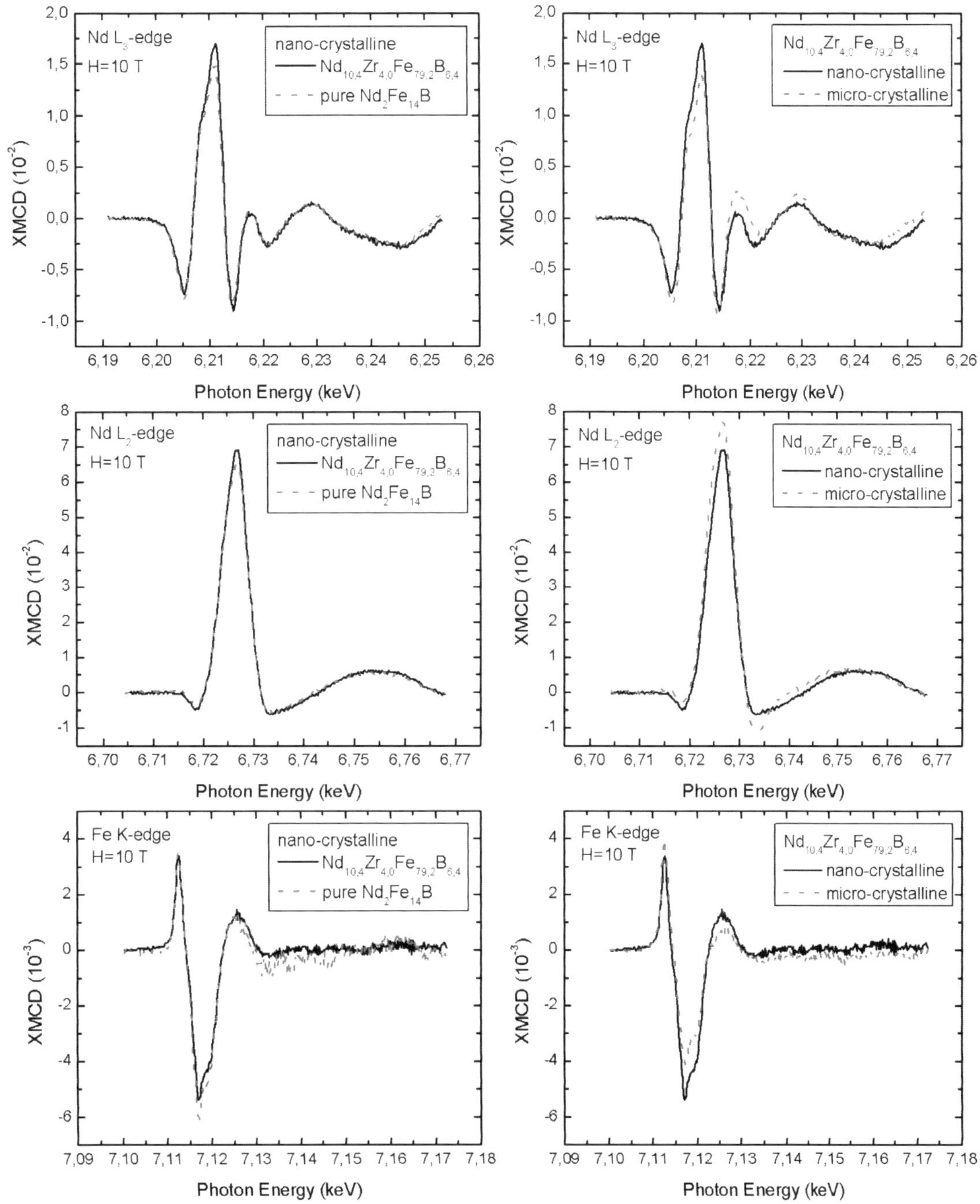


**Figure 2**. Comparative XMCD spectra measured at Nd-$L_{3,2}$ and $K$-Fe absorption edges in a magnetic field of 10 T for doped $Nd_{10,4}Zr_{4,0}Fe_{79,2}B_{6,4}$ and pure $Nd_2Fe_{14}B$ nano-crystalline samples (left panel), and for nano- and micro-crystalline samples of $Nd_{10,4}Zr_{4,0}Fe_{79,2}B_{6,4}$ (right panel).

### *1.1. Effect of* Zr *addition to the* $Nd_2Fe_{14}B$ *on XMCD spectra of nano-crystalline samples.*

XMCD spectra obtained for nano-crystalline samples at Nd-$L_{3,2}$ absorption edges have a greater amplitude for Zr-doped $Nd_{10,4}Zr_{4,0}Fe_{79,2}B_{6,4}$ than for undoped $Nd_2Fe_{14}B$ whereas the opposite effect is observed at the $K$-edge of Fe (Fig. 2, left panel). The highly structured XMCD spectra at the $L_3$ absorption edge of Nd are due to a variety of electronic transitions involved. The first minimum

Journal of Physics: Conference Series **747** (2016) 012039 doi:10.1088/1742-6596/747/1/012039

(prepeak) corresponds to the quadrupolar transition $2p_{3/2} \rightarrow 4f_{5/2}$, and the main maximum – to dipolar transition $2p_{3/2} \rightarrow 5d_{5/2}$ [7]. The oscillating structure afterwards, apparently, corresponds to the transitions to hybridized Nd-Fe *p-d* band and/or manifests multiple scattering of a photoelectron wave from the nearest neighbors, so-called magnetic EXAFS. XMCD signal at the $L_2$ absorption edge of Nd is characterized by the pronounced main maximum, corresponding to dipolar transition $2p_{1/2} \rightarrow 5d_{3/2}$, and much weaker quadrupolar contributions [7]. The shape of the Fe *K*-edge XMCD signal could also be disentangled into quadrupolar and dipolar transitions, but this assignment is usually quite ambiguous.

Thus, the XMCD results indicate that doping of $Nd_2Fe_{14}B$ phase by Zr affects the magnetic state of both Nd and Fe sublattices via changes the overall electronic subsystem, in particular, the 5*d* states of Nd and the 4*p* states of Fe.

*1.2. Effect of crystal grain's size of* $Nd_{10,4}Zr_{4,0}Fe_{79,2}B_{6,4}$ *alloys on XMCD spectra.*
According to the TEM data an average grain size was estimated as 5-10 micron and 10-20 nm in micro- and nano- crystalline samples. Comparing XMCD spectra for micro- and nano- crystalline $Nd_{10,4}Zr_{4,0}Fe_{79,2}B_{6,4}$ (Fig. 2, right panel), one can see that for nano-crystalline sample the signals are greater at Nd-$L_3$ and *K*-Fe edges and smaller at Nd-$L_2$ edge. This is apparently due to the different magnetic state of Nd and Fe ions in the bulk and at grain boundaries, since the transition to nanostructured state leads to growing contribution of ions at the grain boundaries.

It should be noted that XMCD spectra change differently under influence of nano-structuring and Zr doping.

## 4. Conclusion

Experimental *K*-Fe and Nd-$L_{3,2}$ XMCD spectra were measured in magnetic field of 10 T for nano- and micro-crystalline magnetic intermetallides $Nd_2Fe_{14}B$ and $Nd_{10,4}Zr_{4,0}Fe_{79,2}B_{6,4}$.

Noticeable differences in the amplitudes of XMCD signals were found for the samples with addition of zirconium and for nano-crystalline samples. This indicates the important role of the both factors in forming of the magnetic state of Nd and Fe ions in $Nd_2Fe_{14}B$-based compounds.

The results are helpful for understanding the magnetic moment formation and nature of the high coercivity state in $Nd_2Fe_{14}B$-based hard magnetic alloys and may be used to improve their functional magnetic properties.

**Acknowledgements**
The European Synchrotron Radiation Facility is acknowledged for providing beamtime in the framework of the project HC-2600. This work was partly supported by the Ministry of Education and Science of the Russian Federation within the framework of the state assignment 3.1540.2014/K and project 14.579.21.0043 (unique number of agreement RFMEFI57514X0043).

## References

[1] Matsuura Y 2006 *JMMM* **303** 344
[2] Bao XQ, Zhu J, Li W, *et al.* 2009*Journal Rare Earths* **27** 843
[3] Sarafidis C, Wilhelm F, Rogalev A, *et al.* 2009 *J. Phys.: Condens. Matt.* **21** 236001
[4] Van der Laan G, Figueroa A I 2014 *Coordin. Chem. Rev.* **277-278** 95
[5] Rogalev A, Wilhelm F 2015 *The Physics of Metals and Metallography***116** 1285
[6] Wilhelm F, Jaouen N, Rogalev A, *et al.* 2007 *Phys. Rev.* B**76** 024425
[7] Bartolomé F, Tonnerre J M, Sève L, *et al.* 1998 *J. Appl. Phys.* **83** 7091

# Synthesis of nanostructured materials based on $YBa_2Cu_3O_{7-y}$ and $BiFeO_3$

**D K Palchaev**[1,2], **Zh Kh Murlieva**[1,3], **M H Rabadanov**[1,5], **S Kh Gadzhimagomedov**[1], **N M-R Alikhanov**[1], **R M Emirov**[1] **and A S Asvarov**[4]

[1] Dagestan State University, Gadzhieva 43-A, 367000 Makhachkala, Russia

[2] South Federal University (branch), Yusupova 51, 367013, Makhachkala, Russia

[3] Dagestan State University of National Economy, Ataeva 5, 367008, Makhachkala, Russia

[4] Institute of Physics, Dagestan Scientific Center of RAS, Kalinina 39-A, 367030 Makhachkala, Russia

E-mail : zariyat@mail.ru

**Abstract**. Methods for creating new nanostructured materials based on $YBa_2Cu_3O_{7-\delta}$ and $BiFeO_3$, results of a study of their structure, morphology and electrical properties are present. Morphology studies show that the micron-sized grains of ceramics are agglomerates of nanoparticles of about 50 nm average size for both $BiFeO_3$ and $YBa_2Cu_3O_{7-y}$. These nanoparticles are responsible for weak ferromagnetism in bismuth ferrite, and for increase of the critical temperature up to 96 K in $YBa_2Cu_3O_{7-\delta}$ nanoceramics.

## 1. Introduction

We report on synthesis of new nanostructured materials based on compounds of $YBa_2Cu_3O_{7-y}$ (YBCO) and $BiFeO_3$ (BFO) and discuss the methods, improving their composition, structure and morphology during the preparation of both the primary powders and ceramics. The structure and morphology of nanopowders and nanostructured ceramics, based on YBCO, were studied by the scanning probe microscope «LEO-1450» with EDX-analyzer «INCA Energy», X-ray diffractometer «PANalytical Empyrean series 2», ASPEX Express with EDX-analyzer. The electrical resistivity of HTS samples was measured by four-probe method, the dielectric constant $BiFeO_3$ was measured using LCR-78110G meter company Goodwill Instrument Co.

## 2. Results and discussion

### *2.1. Nanostructured materials based on* $YBa_2Cu_3O_{7-y}$

The nanostructured superconducting materials with different densities were prepared from nanopowders. The preparation method differs from reported elsewhere [1] both in the burning of nitrite-organic precursors and thermal treatment, and in method of their compactification. Different density of ceramic achieved without additives of organic binders. An optimal oxygen saturation of ceramics in the process of their sintering was reach due to the high specific surface of nanopowders.. The new method of nanopowder fabrication is characterized by the possibility to control the rate of

[5] To whom any correspondence should be addressed.

combustion of thickened sediment arising after evaporation of the liquid solution of precursors. This technology provides definite size distribution of particles in the powder by selecting the combustion rate.

Figure 1 shows the diffraction patterns for two HTSC ceramics samples containing 95% of $YBa_2Cu_3O_{6.9}$ phase, produced from powders, thermally treated at 350 °C for 1 hour ( Fig. 1a) and at 910 °C for 20 hours (Fig. 1b), respectively. After pressing powders were sintered at 920 °C for 1 hour. There is also shown the diffraction pattern of $YBa_2Cu_3O_{7-y}$ from database PAN-ICSD (Fig. 1c).

Copper oxide phase (5%) in the crystalline state (Fig. 1a and 1b), indicates that the $Y_2BaCuO_5$ oxide in X-ray amorphous state may also exist. After prolonged the heat treatment a recrystallization of this phase (211) observed. Morphology of nanostructured ceramics $YBa_2Cu_3O_{7-y}$ with densities of 2.5 g/cm$^3$ and 6 g/cm$^3$ shown in Figure 2a and 2b, respectively. Both highly porous and dense ceramics consist of micron-size grains, which are agglomerates of nanoparticles. According to diffraction studies, the average size of nanoparticles is about 50 nm. Agglomerates in the dense ceramics ( Fig. 2b) have grown together. The uniform distribution of the nanoparticles of $Y_2BaCuO_5$ in a matrix of $YBa_2Cu_3O_{7-y}$ promotes pinning of the magnetic flux [2] and thus increases the critical current density.

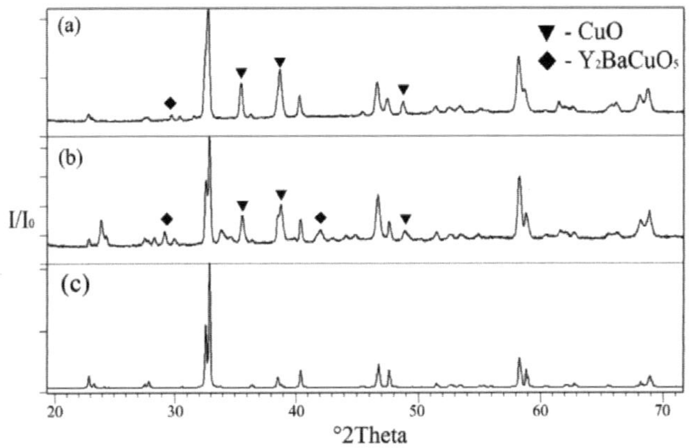


**Figure 1**. Diffraction patterns of ceramics made of thermally treated powders: (a) at 350 °C; (b) at 910 °C; (c) structure $YBa_2Cu_3O_{7-y}$ from the database PAN-ICSD.

Appropriate thermal treatment of nanopowder, in one step, allows to produce nanostructured ceramics samples with different densities up to X-ray density. We used the same technique, but different quantity of glycerol, to synthesize ceramics samples with densities of 3 g/cm$^3$ and 5.7 g/cm$^3$ [3].

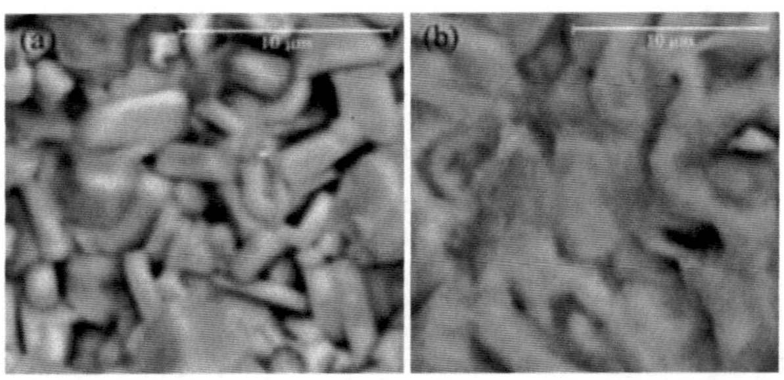


**Figure 2.** Morphology of nanostructured ceramics $YBa_2Cu_3O_{7-y}$: (a) porous; (b) dense.

Various glycerol content leads to different combustion rates of a "gel" which remains after evaporation of aqueous solutions of nitrates Y, Ba and Cu. As one can see, the amount of glycerol in the initial solution, other conditions being equal, allows us to control the value of nanostructured ceramics density. To obtain microcrystalline ceramics in one step it is enough to increase the temperature and duration of nanopowder heat treatment, but the increase in particle size reduces the oxygen content in the ceramics. Optimal oxygen concentration may be achieved by preliminary destruction of nanopowder agglomerates before sintering. In this case, the sintering temperature of the powder increased by about 5 degrees. Figure 3 shows the temperature dependence of the resistivity of nanostructured ceramics $YBa_2Cu_3O_{7-y}$ made from nanopowders, heat treated at 350 °C - (▲) and 910 °C - (●) respectively. It turned out that the transition into the superconducting state, both for dense and porous ceramics (Fig. 3) starts at $T_c$ ~ 96 K. The absolute values of the electrical resistance of high-porous ceramics are an order of magnitude greater. For this ceramics, the temperature coefficient of resistance is also higher, and the superconducting transition stretched up to 85 K, apparently due to the small area of contacts between the grains. Because the increasing Tc does not depend on the porosity this effect may be explained by the fact, that the size of the nanoparticles is the same for the dense and porous ceramics. Similar results were obtained for the nanostructured ceramics with the densities of 3 g /cm3, and 5.7g /cm$^3$ [3] with a particles sizes about 60 nm.

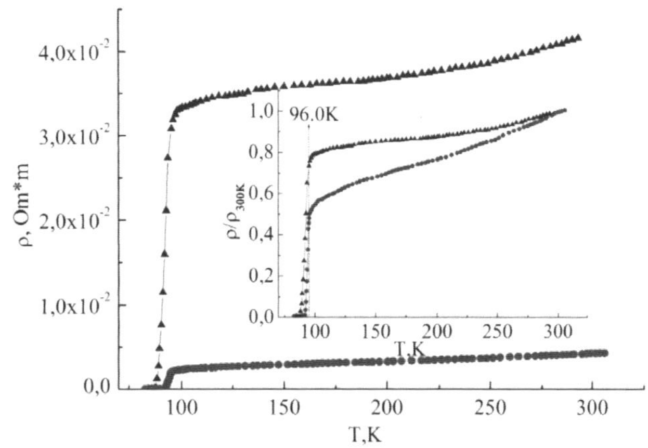


**Figure 3**. The electrical resistance of nanostructured ceramics made from $YBa_2Cu_3O_{7-y}$ nanopowders, precalcined at 350°C - (▲) and 910°C - (●)

### 2.2. *Nanostructured materials based on $BiFeO_3$*

To implement the linear magnetoelectric effect in bismuth ferrite the crystallite size should not exceed the size of the space-modulated spin cycloid with a period of 62 nm. Rabadanov et al. [4] developed a method (burning of bismuth nitrate and iron nitrate with the addition of glycine), which allows one to obtain the nanopowder, consisting of stoichiometric $BiFeO_3$ crystals only (Fig. 4a). Thermal treatment of the powder at temperatures higher than 600 °C gives rise to side phases, which are undesirable in the ceramics manufacture [5]. Optimization of this technology allows to obtaining powders having a particle size from 35 nm to 45 nm without increase of side phase after calcination at 600°C (Fig. 4b). Phase analysis of powder before thermal treatment shows the presence of about 6% impurities, representing probably a mixture of $Bi_{25}FeO_{39}$ and $Bi_2Fe_4O_9$. After calcination of the powder at temperature 600°C, the amount of secondary phases reduced to 4%, and the crystallite size calculated from Scherrer-Wulff formula, increases by approximately 10-15%.

The morphology of nanopowder obtained by optimized technology (Fig. 5a) indicates that powder consists of agglomerates of bismuth ferrite nanocrystals. Figure 5c presents diffraction pattern of ceramics obtained from this powder. As can be seen, all the peaks of side phases are suppressed. The density of the synthesized ceramics is close to the theoretical value and is approximately of 7.8 g /cm$^3$.

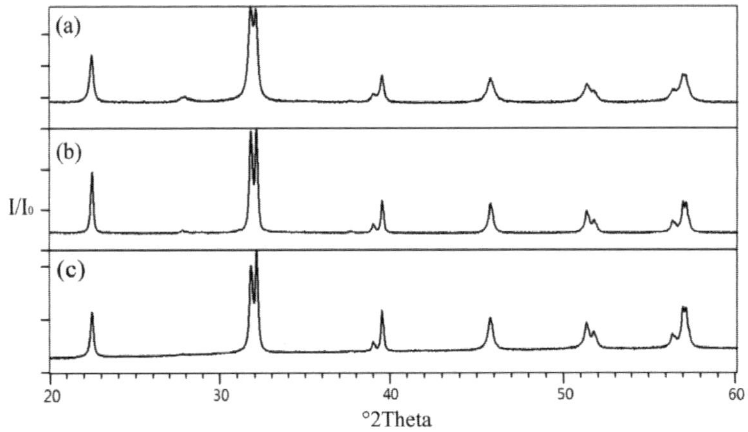


**Figure 4**. Diffraction patterns of $BiFeO_3$ nanopowders: (a) initial; (b) after thermal treatment at 600 °C; (c) nanoceramics

The morphology analysis shows that ceramics is dense and homogeneous (Fig. 5b). There are also clearly seen pores of faceted shape in ceramics grains. This indicates that grains are agglomerates of nanoparticles. This fact agrees with crystallite size evaluation from diffraction data (Fig. 1c) which gives size of ~ 50 nm, while the large grains are of micron size. Faceting agglomerates probably, acquire during their compression. High level of porosity in the ceramic $BiFeO_3$ is a serious obstacle for most practical applications of this material. According [6] the grain growth and compaction occur simultaneously during the sintering, so it is important to find the sintering process mode, in which the compacting will prevail over the increase of the crystallite size.

**Figure 5**. The surface morphology studied by the scanning electron microscope LEO - 1450: a) - nanopowder; b) - nanoceramics

Figure 6 shows the frequency and temperature dependences of the dielectric constant for nanoceramics and cold-pressed sample of the same powder. The anomaly on the $\varepsilon = f(t)$ dependence in the temperature range of 20 – 150 °C (Fig. 6b) is probably due to the presence of adsorbed moisture. Abnormal behavior of heat capacity at the same temperatures [7] also indicates the presence of hydrated moisture in in the cold-pressed samples. This anomaly is absent in a ceramic sample. The values of ε' for ceramics are slightly higher, this may be caused by increased interaction between the nanoparticles which are in the ferroelectric phase and by the absence of moisture as well. Hydrated moisture may improve the electrical conductivity [6], which complicates the use of bismuth ferrite as a ferroelectric.

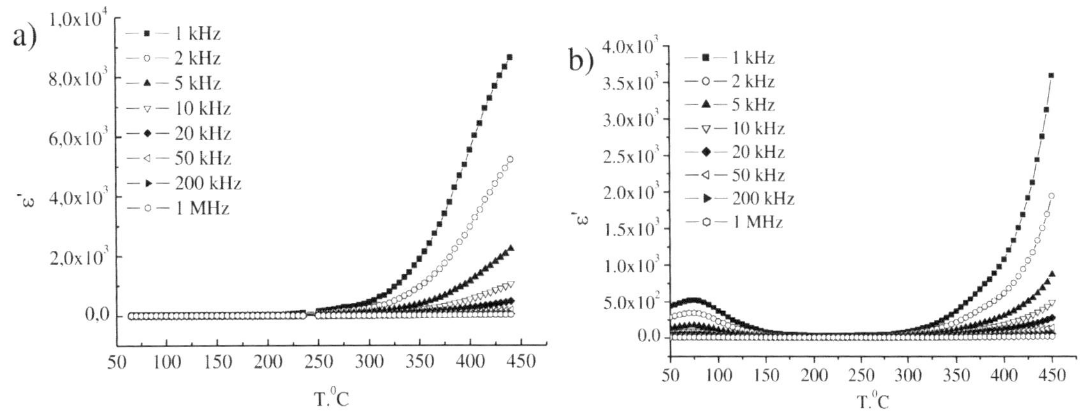


**Figure 6**. The real part of the dielectric constant: a) nanoceramics; b) cold pressed

**Acknowledgments**

The work was supported by the State tasks № 2560 and 16.1103.2014 / K.

**References**

[1] Rabadanov M, Palchaev D, Khidirov Sh, Murleva Zh, Samudov Sh, Ahmedov Sh and Asvarov A 2013 *Patent No. 2486161.*

[2] Pavan Kumar Naik S, Missak Swarup Raju P, Rajasekharan T and Seshubai V 2014 Growth Mechanism in Infiltration Growth Processed YBCO Composites Through Quench Studies. *J. Supercond. Nov. Magn.* **27** 1211.

[3] Gadzhimagomedov S, Palchaev D, Rabadanov M, Murleva Zh, Shabanov N, Palchaev N, Murlev E and Emirov R 2016 $YBa_2Cu_3O_{7-\delta}$ Based Ceramic Materials Manufactured from Nanopowders *J. Technical Physics Letters* **42** **1** 4–7.

[4] Rabadanov M, Palchaev D, Akhmedov Sh, Faradzheva M, Murleva Zh, Kallaev S and Sadykov S 2015 *Patent No. 2556181.*

[5] Palchaev D, Faradzheva M, Sadykov S, Rabadanov M, Murlieva Zh, Kallaev S, Tabit A F A and Emirov R 2014 Peculiarities of Dielectric Properties of Nanocrystalline Bismuth Ferrite *J.Tech. Phys. Lett.* **40** **11** 961–64.

[6] Bernardo M 2014 Synthesis, microstructure and properties of $BiFeO_3$-based multiferroic materials: A review *Bol. Soc. Esp. Ceram. Vidr.***53** **1** 1–14.

[7] Kallaev S, Omarov Z, Palchaev D, Rabadanov M, Murlieva Zh, Faradzheva M and Sadykov S 2015 Heat Capacity of Nanocrystalline Bismuth Ferrite *J. High Temperature* **53** **4** 636–39.

# Influence of synthesis conditions on the crystal structure of the powder formed in the $ZrO_2$ - $Ce_2O_3/CeO_2$ system.

**V V Popov**[1,2,4], **A P Menushenkov**[1], **R M Khubbutdinov**[1], **R D Svetogorov**[2], **Ya V Zubavichus**[2], **A S Sharapov**[1] **and V V Kurilkin**[3]
[1] National Research Nuclear University MEPhI (Moscow Engineering Physics Institute), Kashirskoe sh. 31, 115409 Moscow, Russia
[2] NRC "Kurchatov Institute", pl. Akademika Kurchatova 1, 123182 Moscow, Russia
[3] Peoples' Friendship University of Russia

E-mail: victorpopov@mail.ru

**Abstract.** Influence of synthesis conditions (type of atmosphere, reduction and oxidation, annealing temperature) on the chemical composition and structure of the compounds formed in the "$ZrO_2$ - $Ce_2O_3$ / $CeO_2$" system has been investigated by X-ray absorption fine structure (XAFS) spectroscopy combined with X-ray diffraction (XRD) and thermogravimetric analysis (TGA). It is revealed that isothermal annealing of precursor at temperatures less than 1000 °C in air leads to formation of $Ce_{0.5}Zr_{0.5}O_2$ powders with cubic fluorite-type structure (*Fm-3m*). Further increase of annealing temperatures above 1000 °C causes decomposition of formed crystal structure into two phases: cubic and tetragonal. Annealing in reduction hydrogen atmosphere causes formation of $Ce^{4+}{}_{2x}Ce^{3+}{}_{2-2x}Zr_2O_{7+x}$ compounds with intermediate valency of cerium, where value of $x$ depends on the reducing conditions and treatment parameters of precursor. Annealing in vacuum at 1400 °C strongly reduces the content of $Ce^{4+}$ in a powder samples and leads to formation of pyrochlore structure (space group *Fd-3m*) with practically $Ce^{3+}$ valence state.

## 1. Introduction

The oxygen-containing cerium compounds are of great interest from the point of view of physics and chemistry due to the ability of cerium cation to change an oxidation state in dependence on the synthesis conditions. Materials based on cerium oxide are of great practical importance for obtaining the catalysts [1], solid oxide fuel cells [2], polishing composition, etc. [3].

The aim of the present work is to study the influence of the synthesis conditions and annealing in reduced atmosphere on chemical composition and structure of the powders formed in "$ZrO_2$-$Ce_2O_3/CeO_2$" system by means of X-ray diffraction, XAFS spectroscopy and TGA.

## 2. Experimental

The starting materials were $Ce(NO_3)_3 \cdot 6H_2O$ (99.99% purity), $ZrOCl_2 \cdot 8H_2O$ (99% purity). Precursor was prepared by co-precipitation of metal salts solution with ammonia $NH_3 \cdot H_2O$ (analytical grade) [4]. The powders of $Ce_{0.5}Zr_{0.5}O_2$ were prepared by calcination of precursor at 600-1400°C for 3 h in air. Oxygen content reducing was made by two ways: isothermal annealing in vacuum at 1400°C for 3h and isothermal annealing in hydrogen at 1000°C for 4h (for more detailed experimental procedure refer [5], [6]).

[4] To whom any correspondence should be addressed.

TGA of powders was measured at temperature range 30 - 1400°C using thermogravimetric analyzer SDT Q600 (TA Instruments) with a heating rate of 10 K/min in air flow. The composition and the oxidation degree of synthesized powders was determined by TGA method in measurements of sample mass during heating to 1000 °C and further isothermal annealing for 4 hours in air flow.

Crystal structure of synthesized powders was studied by X-ray powder diffraction at "Structural Materials Science" beamline of the Kurchatov synchrotron radiation source. Measurements were carried out in the transmission mode at $\lambda$ = 0.68886 Å, using XY detector Imaging Plate Fuji Film BAS-5000. The Rietveld full-profile analysis of X-ray diffraction patterns was performed with the Jana2006 software [7].

The X-ray absorption spectra (XANES) were measured above the $L_3$-Ce edge at i811 beamline of MAX-lab (Lund, Sweden) in transmission mode at room temperature. XANDA software pack [8] was used to fit the near-edge region of spectra.

## 3. Results and discussion

It is revealed that isothermal annealing of precursor at 600-1000°C temperatures in air leads to formation of $Ce_{0.5}Zr_{0.5}O_2$ powders with FCC fluorite-type structure (space group *Fm-3m*) (Figure 1). Further increase of annealing temperatures above 1000 °C causes decomposition of formed crystal structure into the two phases: cubic $Ce_{0.55}Zr_{0.45}O_2$ and tetragonal $Ce_{0.07}Zr_{0.93}O_2$ (Figure 2).

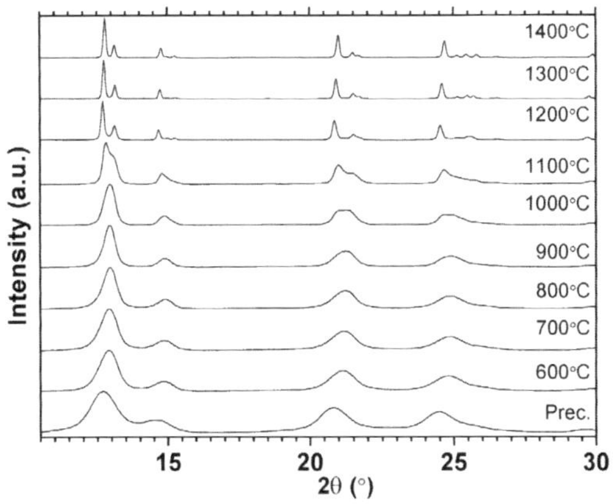


**Figure 1** XRD patterns of $Ce_{0.5}Zr_{0.5}O_2$ powders obtained by calcination of the precursor at different temperatures (shown on the corresponding curves)

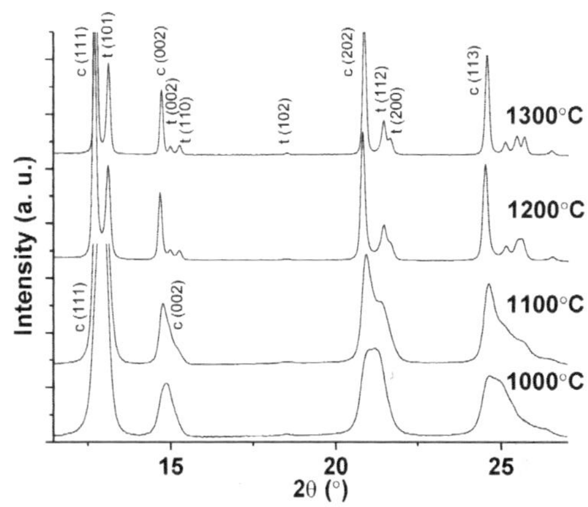


**Figure 2** XRD patterns of $Ce_{0.5}Zr_{0.5}O_2$ powders in a narrower 2θ range.

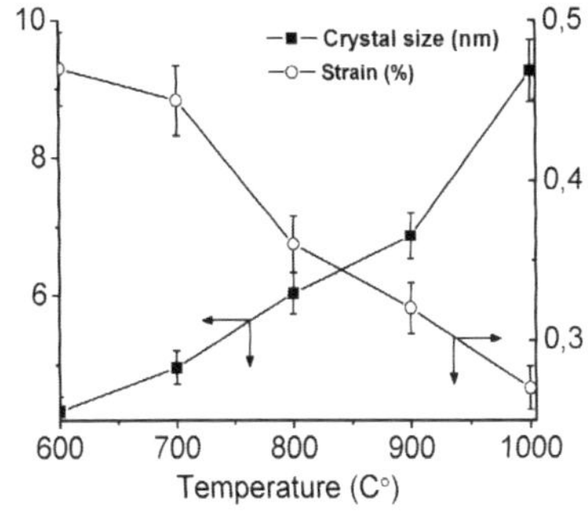


**Figure 3** Crystal size and strain for $Ce_{0.5}Zr_{0.5}O_2$ compound with cubic fluorite-type structure at 600-1000 °C annealing temperatures.

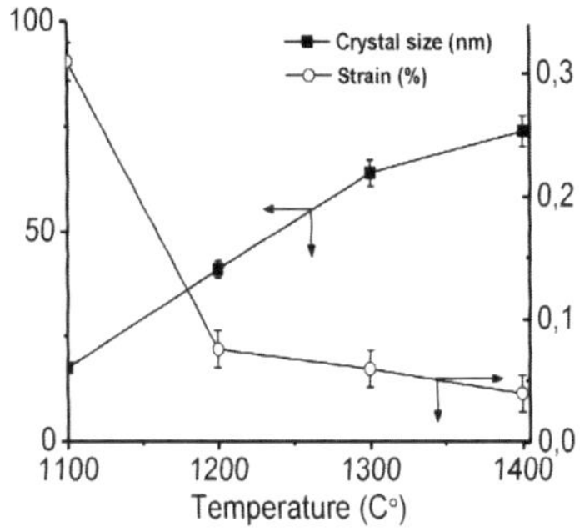


**Figure 4** Crystal size and strain for cubic phase at 1100 – 1400 °C

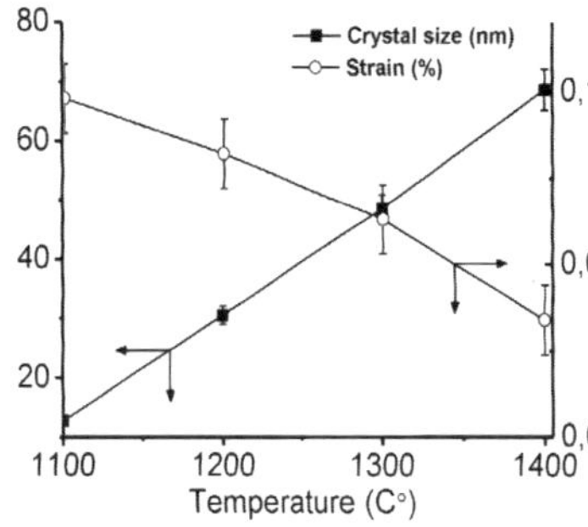


**Figure 5** Crystal size and strain for tetragonal phase at 1100-1400 °C.

**Table 1.** . Structural parameters of $Ce^{4+}{}_{2x}Ce^{3+}{}_{2\text{-}2x}Zr_2O_{7+x}$ reduced samples.

| Sample | Structure | Phase fraction (%) | CSL (nm) | Strain (%) | Compound (TGA) | Compound (XANES) |
|---|---|---|---|---|---|---|
| precursor / $H_2$ 1000°C | - | - | - | - | $Ce^{4+}{}_{0.984}Ce^{3+}{}_{1.016}Zr_2O_{7.508}$ | $Ce^{4+}{}_{0.693}Ce^{3+}{}_{1.307}Zr_2O_{7.347}$ |
| Air 1100°C / $H_2$ 1000°C | *P4₂/nmc*<br>*Fm-3m* | ~ 30<br>~ 70 | 16<br>23 | 0.140<br>0.095 | $Ce^{4+}{}_{0.924}Ce^{3+}{}_{1.076}Zr_2O_{7.462}$ | - |
| Air 1400°C / $H_2$ 1000°C | *P4₂/nmc*<br>*Fm-3m* | ~ 30<br>~ 70 | 92<br>110 | 0.014<br>0.069 | $Ce^{4+}{}_{0.488}Ce^{3+}{}_{1.512}Zr_2O_{7.244}$ | $Ce^{4+}{}_{0.356}Ce^{3+}{}_{1.644}Zr_2O_{7.178}$ |
| Air 1100°C / Vac 1400°C | *Fd-3m* | ~ 100 | 155 | 0.090 | $Ce^{4+}{}_{0.206}Ce^{3+}{}_{1.794}Zr_2O_{7.103}$ | $Ce^{4+}{}_{0.135}Ce^{3+}{}_{1.865}Zr_2O_{7.068}$ |

The increasing of synthesis temperature from 600 to 1000°C leads to the narrowing of the main peaks, due to the increase in the size of the coherent scattering length (CSL) and the decrease in microstrain values (Figure 3). Further increase in temperature at range 1100-1400 °C causes also an increase of CSL and a decrease in the microstrain for both types of formed phases (Figure 4 and Figure 5).

Annealing in reduction hydrogen atmosphere causes formation of $Ce^{4+}{}_{2x}Ce^{3+}{}_{2\text{-}2x}Zr_2O_{7+x}$ compounds with intermediate valency of cerium, where value of $x$ depends on the reducing conditions and treatment parameters of precursor (Figure 6 and Table 1). Annealing in vacuum at 1400 °C strongly reduces the content of $Ce^{4+}$ in a powder samples and leads to formation of pyrochlore structure (space group *Fd-3m*) with practically $Ce^{3+}$ valence state.

Valence state of cerium for the synthesized powders $Ce^{4+}{}_{2x}Ce^{3+}{}_{2\text{-}2x}Zr_2O_{7+x}$ was obtained from XANES spectroscopy at $L_3$ absorption edge of cerium, corresponding to electron dipole transition from the domestic level $2p_{3/2}$ to valence $5d_{5/2}$ level (Figure 7). As seen from Figure 7 the powder obtained by heat treatment in air at 1400 °C/3h has the $L_3$-Ce XANES spectrum in the form of split "white line" corresponding to compounds containing $Ce^{4+}$ cations, particularly to $CeO_2$ [9]. Variation of the absorption peak for the reduced samples were seen owing to the excitation of two possible channels due to the difference in the electronic structure of the cations $Ce^{3+}$ ($4f^1$) and $Ce^{4+}$ ($4f^0$). Herewith the decrease in the degree of oxidation of cerium cations causes shifting the position of the absorption edge to lower energies (Figure 7). It should be noted that the variation of the XANES-spectra for reduced samples correlates qualitatively with the chemical compositions of the samples determined by thermogravimetric method (table 1).

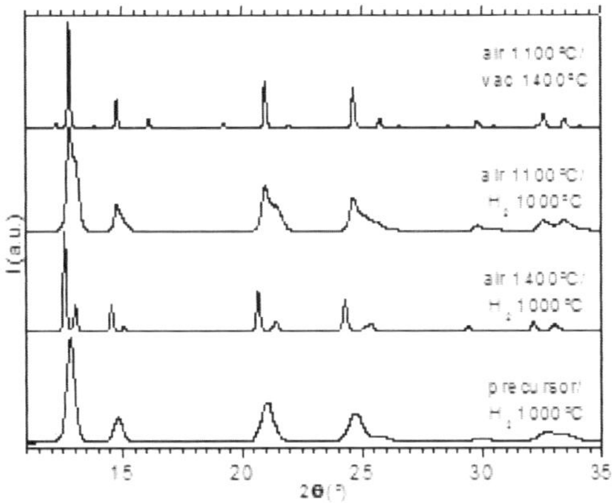


**Figure 6** XRD patterns of reduced samples obtained under different conditions.

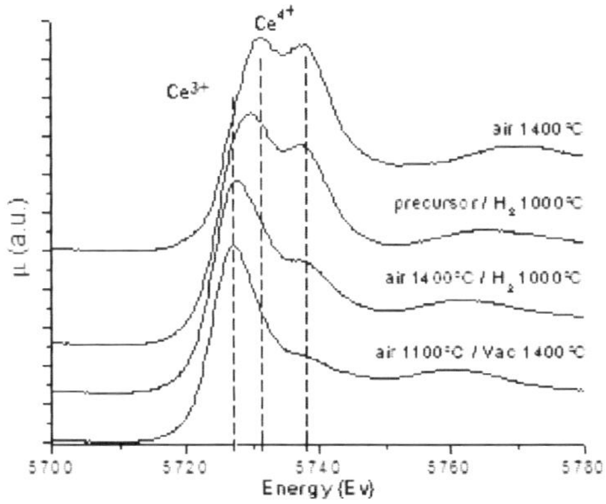


**Figure 7** XANES-spectra of reduced samples obtained under different conditions.

## 4. Conclusion

Isothermal annealing of mixed hydroxides of cerium and zirconium at 600-1000 °C in air leads to formation of $Ce_{0.5}Zr_{0.5}O_2$ powders with FCC fluorite-type structure (space group *Fm*-3*m*). Increase of annealing temperatures above 1000 °C causes decomposition of formed crystal structure into the two phases: cubic $Ce_{0.55}Zr_{0.45}O_2$ and tetragonal $Ce_{0.07}Zr_{0.93}O_2$. It is shown that type and parameters of crystal structure of obtained powders strongly depends on the reducing conditions and treatment parameters of precursor. It has been found that the reduction in vacuum or in hydrogen atmosphere causes formation of $Ce^{4+}{}_{2x}Ce^{3+}{}_{2-2x}Zr_2O_{7+x}$ compounds with intermediate valency of cerium. Also shown, that the reduction in vacuum leads to formation of a compound with a minimum content of $Ce^{4+}$ which has the pyrochlore-type structure (space group *Fd*-3*m*).

**Acknowledgements**

The authors thank S. Carlson (MAX IV lab.), A.A. Ivanov (NRNU MEPhI) for their help in experiments. This work was partly supported by the Ministry of Education and Science of the Russian Federation within the framework of the state assignment 3.1540.2014/K and by Russian Foundation for Basic Research (grant No. 14-22-01002-ofi_m).

**References**

[1] Kašpar J, Graziani M, Fornasiero P 2000 *Handbook on the physics and chemistry of rare earths: The role of rare earths in catalysis* vol **29**, ed K A Gschneidner Jr and L Erwin (Amsterdam: Elsevier Science) 159.

[2] Venâncio S A, de Miranda P E V 2011 Synthesis of $CeAlO_3/CeO_2$-$Al_2O_3$ for use as solid oxide fuel cell functional anode materia *Ceram. Inter.* **37** 3139-3152.

[3] Popov V V, Menushenkov A P, Zubavichus Ya V *et al* 2016 Effect of the Synthesis Conditions on the Crystal, Local, and Electronic Structure of $Ce^{3+}{}_{(1-x)}Ce^{4+}{}_{(x)}AlO_{(3+x/2)}$ *Russ. J. Inorg. Chem.* **61** 225-231

[4] Popov V V 2015 Formation Regularities of Dispersed Hydrated Oxide Systems *Russ. J. Inorg. Chem.* **60** 420-427.

[5] Popov V V, Menushenkov A P, Zubavichus Ya V *et al* 2013 Characteristic Features of the Nanocrystalline Structure Formation in $Ln_2Hf_2O_7$ (Ln = Gd, Dy) compounds *Russ. J. Inorg. Chem.* **58** 1400-1407.

[6] Popov V V, Zubavichus Ya V, Menushenkov A P *et al* 2015 Lanthanide Effect on the Formation and Evolution of Nanocrystalline Structures in $Ln_2Hf_2O_7$ Compounds (Ln = Sm–Dy) *Russ. J. Inorg. Chem.* **60** 16-22.

[7] Petricek V, Dusek M, Palatinus L 2014 Crystallographic computing system JANA2006: General features *Z. Kristallogr.* **229** 345-352.
[8] Klementiev K V 2001 VIPER for Windows *J. Phys. D: Appl. Phys.* **34** 209
[9] Safonova O V, Guda A A, Paun C *et al* 2014 Electronic and Geometric structure of $Ce^{3+}$ Forming Under Reducing Conditions is Shaped ceria Nanoparticles Promoted by Platinum *J. Phys. Chem.* **118** 1974-1982.

# Features of crystal and local structures of complex oxides formed in the binary system $Dy_2O_3$-$ZrO_2$

**V V Popov**[1,2,3], **A P Menushenkov**[1], **A A Yastrebtsev**[1], **Ya V Zubavichus**[2], **R D Svetogorov**[2] **and N A Kolyshkin**[2]

[1] National Research Nuclear University MEPhI (Moscow Engineering Physics Institute), Kashirskoe sh. 31, 115409 Moscow, Russia

[2] NRC "Kurchatov Institute", pl. Akademika Kurchatova 1, 123182 Moscow, Russia

E-mail: victorvpopov@mail.ru

**Abstract**. The process of $Dy_2Zr_2O_7$ ($r_{Dy3+}/r_{Zr4+}$ = 1.426) nanocrystals formation and evolution upon calcinations up to 1400 °C has been investigated by means of synchrotron radiation X-ray diffraction (XRD) and Raman spectroscopy. We observed that isothermal calcination of the X-ray amorphous precursor first leads to the formation of oxide nanocrystalline powders with a defect fluorite structure at 800 °C/3 h. The nucleation and growth of pyrochlore phase take place in nanodomains with pyrochlore-type superstructural ordering of cations and anions inside a well crystalline fluorite matrix at ≥ 1200 °C. This is perhaps the first XRD study that reports the presence of a weak pyrochlore-type superstructure in the $Dy_2Zr_2O_7$.

## 1. Introduction

Complex oxides formed in the binary systems $Ln_2O_3$–$MO_2$ (Ln = lanthanides, M = Zr, Hf) intensively studied in recent years [1–4]. The formation of different types of crystal structure of these compounds depends on the synthesis conditions, types of $Ln_2O_3$ and $MO_2$ oxides, temperature, etc. [5]. One of the main factors that determine the type of the resulting crystal structure is the cation radius ratio $r_{Ln3+}/r_{M4+}$ [1]. The "pyrochlore – fluorite" phase transition takes place at $r_{Ln3+}/r_{M4+}$ ~ 1.46 [1]. According to updated radius-ratio rules for $Ln_2M_2O_7$ the radius-ratio limits are as follows: disordered fluorite < 1.21 < δ-phase < 1.42–1.44 < pyrochlore < 1.78–1.83 < monoclinic pyrochlore < 1.92 [6]. The most interesting compounds are located near the lower boundary of "fluorite - pyrochlore" solid state phase transition. These compounds reveal maximum ionic conductivity [2] and the best radiation-tolerance properties [7]. The phase transition from an ordered pyrochlore to a disordered defect fluorite structure is the rare example of simultaneous disorder in both cation and anion sub-lattice [8].

Dysprosium zirconate $Dy_2Zr_2O_7$ ($r_{Dy3+}/r_{Zr4+}$ = 1.426) is one of such compound with structure exactly between pyrochlore and fluorite phase (or δ-phase). It has been reported, based on XRD studies, that $Dy_2Zr_2O_7$ has a defect fluorite structure with complete disorder of the metal ions and oxygen vacancies [9, 10]. Recent theoretical work [11] showed that the "order-disorder" transformation temperature (793 K) in $Dy_2Zr_2O_7$ is too low for the formation of ordered pyrochlore structure. However few years ago Sayed et al. [12] using Raman spectroscopy experimentally found the formation of a weak pyrochlore-type ordering in the $Dy_2Zr_2O_7$ pellet. On the other hand, in [13,

[3] To whom any correspondence should be addressed.

14] have been shown that $Dy_2Hf_2O_7$ ($r_{Dy3+}/r_{Hf4+}$ = 1.446) manifested a weakly tendency to form pyrochlore structure in ceramic pellets.

The aim of the present work is to study the formation and evolution of crystal and local structures in "boundary" compound $Dy_2Zr_2O_7$, upon isothermal calcination of an amorphous precursor by combined X-ray diffraction (XRD) using synchrotron radiation (information about cation ordering) and Raman spectroscopy (anion ordering).

## 2. Experimental

The starting materials were $Dy(NO_3)_3{\cdot}5H_2O$ (99.5% purity) and $ZrOCl_2{\cdot}8H_2O$ (99% purity). Precursor (mixed Dy-Zr hydroxide) was prepared by co-precipitation of metal salts solution with ammonia $NH_3{\cdot}H_2O$ (analytical grade) [15]. The $Dy_2Zr_2O_7$ powders were prepared by calcination of precursor at 600-1400 °C for 3 h in air. The $Dy_2Zr_2O_7$ powder (800 °C/3 h) was further compacted as pellet and sintered at 1450 °C in air for 4 h to get the final ceramic material. A more detailed experimental procedure was described in Refs [14–17].

The thermal behaviour of the dried powders was studied from 100 to 1400 °C using thermogravimetric analysis and differential scanning calorimetry (TG/DSC, Netzsch STA 409 PC Luxx) with a heating rate of 10 K $min^{-1}$ in an argon flow of 30 mL/min in corundum crucibles.

Crystal structure of $Dy_2Zr_2O_7$ synthesized powders and pellet was studied by X-ray powder diffraction at "Structural Materials Science" beamline of the Kurchatov synchrotron radiation source. Measurements were carried out in the transmission mode at $\lambda$ = 0.68886 Å. The Rietveld full-profile analysis of X-ray diffraction patterns was performed with the Jana2006 software [18].

FT–Raman spectra were recorded at research analytical center JSC "VNIIChT" on the Nicolet iS50 FT-IR spectrometer (Thermo Scientific) equipped with iS50 Raman module with a laser excitation wavelength of 1064 nm. The spectra were measured in wave number range from 100 to 3700 $cm^{-1}$ and were averaged out of 256 scans with a time interval of 2 s and a resolution of 4 $cm^{-1}$.

## 3. Results and discussion

The freshly washed precipitate of mixed Dy-Zr hydroxide was strongly hydrated (~ 95% water). The XRD study showed that both wet and dry powders were amorphous. However, XRD pattern indicates first broad maximum and thus the presence of near order in the precursor structure.

According to TG-analysis, the dry precursor had formula $Dy_2O_3{\cdot}ZrO_2{\cdot}12.7\ H_2O$. Significant parts of weight loss up to 344 °C (~ 63% of total) and above 900 °C (~ 10% of total) indicate a high degree of hydration and strong retention of $OH^-$-groups in the crystal lattice of the synthesized samples.

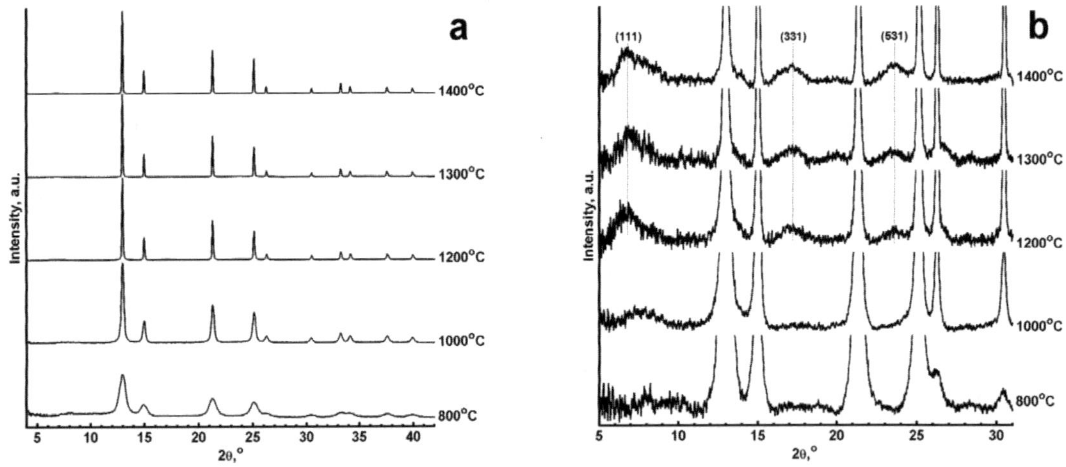


**Figure 1.** XRD patterns of $Dy_2Zr_2O_7$ powders obtained by calcination of the precursor at different temperatures (shown on the corresponding curves) for 3 h in the air: (a) general pattern, (b) in a narrower range. In brackets are the reflection indices of the pyrochlore phase.

It has been found that primary crystallization of precursor occurs at 800 °C/3 h. At 800 °C the powder appears to be nanocrystalline (coherent scattering length – CSL – 8 nm) with a fluorite structure (*Fm-3m*) in accordance with XRD data. A further increase in the calcination temperature leads to the narrowing of diffraction peaks (Figure 1a) owing to the increase of CSL and the decrease of microstrain values (ε) (Table 1). The diffraction patterns for $Dy_2Zr_2O_7$ strongly change at calcination temperature higher than 1200 °C. The XRD patterns of the powders synthesized at temperatures ≥ 1200 °C/3 h show the appearance of superstructure peaks (111), (331), and (531), which point to the onset of pyrochlore cationic ordering (*Fd-3m*) (Figure 1b). This is perhaps the first synchrotron radiation XRD study that reports the presence of a weak pyrochlore-type superstructure in the $Dy_2Zr_2O_7$.

Figure 1b shows that the superstructure reflections remain significantly broader than the basic peaks characteristic of fluorite. The XRD quantitative analysis (Table 1) enables to suggest that, upon the calcinations of $Dy_2Zr_2O_7$ the cationic ordering formation of the pyrochlore phase presumably occus in separate regions (nanodomains) distributed inside of bulkier crystals with the fluorite structure. Similar results were obtained earlier for $Ln_2Hf_2O_7$ (Ln = Sm – Dy) [19] and $Gd_2Zr_2O_7$ [20].

**Table 1.** Temperature dependence of atomic crystal structure parameters of $Dy_2Zr_2O_7$ powders.

| | | Temperature, °C | | | | |
|---|---|---|---|---|---|---|
| | | 800 | 1000 | 1200 | 1300 | 1400 |
| Crystal structure | Parameter | | | | | |
| *Fm-3m* | CSL, nm | 8 | 21 | 217 | - | 302 |
| | ε, % | 0.599 | 0.165 | 0.497 | - | 0.108 |
| *Fd-3m* | CSL, nm | - | - | 4 | 5 | 8 |
| | ε, % | - | - | 7.461 | 6.543 | 9.239 |
| | *a*, Å | 5.2388 | 5.2338 | 5.2318 | 5.2344 | 5.2304 |

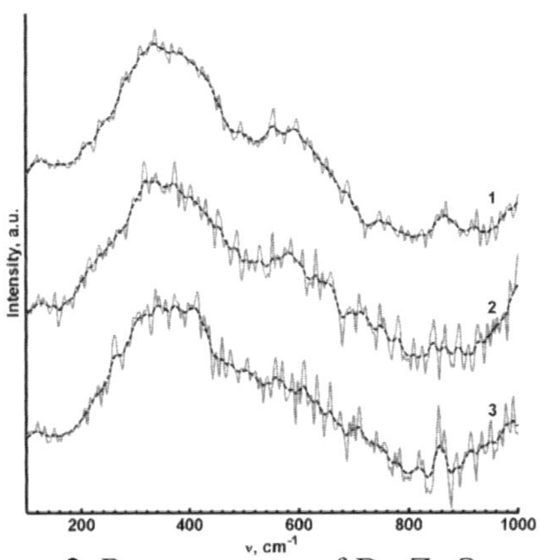

**Figure 2.** Raman spectra of $Dy_2Zr_2O_7$ powders (1, 2) and ceramic pellet (3) prepared by calcination at different temperatures: 1 – 1200; 2 – 1400; 3 – 1450 °C

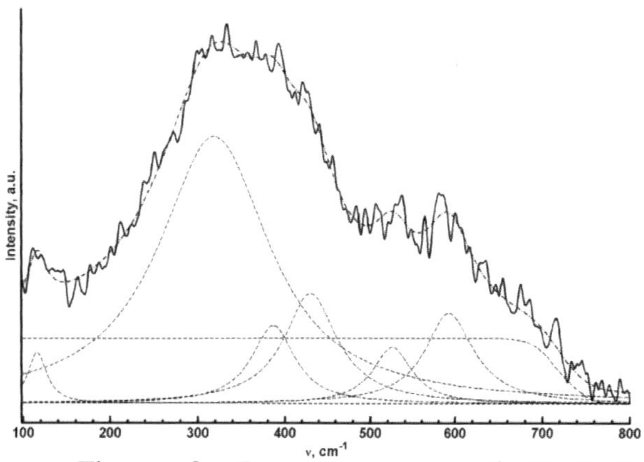

**Figure 3.** Raman spectrum of $Dy_2Zr_2O_7$ (1450°C/4 h): solid line – experimental spectrum; dashed lines approximation curves.

Raman spectroscopy data are represented in Figure 2. One can see that the Raman spectra of $Dy_2Zr_2O_7$ pellet (1450 °C/4 h) include few broaden modes. The similar spectra were represented at [12]. Quantitative analysis of the received Raman spectrum of $Dy_2Zr_2O_7$ pellet showed five of the six pyrochlore structure modes: $E_g$ (386 cm$^{-1}$), $A_{1g}$ (525 cm$^{-1}$) and $3F_{2g}$ (319 main, 430, 593 cm$^{-1}$). The Raman spectroscopy is sensitive to "oxygen-cation" vibrations and gives information about the local structure ordering of the anionic sublattice. Thus our Raman spectroscopic data shown that $Dy_2Zr_2O_7$

ceramics reveal a tendency to form the weak pyrochlore-type ordering in anion sublattice. In addition to our XRD data (Figure 1b) this result points to the both the cation and anion pyrochlore-type ordering in the nanodomains inside the well crystalline fluorite matrix.

## 4. Conclusion

The evolution of $Dy_2Zr_2O_7$ crystalline and local structures upon calcination of initial mixed hydroxide has been studied by a complex of XRD and Raman spectroscopy methods. It has been shown that isothermal calcination of the X-ray amorphous precursor at temperatures $\geq$ 800°C first leads to the formation of oxide nanocrystalline powders with a defect fluorite structure and reveals an increase in coherent scattering lengths and a decrease in microstrain values upon calcination temperature increase. Heat treatment at temperature $\geq$ 1200 °C initiates nucleation and growth of nanodomains with pyrochlore-type superstructural ordering of cations and anions inside a microcrystalline fluorite matrix.

**Acknowledgements**

The authors thank N.A. Tsarenko (JSC "VNIIChT") for her help in Raman experiments. This work was partly supported by Ministry of Education and Science of the Russian Federation within agreement No. 14.587.21.002 and in the framework of the state assignment 3.1540.2014/K.

**References**

[1] Subramanian M A, Aravamudan G and Subba Rao G V 1983 Oxide pyrochlores – a review *Prog. Solid State Chem.* **15** pp 55–143

[2] Yamamura H, Nishino H, Kakinuma K *et al* 2003 Electrical conductivity anomaly around fluorite–pyrochlore phase boundary *Solid State Ionics* **158** 359–65

[3] Shimamura K, Arima T, Idemitsu K *et al* 2007 Thermophysical properties of rare-earth-stabilized zirconia and zirconate pyrochlores as surrogates for actinide-doped zirconia *Int. J. Thermophys.* **28** 1074–84

[4] Blanchard P E R, Liu S, Kennedy B J *et al* 2012 Does local disorder occur in the pyrochlore zirconates? *Inorg. Chem.* **51** 13237–44

[5] Andrievskaya E R 2008 Phase equilibria in the refractory oxide systems of zirconia, hafnia and yttria with rare-earth oxides *J. Eur. Ceram. Soc.* **28** 2363–88

[6] Stanek C R, Jiang C, Uberuaga B P *et al* 2009 Predicted structure and stability of $A_4B_3O_{12}$ δ-phase compositions *Phys. Rev. B* **80** 174101-1–174101-11

[7] Sickafus K E, Grimes R W, Valdez J A *et al* 2007 Radiation-induced amorphization resistance and radiation tolerance in structurally related oxides *Nature Mater.* **6** 217–23

[8] Blanchard P E R, Liu S, Kennedy B J *et al* 2013 Investigating the order–disorder phase transition in $Nd_{2-x}Y_xZr_2O_7$ *via* diffraction and spectroscopy. *Dalton Trans.* **42** 14875–82

[9] Qiang X, Wei P, Jingdong W *et al* 2005 Preparation and thermophysical properties of $Dy_2Zr_2O_7$ ceramic for thermal barrier coatings *Mater. Letters.* **59** 2804–8

[10] Tong Y, Yu Z, Lu L *et al* 2008 Rapid preparation and characterization of $Dy_2Zr_2O_7$ nanocrystals *Mater. Res. Bull.* **43** 2736–41

[11] Jiang C, Stank C R, Sickafus K E *et al* 2009 First-principles prediction of disordering tendencies in pyrochlore oxides *Phys. Rev. B* **79** 104203-1–104203-5

[12] Sayed F N, Grover V, Bhattacharyya K *et al* 2011 $Sm_{2-x}Dy_xZr_2O_7$ Pyrochlores: probing order-disorder dynamics and multifunctionality *Inorg. Chem.* **50** 2354–65

[13] Mandal B P, Garg N, Sharma S M *et al* 2006 Preparation, XRD and Raman spectroscopic studies on new compounds $RE_2Hf_2O_7$ (*RE* = Dy, Ho, Er, Tm, Lu, Y): Pyrochlores or defect-fluorire? *J. Solid State Chem.* **179** 1990–4

[14] Popov V V, Menushenkov A P, Zubavichus Ya V *et al* 2016 Features of crystal and local structures of compounds formed in "$Dy_2O_3$-$HfO_2$" system *Russ. J. Inorg. Chem.* **61** (in press)

[15] Popov V V 2015 Formation Regularities of Dispersed Hydrated Oxide Systems

*Russ. J. Inorg. Chem.* **60** 420–7

[16] Popov V V, Menushenkov A P, Zubavichus Ya V *et al* 2013 Characteristic features of the nanocrystalline structure formation in $Ln_2Hf_2O_7$ (Ln = Gd, Dy) compounds *Russ. J. Inorg. Chem.* **58** 1400–7

[17] Popov V V, Petrunin V F, Korovin S A *et al* 2011 Formation of nanocrystalline structures in the $Ln_2O_3$-$MO_2$ systems (Ln = Gd, Dy; M = Zr, Hf) *Russ. J. Inorg. Chem.* **56** 1538–44

[18] Petricek V, Dusek M and Palatinus L 2014 Crystallographic computing system JANA2006: General features *Z. Kristallogr.* **229** 345–52

[19] Popov V V, Zubavichus Ya V, Menushenkov A P *et al* 2015 Lanthanide effect on the formation and evolution of nanocrystalline structures in $Ln_2Hf_2O_7$ compounds (Ln = Sm–Dy) *Russ. J. Inorg. Chem.* **60** 16–22

[20] Popov V V, Zubavichus Ya V, Menushenkov A P *et al* 2014 Short- and long-range order balance in nanocrystalline $Gd_2Zr_2O_7$ powders with a fluorite–pyrochlore structure *Russ. J. Inorg. Chem.* **59** 279–285

Journal of Physics: Conference Series **747** (2016) 012043 doi:10.1088/1742-6596/747/1/012043

# $La_2Hf_2O_7$ crystal and local structure changes on the fluorite – pyrochlore phase transition

**V V Popov**[1,2,3], **A P Menushenkov**[1], **A A Yastrebtsev**[1] **and Ya V Zubavichus**[2]

[1] National Research Nuclear University MEPhI (Moscow Engineering Physics Institute), Kashirskoe sh. 31, 115409 Moscow, Russia

[2] NRC "Kurchatov Institute", pl. Akademika Kurchatova 1, 123182 Moscow, Russia

E-mail: victorvpopov@mail.ru

**Abstract**. The process of $La_2Hf_2O_7$ ($r_{La3+}/r_{Hf4+}$ = 1.63) nanocrystals formation and evolution upon calcinations up to 1400 °C has been investigated by means of synchrotron radiation X-ray diffraction (XRD) and Raman spectroscopy. It has been shown that isothermal calcination at 800 °C/3h of the X-ray amorphous precursor firstly leads to the formation of oxide nanocrystalline powders with a defect fluorite structure. In the temperature range 900 – 1000 °C we observed the nucleation and growth of pyrochlore nanodomains inside a well crystalline fluorite matrix. The pyrochlore-type superstructural ordering of cations and anions appears at calcinations temperature higher than 1000 °C.

## 1. Introduction

The compounds and solid solutions formed in «$Ln_2O_3$ - $MO_2$» systems (*Ln* – lanthanides, *M* – titanium subgroup elements) has intensively investigated due to their important scientific interest (order-disorder phase transitions [1, 2]; geometrically frustrated magnets [3], etc.) and technological applications (thermal barrier coatings [4]; solid electrolytes in high temperature solid oxide fuel cells [5]; neutron absorbing materials [6]; nuclear waste storage materials [7], etc.). The application functionality of materials depends on the crystal and local structure of used materials. The structural phase transition from the order *Fd-3m* pyrochlore structure to the disorder *Fm-3m* defect fluorite structure involves the randomization of the oxygen ions among the 48*f*, 8*b*, and 8*a* sites, and the cations between the 16*c* and 16*d* sites [8].

Lanthanides hafnates are the less investigated in the $Ln_2M_2O_7$ family. During some years we actively investigated the specific features of the structure formation in "boundary" compounds $Ln_2Hf_2O_7$ (*Ln* = Sm – Dy) with the cation ratio $r_{Ln3+}/r_{M4+}$~1.46 [9–11]. The aim of the present work is to study the formation and evolution of crystal and local structures in $La_2Hf_2O_7$ ($r_{La3+}/r_{Hf4+}$ = 1.63), which has the most thermodynamically stable pyrochlore phase among $Ln_2Hf_2O_7$ group, upon isothermal calcination of an amorphous precursor. We simultaneously observed the processes of cation ordering using synchrotron radiation X-ray diffraction and anion ordering using Raman spectroscopy upon the "fluorite - pyrochlore" phase transition.

## 2. Experimental

[3] To whom any correspondence should be addressed.

The starting materials were $La(NO_3)_3{\cdot}6H_2O$ (99.96% purity) and $HfOCl_2{\cdot}8H_2O$ (99.7% purity). Precursor (mixed La-Hf hydroxide) was prepared by co-precipitation of metal salts solution with ammonia $NH_3{\cdot}H_2O$ (analytical grade) [12]. The $La_2Hf_2O_7$ powders were prepared by calcination of precursor at 600-1400 °C for 3 h in air. A more detailed experimental procedure was described in Refs [9–11].

The thermal behaviour of the dried powders was studied from 100 to 1400 °C using thermogravimetric analysis and differential scanning calorimetry (TG/DSC, Netzsch STA 409 PC Luxx) with a heating rate of 10 K min$^{-1}$ in an argon flow of 30 mL/min in corundum crucibles.

Crystal structure of $La_2Hf_2O_7$ synthesized powders was studied by X-ray powder diffraction at "Structural Materials Science" beamline of the Kurchatov synchrotron radiation source. Measurements were carried out in the transmission mode at $\lambda$ = 0.68886 Å. The Rietveld full-profile analysis of X-ray diffraction patterns was performed with the Jana2006 software [13].

FT–Raman spectra were recorded at research analytical center JSC "VNIIChT" on the Nicolet iS50 FT–IR spectrometer (Thermo Scientific) equipped with iS50 Raman module with a laser excitation wavelength of 1064 nm. The spectra were measured in wave number range from 100 to 3700 cm$^{-1}$ and were averaged out of 256 scans with a time interval of 2 s and a resolution of 4 cm$^{-1}$.

## 3. Results and discussion

The freshly washed precipitate of mixed La-Hf hydroxide was strongly hydrated (~ 95% water). The XRD study showed that both wet and dry powders were amorphous. However, XRD pattern indicated the presence of the first broad maximum which pointed to origin of the local order in the precursor structure.

According to TG-analysis, the dry precursor had formula $La_2O_3{\cdot}HfO_2{\cdot}10.7\ H_2O$. Significant parts of weight loss up to 300 °C (~ 50% of total) and above 950 °C (6% of total) indicate a high degree of hydration and strong retention of $OH^-$-groups in the crystal lattice of the synthesized samples.

It has been found that primary crystallization of precursor occurs between 700-800 °C/3 h. At 800 °C the powder appears to be nanocrystalline (coherent scattering length – CSL – 35 nm). Its XRD patterns show reflections corresponding to a fluorite structure (*Fm-3m*) with a unit cell parameter (a) of ~5.33 Å. A further increase in the calcination temperature leads to the narrowing of diffraction peaks (Figure 1a) owing to the increase of CSL values and the decrease in microstrain values ($\varepsilon$) (Table 1).

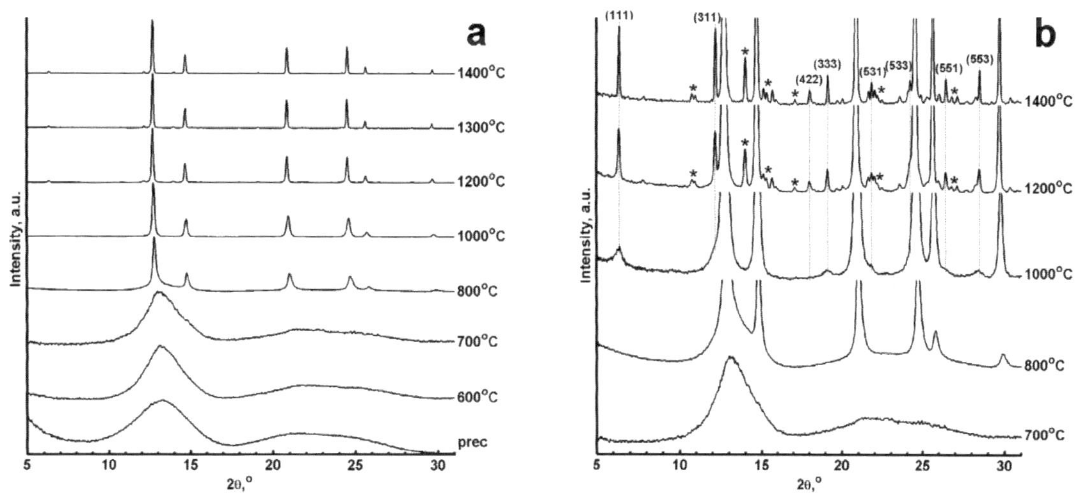


**Figure 1.** XRD patterns of $La_2Hf_2O_7$ powders obtained by calcination of the precursor at different temperatures (shown on the corresponding curves) at 3 h in the air: (a) general pattern, (b) in a narrower range. In brackets are the reflection indices of the pyrochlore phase. * - impurity of $HfO_2$.

The crystallization pattern for $La_2Hf_2O_7$ starts qualitatively changing at temperature ~1000 °C. The XRD patterns of the powders synthesized at temperatures ≥ 1000 °C/3 h show the appearance of superstructure peaks (111) (main), (311), (531) and others, which point to the onset of pyrochlore cationic ordering (*Fd-3m*). From Figure 1b, it can be noted that the superstructure reflections remain significantly broader than the basic peak characteristic of fluorite. The XRD quantitative analysis (Table 1) enables to suggest that, in the case of $La_2Hf_2O_7$ sample calcinations in temperature range 900-1000 °C, the cationic ordering in the pyrochlore phase presumably occurs in separate regions (nanodomains) distributed over bulk crystals with the fluorite structure. It should be noted, that the observed "amorphous→ fluorite → pyrochlore" phase transformation is consistent with Ostwald's step rule. Similar results were obtained earlier for $Ln_2Hf_2O_7$ (Ln = Sm – Dy) [11] and $Gd_2Zr_2O_7$ [14].

**Table 1.** Temperature dependence of atomic crystal structure parameters of $La_2Hf_2O_7$ powders.

| | | Temperature, °C | | | | |
|---|---|---|---|---|---|---|
| | | 800 | 1000 | 1200 | 1300 | 1400 |
| Crystal structure | Parameter | | | | | |
| *Fm-3m* | CSL, nm | 35 | 115 | 149 | 445 | 839 |
| | ε, % | 1.258 | 1.083 | 0.242 | 0.032 | 0.022 |
| *Fd-3m* | CSL, nm | - | 17 | 72 | 375 | 462 |
| | ε, % | - | 0.317 | 0.386 | 0.743 | 0.610 |
| *a*, Å | | 5.3328 | 10.7253 | 10.7531 | 10.7644 | 10.7546 |
| | % wt. $HfO_2$ (XRD / Raman) | - | 2.4 / 1.2 | 4.1 / 4.4 | 4.8 / 5.1 | 6.2 / 6.1 |

The results of Raman spectroscopy investigations are presented in Figure 2. As shown in Figure 2, all the six pyrochlore modes $E_g$ (321 cm$^{-1}$), $A_{1g}$ (499 cm$^{-1}$) and $4F_{2g}$ (304 main, 401, 520 and 755 cm$^{-1}$) are observed in the Raman spectra of $La_2Hf_2O_7$ powders prepared at temperature ≥ 1000°C. These results are in good agreement with [15].

Quantitative analysis of the obtained Raman spectra showed that the process of anion ordering of pyrochlore type in $La_2Hf_2O_7$ powders begins at a lower temperature than the process of cation ordering (according to XRD data). However at calcinations temperature higher than 1100 °C both the cation and anion sublattice of $La_2Hf_2O_7$ became fully ordered. So above this temperature the "fluorite - pyrochlore" phase transformation is fully completed. As shown in Figure 3, the Raman spectra of $La_2Hf_2O_7$ (1400 °C/3h) clear indicate the presence of monoclinic $HfO_2$ impurity. The $HfO_2$ contents calculated from Raman spectra were in good agreement with XRD data (Table 1).

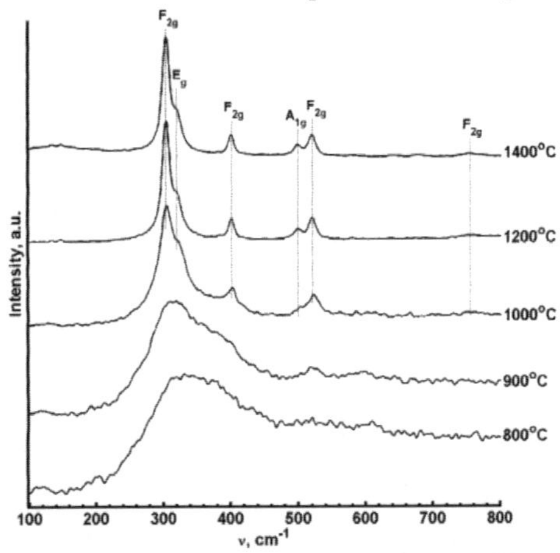


**Figure 2.** Raman spectra of $La_2Hf_2O_7$ powders obtained by calcination of the precursor at different temperatures

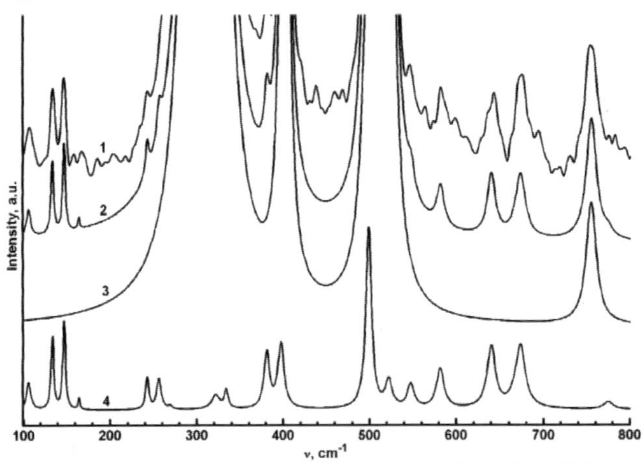


**Figure 3.** Raman spectra of powders $La_2Hf_2O_7$ (1400 °C/3 h) (1) and *m*-$HfO_2$ phase (4); approximation curves: summary (2) and $La_2Hf_2O_7$

(shown on the corresponding curves) for 3 h in the air. (*Fd*-3*m*) (3).

## 4. Conclusion

The evolution of $La_2Hf_2O_7$ crystalline and local structures upon calcination of initial mixed hydroxide has been studied simultaneously by using of synchrotron radiation XRD and Raman spectroscopy methods. It has been shown that isothermal calcination of the X-ray amorphous precursor at temperatures ≥ 800 °C first leads to the formation of oxide nanocrystalline powders with a defect fluorite structure and reveals an increase in coherent scattering lengths and a decrease in microstrain values upon calcination temperature increase. Heat treatment at temperature ≥ 1000 °C initiates nucleation and growth of nanodomains with pyrochlore-type superstructural ordering of cations and anions inside a microcrystalline fluorite matrix. At calcinations temperature higher than 1100 °C both the cation and anion sublattice of $La_2Hf_2O_7$ became fully ordered and the "fluorite - pyrochlore" phase transformation is fully completed.

**Acknowledgements**
The authors thank N.A. Tsarenko (JSC "VNIIChT") for her help in Raman experiments and the Russian Science Foundation (grant 14-22-00098) for financial support.

**References**
[1] Jiang C, Stank C R, Sickafus K E *et al* 2009 First-principles prediction of disordering tendencies in pyrochlore oxides *Phys. Rev. B* **79** 104203-1–104203-5
[2] Blanchard P E R, Liu S, Kennedy B J *et al* 2013 Investigating the local structure on lanthanoid hafnates $Ln_2Hf_2O_7$ via diffraction and spectroscopy *J. Phys. Chem.* C **117** 2266-73
[3] Gardner J S, Gingras M J P, Greedan J E, 2010 Magnetic pyrochlore oxides *Rev. Modern Phys.* **82** 53–107
[4] Pan W, Phillpot S R, Wan C *et al* 2012 Low thermal conductivity oxides *MRS Bull.* **32** 917–22
[5] Yamamura H, Nishino H, Kakinuma K *et al* 2003 Electrical conductivity anomaly around fluorite–pyrochlore phase boundary *Solid State Ionics* **158** 359–65
[6] Risovany V D, Zakharov A V, Muraleva E M *et al* 2006 Dysprosium hafnate as absorbing material for control rods *J. Nucl. Mater.* **355** 163–70
[7] Ewing R C, Weber W J, Lian J 2004 Nuclear waste disposal - pyrochlore ($A_2B_2O_7$): Nuclear waste form for the immobilization of plutonium and "minor" actinides *J. Appl. Phys.* **95** 5949–71.
[8] Nachimuthu P, Thevithasan S, Stutthanandan V *et al* 2005 Near-edge X-ray absorption fine-structure study of ion-beam-induced phase transformation in $Gd_2(Ti_{1-y}Zr_y)_2O_7$ *J. Appl. Phys.* **97** 033518-1–033518-5
[9] Popov V V, Petrunin V F, Korovin S A *et al* 2011 Formation of nanocrystalline structures in the $Ln_2O_3$-$MO_2$ systems (Ln = Gd, Dy; M = Zr, Hf) *Russ. J. Inorg. Chem.* **56** 1538–44
[10] Popov V V, Menushenkov A P, Zubavichus Ya V *et al* 2013 Characteristic features of the nanocrystalline structure formation in $Ln_2Hf_2O_7$ (Ln = Gd, Dy) compounds *Russ. J. Inorg. Chem.* **58** 1400–7
[11] Popov V V, Zubavichus Ya V, Menushenkov A P *et al* 2015 Lanthanide effect on the formation and evolution of nanocrystalline structures in $Ln_2Hf_2O_7$ Compounds (Ln = Sm–Dy) *Russ. J. Inorg. Chem.* **60** 16–22
[12] Popov V V 2015 Formation Regularities of Dispersed Hydrated Oxide Systems *Russ. J. Inorg. Chem.* **60** 420–7
[13] Petricek V, Dusek M and Palatinus L 2014 Crystallographic computing system JANA2006: General features *Z. Kristallogr.* **229** 345–52
Popov V V, Zubavichus YaV, Menushenkov A P *et al* 2014 Short- and long-range order balance in

nanocrystalline $Gd_2Zr_2O_7$ powders with a fluorite–pyrochlore structure *Russ. J. Inorg. Chem.* **59** 279–85
[14] Kumar S, Gupta H C 2012 First principles study of dielectric and vibration properties of pyrochlore hafnates *Solid State Sci.* **14** 1405–11

# Investigation of the hysteresis losses in CC tapes after laser filamentation

**M Osipov[1,2,3], E Splavnik[1], S Pokrovskii[1,2], A Andreev[1], N Mineev[1,2], D Abin[1,2] and I Rudnev[1,2]**

[1] National Research Nuclear University MEPhI (Moscow Engineering Physics Institute), Kashirskoe sh. 31, 115409 Moscow, Russia
[2] International Laboratory of High Magnetic Fields and Low Temperatures, Gajowicka 95, 53-421 Wroclaw, Poland

E-mail address: maosipov1@mephi.ru

**Abstract**. Study of AC loss has been conducted on the 4 mm wide CC-tapes manufactured by SuperOx. Filamentation was carried out by laser cutting of the commercially available tapes with a copper coating. Also for comparison, original tapes and commercial multifilament tapes (manufactured by SuperOx by using chemical etching method) were studied. Losses were obtained from the magnetization curves measured on a vibration sample magnetometer in the temperature range from 4 K to 77 K. In addition, the current carrying characteristics of tapes were studied, and the effect of filamentation on the critical current value were examined. We present an analysis of the experimental results, as well as a comparison of data for different types of samples. The possibility of decreasing the energy losses due to filamentation tapes was demonstrated.

## 1. Introduction

Due to the extensive development of high-temperature superconductivity important task of the efficient use of HTS materials in the energy sector is a problem of reducing the alternating current energy losses in coated conductors tapes (CC-tapes). This problem can be solved by various methods. So, to reduce the losses one can use a special laying method in the cable [1], [2]. Work [3] presents the effect of the tape size and the distance between the individual layers on the losses. The calculation results indicate low expediency of using tapes less than 4 mm width. Thus, it is necessary to develop methods to reduce AC losses for CC-tapes with optimal width.

There are several basic approaches, one of which is based on the separation of the HTS layer into several filaments. Multi-filamentation of the CC-tape can be performed in various ways. For example, work [4] presents the filamentation technique with additional virtual cross-section of the buffer layer. Filamentation can also be performed by the deformation, partial removal of the buffer layer [5], or coating of buffer layer with nanolayer [6]. All this leads to disruption of the superconductor structure above the modified part of the buffer layer. Filamentation can be performed without modification of the substrate or buffer layer. In particular, in work [8] authors proposed a method for producing CC-tape with a zigzag pattern of filaments obtained using the three-step photolithography and chemical etching. The tape filamentation can be done, for example, by laser irradiation of the CC-tapes with

[3] To whom any correspondence should be addressed.

copper coating [9]. This approach is relatively simple since only the post-processing of finished CC-tape is required.

Work [10] describes a complex method of filamentation, leading to record level of AC losses reduction. Non-stabilazed tape was annealed in oxygen after the filamentation process, which leads to oxidation of the buffer layer between the adjacent filaments and isolates them from each other.

Among the methods described above the laser filamentation of copper coated tapes is one of the most promising due to its simplicity and scalability. Also, this technique allows one to tune filamentation parameters for specific tasks.

## 2. Experimental technique and discussion

### 2.1. *Samples*

In our study we used commercially available 4 mm width GdBCO tapes manufactured by SuperOx. The tapes of 0.11 mm thickness on Hastelloy C276 substrate had no isolation. They were copper coated with 50 microns plating thickness. The critical current was about 150 A. The tapes were filamented using laser cutting technique. In addition, we investigated SuperOx produced tapes without copper coating which were divided into 4 filaments by chemical etching.

### 2.2. *Laser specification.*

The tapes were divided into four filament by laser cutting performed using ytterbium pulsed fiber laser with 1.068 micron wavelength and 50 microns focus diameter. Maximal power per pulse of 1 mJ, pulse frequency range from 20 to 100 kHz and beam moving speed up to 20,000 mm/sec and were used.

### 2.3. *Selection of a laser power*

We used four-probe method for critical current measurements and the scanning Hall magnetometry to assess the impact of laser radiation on the superconducting layer. To obtain the laser parameters suitable for the filamentation, the tapes were cut in transverse direction and the critical current was measured. The dependence of the critical current on the laser incision parameters is shown in Figure 1. As a result, we picked up minimal energy of laser pulse, which destroys superconducting layer and provides minimal substrate damage and distortion of the tape.

To check the quality of the laser cut we used a home-made Hall magnetometer [11]. After the laser cutting (the pulse energy 1 mJ, the pulse frequency 20 kHz, the beam moving speed 50 mm/s.) the sample was cooled in liquid nitrogen and magnetized using permanent magnets. Then the picture of remanent flux was measured (Figure 2) and converted into a current distribution by inversing of Biot–Savart equation (Figure 3). Laser cut is clearly visible in Figure 3. There is no current flow through the cut.

### 2.4. *Multifilamentization technique.*

After the laser parameters were selected, tape was filamented into 4 stripes. To estimate the filamentation effect, the direct measurements of the critical current was carried out for both non-filamented and filamented tapes. The tapes chemically filamented by SuperOx were also measured for comparison. The critical current decreased from 150 A to 140 A under the laser filamentation. Such weak decrease (less than 10%) indicates a small cutting width leading to small change of the superconductor cross-section. The critical current of chemically filamented tape dropped to 40 A.

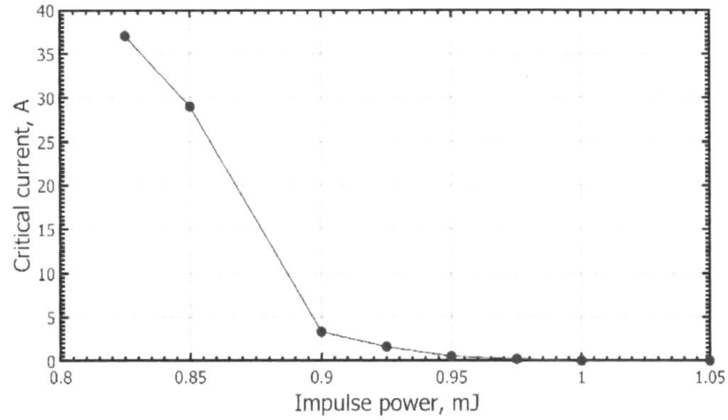


**Figure 1.** The dependence of the critical current obtained after the laser incision on the laser pulse energy. The pulse frequency - 20 kHz, the beam moving speed - 50 mm/sec.

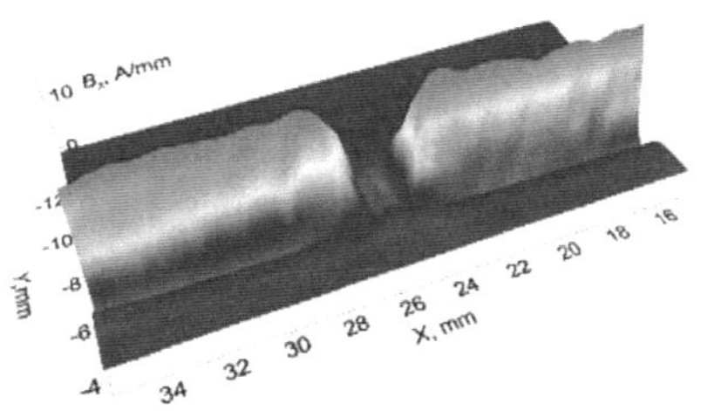


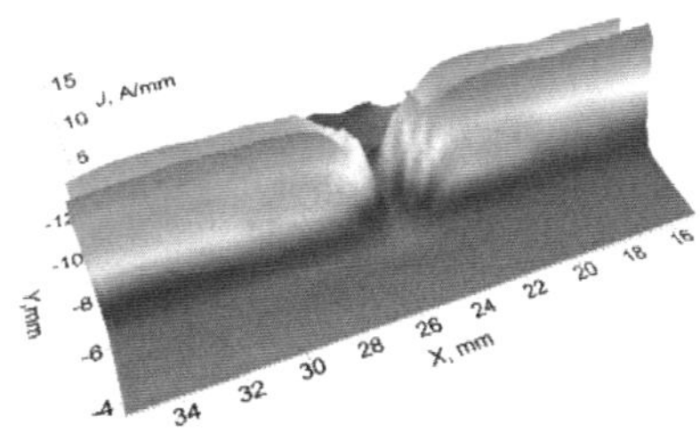


**Figure 2.** Distribution of the trapped field near the transverse cut at temperature $T = 77$ K.

**Figure 3.** Calculated current distribution near the transverse cut at temperature $T = 77$ K.

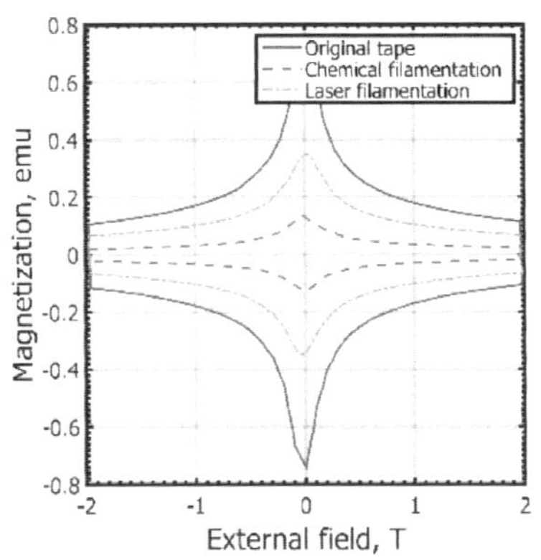


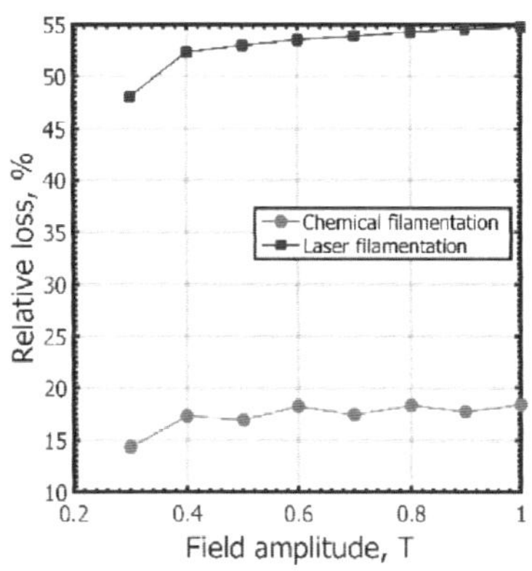


**Figure 4.** An example of the magnetization curve for all three samples at $T = 50$ K.

**Figure 5.** Dependence of the relative losses on the integration region at $T = 50$ K.

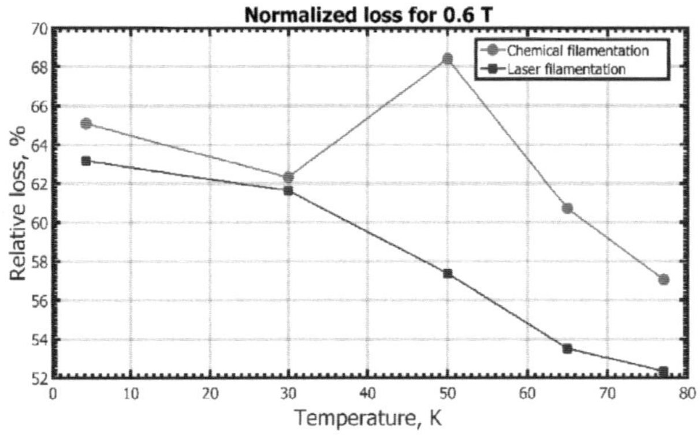


**Figure 6.** Temperature dependence of the relative losses normalized by the relative change of critical current. Data are presented for the integration region of 0.6 T. Normalization factors are 140/150 and 40/150 respectively for laser and chemically filamented tapes.

On the next step we studied filamentation influence on AC losses estimated as hysteretic losses per cycle multiplied by operating frequency. The hysteretic losses were evaluated from the magnetization curves measured on a vibration magnetometer at various temperatures in external magnetic field produced by 15 T superconducting magnet Oxford Instruments. Measurements were performed with sweep rate of 0.5 T/min. Each curve was measured from zero field up to 2 T, then to -2 T and again to zero. The curves obtained at $T = 50$ K are shown in Figure 4 as an example. Hysteretic losses per cycle can be evaluated according to the formula:

$$P_h = C \times A \times \mu_0 \times \oint H_e dM = -C \times A \times \mu_0 \times \oint M dH_e$$

where $H_e$ – is applied magnetic field, $M$ - magnetization, $C \times A$ - the effective cross section of the superconductor, $A$ represents a geometric section, and $C$ - the effective coefficient, at a low frequency $C = 1$ [12]. The region of integration is chosen the same as an amplitude of the external field affecting the superconductor (or tape self-field, in the case of AC current loss in zero external field). The hysteretic losses calculated for several fields from the magnetization curves presented in Figure 4 are shown in Figure 5. The losses of filamented tapes were normalized to the losses of the non-filamented one. The relative losses of two samples with identical sizes can be obtained as the ratio of their hysteresis loops areas. So, there is no need to calculate the coefficients before the integral. Example of dependence of the relative losses on the field amplitude for the temperature $T = 50$ K is shown in Figure 5. The graph shows the losses reduction for both types of filamentation, in comparison with the non-filamented tape. We obtained that the losses increase with field, therefore in low fields, or in self-field for AC current, the filamentation is more efficient.

Due to the differences of the critical currents of the laser and is chemically filamented tapes, it makes sense to normalize losses on the relative change of critical current caused by the filamentation procedure. Normalization factors are 140/150 and 40/150 respectively for laser and chemically filamented tapes. Figure 6 shows the normalized loss as function of temperature. This dependence shows that the relative losses of sample filamented by laser less than the losses of the chemically filamented sample. In addition, for both samples the relative losses tend to decline at higher temperatures. The only exception is the point at $T = 50$ K for the chemically filamented sample for which the loss growth can be attributed to the error of the experiment.

## 3. Conclusion

Effect of laser filamentation on hysteretic losses in CC-tapes was studied. Optimal parameters of laser cutting of tapes with copper coating were chosen experimentally. Measurements of hysteretic losses per cycle demonstrated the possibility of AC losses reducing via tapes filamentation.

**Acknowledgements**
This work was partly supported by the Ministry of Education and Science of the Russian Federation within the framework of the state assignment 3.1540.2014/K and by the Competitiveness Program of NRNU MEPhI. We express our gratitude to SuperOx Company for the cooperation and providing with CC-tapes.

**References**

[1] Clem J R 2008 Field and current distributions and ac losses in a bifilar stack of superconducting strips *Phys. Rev. B* **77** 1–7
[2] Terzieva S, Vojenčiak M, Pardo E, Grilli F, Drechsler A, Kling A, Kudymow A, Gömöry F and Goldacker W 2010 Transport and magnetization AC losses of ROEBEL assembled coated conductor cables: measurements and calculations *Supercond. Sci. Technol.* **23** 14023
[3] Malozemoff A. P, Snitchler G and Mawatari Y 2009 Tape-width dependence of AC losses in HTS cables *IEEE Trans. Appl. Supercond.* **19** 3115–8
[4] Zhang Y, Duckworth R C, Ha T T, List F A., Gouge M J, Chen Y, Xiong X, Selvamanickam V and Polyanskii A 2011 AC loss reduction in filamentized YBCO coated conductors with virtual transverse cross-cuts *IEEE Trans. Appl. Supercond.* **21** 3301–6
[5] Majkic G, Kesgin I, Zhang Y, Qiao Y, Schmidt R and Selvamanickam V 2011 AC Loss Filamentization of 2G HTS Tapes by Buffer Stack Removal *IEEE Trans. Appl. Supercond.* **21** 3297–300
[6] Goyal A, Superconductor films with improved flux pinning and reduced AC losses // US Patent № US 7,919,435 B2, 2011
[7] Cai X, Kesgin I, Schmidt R, Chen Y and Selvamanickam V 2013 Completely Etch-Free Fabrication of Multifilamentary Coated Conductor Using Inkjet Printing and Electrodeposition *IEEE Trans. Appl. Supercond.* **23** 6603005–6603005
[8] Abraimov D, Gurevich A, Polyanskii A, Cai X Y, Xu A, Pamidi S, Larbalestier D and Thieme C L H 2008 Significant reduction of AC losses in YBCO patterned coated conductors with transposed filaments *Supercond. Sci. Technol.* **21** 082004
[9] Levin G, Murphy J and Haugan T 2013 AC Losses of Copper Stabilized Multifilament YBCO Coated Conductors *Appl. Supercond.* **23** 6600604
[10] Kesgin I, Levin G A., Haugan T J and Selvamanickam V 2013 Multifilament, copper-stabilized superconductor tapes with low alternating current loss *Appl. Phys. Lett.* **103** 252603
[11] Rudnev I , Pokrovskiy S and Podlivaev A 2012 An Application of Magnetic Measurements for Study of Local Transport Characteristics of Modern Superconducting Materials *IEEE Trans. Appl. Supercond.* **22** 9001304–9001304
[12] Wang Y 2013 Fundamental Elements of Applied Superconductivity in Electrical Engineering (Singapore: Wiley)

Journal of Physics: Conference Series **747** (2016) 012045 doi:10.1088/1742-6596/747/1/012045

# Regulation of nanoparticle impact on the growth of $MoSe_x$ films during pulsed laser evaporation of $MoSe_2$ target

**R I Romanov**[1], **V Y Fominski**[1,4], **A G Gnedovets**[2], **S N Grigoriev**[3] **and M A Volosova**[3]

[1] National Research Nuclear University MEPhI (Moscow Engineering Physics Institute), Kashirskoe sh. 31, 115409 Moscow, Russia
[2] Baikov Institute of Metallurgy and Materials Science RAS, Leninskii pr. 49, 119991 Moscow, Russia
[3] Moscow State University of Technology “STANKIN”, Vadkovskii per. 3a, 127005 Moscow, Russia

E-mail: vyfominskij@mephi.ru

**Abstract.** Irradiation of $MoSe_2$ target with intensive laser pulses caused the formation of micro- and nanoparticles. The particles were observed on the target surface and in $MoSe_x$ films prepared by deposition of the laser-induced plume. Content of nanoparticles on the film surface was markedly larger than that of microparticles. Transport of the plume in vacuum and in a buffer helium (He) gas was studied. The He pressure was high enough to provide effective atom scattering and plume deceleration. For a medium target-substrate distance, the structure of $MoSe_x$ film was formed due to intensive deposition of atomic flux scattered in the buffer gas. Impact of nanoparticles on the structure was negligible. For a large distance, deposition of the nanoparticles from the plume was assisted by co-deposition of drifting atomic flux. Such deposition resulted in the formation of relatively smooth film containing nanoparticles transferred with plume, as well as the growth of spherical Se particles on the substrate.

## 1. Introduction

Sen et al. [1], Fominski et al. [2], and Hu et al. [3] have observed the nanoparticles formation during pulsed laser deposition (PLD) of transition metal dichalcogenides (TMDs, $Mo/W/S_2/Se_2$) films. Grigoriev et al. [4] revealed the important influence of Mo nanoparticles on the structure of the PLD $MoSe_x$ films and their catalytic properties in hydrogen evolution reaction. The possibility of a positive effect of nanoparticles on the tribological properties of gradient Mo-Se-Ni-C coatings obtained by PLD was proposed by Grigoriev et al. [5].

At present, an interest to study the mechanisms of nanoparticles formation under pulsed laser irradiation/ablation and the impact of nanoparticles on the structural and functional properties of thin films increases (e.g., Cao et al. [6]). The aim of this work was to investigate the formation of nanoparticles during pulsed laser irradiation of $MoSe_2$ target and the transport of these particles from the target to $MoSe_x$ films formed due to the deposition of the laser-induced plume. The transport and deposition of the laser plume were studied in vacuum and in a buffer gas with a relatively high pressure. Inert helium (He) was used as a buffer gas to change the dynamics of atomic particles of the

[4] To whom any correspondence should be addressed.

laser plume. The transport of the laser plume at the different conditions was monitored by measuring the time-of-flight signals for the pulsed laser plasma.

## 2. Experimental details

An electro-optically Q-switched yttrium-aluminium-garnet laser with wavelength of 1.06 μm and a pulse duration of 15 ns was used for PLD. For irradiation of $MoSe_2$ target, a relatively high laser fluence of ~$8\times10^8$ W/cm$^2$ was used. The deposition chamber was evacuated to a pressure of ~$10^{-4}$ Pa. Hereinafter, PLD at a pressure of ~$10^{-4}$ Pa will be referred to as PLD in vacuum. In some experiments, the buffer He gas with pressure of 30 Pa was introduced into the chamber after evacuation. This option will be denoted as PLD in He gas. The laser plume was directed along the normal toward the substrate surface. Si substrates were placed at 5 or 10 cm from the target.

To control the transport of ionized laser plume in various conditions, time-of-flight ion signals were detected in places of the target location. Surface morphology of the target and the deposited films was characterized by scanning electron microscopy (SEM), energy dispersive X-ray spectroscopy (EDS) and micro-Raman spectroscopy (MRS).

The $MoSe_2$ target for PLD was manufactured by means of cold compacting technology under the pressure of 600 MPa using $MoSe_2$ powder obtained by self-propagating high-temperature synthesis. The $MoSe_2$ powder was synthesized from Mo powder of 99.93% purity and particle size of 1–4 μm, and Se powder of 99.997% purity. The pressed target was composed of $MoSe_2$ microcrystals having a lamellar shape. EDS measurements showed that the elements (Mo and Se) were fairly uniformly distributed throughout the volume of the microcrystals. Chemical composition of the synthesized microcrystals was equal to a stoichiometric value. Prior to pulsed laser irradiation, the surface of the pressed $MoSe_2$ target was treated mechanically (grinding, polishing) and washed with the alcohol.

## 3. Results and discussion

Figure 1 shows SEM images of the surface of a polished $MoSe_2$ target before and after pulsed laser irradiation under vacuum conditions. Mechanical treatment of the target removed the microcrystalline morphology/roughness and caused the enrichment of the surface layer with Se. Mean value of ratio $x$ = Se/Mo was ~10. Pulsed laser irradiation caused melting of the surface and Se content ($x$~1.7) in the surface layer decreased. Mo- and Se-enriched nano- and micro-particles were observed on the surface. The possible mechanisms of such particle formation were discussed in [2,4]. On the next step of pulsed irradiation, these particles could be captured by vapor flux and transferred from the target to $MoSe_x$ film.

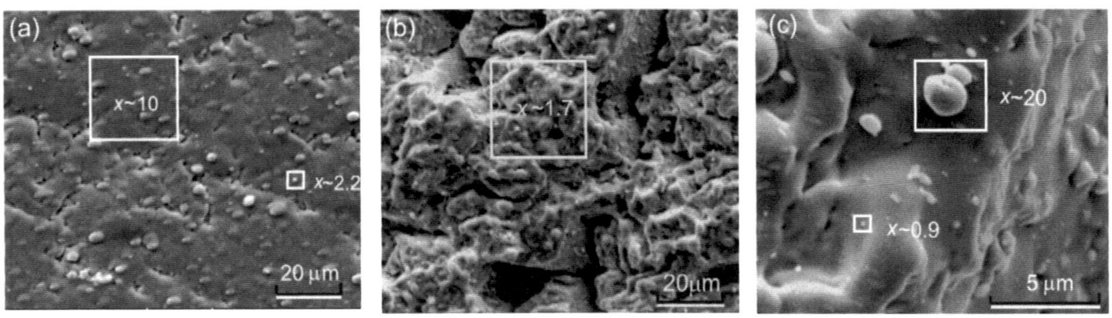


**Figure 1**. SEM images of the surfaces: (*a*) polished $MoSe_2$ target; (*b*) and (*c*) the same target after pulsed laser irradiation.

Figure 2 shows ion signals measured at distances 5 and 10 cm for pulsed laser plasma initiated in vacuum and in He gas at 30 Pa. For a distance of 5 cm, presence of buffer He gas leads to a decrease of ion pulse intensity but a large fraction of the ion beam effectively penetrates trough the gas.

Calculations with models of Fominski et al. [7] show that the atomic fraction of the plume can reach the substrate, but the energy of atoms is relatively low and atoms fall on the surface at various angles.

For a distance of 10 cm and a pressure of 30 Pa, only very small fraction of ions penetrated through the gas and conserved their energy (Figure 2(*b*)). Main portion of the ion beam has arrived to the detector in "a long tail" of the ion pulse. In this case, the film is assumed to grow due to the deposition of the flow of drifting atoms and ions. Drift velocity of deposited atoms is caused by a movement of shock wave that arises in the He gas at a pressure of 30 Pa.

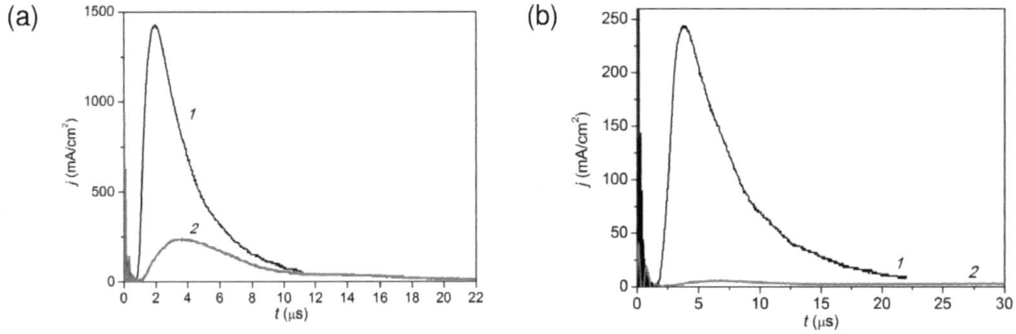


**Figure 2**. Ion signals of pulsed laser plasma measured in (*1*) vacuum and (*2*) He at 30 Pa for two distances from the target: (*a*) 5 cm; (*b*) 10 cm.

Results of SEM studies of the $MoSe_x$ films prepared by PLD at various conditions are shown in Figures 3 and 4. For PLD at a distance of 5 cm (Figure 3), the vacuum deposited film consisted of dense matrix that contained spherical nanoparticles. The mean value of ratio $x$ was 1.2. The main content of the particles had the size in a range ~5 – 50 nm and the ratio $x$ for particles varied in a range ~ 0.8 – 2.5. For PLD in He, the film thickness and the Se concentration were larger than that for the vacuum deposited film. This film had a granulated morphology and the average size of the granules was ~50 nm. The growth of the granulates was probably initiated by scattered atomic flux deposition on the surface of smaller nanoparticles. When $x$ exceeds 3 in the $MoSe_x$ film, the deposition of excess selenium could make contribution to processes of the granules formation.

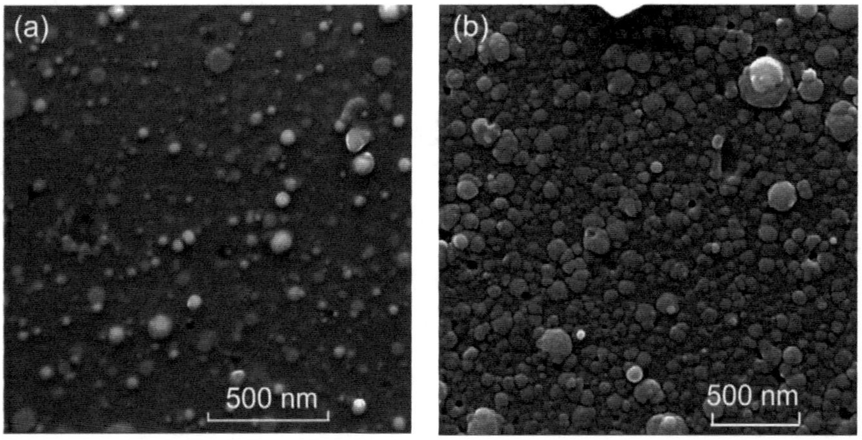


**Figure 3**. SEM images of the surface of $MoSe_x$ films deposited at a distance of 5 cm in (*a*) vacuum and (*b*) He gas at 30 Pa.

Transport of nanoparticles ejected from the target surface or formed in the laser plume is described by a drag model which is given by $X(t) = L_p[1 - \exp(-t/\tau_p)]$, where $L_p = v_0\tau_p$ is the nanoparticle stopping distance, $v_0$ is the initial nanoparticle velocity, and $\tau_p$ is the relaxation time. In the free-molecular regime (e.g., Smirnov [8]), the relaxation time is $\tau_p = (3/8)(M_p/R_p^2P)(kT/2\pi m)^{1/2}$, where $k$ is Boltzmann constant; $R_p$ and $M_p$ are the nanoparticle radius and mass; $m$, $T$ and $P$ are the molecule mass, temperature and pressure of the ambient gas, respectively. For 10 nm sized nanoparticles of Mo and Se in He gas at a pressure of 30 Pa, estimates give the values of the relaxation time $\tau_p \sim 8.5\times10^{-4}$ s and $4.0\times10^{-4}$ s, respectively. Thus, nanoparticles having this size and larger with the initial velocity in the range ~0.1 – 1 km/s could effectively penetrate through the gas and bombard the substrate located at a distance of 5 – 10 cm from the target.

For PLD in vacuum at a distance of 10 cm, the deposited film possessed the specific surface morphology modulated at the nanoscale. Figure 4 (*a*) shows that the sizes of "nanoparticles" were distributed mainly in the range ~5 – 50 nm. The film deposited in He gas was relatively smooth. The content of "nanoparticles" with the sizes 5 – 50 nm was not so large, but the content of spherical nanoparticles with the sizes 50 – 100 nm has increased essentially. According to EDS measurements, these larger particles were enriched with Se. The formation of nanostructure of the vacuum-deposited film was probably induced by self-sputtering of the film by energetic atoms (neutrals and ions). The Mo-enriched nanoparticles were not sputtered so effectively as the film matrix and the inclusions of Mo-enriched nanoparticles caused the corrugate surface formation. Figure 5 (*a*) shows SEM images (side view) of the $MoSe_x$ film illustrating nanostructure character of this film.

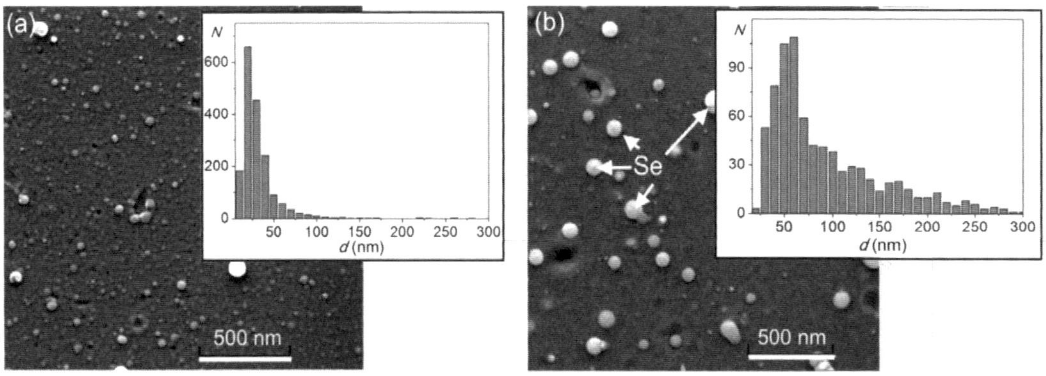


**Figure 4**. SEM images of the surface of $MoSe_x$ films deposited at a distance of 10 cm in (*a*) vacuum and (*b*) He gas at a pressure of 30 Pa. Inserts show the particle lateral size distributions measured on an area of 5×5 μm$^2$.

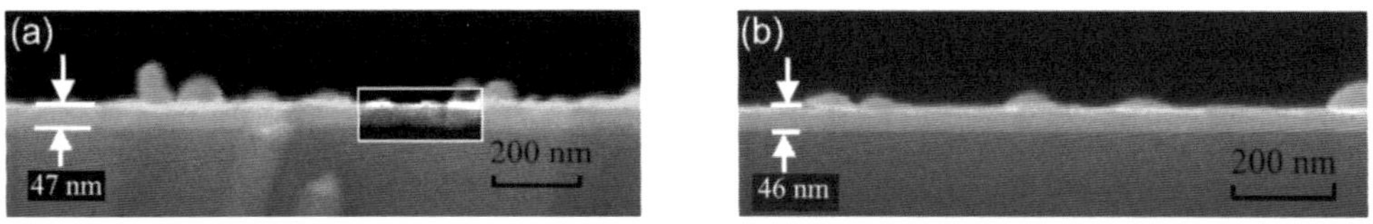


**Figure 5**. Cross-section SEM images of the $MoSe_x$ films deposited on Si substrate at a distance of 10 cm in (*a*) vacuum and (*b*) He gas at a pressure of 30 Pa.

For PLD in He gas at a distance of 10 cm, the buffer gas prevents an intensive bombardment of the film with energetic atoms. Figures 4 (*b*) and 5 (*b*) show that in the case of drifted atoms deposition, the smooth matrix could be grown and the small Mo-enriched nanoparticles would be captured in it. In such conditions, the particles do not essentially influence on the film morphology. The EDS measurement has shown that the ratio $x$ was ~2. Selenium was concentrated in isolated spherical Se nanoparticles and the $MoSe_x$ film was enriched with molybdenum. MRS studies have shown (results not presented) that selenium condensation resulted in appearance of intense line at 234 $cm^{-1}$ in the Raman spectrum corresponding to nanocrystalline Se phase.

## 4. Conclusions

In this work, the formation of nano- and micro-particles on the surface of the laser-irradiated $MoSe_2$ target was observed. The particles had various composition and they were enriched with Mo or Se. For the condition of PLD used in this work, the transfer of particles from the target to the substrate was assisted with vapor flux transport and co-deposition that occurred on the substrate. The formation of the composition and morphology of the $MoSe_x$ films deposited on the substrate strongly depends on the dynamic characteristics of the vapor flux.

For PLD in vacuum, the bombardment of the growing films with high energy atoms induced sputtering of Se-enriched phase and the consequent nanostructuring of the film was developed. The buffer He gas has changed the dynamic of the laser plume but it could not prevent the deposition of the vapor flux even at a high pressure and a large distance from the target. For PLD at a medium target-substrate distance, the deposition of the scattered vapor flux causes granulated morphology formation. For PLD at a large distance, the Se vapor condenses on the substrate and larger Se particles are formed. The deposited Mo nanoparticles are captured in the bulk of the grown film due to the deposition of drifted vapor flux.

## Acknowledgments

This work was partly supported by the Ministry of Education and Science of the Russian Federation within the framework of the state assignment (including the work of "Research Organization" for VYF) and by the Competitiveness Program of NRNU MEPhI.

## References

[1] Sen R, Govindaraj A, Suenaga K, Suzuki S, Kataura H, Iijima S and Achiba Y 2001 *Chem. Phys. Lett.* **340** 242–8
[2] Fominski V Y, Grigoriev S N, Gnedovets A G and Romanov R I 2013 *Technical Physics Lett.* **39** 312–5
[3] Hu J J, Zabinski J S, Bultman J E, Sanders J H and Voevodin A A 2008 *Cryst. Growth Des.* **8** 2603–5
[4] Grigoriev S N, Fominski V Yu, Romanov R I, Volosova M A and Shelyakov A V 2015 *Thin Solid Films* **592** 175–81
[5] Grigoriev S N, Fominski V Yu, Romanov R I and Gnedovets A G 2014 *Thin Solid Films* **556** 35–43
[6] Cao Y and Shin Y C 2015 *Appl. Surf. Sci.* **357** 74–85
[7] Fominski V Yu, Romanov R I, Gnedovets A G and Nevolin V N 2010 *Technical Physics* **55** 1509–16
[8] Smirnov B M 2011 *Physics-Uspekhi* **54** 691-721

Journal of Physics: Conference Series **747** (2016) 012046 doi:10.1088/1742-6596/747/1/012046

# Structural and electrochemical properties of W-Se-O layers prepared by pulsed laser pre-deposition and thermal post-treatment

**A A Soloviev[1], S N Grigoriev[2], D V Fominski[1,3], M A Volosova[2] and R I Romanov[1]**

[1] National Research Nuclear University MEPhI (Moscow Engineering Physics Institute), Kashirskoe sh. 31, 115409 Moscow, Russia
[2] Moscow State University of Technology "STANKIN", Vadkovskii per. 3a, 127005 Moscow, Russia

E-mail: dmitryfominski@gmail.com

**Abstract**. Novel nanostructured thin-film coatings containing $WSe_2$ and $WO_3$ nanosheets were prepared through a simple and highly reproducible method. Pulsed laser pre-deposition of W-Se-O films on a microcrystalline graphite substrate by ablation of $WSe_2$ target in a mixture of Ar and $O_2$ gases was followed by thermal post-treatment. The influence of pre-deposition and post-treatment conditions on the structure, morphology and chemical composition of the W-Se-O films was studied. Thermal annealing at appropriate conditions of pre-deposited amorphous W-Se-O films resulted in the formation of crystalline structure consisted of nanocrystals of $WSe_2$ and $WO_3$ phases. Such structural modification significantly altered the electrochemical properties of the thin-film coatings with consequences, in particular, on their catalytic activity toward hydrogen evolution reaction in an acid solution.

## 1. Introduction

Recently, two-dimensional (2D) or quasi-two-dimensional (Q2D) layered transition metal dichalcogenides (TMDs) have received great attention as potential catalysts for hydrogen evolution reaction (HER) in acid solutions (e.g., Pumera et al. [1]). The active sites for HER are located along the edges of the molecular layers consisted of atomic chalcogen-metal-chalcogen sandwiches. Therefore, ultra-thin TMD nanocrystals possessing sheet-like shape are promising materials for obtaining effective catalysts of HER [2].

Unfortunately, in accordance with conclusions of Mao et al. [3] and Vrubel et al. [4], the catalytic activity of such TMD nanosheets is still limited due to the inevitable agglomeration and stacking of the liquid exfoliated nanosheets. TMD nanosheets coupled with highly conductive nanomaterials, e.g., carbon nanotubes and graphene, have been proposed by Tang et al. [5] and Mao et al. [3] to enhance the catalytic activity.

Transition metal oxides also have a layered packing of atoms in the crystal lattice and when dimensions of the oxide crystals decreased due to ultra-thin 2D/Q2D nanosheets formation the physicochemical properties of the oxides can be changed. Zhuiykov et al. [6] and Ketpang et al. [7] revealed that in contrast to "bulk" metal oxides samples, nanostructured oxides exhibit good

[3] To whom any correspondence should be addressed.

conductivity and improved catalytic properties in HER. Electrical properties of some metal oxide films, such as $WO_3$ films, can be modified significantly due to interaction with hydrogen [8].

For $WO_3$-S system, Wu et al. [2] revealed interesting effects when chalcogen atoms chemically interacted at elevated temperatures with $WO_3$ ultra-small particles. During the annealing of the mixture consisted of $WO_3$ and S particles, $WS_2$ shells were produced on the $WO_3$ particles surface and then peeled off to form $WS_2$ nanosheets. After the outer layer (i.e., $WS_2$ shell) peels away, $WO_3$ particles became smaller and more active than the ones with larger size, and it became easier for S to coat their outer surface to form new $WS_2$ shell. Mechanical activation of ball milling was essential factor to get close contact between S and $WO_3$ and facilitate the shell formation due to sulfurization. For $WS_2$ nanosheets catalyst fabrication, additional/special mechanical and chemical procedures should be carried out [2].

It is of great interest the preparation of the nanosheet-based thin-film catalyst by using the specific processes of phase transformation that are characteristic of chalcogenide-oxide system. The aim of this work was to develop a novel thin-film material containing nanosheets of tungsten diselenide ($WSe_2$) and $WO_3$ and to study the HER catalytic properties of this thin-film material deposited on a graphite substrate.

## 2. Experimental details

For preparation of $WSe_2/WO_3$ layers, pulsed laser deposition (PLD) of W-Se-O films was carried out, and then the deposited films were subjected to thermal post-treatment (annealing). An yttrium-aluminium-garnet laser (wavelength 1.06 μm) was used for PLD with a pulse duration of 15 ns, a repetition rate of 25 Hz, and a pulse energy of 30 mJ. The fluence in the laser spot was ~8 J/cm$^2$. The $WSe_2$ target for PLD was manufactured by means of cold compacting technology under the pressure of 600 MPa using $WSe_2$ powder obtained by self-propagating high-temperature synthesis.

A deposition chamber was evacuated to a pressure of ~$10^{-4}$ Pa. The deposition of the films was carried out in vacuum (residual gas pressure was ~$10^{-4}$ Pa) and at various pressures of buffer/reactive gases. Buffer argon (Ar), reactive oxygen ($O_2$), and mixture of Ar and $O_2$ gases was introduced into the chamber after evacuation. The W-Se-O films were deposited at room temperature on substrates made from polished graphite. The obtained W-Se-O films were annealed at 450 °C and 550 °C for 30 min in Ar at a pressure of $10^5$ Pa.

Surface morphology of the substrate and deposited films, structure and chemical composition were characterized by scanning electron microscopy (SEM), energy dispersive X-ray spectroscopy (EDS) and micro-Raman spectroscopy (MRS, laser wavelengths were 488 and 647 nm). The catalytic activity of the prepared coatings on the graphite substrates was studied in 0.5M $H_2SO_4$ solution at room temperature with a typical three-electrode electrochemical cell setup. Potential was measured relative to the standard hydrogen electrode.

## 3. Results and discussions

Figure 1 shows surface morphology of the polished graphite substrate prior the film deposition. The pristine substrate consisted of randomly oriented graphite sheets having the sizes of several micrometers.

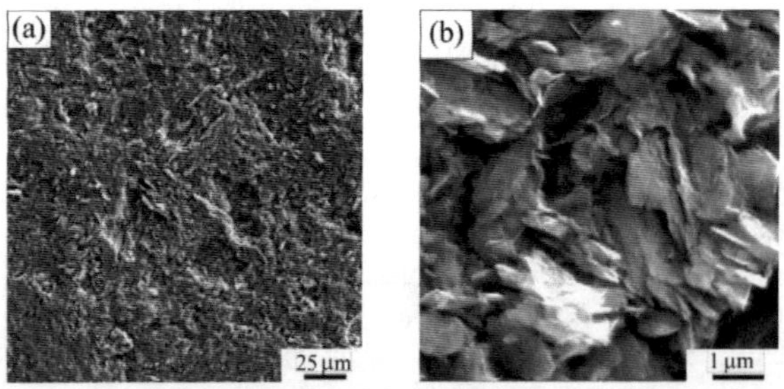


**Figure 1**. SEM images with two magnifications for the

polished graphite substrate.

Surface morphology and composition of the pre-deposited W-Se-O films strongly dependent on the pressure and composition of the used gas. The deposition in the residual gas resulted in the formation of W-enriched films ($WSe_{0.3}$) which possessed relatively smooth surface and contained ball-shaped nanoparticles (Figure 2*a*). These nanoparticles were made of β-W [9] and they were ejected from the $WSe_2$ target due to pulsed laser irradiation. Incorporation of oxygen from the residual gas into the bulk of the films was negligible.

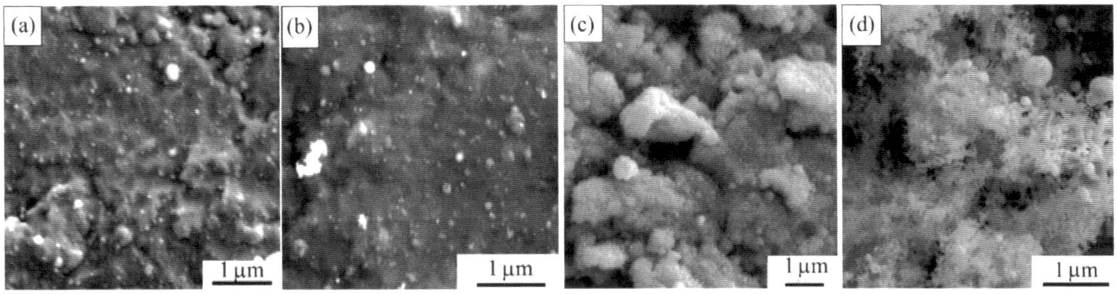


**Figure 2**. SEM images for W-Se-O films prepared by PLD in (a) vacuum conditions; (b) $O_2$ gas; (c) $Ar+O_2$ gas mixture at a pressure of 5 Pa; (d) $Ar+O_2$ mixture at a pressure of 20 Pa.

The use of $O_2$ gas for PLD at a pressure of 7 Pa did not alter the morphology of the films but the oxygen addition was realized and $WSe_{0.5}O_2$ film was formed (Figure 2*b*). As a rule, the PLD of TMDs in a gas should result in a significant increase in the chalcogen atoms content for TMD films [10]. The increase of chalcogen atoms content is caused by the decrease of energy for condensed atoms with consequence, in particular, on preferential sputtering of chalcogen atoms. In fact, a large increase of Se/W ratio was detected for the W-Se-O films when the deposition was performed in $Ar+O_2$ gas mixture with relatively lower $O_2$ concentration (~10%). For such gas mixture composition at a pressure of 5 Pa, $WSe_2O_{0.8}$ film was prepared. The comparison of the films prepared in $O_2$ and $Ar+O_2$(10%) mixture shows that in the case of a high $O_2$ content in the gas, Se atoms could be removed from the W-Se-O film due to volatile selenium oxide formation and/or Se-O molecules formation in collisions of Se atoms with $O_2$ molecules. The $WSe_2O_{0.8}$ film consisted of round-shaped nanoparticles having 50 – 100 nm sizes (Figure 2*c*).

SEM and EDS studies of the films prepared by PLD at higher pressure of $Ar+O_2$(50%) gas mixture (the pressure was ~20 Pa) showed that the concentration of oxygen in the films has increased significantly and the porous/sponge-like $WSeO_3$ films were grown (Figure 2*d*). These films had nonuniform distribution and low adhesion on the substrate. Thermal post-treatment of the prepared films at a temperature of 450 °C caused insignificant modification of their surface morphology (Figure 3) and Se content has decreased to some extent.

**Figure 3**. SEM images for various W-Se-O films after annealing at a temperature of 450 °C. SEM images of the pre-deposited W-Se-O films are shown in Figure 2 (a), (b), (c), and (d), respectively.

Journal of Physics: Conference Series **747** (2016) 012046 doi:10.1088/1742-6596/747/1/012046

Figure 4 shows MRS spectra measured for various W-Se-O films before and after annealing at 450 °C. Laser wavelength was 647 nm. MRS spectrum of the polished graphite substrate had two narrow G and D peaks at ~1580 $cm^{-1}$ and ~1360 $cm^{-1}$, respectively (results not presented). Deposition of the W-Se-O films caused the appearance of several broadened lines in the range of wave numbers from 200 $cm^{-1}$ to 300 $cm^{-1}$. The most intense line was near 252 $cm^{-1}$. The position and shape of this line indicated the formation of a disordered structure, in which a short-range order as in 2H-$WSe_2$ phase and/or pure Se phase could be arranged. After annealing at 450 °C, MRS spectra contained two broad bands, the first one at low frequency in the range 300–700 $cm^{-1}$ and the second one at 800–1000 $cm^{-1}$. Such spectra are characteristic of quasi-crystalline $WO_3$ films [11].

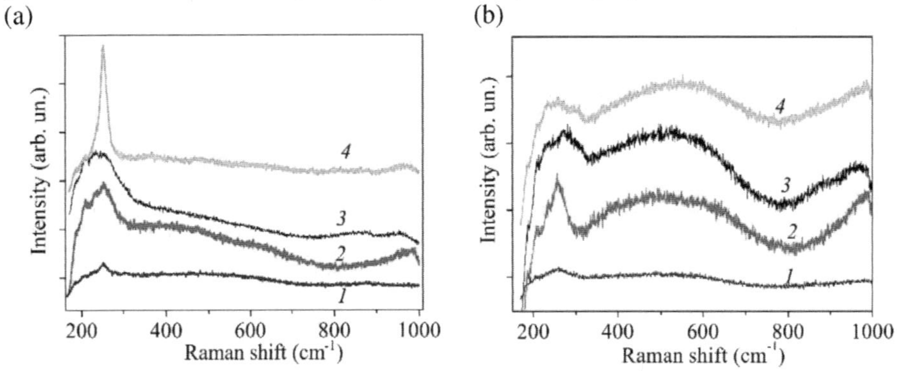


**Figure 4**. MRS spectra measured (a) before and (b) after annealing at 450 °C for W-Se-O films prepared by PLD in vacuum (*1*), $O_2$ gas (*2*), (c) Ar+$O_2$ gas mixture at a pressure of 5 Pa (*3*) and Ar+$O_2$ mixture at a pressure of 20 Pa (*4*).

Annealing of the $WSe_2O_{0.8}$ film at a higher temperature caused significant reconstruction of the film and a "flower-like" morphology was observed by SEM (Figure 5). "Buds" of the flowers consisted of petals with a sheet-like structure. Dimensions of the sheets did not exceed 200 nm, and thicknesses of the sheets were several nanometers.

**Figure 5**. SEM images with two magnifications for W-Se-O film containing $WSe_2$ and $WO_3$ nanosheets.

For annealed samples, MRS studies (laser wavelength was 488 nm) showed that the intensity of the line for the $WSe_2$ nanophase at 257 $cm^{-1}$ increased and new lines at 727, 813, and 963 $cm^{-1}$ had arisen (Figure 6). The new lines correspond to $WO_3$ phase and the specific line at 963 $cm^{-1}$ reveals the nanometer dimensions of $WO_3$ phase because this line corresponds to W=O boundary bonds in the $WO_3$ nanocrystals.

Measurements of the cathodic polarization curves showed that only post-treatment at 550 °C resulted in a significant increase in the HER activity for W-Se-O films. The cathodic current density increased up to 50 mA/cm$^2$ at a potential of 100 mV. For comparison, the current density for the graphite cathode was about 7 mA/cm$^2$. Tafel curves could be extrapolated as linear dependencies in a relatively short range of cathodic currents. For the graphite cathode covered with the $WSe_2/WO_3$ nanosheets layer and without it, the Tafel slopes were ~90 mV/dec and 340 mV/dec, respectively. It should be noted that cathodic pretreatment at -500 mV was carried out to activate the hydrogen production. An improvement of the catalytic properties of the $WSe_2/WO_3$ catalysts could be achieved by optimizing their chemical composition and varying the annealing conditions.

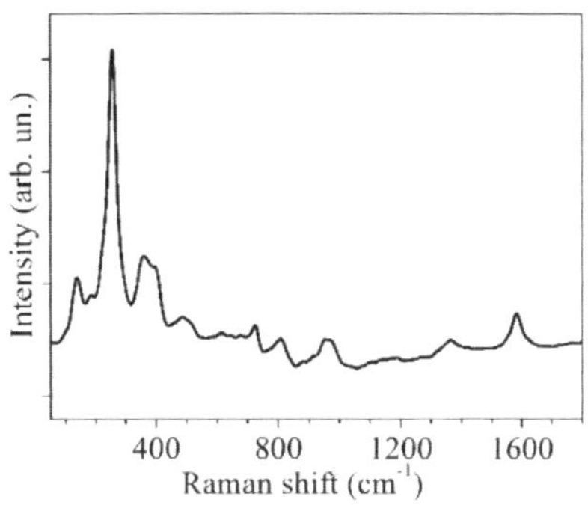


**Figure 6**. MRS spectrum for $WSe_2/WO_3$ nanosheets containing films prepared on the graphite substrate.

## 4. Conclusion

PLD of W-Se-O films in $Ar+O_2$ gas mixture followed by thermal annealing at a quite high temperature resulted in the formation of efficient catalysts for HER. These thin-film coatings had perpendicularly oriented few-layered $WSe_2$ and $WO_3$ sheets that uniformly covered graphite microcrystals. The cathodic electrochemical treatment of the nanosheets containing structure could initiate phase transformation/restructuring under hydrogen interaction with the $WSe_2$ and $WO_3$ phases.

**Acknowledgments**
This work was supported by Russian Science Foundation (project No 14–22–00098).

## References

[1] Pumera M, Sofer Z and Ambrosi A 2014 *J. Mater. Chem.* A **2** 8981–8987
[2] Wu Z, Fang B, Bonakdarpour A, Sun A, Wilkinson D P and Wang D 2012 *Appl. Catal.* B **125** 59–66
[3] Mao S, Wen Z, Ci S, Guo X, Ostrikov K and Chen J 2014 *Small* **11** 414–419
[4] Vrubel H, Moehl T, Grätzel M and Hu X 2013 *Chem. Commun.* **49** 8985–8987
[5] Tang H, Dou K, Kaun C-C, Kuang Q and Yang, S 2014 *J. Mater. Chem.* A **2** 360–364
[6] Zhuiykov S and Kats E 2014 *Nanoscale Res. Lett.* **9** 401
[7] Ketpang K, Kim M, Kim S and Shanmugam S 2013 *Inter. J. Hydrogen Energy* **38** 9732–9740
[8] Fominski V Y, Grigoriev S N, Romanov R I, Zuev V V and Grigoriev V V 2012 *Semiconductors* **46** 401–9
[9] Fominski V Yu, Grigoriev S N, Celis J P, Romanov R I and Oshurko V B 2012 *Thin Solid Films* **520** 6467–6483
[10] Grigoriev S N, Fominski V Yu, Gnedovets A G and Romanov R I 2012 *Appl. Surf. Sci.* **258** 7000 – 7007
[11] Filipescu M, Orlando S, Russo V, Lamperti A, Purice A, Moldovan A and Dinescu M 2007 *Appl. Surf. Sci.* **253** 8258–8262

# The effect of temperature on pinning mechanisms in HTS composites

**A P Sotnikova**[1,3] **and I A Rudnev**[1,2]

[1] National Research Nuclear University MEPhI (Moscow Engineering Physics Institute), Kashirskoe sh. 31, 115409 Moscow, Russia
[2] International Laboratory of High Magnetic Fields and Low Temperatures, Gajowicka 95, 53-421 Wroclaw, Poland

E-mail: apsotnikova@mail.ru

**Abstract.** Pinning mechanism in samples of second generation tapes (2G) of high-temperature superconductors (HTS) was studied The critical current and the pinning force were calculated from the magnetization curves measured in the temperature range of 4.2 – 77 K in magnetic fields up to 14 Tesla using vibration sample magnetometer. To determine the pinning mechanism the dependences of pinning force on magnetic field were constructed according to the Dew-Hughes model and Kramer's rule. The obtained dependences revealed a significant influence of the temperature on effectiveness of different types of pinning. At low temperatures the 2G HTS tapes of different manufacturers demonstrated an equal efficiency of the pinning centers but with temperature increase the differences in pinning mechanisms as well as in properties and effectiveness of the pinning centers become obvious. The influence of the pinning mechanism on the energy losses in HTS tapes was shown.

## 1. Introduction

Modern 2G HTS tapes produced by a number of commercial manufacturers such as American Superconductors, SuperPower, SuNaM. *et al.* have different transport properties, especially under applying magnetic fields *H*. Differences in values of the critical current $J_c$ and in $J_c(H)$ dependences are associated with different types of artificial pinning centers in the tapes [1, 2]. Determination of pinning type (volume, surface, point) in the tapes and analysis of pinning evolution with temperature are still actual problems studied in this paper.

## 2. Materials and methods

Industrial tapes with artificial pinning centers manufactured by AMSC (magnetic substrate), Super Power (non-magnetic substrate) and tape without artificial pinning centers manufactured by SuNam were selected for study. Magnetization measurements were carried out at the International Laboratory of High Magnetic Fields and Low Temperatures in Wroclaw by use of the vibration sample magnetometer. The temperature and magnetic field varied in the ranges $T$ = 4,2 – 100 K and $H$ = 0 – 14 T. The field was applied perpendicular to the tape plane. The dependences of the critical current on external magnetic field were calculated from magnetization curves using Bean model.

[3] To whom any correspondence should be addressed.

## 3. Results and discussion

The examples of the magnetization curves measured for AMSC tape in the temperature range from 4.2 K to 77 K are shown on Figure 1. As seen, width of the curves decreases as temperature increases indicating a decrease the critical current with temperature. According to the Bean model the hysteresis width is proportional to the critical current of a sample. Field dependences of the critical current densities obtained from the magnetization curves using the Bean model are presented in Figure 2.

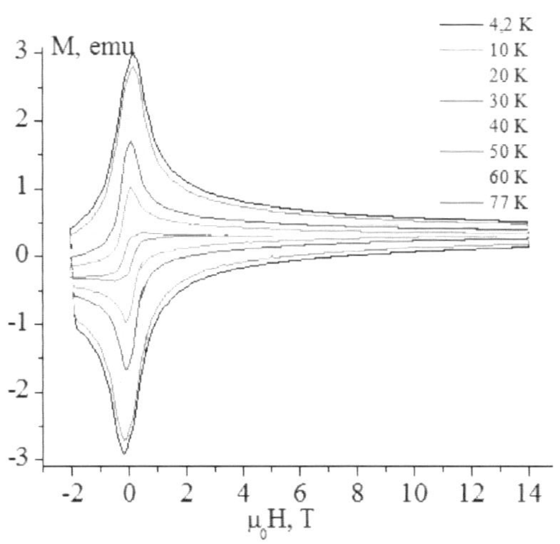


**Figure 1.** Magnetization curves from $T = 4.2$ K to 77 K for AMSC sample of 4x4x0.1 mm size.

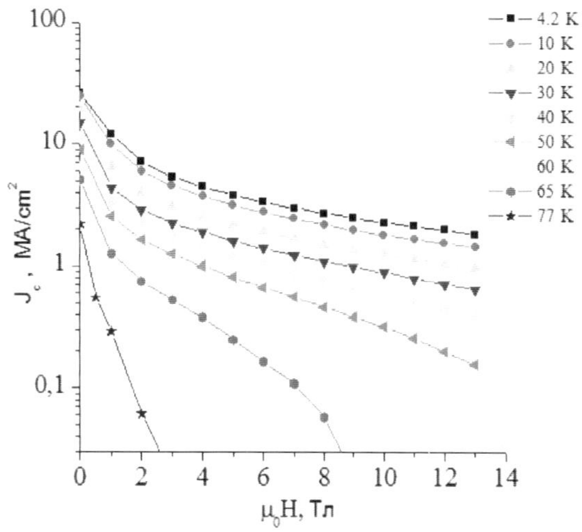


**Figure 2.** Field dependence of the critical current density for AMSC sample at a wide temperature range.

The pinning force was calculated as product of the critical current density and magnetic field values. Analysis of pinning mechanism was carried out in the frame of Dew - Hughes model for the normalized pinning force [3]:

$$F_p \sim F_{po} \times b^n \times (1 - b)^m \qquad (1)$$

where $b = \frac{H}{H^*}$, $F_{po}$ – the maximal pinning force, $H^*$ is the magnetic field at which the critical current density is equal to zero. Combination on $n$ and $m$ parameters depends on the pinning mechanism. It refers to a surface pinning if $n = 0.5$ and $m = 2$, to a point pinning when $n = 1$ and $m = 2$, and to a volume pinning for $n = 0$ and $m = 2$.

A scalable field $H_k$ proportional to $H^*$ is commonly used instead of $H^*$. The former can be easily found from so-called Kramer's plots representing a function $j_c^{0.5} \times H^{0.25}$ *vs* $H$ [4]. These Kramer plots are in general linear with field. To find a magnetic field at which the critical current is equal to zero, the obtained dependences were approximated by linear functions (Fig. 3). For each sample the pinning force was normalized by its maximal value.

Obtained field dependences of the pinning force (Fig. 4 (a)) showed that the surface pinning dominates at low temperatures 4,2 - 20 K in all investigated samples. At low temperatures all samples demonstrated high critical currents, but as temperature increases an efficiency of the surface pinning centers is reduced and difference in pinning mechanisms comes into effect (Fig. 4(b)). It was found that from low to hight temperatures the pining type in AMSC and SUNAM tapes changes respectively from surface to volume and from surface to point. At the same time, temperature does not affect the pinning type of the SP tapes. This can be explained by the different magnetic properties of substrates and properties of pinning centers [5, 6].

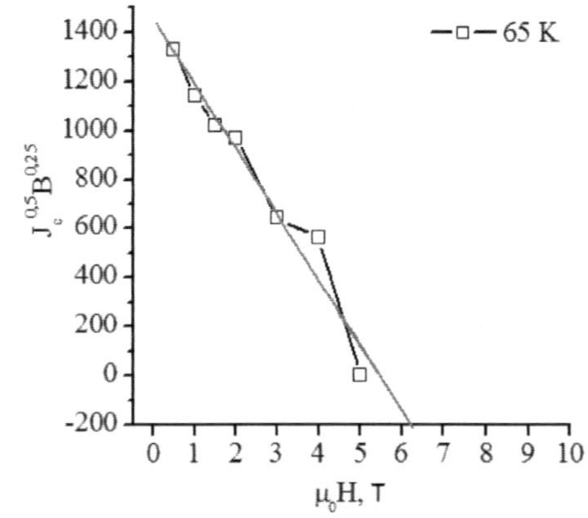


**Figure 3.** Sample of Kramer's plot for $T = 65$ K.

Analysis of the field dependences of the pinning force revealed a significant influence of temperature on the effectiveness of the different types of pinning in the tapes. At low temperatures the efficiency of the pinning centers in all samples is approximately the same, but at high temperatures there is the obvious difference in both efficiency of the centers and pinning mechanisms.

Field dependence of the energy losses at different temperatures was also researched. Losses were determined by calculating of the hysteresis loop area. To compare field dependences of the losses for different tapes, they were normalized to maximal value of the losses calculated for each sample. Obtained curves are presented in Figure 5. Field dependences of the normalized losses coinciding at low temperatures (Fig. 5 (a)) gradually diverge each other with temperature increase and at 77 K the difference is most pronounced (Fig. 5 (b)).

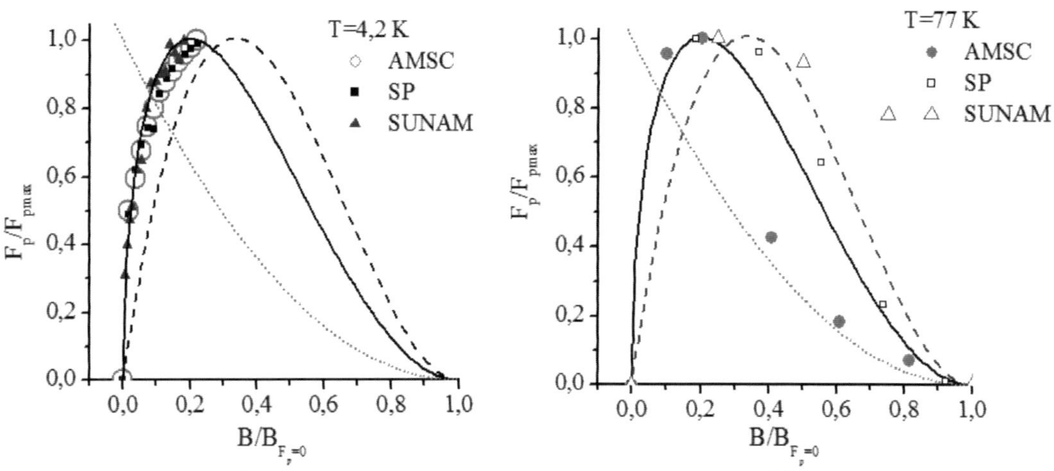


**Figure 4.** Dependences of normalized pinning force $F/F_{pmax}$ on scaled magnetic field $B / B_{Fp=o}$ at 4,2 K (a) and at 77 K (b). The lines correspond to: solid – surface pinning, dashed – point pinning, dotted – volume pinning.

Study of the losses has also indicated that low-temperature efficiency of the pinning centers is approximately the same in all samples, but it differently changes with temperature in different tapes.

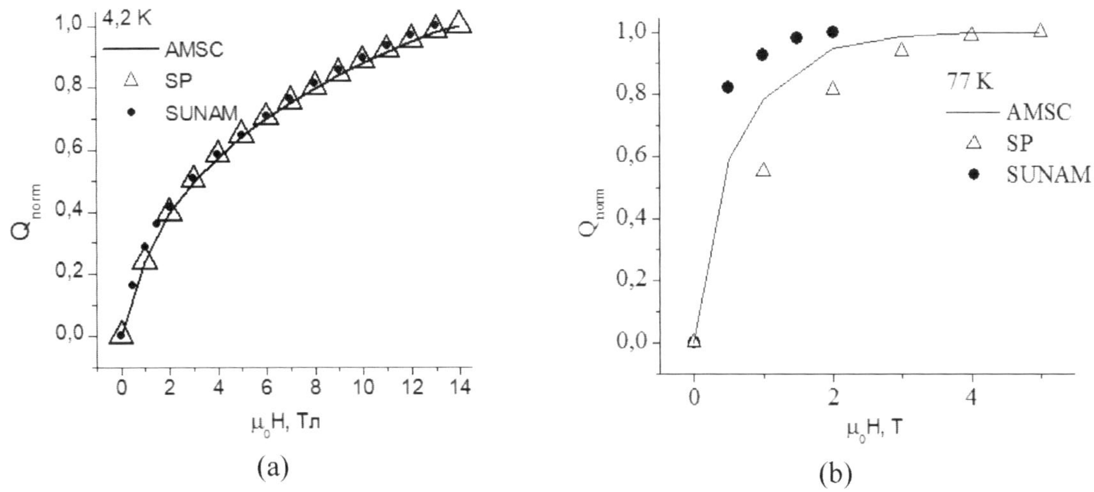


**Figure 5.** Dependences of normalized losses on magnetic field in HTS tapes at 4.2 K (a) and 77 K (b).

## 4. Conclusions

Obtained results demonstrated a significant influence of temperature on pinning in tapes with different landscapes of the pinning centers. The surface pinning dominates in all investigated samples at low temperatures and the efficiency of the pinning centers is approximately the same. At high temperatures the obvious differences in magnetic properties of the samples is caused by change of dominated pinning mechanism which becomes the volume pinning in AMSC tapes and the point pinning in SUNAM tapes. The type of pinning in SP tapes does not changed by temperature.

**Acknowledgments**
The work was done with the financial support of the Ministry of Education and Science of the Russian Federation (the Agreement №14.575.21.0050 (RFMEFI57514X0050)).

**References**

[1] Blanca G R S, De Vero J C, Garcia W O, Sarmago R V 2013 *Physica C* **484** 74-76
[2] Feldmann D M, Holesinger T G, Maiorov B, Zhou H, Foltyn S R, Coulter J Y, Apodoca I 2010 *Supercond. Sci. Technol* **23** 115016
[3] Dew-Hughes D 1987 *Philosophical Magazine* **55** 459-479
[4] Kramer E J 1973 *J. Appl. Phys.* **44** 1360
[5] Kashurnikov V A, Rudnev I A, Zyubin M V 2002 *Journal of Experimental and Theoretical Physics* **94** 377-386
[6] Kashurnikov V A, Rudnev I A, Zyubin M V 2001 *Supercond. Sci. Technol* **14** 695-698

Journal of Physics: Conference Series **747** (2016) 012048 doi:10.1088/1742-6596/747/1/012048

# Variation of critical current and *n*-value of 2G HTS tapes in external magnetic fields of different orientation

**V V Sychugov[1,4], P N Degtyarenko[1,2], A V Ovcharov[1], S V Shavkin[1], V S Kruglov[1,3], A L Vasiliev[1], P V Volkov[1] and Yu M Chesnokov[1]**
[1] National Research Centre "Kurchatov Institute", Akademika Kurchatova pl. 1, 123182 Moscow, Russia
[2] Joint Institute for High Temperatures of Russian Academy of Sciences, Izhorskaya st. 13 Bd.2, 125412 Moscow, Russia
[3] National Research Nuclear University MEPhI (Moscow Engineering Physics Institute), Kashirskoe sh. 31, 115409 Moscow, Russia

E-mail: sychugovvv@gmail.com

**Abstract.** The in-field orientation dependence of critical current and *n*-value in second generation high temperature superconductive tapes was investigated. The samples were manufactured by Metalorganic Chemical Vapour Deposition method with $BaZrO_3$ inclusions (SuperPower Inc.) and Pulsed Laser Deposition method (Bruker HTS). For samples of each kind of fabrication techniques we observed higher critical current value in the case of external magnetic field aligned along (or nearby) *c*-axis direction in comparison with one aligned along *ab*-plane. We analysed possible reasons for this effect. Angular dependences of the critical current and *n*-value were investigated. The microstructure images of superconductive layer of studied samples show tilt of $BaZrO_3$ nanorods in MOCVD sample and high density of structural defects for PLD sample.

## 1. Introduction

The development of high temperature superconductor tapes of second generation (2G HTS) allows implementation of the superconductors in the wider application areas. An extended research of the superconductivity phenomenon is necessary for development of more powerful conductors. It was proved [1-3] that addition of nanoparticles to superconducting material leads to great enhancing of flux pinning. In this paper in-field critical current behavior of 2G HTS tapes prepared by a Metalorganic Chemical Vapour Deposition (MOCVD) with the $BaZrO_3$ (BZO) doping of (GdY)BCO superconductive layer [4] and by Pulsed Laser Deposition (PLD) [5] without doping was investigated. We analyze reasons for the observed behavior and influence of deposition technologies on the microstructure of YBCO/(GdY)BCO layer.

## 2. Experimental details

The microstructural and electrical properties of two types of commercial 2G HTS tapes produced by SuperPower Inc. (MOCVD technology with advanced pinning (AP), called below "MOCVD sample")

[4] To whom any correspondence should be addressed.

Journal of Physics: Conference Series **747** (2016) 012048 doi:10.1088/1742-6596/747/1/012048

and Bruker HTS (PLD technology, called below "PLD sample") (see Table 1) have been investigated. We used the same equipment and the methodology as in [6] to investigate the influence of $BaZrO_3$ nanoparticles embedded into the HTS layer on the critical current value in external magnetic fields up to 1T at different orientations. Standard four-probe measuring method was used. *n*-value was calculated for each Voltage-Current Curve. All experiments were performed in the maximum Lorentz force configuration.

**Table 1.** The experimental samples.

| | SuperPower | Bruker |
|---|---|---|
| Technology | MOCVD | PLD |
| Width | 4 mm | 4 mm |
| Thickness | 0.1 mm | 0.2 mm |
| $I_{c0}$ (77K, SF) | 104 A | 69 A |
| Stabilization | Copper | Copper |
| HTS layer | GdYBCO | YBCO |
| Doping | $BaZrO_3$ | ----- |

Detailed microstructural analysis was performed using scanning/transmission electron microscope TITAN 80-300 TEM/STEM (FEI, US) with Probe-Cs corrector and EDX spectrometer (EDAX, US) operated at 300 kV in NRC "Kurchatov Institute". Platelets were prepared by standard FIB procedure in a Helios (FEI, US) SEM/FIB dual beam microscope.

## 3. Results and discussion

Figures 1a and 1d show in-field behavior of the 2G HTS tapes, the external magnetic field ranges from 0.0625 T to 1 T, orientation angle ranges from -20 to 180 degrees. As seen in Figure 1a there are pronounced peaks on the critical current curves for both MOCVD and PLD samples. The angular dependence of the critical current is not symmetrical for the MOCVD sample. In this sample the main peak is not aligned along *c*-axis but deflects from it at about 10° degrees. Figure 1b shows that up to the magnetic field of 1 T both PLD and MOCVD samples do not reach maximum of flux pinning force that means the flux pinning force peak located in the higher magnetic field.

As seen on Figures 1a and 1d the maximum of critical current in both samples for the magnetic field along the *c*-axis direction exceeds the one along *ab*-planes. The ratio for $B = 1$ T is $I_{ab}(1\ T)/I_c(1\ T)=0.79$ for PLD sample, and is $I_{ab}(1\ T)/I_c(1\ T)=0.53$ for the MOCVD sample. Therefore, it is clear that even in 77K/1T the BZO columnar structures pin magnetic flux stronger than intrinsic pinning between *ab*-planes, probably because the intrinsic pinning is disturbed by columnar structures. Concerning the PLD sample with no inclusions we suppose that high density of the intrinsic defects such as stacking faults, grain boundaries and twin-boundaries are responsible for such in-field critical current angular dependence.

Figure 1f shows that for the MOCVD sample there are two distinct peaks in *n*-value corresponding to the magnetic field aligned along the *c*-axis and parallel to the *ab*-plane. The *n*-value behavior for the PLD sample differs. For the field value as low as 0.0625 T *n*-value has two distinct peaks for *n*-value angular dependence. There is just one *n*-value peak for the field range from 0.0625 T to 1 T, and there are no pronounced peaks in the *n*-value for the field of 1 T.

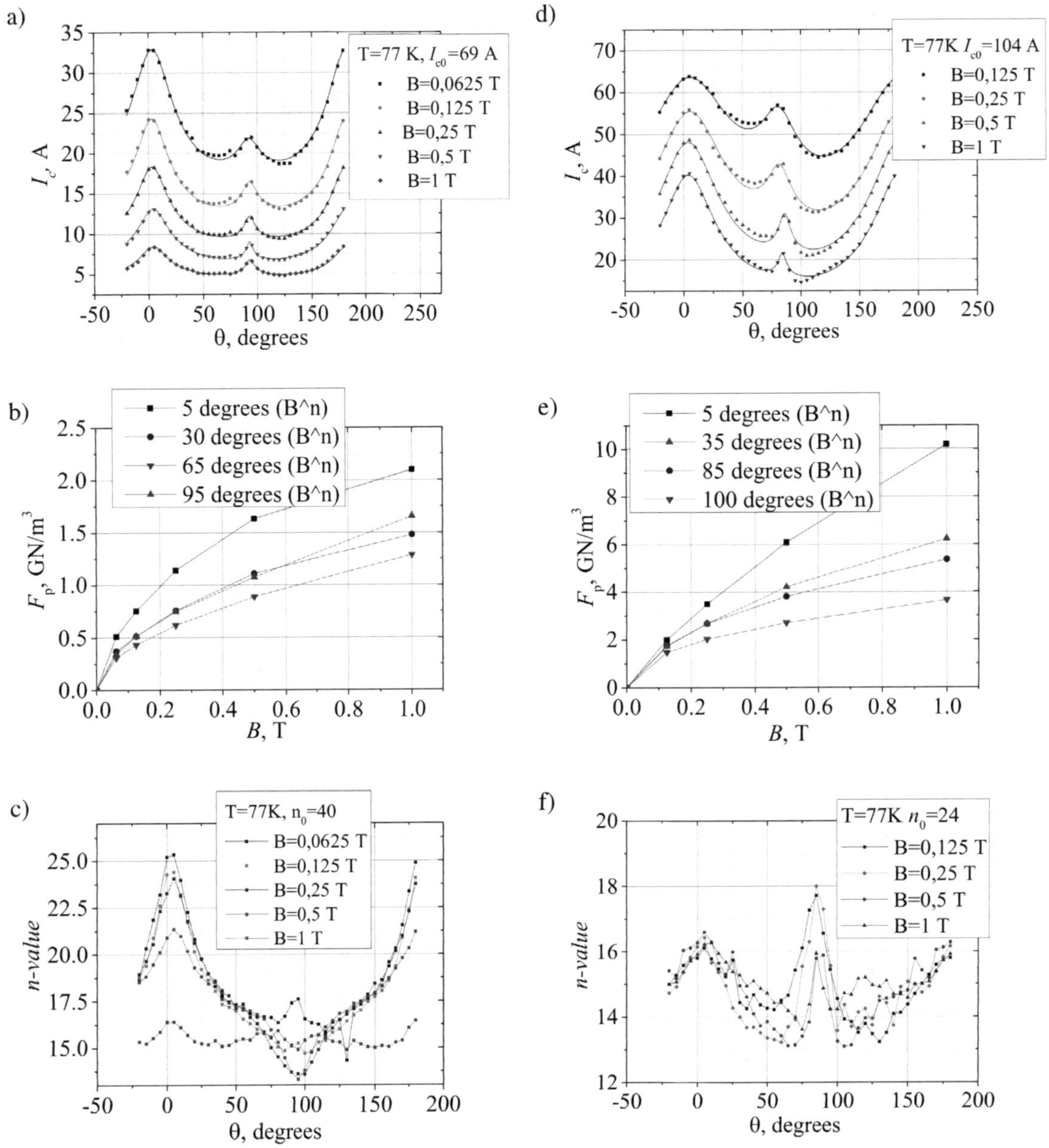


**Figure 1.** Field orientation dependence of the critical current, *n*-value (maximum Lorentz force configuration) θ=**B^n** (normal to the sample surface). Field dependence of the flux pinning force. (a)-(c) PLD sample, (d)-(f) MOCVD sample. For plots b) and e) plots were chosen as follows: local maxima of the critical current value, one local minimum and one point is equal to the local maximum for the *ab*-plane direction of magnetic field.

We used approximation model described in [7] to fit the in-field critical current data. Good agreement with periodical model with two peaks within 5% error margin was obtained.

Two TEM images of the YBCO layer cross-section of the PLD sample are presented in Figure 2. No inclusions were observed in the layer, but only the columnar YBCO grains with high density of grain boundaries and threading dislocations, which are aligned pinning centers along *c*-axis, which probably could cause high critical current value in magnetic field aligned along *c*-axis. The exact reason for the increased critical current values needs further investigation.

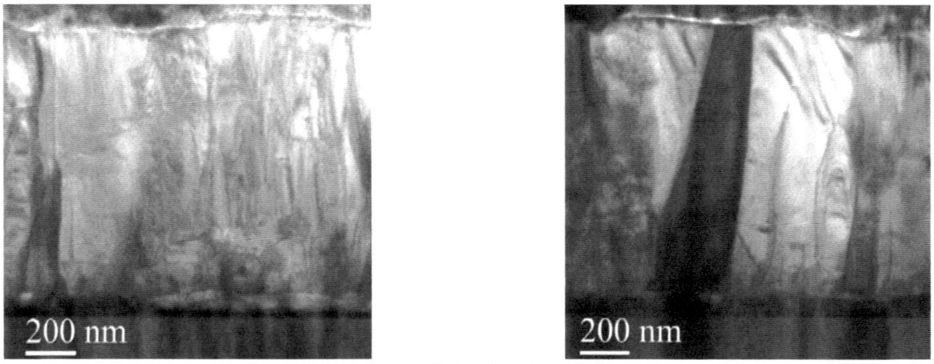


**Figure 2.** TEM images of the PLD sample cross section.

In contrast, distinct columnar $BaZrO_3$ nanorods are clearly visible in the cross-section images of MOCVD sample shown in Figures 3 and 4. The lateral dimension of the BZO nanorods varies from 3 to 7 nm. Thus these nanorods should provide strong pinning for the magnetic flux aligned in parallel with main nanorods direction. It is noticeable from Figures 3 and 4 that $BaZrO_3$ nanorods have an average tilt of about 11° degrees relative to the substrate normal which is in good conformance with asymmetrical shape of $I_c$ graph shown in Figure 1d. The direction of the columnar defects coincides with the maximum of critical current. Thereby $BaZrO_3$ nanorods make main contribution to pinning force for MOCVD sample in the direction of the *c*-axis of the sample.

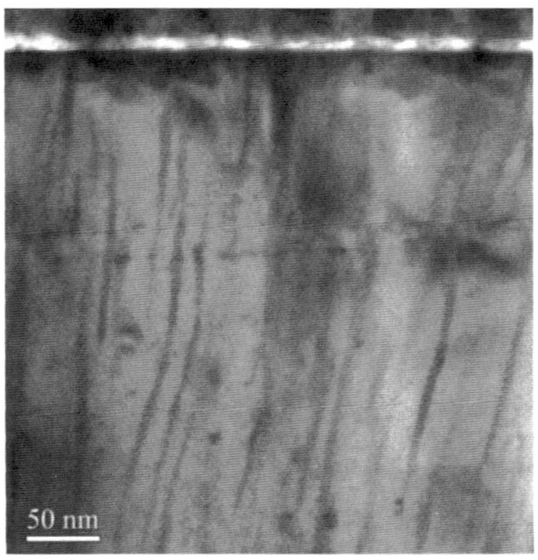


**Figure 3.** TEM image of the MOCVD sample cross section.

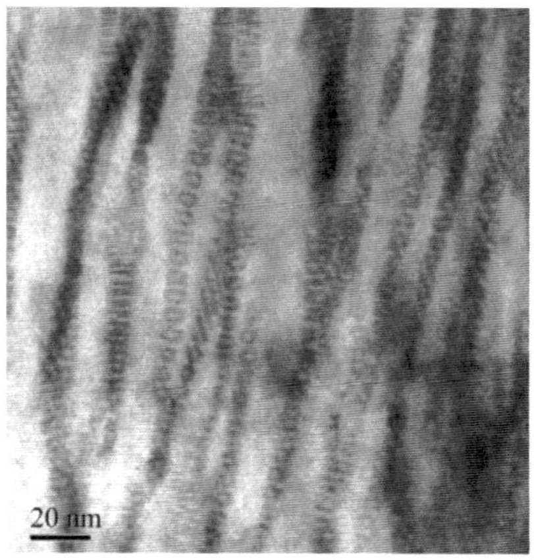


**Figure 4.** The enlarged TEM image of the MOCVD sample cross section. Dimensions of BaZrO3 nanorods are 3 to 7 nm.

## 4. Conclusions

Critical current and *n*-value of 2G HTS samples prepared by MOCVD techniques with $BaZrO_3$ doping centers and PLD techniques were systematically investigated at applied magnetic field up to 1 T. The orientation dependence of the critical current value of such samples showed two distinct peaks. In the MOCVD sample the main critical current peak is not aligned along *c*-axis but slightly deflects from it at about 11° degrees. The microstructural study of the samples showed that there are $BaZrO_3$ columnar nanorods which are tilted at the same value relative to the substrate normal. For the PLD samples TEM images show high density of grain boundaries which aligned in normal direction to the sample surface. Corresponded orientation dependence of critical current is symmetrical. For both kinds of samples *n*-value orientation dependence were calculated from Voltage-Current Curves. Pronounced differences in *n*-values demonstrate in general that vortex matter is essentially different for these two kinds of technologies.

**Acknowledgments**
The authors would like to thank management of NRC "Kurchatov institute" for support of this work. The TEM measurements were performed on equipment of Resource Center for electronic microscopy "Nanozond" in NRC "Kurchatov Institute".

**References**
[1] Malmivirta M, Yao L D, Huhtinen H, Palonen H, van Dijken S and Paturi P 2014 *Thin Solid Films* **562** 554-60
[2] Ding F, Gu H, Zhang T, Wang H, Qu F, Dai S and Cao J 2012 *Journal of Alloys and Compounds* **513** 277-81
[3] Sueyoshi T, Tokita Y, Fujiyoshi T, Mitsugi F and Ikegami T 2015 *Physics Procedia* **65** 137-40
[4] Majkic G *et al.* 2013 *Applied Superconductivity, IEEE Transactions on* **23**(**3**), 6602605
[5] Abraimov D, Ballarino A, Barth C, Bottura L, Dietrich R, Francis A and Rossi L 2015 *Superconductor Science and Technology* **28**(**11**) 114007
[6] Shuvalov D B, Barkalov K E, Lymar A V, Sychugov V V, Degtyarenko P N, Shavkin S V and Kruglov V S 2015 *Physics Procedia* **71** 417-22
[7] Hilton D K, Gavrilin A V and Trociewitz U P 2015 *Superconductor Science and Technology* **28**(**7**) 074002

Journal of Physics: Conference Series **747** (2016) 012049 doi:10.1088/1742-6596/747/1/012049

# The method of measuring the temperature dependence of the critical current of the HTS-2

**D S Yashkin[1,2,3], E P Krasnoperov[1], P A Brazhnik[1] and V S Kruglov[1,2]**
[1] National Research Centre "Kurchatov Institute", Kurchatov sq. 1, 123182 Moscow, Russia
[2] National Research Nuclear University MEPhI (Moscow Engineering Physics Institute), Kashirskoe sh. 31, 115409 Moscow, Russia

E-mail: dsjashkin@yandex.ru

**Abstract**. A new method of measurement of $I_c(T)$ dependence for the HTS-2 tape is realized. In this method, the metal substrate of the tape is used as both a heater and a thermometer. Index $n(T)$ was calculated from the measured current-voltage characteristics (CVC). For the tape manufactured by SuperPower Inc. the data of $n$-index temperature dependence in the range from 77 K up to 90 K are presented.

## 1. Introduction

The critical current $I_c$ is one of the main characteristics of superconductors for their application. There is a lot of data on the HTS-2G critical current measurement at the liquid nitrogen temperature [1,2]. But published results on the $I_c(T)$ dependence at elevated temperatures are much more modest because of the complication in measurement procedure. Cryocoolers are often used for tape cooling [3]. In this case, the tape is placed in the vacuum vessel and absence of a heat-exchange gas causes a drawback of the method. Along the tape there is a temperature gradient and local thermometer cannot correctly measure a temperature of long sample. Besides, there is a high probability of the tape's burnout at large currents. In the case of gas or liquid cooling the tape's burnout is less likely but the thermometric problem is still actual. In this paper a new method for measurement of $I_c(T)$ dependence in the superconducting tapes is proposed.

## 2. Experiment

A layered structure of the HTS-2 tape is shown in the right part of Figure 1. The superconductive layer (YBCO – 1 μm) is separated by dielectric buffer layers (0.2 μm total thickness) from the substrate (Hastelloy - 50 μm). These layers are covered by protective silver layer (Ag – 2 μm) and stabilizing copper layer Cu (20 μm) at both sides. Being closed at the edges, the copper layer is a single coating.

After removal of copper from the lateral edges, the substrate and the superconducting layer proved to be galvanically isolated but the critical current of tapes at $T = 77.4$ K fell down from 114 to 109 A. Probably, that is caused by removing of a portion of the superconductor. After the isolating the metal layers at substrate side can be used as a heating element. The resistance of the Hastelloy layer is several orders higher than that of copper and silver layer [5], so the latter bridges it. If the temperature dependence of Cu-Ag layer resistance $R(T)$ is calibrated it can be used as a thermometer (hot wire

[3] To whom any correspondence should be addressed.

method – [4]). Electrical connections of tape are shown on the left side of Figure 1. In this structure the thermometer is placed along the tape surface. This thermometer is convenient in the temperature range from 50 to 250 K. We found that $R(T)$ dependencies for two pieces of the tape are practically linear in the range of 75-100 K, and their slopes $dR / dT$ differ by less than 5%.

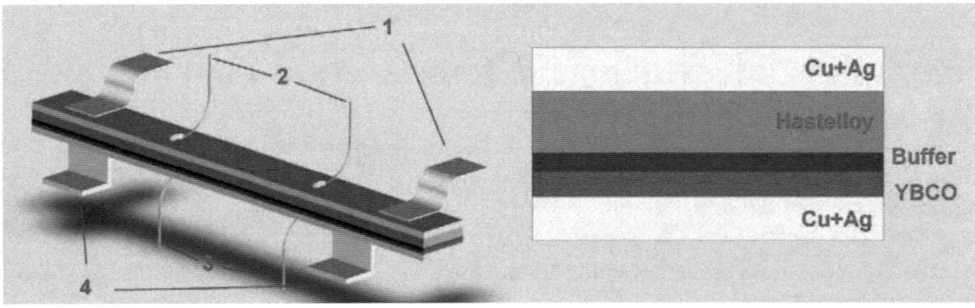


**Figure 1**. (left side) the CVC measurement diagram: 1 – the thermometer current leads, 2 – the thermometer voltage contacts, 3 – the superconductor voltage contacts, 4 – the superconductor current leads; (right side) the tape layers structure.

The thermometer readings are consistent with the superconducting layer temperature if heat flux is small. Indeed, the temperature difference $\Delta T$ between the superconductor layer and the thermometer depends on the transverse heat flux $q$ as

$$\Delta T = q\left(\frac{h_{Hast}}{\kappa_{Hast}} + \frac{1}{\kappa_{surf\_buf}}\right)$$

where $h_{\mathrm{Hast}}$ - the thickness, $\kappa_{\mathrm{Hast}}$ - thermal conductivity of the Hastelloy layer, $\kappa_{\mathrm{surf\_buf}}$ - thermal conductivity of the buffer layer per unit area. According to [5, 6] $\kappa_{\mathrm{Hast}} = 0{,}06$ W / cm • K, $\kappa_{\mathrm{surf\_buf}} = 270$ W / cm$^2$ • K and $h_{\mathrm{Hast}} = 50$ μm. So when the transverse heat flow is less than 1 W/cm$^2$, the temperature difference between the superconducting layer and the thermometer does not exceed 0.1 K.

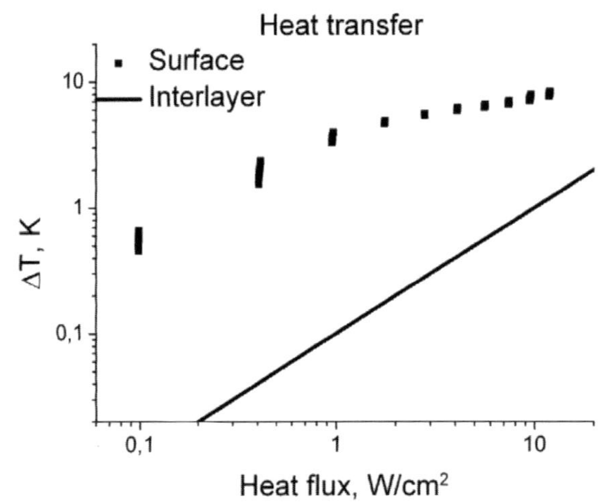


**Figure 2.** (■) – the tape surface overheat over the nitrogen boiling temperature and (—) – the difference between temperatures of YBCO layer and thermometer.

Conventional method to determine $I_c$ is direct measurement CVC in liquid nitrogen. The dependencies of the temperature jump on the tape's surface and interlayer temperature difference (YBCO – thermometer, $\Delta T$) on transverse heat flow are shown on the Figure 2. When the surface temperature exceeds the boiling point of liquid nitrogen up to 3 K the transverse heat flow causes negligible temperature gradients in the cross section of the tape. In the case of overheating the surface of about 5 K the boiling regime changes. At higher temperatures the heat flux substantially distorts the temperature distribution.

At temperatures above 80 K (up to $T_c$) measurements were carried out in evaporated nitrogen gas. In this case, the measuring cell was placed in Styrofoam box turned upside-down and filled with cotton wool to prevent the convective flow of nitrogen gas. In such an arrangement, known as the "inverted Dewar" [7], the temperature field is fairly uniform, and substantial overheating above the nitrogen boiling temperature is achieved with small heating power. On the other hand, the possibility of tape burnout significantly reduced in the ambient gas compared to vacuum.

## 3. Results

Figure 3 shows the results of measurement of current-voltage characteristics of the tape at various temperatures in ambient gas. The voltage generated by the thermoelectric power was excluded during the processing the CVC.

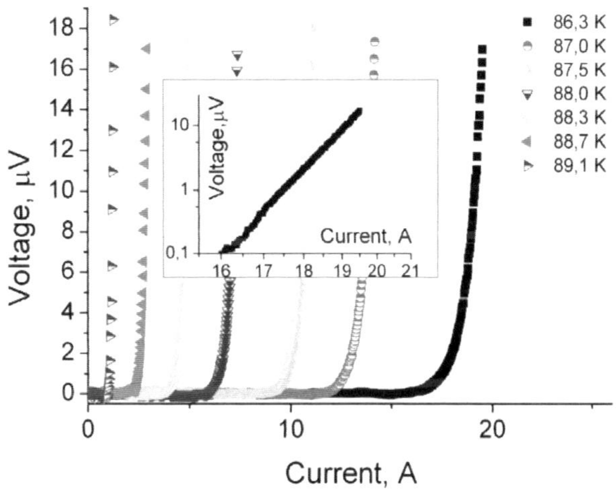


**Figure 3.** The HTS current-voltage curves at differ temperatures. Inset – one of the curves in logarithmic scale.

Using an electric field level 1 μV/cm as the criterion of the critical current – the $I_c(T)$ dependence was determined. As shown in Figure 4, this dependence is linear and slightly flattens in vicinity of the critical temperature. It confirms well the known data [3, 8].

As a rule, the CVC is described by a power dependence

$$V = V_0 \left( \frac{I}{I_c} \right)^n$$

where $n$ is the the CVC index [9]. Linearity of CVC in a logarithmic scale, presented in the inset of Figure 3, demonstrates a good agreement with the CVC power-law approximation. Therefore several CVCs obtained at differ temperatures allow to obtain $n(T)$ dependence. This dependence is shown in

Figure 4. In the temperature range 78-86 K the CVC index remains practically unchanged. As the temperature increases further up to $T_c$ the sharp decrease in $n$ is observed.

The superconducting transition of $R(T)$ is also shown in Figure 4 as an illustration. For the measured tape the critical temperature, corresponding to $R = 0$, is equal to 89.8 K.

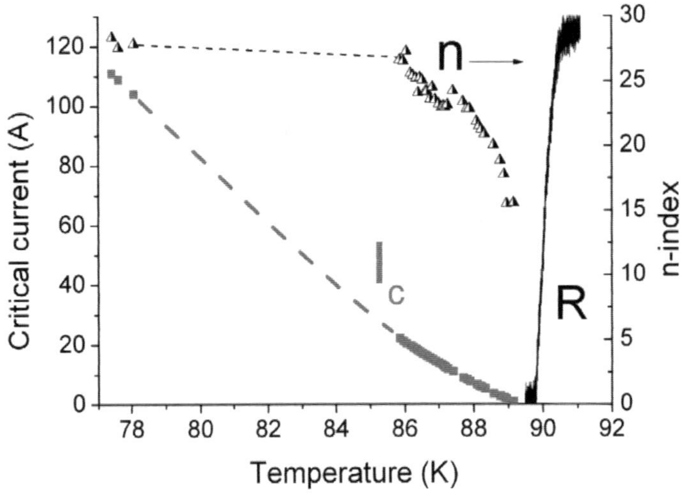


**Figure 4.** The temperature dependencies of critical current (■), $n$-index (▲). The left part corresponds the measurement in liquid nitrogen, the right part – in ambient gas.

## 4. Conclusion

An original method of CVC measurement with precise temperature control was developed for the HTS-2 tape. In this method the normal layer was employed as both a heater and a thermometer.

## References

[1] Xiang Wang, Alan Dibos, and J Z Wu 2008 *Phys. Rev. B* **77**, 144525

[2] Sung Hun Wee, Amit Goyal, Eliot D Specht, Claudia Cantoni, Yuri L Zuev, Selvamanickam V, and Sy Cook 2010 *Phys. Rev.* B **81** 140503(R)

[3] Hae-Yong Park, A.-Rong Kim, Minwon Park et al. 2010 *IEEE Trans. Appl. Supercond.* **20(3)** 2122

[4] Korotkov V S, Krasnoperov E P, and Yashkin D S 2015 *High Temperature* **53(5)** 768-770

[5] Iwasa Y 2009 *Case Studies in Superconducting Magnets* (New-York: Springer Science+Business Media)

[6] Antognazza L, Decroux M, Badel A and Abplanalp M 2013 *IEEE Trans. Appl. Supercond.* **23(3)** 900604

[7] Zavaritskii N V 1974 *JETP Lett*, **19(4)** 126-128

[8] Falorio I, Young E A and Yang Y 2012 *Physics Procedia* **3** 1462 – 67

[9] Woo Seung Lee, Seokho Nam, Jinsub Kim, Jeyull Lee, and Tae Kuk Ko 2015 *IEEE Trans. Appl. Supercond* **25(3)** 8000904

# Comparative studies of monoclinic and orthorhombic $WO_3$ films used for hydrogen sensor fabrication on SiC crystal

**V V Zuev[1], S N Grigoriev[2], R I Romanov[1], V Y Fominski[1,4], M A Volosova[2] and M V Demin[3]**

[1] National Research Nuclear University MEPhI (Moscow Engineering Physics Institute), Kashirskoe sh. 31, 115409 Moscow, Russia
[2] Moscow State University of Technology "STANKIN", Vadkovskii per. 3a, 127005 Moscow, Russia
[3]Immanuel Kant Baltic Federal University, Nevskogo 14 A, 236041 Kaliningrad, Russia

E-mail: vyfominskij@mail.ru

**Abstract**. Amorphous $WO_x$ films were prepared on the SiC crystal by using two different methods, namely, reactive pulsed laser deposition (RPLD) and reactive deposition by ion sputtering (RDIS). After deposition, the $WO_x$ films were annealed in an air. The RISD film possessed a m-$WO_3$ structure and consisted of closely packed microcrystals. Localized swelling of the films and micro-hills growth did not destroy dense crystal packing. RPLD film had layered β-$WO_3$ structure with relatively smooth surface. Smoothness of the films were destroyed by localized swelling and the micro-openings formation was observed. Comparative study of m-$WO_3$/SiC, Pt/m-$WO_3$/SiC, and β-$WO_3$/SiC samples shows that structural characteristics of the $WO_3$ films strongly influence on the voltage/current response as well as on the rate of current growth during $H_2$ detection at elevated temperatures.

## 1. Introduction

Recently, tungsten oxide ($WO_3$) films have been widely used for the development of resistive and semiconductor sensors of explosive and environmentally harmful gases (Anisimov et al. [1], Trinchi et al. [2], Kandasamy et al. [3], and ets.). For field-effect semiconductor sensors (e.g., metal - metal-oxide - semiconductor, MOS), $WO_3$ films are deposited on the semiconductor and, as a rule, are coated with a thin film of catalytic metal (Pt, Pd). The use of a silicon carbide (SiC) crystal in such structure provides a significant increase in the operation temperatures, the improved radiation and chemical resistance of sensors. Nakagomi et al. [4], Yamamoto et al. [5] and a number of other authors have demonstrated that $WO_3$ films can be formed using a variety of the chemical and physical methods. Among the physical vapor deposition methods, reactive pulsed laser deposition (RPLD) and reactive deposition by ion sputtering (RDIS) have certain advantages and they are more often used. Filipescu et al. [6], Fominski et al. [7], and Dellasega et al. [8] have shown that the RPLD of $WO_x$ films exhibits some interesting features influenced by various conditions such as the laser fluence, the reactive/buffer gas pressure, and the post-treatment conditions.

[4] To whom any correspondence should be addressed.

The aim of this work was to study features of the formation of $WO_3$ films by using RDIS and RPLD methods for pre-deposition and conventional thermal annealing for post-treatment. It seems important to perform comparative studies of the structure and properties of $WO_3$ films grown by these methods on the SiC crystal with determination of the factors that have an important effect on the properties of MOSiC sensors for prolonged hydrogen detection at high temperatures.

## 2. Experimental details

For $WO_x$ films preparation, a 6H-SiC substrate was placed at a distance of 3 cm from a W target in molecular oxygen at a pressure of 5 Pa. In case of RPLD the target was irradiated by nanosecond laser pulses (energy density was ~9 J/cm$^2$). In case of RDIS a high-voltage pulsed bipolar voltage was applied to the target, which provided W sputtering under ion bombardment, and bombardment of the $WO_x$ film with high-energy ions of gas-discharge plasma. The RDIS $WO_3$ film was not sensitive enough to $H_2$, that is why an additional Pt thin film was deposited on the surface of the $WO_x$ film by PLD. After deposition, the films were annealed in air at 550 °C for 1 h. The prepared samples were used for $H_2$ detection in the range 20–400 °C for more than 1 year.

The prepared samples were characterized by scanning electron microscopy (SEM), energy dispersive X-ray spectroscopy (EDS), and atomic force microscopy (AFM). Structure of the films was studied by X-ray diffraction method (XRD). To measure current–voltage characteristics (CVCs) of MOSiC samples, the first electrode touched the Pt or $WO_3$ film surface, while the second one touched the back surface of the SiC substrate, on which an ohmic contact was preliminary obtained by means of Pt deposition and annealing at high temperature. The diameter of the contact area for the probing electrode was no more than 300 μm. The CVCs were measured in air and in air+$H_2$ (2%) mixture at a temperature of 350 °C.

## 3. Results and discussion

Post-treatment of the prepared samples in air caused crystallization of as-deposited sub-stoichiometric $WO_x$ films and their chemical composition approached stoichiometric ($x$~3). Figure 1 shows SEM images of the RDIS $WO_3$ film covered with Pt on the SiC substrate after annealing post-treatment.

**Figure 1.** SEM images of

**Figure 2.** SEM images of

annealed $Pt/WO_3/SiC$ sample prepared by RDIS. annealed $WO_3/SiC$ sample prepared by RPLD.

The RDIS $WO_3$ film possessed crystalline structure which consisted of closely packed microcrystals. Localized swelling of the $WO_3$ films caused the micro-hills (diameter of 2 μm) formation. EDS studies have shown that this process does not destroy dense microcrystal packing (Figure 3). Uniform W and O distributions on the surface of the SiC substrate preserved. Surface of the RDIS films was covered by individual Pt nano-islands. According to XRD measurements, this film contained two monoclinic (m-$WO_3$) phases (Figure 4).

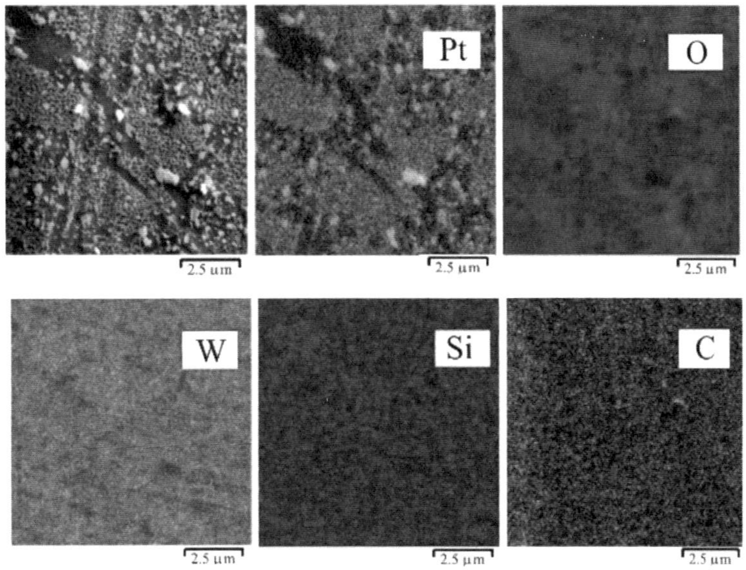


**Figure 3.** SEM image and EDS maps of the elements for $Pt/WO_3/SiC$ sample.

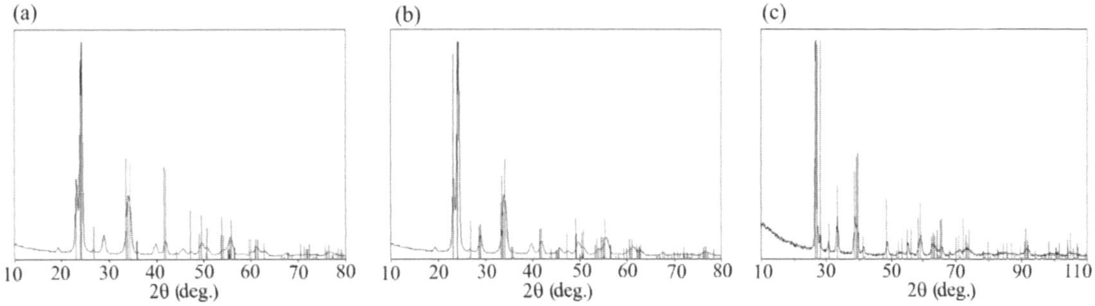


**Figure 4**. XRD spectra measured for (*a*), (*b*) Pt/m-$WO_3$/SiC and (*c*) β-$WO_3$/SiC samples. Tabulated/line spectra for (*a*) monoclinic *mP*16, (*b*) monoclinic *mP*32, and (*c*) orthorhombic *oP*32 phases are shown.

The RPLD $WO_3$ film possessed relatively smooth surface which was destroyed by localized swelling (Figure 2). Lateral size of the formed hills was ~1 – 7 μm. Large-sized hills could collapse down to SiC substrate and openings appeared in the formed craters. The formation of openings has caused nonuniform distribution of W, O, Si, and C on the EDS maps (Figure 5). There were ~200 micro-openings on an area of $200\times200$ $\mu m^2$. XRD studies have shown (Figure 3) that the RPLD $WO_3$ film had orthorhombic (β-$WO_3$) crystal structure with strong texture. For the layered crystalline structure of β-$WO_3$, W-containing atomic planes were parallel to each other as well as to the surface of the SiC surface. Such type of atomic layer package resulted in a specific character of film destruction in local areas containing micro-hills. The insert in Figure 2 shows nanosheets that formed due to cracking of the β-$WO_3$ film. AFM studies of local areas with smooth surface on the RPLD $WO_3$ film have shown (results not presented) that this film was nanostructured and ~~it~~ consisted of slightly disoriented nanocrystals with sizes ~100 nm.

The character of stationary CVCs has shown that currents are far higher at positive voltage on the Pt or $WO_3$ contact than at negative. $H_2$ addition to air resulted in the current increase and the response depended on the temperature and the concentration of $H_2$. At elevated temperatures (~350 °C), the most distinct response to $H_2$ from the Pt/m-$WO_x$/SiC and β-$WO_3$/SiC samples was achieved at registration of change in voltage/current for the reverse branch of CVCs (Figure 6).

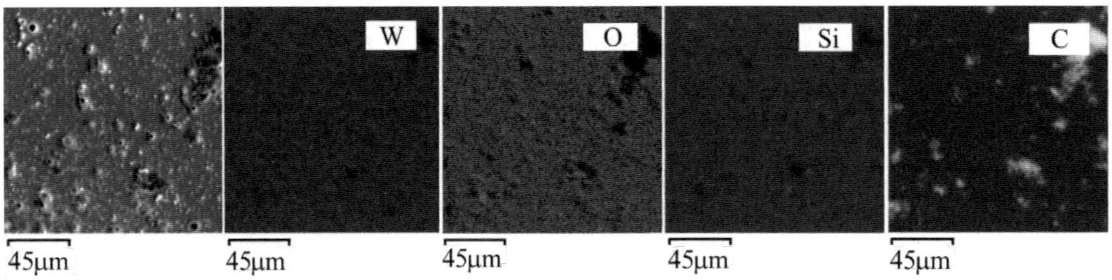


**Figure 5.** SEM image and EDS maps of elements for the $WO_3$/SiC sample.

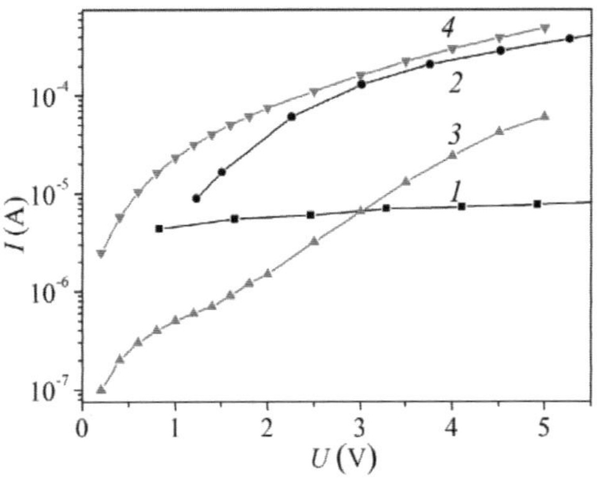


**Figure 6**. Reverse branches of CVCs for (*1*, *2*) Pt/m-$WO_3$/SiC and (*3*, *4*) β-$WO_3$/SiC samples measured (*1*, *3*) in air and (*2*, *4*) in the mixture of air and $H_2$

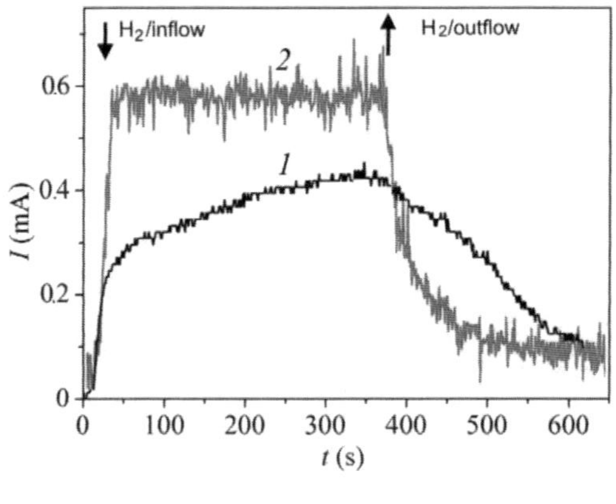


**Figure 7.** Electric current change on time during inflow and outflow of $H_2$ for (*1*) Pt/m-$WO_3$/SiC and (*2*) β-$WO_3$/SiC samples. Voltage was –5 V; temperature was 350 °C.

For Pt/m-$WO_3$/SiC sample, the voltage shift of the reverse branch at a current of ~10 μA reached ~5 V. The voltage response of β-$WO_3$/SiC sample was not so large. However, this sample can effectively detect $H_2$ without using Pt layer and the current response at a low voltages (< 3 V) was larger than that for Pt/m-$WO_x$/SiC sample. Moreover, for β-$WO_3$/SiC sample in comparison with Pt/m-$WO_3$/SiC sample, the current pulse has grown faster under $H_2$ inflow and decreased sharply when $H_2$ was removed (Figure 7). For these samples, the difference in the responses to $H_2$ may be due to the differences in the structure and morphology of the m-$WO_3$ and β-$WO_3$ films. For Pt/m-$WO_3$/SiC sample, relatively slow kinetics of transport (e.g., diffusion) of atomic hydrogen from the film surface through the dense m-$WO_3$ film matrix determines the response of the sensors. The catalytic Pt is necessary for the $H_2$ dissociation that has not occurred effectively on the m-$WO_3$ film surface.

For β-$WO_3$/SiC sample, $H_2$ transport through openings is possible. In the openings, edges of the nanosheets of the layered β-$WO_3$ structure could exhibit improved catalytic properties. The layered structure of β-$WO_3$ film (layers are oriented perpendicular to the axis of openings) facilitates an effective transport into the bulk of β-$WO_3$ film. At the same time, fast penetration of hydrogen through the openings to the interface between β-$WO_3$ layer and the SiC substrate could change an energy barrier and initiate the sensor response to $H_2$.

## 4. Conclusions

Sensor on MOSiC platform with the compact m-$WO_3$ film needs the Pt film deposition for activation of $H_2$ dissociation. The functional layers of Pt/$WO_3$/SiC sensor preserves solidity and durability and this sensor has larger voltage response. The sensor with the β-$WO_3$ film has fairly good properties without using Pt and this sensor exhibited obviously enhanced current response and fast response-recovery speed. The sensitivity of the β-$WO_3$/SiC detector may be caused by unique structural characteristics of the β-$WO_3$ film consisted of loosely stacked ultra-thin crystalline layers. Such texture of the β-$WO_3$ film and openings formation in it facilitate $H_2$ dissociation and effective hydrogen transport. The structure of the β-$WO_3$ film preserves during prolonged exploitation of the β-$WO_3$/SiC sensor.

## Acknowledgments

This work was partly supported by the Ministry of Education and Science of the Russian Federation within the framework of the state assignment (including the work of "Research Organization" for VYF) and by the Competitiveness Program of NRNU MEPhI.

## References

[1] Anisimov O V, Gaman V I, Maksimova N K, Naiden Yu P, Novikov V A, Sevast'yanov E Yu, Rudov F V and Chernikov E V 2010 *Semiconductors* **44** 366–71
[2] Trinchi A, Kandasamy S and Wlodarski W 2008 *Sens. Actuators* B **133** 705–16
[3] Kandasamy S, Trinchi A, Wlodarski W, Comini E and Sberveglieri G 2005 *Sens. Actuators* B **111–112** 111–6
[4] Nakagomi S, Okuda K and Kokubun Y 2003 *Sens. Actuators* B 96 364–71
[5] Yamamoto S, Inouye A and Yoshikawa M 2008 *Nucl. Instrum. Methods Phys. Res.* B **266** 802–6
[6] Filipescu M, Orlando S, Russo V, Lamperti A, Purice A, Moldovan A and Dinescu M 2007 *Appl. Surf. Sci.* 253 8258–62
[7] Fominski V Y, Grigoriev S N, Romanov R I, Zuev V V and Grigoriev V V 2012 *Semiconductors* **46** 401–9
[8] Dellasega D, Pietralunga S M, Pezzoli A, Russo V, Nasi L, Conti C, Vahid M J, Tagliaferri A and Passoni M 2015 *Nanotechnology* **26** 365601

Journal of Physics: Conference Series **747** (2016) 012051 doi:10.1088/1742-6596/747/1/012051

# Comparison of particle size measurements of some aqueous suspensions by laser polarimetry and dynamic light scattering

**S N Chirikov**[1]

National Research Nuclear University MEPhI, Kashirskoe sh. 31, Moscow, 115409, Russia

E-mail: SNChirikov@mephi.ru

**Abstract**. The results of the size distributions measurements of the particles of aqueous suspensions of ZnO, CuO, $TiO_2$, and $BaTiO_3$ by methods of laser polarimetry and dynamic light scattering are considered. These measurements are compared with the results obtained by electron microscopy. It is shown that a laser polarimetry method gives more accurate results for size parameter values more than 1–2.

## 1. Introduction

Laser techniques allow to determine the particle size of dispersed media remotely, efficiently, and with a high sensitivity [1]. Typically, the size determination is carried out on the assumption that the medium particles are spherical. Therefore, for media with irregular particles, data from these measurements are approximate. Measurement of the scattering matrix (4×4) whose elements are dependent on the refractive index, particle size and shape, in principle allows to determine the characteristic size of the particles considering their shapes. Accounting for the specific features of particle shape is a complex task requiring significant computing resources and practically difficult to realize. Research efforts are currently focused on finding the representative particles that are able to mimic real scattering properties of dispersed media and shape, which can be characterized by two or three parameters [2–5]. Spheroids (ellipsoids of revolution) are among these representative particle shapes. This paper presents the results of determination of particle size distributions of some suspensions obtained by laser polarimetry (LP) and dynamic light scattering (DLS).

## 2. Materials and equipment

Aqueous suspensions of ZnO, CuO, $TiO_2$, and $BaTiO_3$ were investigated, and the electron microscopy (EM) images of the particles are shown in figure 1. The effective area of the particles and the ratio of the transverse dimensions were determined from microphotographs. Histograms of particle size distribution are shown in figure 2 (dark bars), where $r$ – radius of a circle with an area equal to the effective area of the particles. The values of the average radius ($r_{av}$) and standard deviation ($\sigma$), which characterizes the width of the distribution, were derived from the histogram. The values obtained are shown in table 1 (columns EM). CuO particles are characterized by strong absorption at a wavelength of 0.63 μm, and the absorption coefficient is $8.98 \cdot 10^4$ cm$^{-1}$. Particles in other suspensions do not absorb radiation.

The scattering matrices were measured by laser polarimeter [6], where a single-mode He-Ne laser with a wavelength of 0.63 μm and a power of 7 mW was used as a radiation source.

[1] To whom any correspondence should be addressed.

**Figure 1.** Microphotographs of particles of (a) ZnO, (b) CuO, (c) $TiO_2$, (d) $BaTiO_3$.

Measurements of the scattering matrices were carried out in the scattering angle range 10°–155°. For a macroscopically isotropic medium that contains identical numbers of randomly oriented scatterers and their mirror-symmetric counterparts, scattering matrix F(4 × 4) has a block-diagonal form [7]. In this case, elements $F_{14}$, $F_{41}$, $F_{24}$, $F_{42}$, $F_{31}$, $F_{32}$, $F_{13}$, and $F_{23}$ are zero and $F_{12} = F_{21}$, $F_{34} = -F_{43}$. The dependence of element $F_{11}$ on angle of scattering θ describes the scattering indicatrix of the nonpolarized radiation. The measured scattering matrix was represented as a weighted sum of the theoretically calculated scattering matrices of model (spheroidal) particles with different sizes and shapes (aspect ratio *a/b*). The values of the corresponding weights that ensured a minimum of mean square deviations of the theoretical and experimental data determined the sought particle size distribution. The scattering matrices of model particles were calculated using the program that is based on the T-matrix method developed by Mishchenko M I [7] for an ensemble of randomly oriented spheroids. The Levenberg–Marquardt algorithm was used to solve the optimization problem.

The NanoTrac particle analyzer (Microtrac company), whose operating principle is based on the dynamic light scattering method, was also used to determine the particle size distribution. Particle sizes are determined by measuring the spectral density width of the backscattered electric field. Heterodyne detection scheme is employed; radiation reflected from the optode tip is used as a reference.

## 3. Results of experiment

When recovering the particle size distribution of suspensions according to the measurement of the scattering matrix, best results were obtained by minimizing the sum of squared deviations of the theoretical values of the matrix elements from the experimental values for the diagonal elements, and not for all elements of a block diagonal matrix. Retrieved size distributions are shown in figure 2 (open bars), and the values of the average radius ($r_{av}$) and standard deviation (σ) are presented in table 1 (columns LP). The results for all except ZnO suspensions can be considered satisfactory. Average radius values are restored with an error not exceeding 30%, for both unimodal and bimodal distributions. The value of σ, which characterizes the width of the distribution, in the case of unimodal distribution is restored with an error not exceeding 10%, and in the case of a bimodal distribution – with an accuracy of 27%.

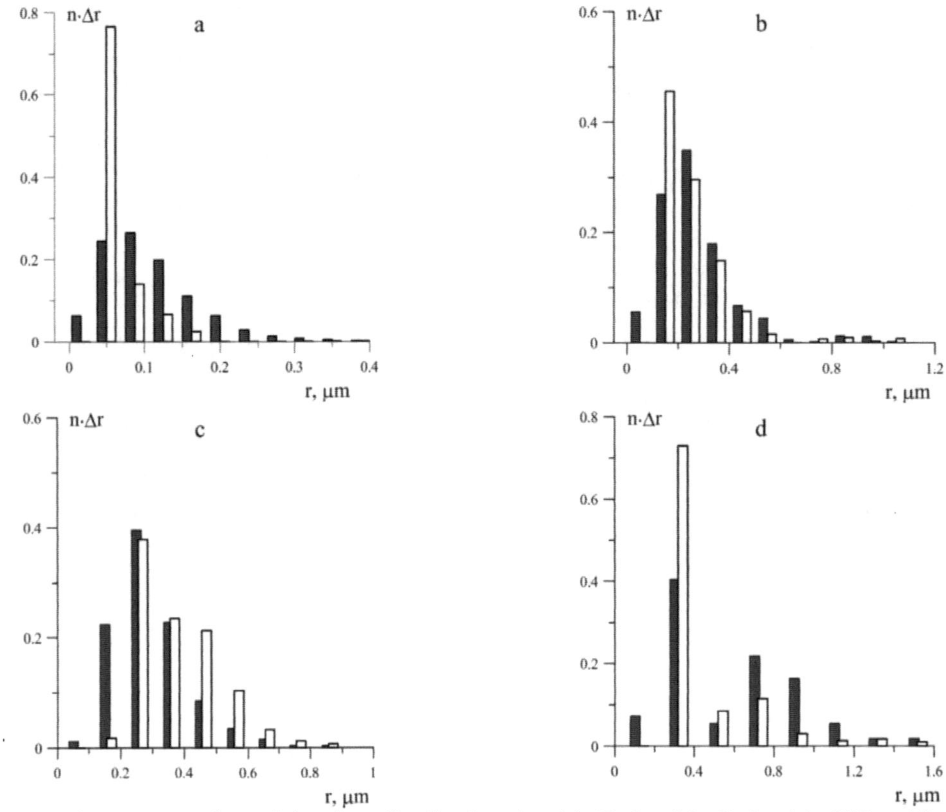


**Figure 2.** Histograms of particle size distribution in (a) ZnO, (b) CuO, (c) $TiO_2$, (d) $BaTiO_3$ suspensions. Dark bars are data obtained by EM. Open bars are data obtained by LP.

**Table 1.** The values of the parameters of the distributions.

| Parameter | ZnO | | | CuO | | | $TiO_2$ | | | $BaTiO_3$ | | |
|---|---|---|---|---|---|---|---|---|---|---|---|---|
| | EM | LP | DLS | EM | LP | DLS | EM | LP | DLS | EM | LP | DLS |
| $r_{av}$, μm | 0.11 | 0.064 | 0.087 | 0.28 | 0.25 | 0.18 | 0.29 | 0.37 | 0.20 | 0.57 | 0.42 | 0.39 |
| σ, μm | 0.06 | 0.03 | 0.02 | 0.16 | 0.15 | 0.07 | 0.12 | 0.13 | 0.02 | 0.33 | 0.24 | 0.04 |
| $(a/b)_m$ | 0.66 | 0.41 | - | 0.71 | 0.49 | - | 0.67 | 0.61 | - | 0.76 | 0.69 | - |

However, for the ZnO suspension, the distortions were observed in the retrieved distribution manifested in increasing the proportion of fine particles and reduction of the width of the distribution. Such distortions of the retrieved distribution are characteristic for the values of the size parameter $X = 2 \cdot \pi \cdot r/\lambda$ smaller than 1–2. Firstly, this is due to the difference in the values of the element $F_{11}$ for irregularly shaped particles and mixtures of spheroids, which can be most easily compensated by adding to the mixture of spheroids a relatively large number of particles of the fine fraction having a small scattering cross section and smooth indicatrix of scattering (dependence $F_{11}(\theta)$). Secondly, dependencies of the matrix elements $F_{33}$, $F_{44}$ on the scattering angle for media with irregular particles are well enough approximated by the similar dependencies for a mixture of spheroids, while the $F_{22}(\theta)$ dependence is approximated worse. Under optimum conditions, the differences between the dependencies $F_{22}(\theta)$ for a mixture of spheroids and irregular particles are larger than the differences of $F_{33}(\theta)$, $F_{44}(\theta)$. Therefore, the main contribution to the sum of squared deviations of the theoretical and experimental values of the elements $F_{22}$, $F_{33}$, $F_{44}$ will be given by a deflection of the element $F_{22}$. Decrease of particle size and increase of absorption result in $F_{22}$ values approaching unity, in a decrease of the differences of this element for particles of different shapes, and hence in reducing the magnitude of this contribution. Note that the average ratio of the transverse dimensions $(a/b)_m$ is restored more accurately when the sum of squared deviations of the calculated and measured values of the element $F_{11}$ is minimized. These values are 0.71, 0.53, 0.65, and 0.73 for suspensions of ZnO, CuO, $TiO_2$, and $BaTiO_3$, respectively.

Data obtained by the method of dynamic light scattering (table 1, columns DLS) indicate greater deviation of the average particle size from the size defined with microphotographs, compared with the method of laser polarimetry for all suspensions except ZnO slurry. The recovery of the particle size distribution based on the width measurements of the spectrum of the scattered light is an ill-conditioned mathematical problem. The noise level greatly affects the widths of the obtained distributions, therefore in the case of weakly scattering suspensions, only the position of the distribution maximum is reliably determined. Due to this fact, smaller values of the widths of the distributions are derived by the dynamic light scattering method. The ability to restore bimodal distribution with a particle size ratio corresponding to the maxima of the modes, 1:2, is a limit for this technique (for spherical particles) [1], therefore bimodality of the particle distribution in $BaTiO_3$ suspension was not detected in these measurements. The error in the size determination by dynamic light scattering method for spherical particles is rather low, not greater than 5–10% [1]. The measurement data even for ZnO suspension, which is characterized by the small size parameter value, indicate a greater error (18%). Possible reasons for these differences are related to the fact that the recovery of particle size distributions in the method of dynamic light scattering is carried out on the assumption that the particles are spherical. However, coefficients of translational and rotational diffusion, scattering cross sections, and backscattered intensity may be different for irregularly shaped particles and spherical particles that can change the field correlation function and the spectral density width of the scattered electric field. Trying to correct the data obtained from the ZnO suspension by dynamic light scattering by changing the diffusion coefficients, scattering cross sections, and the backscatter intensity for spherical particles, corresponding values for spheroids did not lead to significant changes in the distribution resulting from the small changes of these parameters. The asymmetry of the particles according to [3] for small size parameter (smaller than 2) also does not lead to a noticeable change in the scattering cross section and the backscattering intensity. Presumably, their change may be caused by the particle's surface roughness, which requires further research.

## 4. Conclusion

Use of the model of spheroidal scatterers to interpret the results of measurements of the scattering matrix of dispersive media containing non-agglomerated irregular particles allows to restore the particle size distribution for the values of the size parameter greater than 1–2 with satisfactory precision. At lower size parameter values, the restored distribution may exhibit the distortions due to increasing of the fine particle proportion. The dynamic light scattering method for small size parameter (smaller than 1–2) allows to determine the average size of suspension particles more accurately.

**Acknowledgements**

This work was also supported by Competitiveness Growth Program of the Federal Autonomous Educational Institution of Higher Professional Education National Research Nuclear University MEPhI (Moscow Engineering Physics Institute).

## 5. References

[1] Xu R 2002 *Particle characterization: light scattering methods* (New York: Kluwer Academic Publishers)

[2] Dubovik O, Sinyuk A, Lapyonok T, Holten B N, Mishchenko M, Yang P, Eck T F , Volten H, Munoz O, Veihelmann B, Zande W J, Leon J F, Sorokin M, and Slutsker I 2006 *Journ. Geophys. Res.: Atmos.* **111** D11208-42

[3] Liu J, Yang P and Muinonen K 2015 *Journ. Quant. Spectrosc. Radiat. Transfer* **161** 136-44

[4] Bi L, Yang P, Kattawar G W, and Kahn R 2009 *Appl. Opt.* **48** 114-26

[5] Bi L, Yang P, Kattawar G W, and Kahn R 2010 *Appl. Opt.* **49** 334-42

[6] Ezhokin E N, Tymper S I, and Chirikov S N 2012 *Optika i Spectroskopiya* **112** 125-31
[7] Mishchenko M I, Travis L D, and Lacis A A 2002 *Scattering, Absorption, and Emission of Light by Small Particles* (Cambridge: Cambridge University Press)

# Investigation of the effect of cyclic laser heating for creating dispersed structures in the austenitic-martensitic alloys based on Fe-Cr-Ni system

**A O Andreev[1], V D Mironov[1], V N Petrovskii[1], A V Orlov[1] and M A Libman[2]**
[1] National Research Nuclear University "MIPhI," Moscow, Russia
[2] FGUP "TsNIIchermet Im. I. P. Bardina", Moscow, Russia

E-mail.: andreev07@mail.ru

**Abstract.** The effect of cyclic laser heating on the formation of the austenite structure in the austenitic-martensitic alloys based on Fe-Cr-Ni system is investigated. It is shown that under the influence of ultra-fast laser heating on the martensite, which was formed during plastic deformation, the reverse martensitic transformation occurs, and austenite with high strength characteristics is formed. Repeated and multiple laser heating effectively grinds areas of austenite to a size close to the large nanoparticles. There is an additional increase in the strength characteristics of austenite as a result of this fragmentation.

## 1. Introduction

Production of metallic materials that combine high strength and plastic properties is one of the main tasks of modern metallurgy. Typically, increasing the strength of the alloys is accompanied by decrease of its ductility. One known exception to this rule is the grinding of the crystal structure down to the nano-crystalline state. Currently, such materials are prepared by methods of crystallization from the amorphous state, mechanical alloying and severe plastic deformation. Neither of these methods provides a nano-crystalline structure in the large items. One of the perspective methods of preparing dispersed structures in the large items is a cyclic thermal processing comprising multiple heating-cooling cycles. In [1] it was shown that the use heating-cooling cycles in the iron-nickel system leads to a sequence of transformations $\alpha \rightarrow \gamma \rightarrow \alpha$ and grinding of α-phase particles.

In the present study we investigated the possibility of the γ-phase dispersed particles formation by cyclic laser heat treatment. The austenitic-martensitic alloys based on Fe-Cr-Ni system were selected for studying. The need to obtain the γ-phase with high strength properties in these alloys is associated with the fact that they are the basis for the creation of "gradient" materials, in which areas of high strength α- and γ-phases are distributed in a specific manner [2-4].

## 2. Materials and research methods

The alloy based on Fe-Cr-Ni system with the following chemical composition: Cr - 16.5%, Ni - 7.0%, C - 0.2%, Fe - base, was selected for the researches [5]. At first, the conditions of obtaining α- and γ-phases were defined for this alloy. Methods of X-ray diffraction analysis, dilatometric and magnetometric studies were used for this purpose. Mainly austenite was present in the alloy after its melting in the open furnace and hot deformation at temperature 1000-1150 °C. The cold plastic

[1] To whom any correspondence should be addressed.

deformation with at least 75% reduction was performed to obtain martensite. Such deformation led to $\gamma \rightarrow \alpha$ transformation with up to 70% α-phase formation. Magnetometric studies showed that complete conversion of martensite formed during cold deformation to austenite was achieved after heating up to temperatures 1025-1075 °C. During subsequent cooling, two scenarios can be implemented. In the first case, isothermal $\gamma \rightarrow \alpha$-transformation was observed upon cooling at room temperature after slow heating. In the second case, after rapid heating (not slower than 300°C/min) and cooling at room temperature, only γ-phase was fixed. It was found that the formation of α- and γ-phases during such cooling did not depend on the heating temperature, provided the cooling was carried out starting from the temperature range of γ-phase existence. These results are consistent with the data presented earlier in [6].

The cyclic ultrafast laser heating of deformed α-phase with temperatures up to 1050 °C was used for fixing the γ-phase at a room temperature and its subsequent grinding. Ten cycles of heating-cooling were applied. The samples for X-ray examinations and the foils for electron microscopy studies were prepared after each cycle of heating-cooling.

An ytterbium fiber laser (wavelength $\lambda$ = 1068 nm) with a maximum power of 5 kW, operating in continuous mode, was used for ultra-fast heating of samples. The laser was connected to the optical scanning head. In the optical head construction two mirrors were provided, which allowed to deflect the laser beam in mutually orthogonal directions. Thus, the maximum laser processing surface was $30 \times 30$ cm$^2$.

Mask-plate (3 mm thick duralumin) was superimposed on the top side of the sample to heat up the sample by laser irradiation in the required zones and thus to create the necessary geometry of thermal influence zones. A stack of the mask-plate and the sample was irradiated by the laser beam from the top side. The laser beam operated in the scanning mode travelled the entire surface of the stack. Certain areas of the sample were protected (shielded) by the mask-plate, hence they were not subjected to laser irradiation, and, consequently, were not heated. Thus, only the required area of the sample was treated by laser radiation and was heated to a predetermined temperature. Different samples were heated from 1 to 10 times. Each sample was cooled in air to a room temperature after the laser heat treatment.

Linear speed of surface scanning by laser beam was constant (30 m/min). The diameter of the laser spot on the sample surface was 0.5 mm, and the distance between the scan paths was 0.25 mm, resulting in 50% degree of overlap between the scan paths. Samples had to be heated up to a temperature of 1050 °C. The laser power used for the first heating of the sample and for the subsequent heatings was different. All processing parameters are listed in table 1, including the temperature to which the surface of the sample was heated.

We controlled the heating temperature of the sample surface with a thermal imager (model T650sc by FLIR Systems). The device allows to register the temperature with an accuracy of 1°C in a temperature range from 0 to 2000 °C.

**Table 1.** Parameters of laser irradiation and temperature of heating

| Quantity of heating cycles | Laser mode | Linear speed of scanning (m/min) | Diameter of the laser spot (μm) | Overlap of scanning paths (%) | Laser power (W) | Temperature of the sample surface heating (°C) |
|---|---|---|---|---|---|---|
| 1 | Continuous, $\lambda$=1068 nm | 30 | 500 | 50 | 1260 | 1050 |
| 2 .. 10 | | | | | 925 | |

X-ray diffraction analysis was performed on a DRON-3 unit with the Fe radiation. The quantity of γ-phase and α-phase and degree of its dispersion were determined for the following cases: in the initial state (after the plastic deformation), and after the first, seventh and tenth heating-cooling cycles. The data are given in table 2. The phase content was determined by the method of homologous pairs, and the degree of dispersion - by line broadening. Electron microscopic studies were performed on a microscope with an accelerating voltage of 120 kV.

### 3. Experimental results and discussion
As a result of X-ray diffraction analysis, it was found that in the initial state (after plastic deformation) the alloy consisted essentially of a single α-phase. The structure of alloy changed dramatically after the first heating-cooling cycle, when up to 70% of γ-phase was formed. The quantity of γ-phase increased slightly after seven and ten heating-cooling cycles, while there was γ-phase's line broadening, indicating an increase in its degree of dispersion (table 2).

**Table 2.** Parameters of structures in the austenitic-martensitic alloys based on Fe-Cr-Ni system after various treatments

| Processing mode | γ-phase content (%) | Line broadening $220_{\alpha}$ γ-phase (degree) | Average size of the γ-phase domains (μm) |
|---|---|---|---|
| Plastic deformation | - | - | - |
| Plastic deformation followed by one cycle of heating-cooling | 75 | 0,75 | 2,0 |
| Plastic deformation followed by seven cycles of heating-cooling | 80 | 0,8 | 1,0 |
| Plastic deformation followed by ten cycles of heating-cooling | 90 | 0,83 | 0,25 |

Electron microscopic studies of the microstructure of the alloy showed that the structure in the initial deformed state was a mixture of highly deformed martensite and a minor amount of residual austenite fragments with sizes up to 0.5 μm (figure 1, a). After one cycle of heating-cooling the microstructure of sample was sufficiently uniform. It represented recrystallized structure of irregularly shaped austenite grains with sizes up to 2 μm having high density of dislocations, twins and a large amount of stacking faults (figure 1, b). Inhomogeneous microstructure was observed in the samples after seven cycles of heating-cooling. The structure was a mixture of highly fragmented austenite resulting from the reverse α→γ transformation, which inherited the defect structure of martensite, and the areas of the initial stage of recrystallization, which were highly fragmented austenite with some fine recrystallized grains with size not greater than 1 μm. Completely recrystallized areas with larger grains (up to 2 μm) with a high dislocation density and a large quantity of stacking faults were also observed. Generally a mixed microstructure of fragmented austenite with small recrystallized grains was dominant (figure 1, c). After ten cycles of heating-cooling (figure 1, d) of the samples we observed, firstly, areas of strongly fragmented austenite resulting from the reverse α→γ conversion, and secondly, the areas of an initial stage of recrystallization, representing the large quantity of fine recrystallized grains with size from about 0.1 μm to about 0.5 μm. These grains are quite evenly spaced in a fragmented austenitic matrix. Also, a small quantity of large recrystallized grains with a high degree of defects was found.

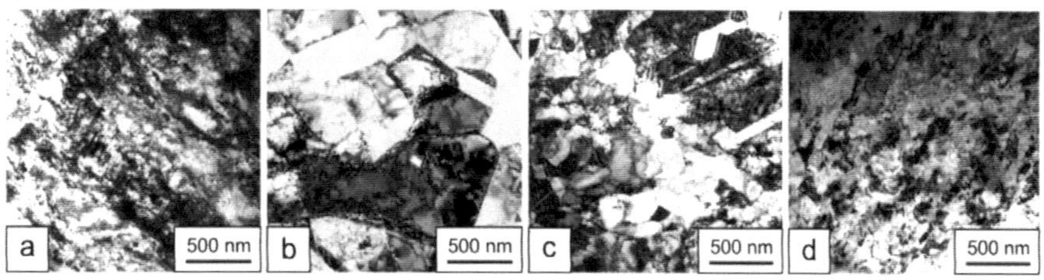


**Figure 1.** The microstructure of alloy after plastic deformation (a); plastic deformation and subsequent heating-cooling cycles: one cycle (b), seven cycles (c), ten cycles (d).

Thus, the results of electron microscopic examination of the microstructure were correlated with the data of X-ray analysis. These results showed that transformation of significant amount of α-phase to γ-phase occurred after one heating-cooling cycle. Grinding of the phase to the large nanoparticles occurred after repeated laser heating-cooling cycles.

Microhardness of γ-phase after one heating-cooling cycle reached 3000 MPa. This result is significantly higher than in the case of the austenite formation in the alloy after melting and hot deformation. This indicates that the strength of the γ-phase generated by laser heating substantially exceeds its typical strength values. Apparently, this is due to inheritance of the defects of deformed martensite in the structure of austenite. Reducing the size of γ-phase with increasing quantity of heating-cooling cycles results in further strengthening while maintaining a high degree of plasticity of austenite.

## 4. Conclusions

Laser cyclic heat treatment of austenitic-martensitic alloys based on Fe-Cr-Ni system leads to fragmentation of particles of γ-phase to the formation of large nanoparticles. In this case, the strength characteristics of austenite improve significantly while maintaining its high ductility.

**Acknowledgments**

This work was financially supported by the Ministry of Education and Science of the Russian Federation (unique identifier PNIER RFMEFI58214X0004).

This work was also supported by Competitiveness Growth Program of the Federal Autonomous Educational Institution of Higher Professional Education National Research Nuclear University MEPhI (Moscow Engineering Physics Institute).

## References

[1] Estrin E I, D'yakonov D L, Libman M A, Podrets M E, Tomchuk A A and Cheretaeva A O 2015 *Inorganic Materials: Applied Research* **6** pp 388-390
[2] Andreev A O, Galkin M P, Libman M A, Mironov V D, Petrovskiy V N and Estrin E I 2014 *MITOM* **1** pp 50-53
[3] Galkin M P, Libman M A and Estrin E I 2014 *Inorganic Materials: Applied Research* **3** pp 25-28
[4] Blinova E N, Glezer A M, Libman M A and Estrin E I 2014 *News of universities. Phisics* **4** pp 8-13
[5] Libman M A and Estrin E I 2014 *Steel* **10** pp 67-68
[6] Saragadze V V and Uvarova A M 2013 *Hardening and properties of austenitic steels* (Ekaterinburg: IFM) p 720

# Subnanosecond and picosecond generation regimes of all-PM Yb-doped fiber lasermode-locked by NOLM

**A A Borodkin[1], D V Khudyakov and S K Vartapetov**
Physics Instrumentation Center of Prokhorov General Physics Institute,Russian Academy of Sciences, Troitsk, Moscow,142190, Russia

E-mail: borodkin_andrey@mail.ru

**Abstract**. We demonstrated two stable pulsed operation regimes of all-polarization maintaining (PM) Yb-doped fiber laser oscillator with pulse duration of 640 and 85 ps. Nonlinear optical loop mirror (NOLM) was used for laser mode-locking. The first operation regime delivered high energy pulses of 5 nJ, and second regime delivered pulses of 0.7 nJ at a common repetition rate 5 MHz. The dynamics of the temporal and spectral parameters of laser pulses was studied using mathematical simulation based on numerical solution of the nonlinear Schrödinger equation. The simulation results showed that in stable regime with pulse duration of 85 ps the pulse could be compressed to 2 ps.

## 1. Introduction

Nowadays ultrashort pulse fiber lasers are widely used in various fields of science, technology and medicine. Due to the easy fabrication, low cost and high repetition rate, fiber lasers are often applied for various technical tasks, such as precision processing of materials, marking, surface treatment, metal cutting. For these tasks lasers with pulse duration in the range of 1 ps - 10 ns are typically used [1,2]. For reaching such a wide range of pulse duration, various stretchers are applied [2,3]. However, stretchers degrade the characteristics of the output pulse and significantly increase the size and cost of the fiber laser. Development of a fiber laser with the possibility of changing the pulse duration by transition from one stable pulse generation regime to another is the important task. The output pulse with duration from 10 ps to a few ns can be obtained in the fiber oscillator with NOLM [4,5]. But none of them are capable of changing the pulse duration significantly. As the additional advantage, fiber lasers with NOLM can be performed on the polarization maintaining fiber. Such fiber allows avoiding instability resulting from the polarization state drift in the fiber due to external influences. Thus such lasers offer the temperature and polarization stability.

In this paper we demonstrate a NOLM mode-locked fiber oscillator with two stable pulsed operation regimes with pulse durations of 640 ps and 85 ps with 5 MHz repetition rate. The transition between modes is performed by means of pump power changing. All elements of the ring fiber laser have normal group velocity dispersion, and fibers are polarization maintaining to achieve polarization and temperature stability of radiation.

## 2. Scheme of the pulsed fiber laser

The scheme of a fiber laser with two stable pulsed operation regimes consisting of elements with normal group velocity dispersion is shown in figure 1. The ring laser is made of PM fibers with a core

[1] To whom any correspondence should be addressed.

diameter of 6 μm and a cladding diameter of 125 μm. The total ring cavity dispersion is 1 ps$^2$. As an active medium we used an ytterbium-doped PM single mode fiber with a core diameter of 6 μm and absorption of 250 dB·m$^{-1}$ at a wavelength of 976 nm. The active fiber was pumped by a single mode laser diode with a maximum average power of 460 mW at a wavelength of 976 nm. The diode radiation was coupled into the cavity using a PM WDM 980/1030 fiber multiplexor based on a PM fiber. Two output channels (through 20% and 50% couplers) were used to couple out radiation from the cavity and to diagnose the laser radiation, respectively. The linear polarization in the cavity was oriented along the slow axis of the PM fiber using a fiber isolator–polarizer. As a mode-locking element we used a NOLM made of a 30 m long ring PM fiber segment. Stable generation of pulses was achieved using a 90/10 coupler (3) in the NOLM loop. Since the resonator scheme consisted only of PM fiber elements, pulsed lasing was stable and did not depend on thetemperature variations and on the optical fibers bends. There was no spectral filter in the laser, which is typically included in the picosecond fiber lasers with NOLM.

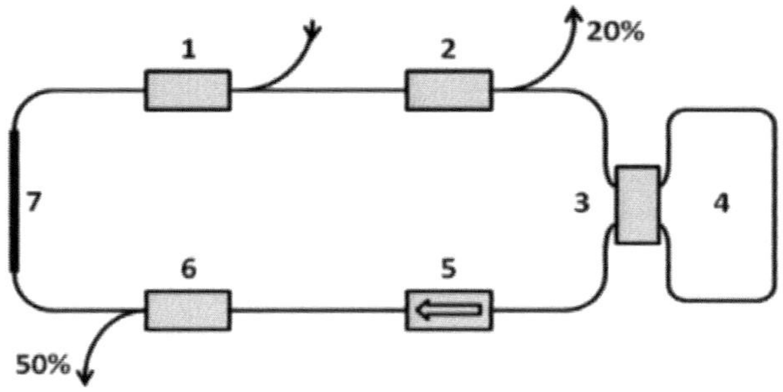


**Figure 1.** Scheme of the fiber laser with two operation regimes: ( 1 ) PM WDM 980/1030 multiplexor; ( 2 ) 20/80 output coupler; ( 3 ) 10/90 coupler; ( 4 ) 30 m long fiber loop; ( 5 ) isolator–polarizer; ( 6 ) 50/50 output coupler; ( 7 ) active Yb fiber.

### 3. Experiment

The threshold of self mode-locking was reached at a pump power of 400 mW. At this pumping level laser operated in a regime of stochastic pulses with pulse-to-pulse instability ~ 15%. After initiation of pulsed lasing, it was possible to decrease the pump diode power to a lower level of 300 mW with stable pulsed operation. The pulse repetition rate at the chosen cavity length was 5 MHz.

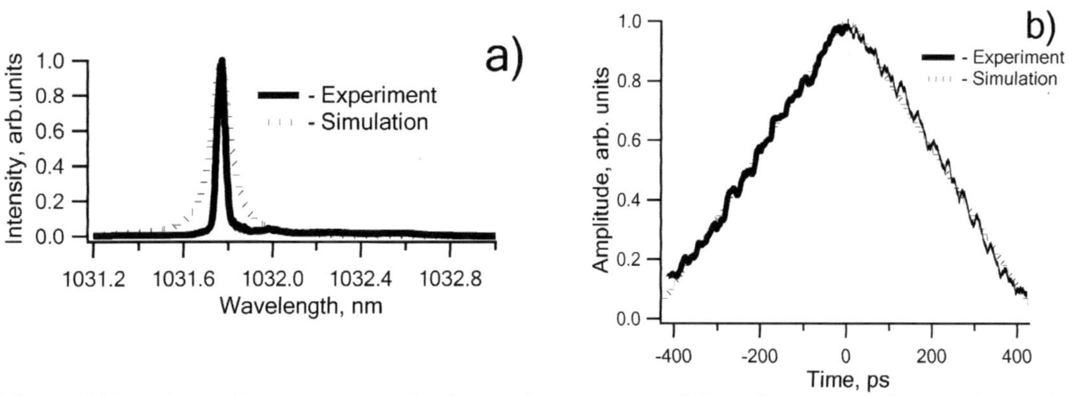


**Figure 2.**Experimentally measured and calculated parameters of the subnanosecond operation regime: (a) spectra; (b) autocorrelation functions of a pulse.

The average laser output power was 25 mW, which corresponds to pulse energy of 5 nJ. The optical output spectrumin the regime of pulsed operation was measured witha Yokogawa AQ6370C spectrometer with a resolution of 0.02 nm (figure 2a). The center wavelength and the FWHM of the

Journal of Physics: Conference Series **747** (2016) 012053 doi:10.1088/1742-6596/747/1/012053

spectrum were 1032 nm and 0.04 nm, respectively.The pulse duration was measured by the method of noncollinear second harmonic generation in a nonlinear crystal,which was used to determine the pulse autocorrelation function(figure2b). The FWHM of the autocorrelation function was 650 ps and the function shape was triangular, whichtestifies that the pulse shape is rectangular.

After decreasing the pump power in the laser below 300 mW, the pulse generation became unstable. With further decrease of the pump power to 80 mW, another stable regime of generation occurs with a lower power radiation output and a shorter pulse duration.

The average laser output power in such stable regime was 3.5mW, which corresponds to a pulse energy of 0.7 nJ, that is 7 times less than in the subnanosecond regime. The optical output spectrum of pulsed operation was measured with a resolution of 0.5 nm (figure3a). The center wavelength and theFWHM of the spectrum were 1033 nm and 4 nm, respectively.The autocorrelation function of the pulse was measured as in the previous case (figure3b). The FWHM of the autocorrelation function was 66ps, and the function shape was close to triangular, whichtestifies that the pulse shape deviated from Gaussian to rectangular.

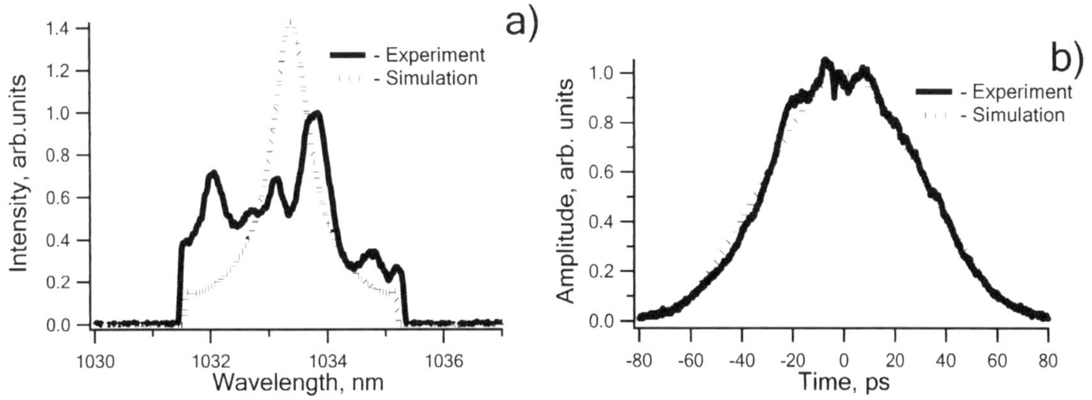


**Figure3.**Experimental and calculated parameters of picosecond operation regime: (a) spectra; (b) autocorrelation functions of a pulse.

## 4. Numerical simulation

The mathematical simulation of the dynamics of temporal and spectral pulse's parameters, obtained in a fiber laser with NOLM, was performed using the nonlinear Schrödinger equation as described in [5]. In simulation, firstly we intentionally chose a longer pulse than in the experiment with a zero chirp. The calculations were continued until the radiation parameters became unchanged during a number of cavity round-trips. After several round-trips, the combined action of the amplifying medium, self-phase modulation, dispersion, and NOLM led to the formation of a stable output spectrum (figure 2a) and pulse (figure 4) with corresponding autocorrelation function (figure 2b). The calculated duration of the output pulse was 640 ps (figure 4). The pulse autocorrelation function obtained in simulation is in good agreement with the autocorrelation function measured by the method of noncollinear second harmonic generation (figure 2b). The pulse shape is rectangular, which agrees with the results of another work on pulsed fiber lasers with NOLM [1].

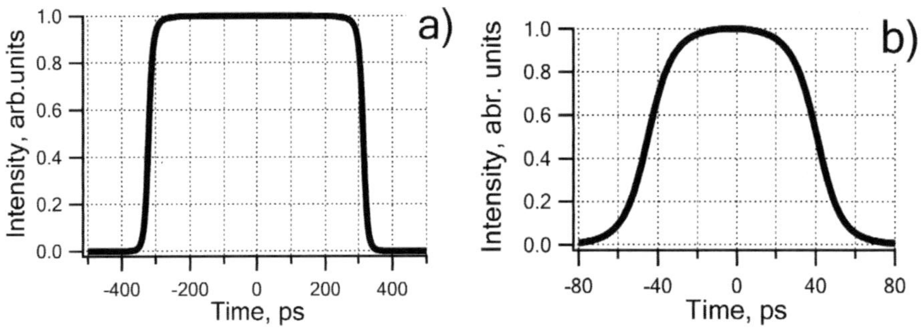


**Figure 4.**Numerically simulated output pulse of the laser:(a) subnanosecond regime; (b) picosecond regime.

To calculate the parameters of laser radiation in picosecond regime, we used the results obtained in the simulation of subnanosecond fiber laser as initial conditions. With pump power decreasing, a stable generation of subnanosecond pulses is transferred to the stable picosecond regime. The pulse autocorrelation and output spectrum obtained in simulation are in good agreement with the experimental data(figure 2). In the picosecond regime, the output pulse has linear chirp and could be compressed. Calculated duration of the compressed pulse is 2 ps (figure 5).

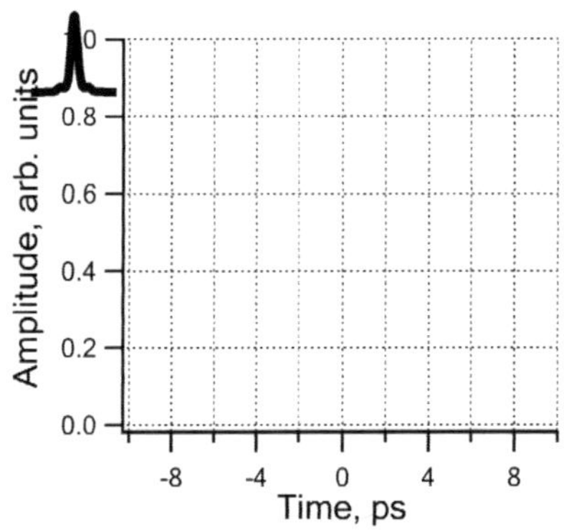


**Figure 5.** Numerically simulated autocorrelation function of the compressed pulse.

## 5. Conclusions

A passively mode-locked ytterbium fiber laser with two stable pulsed operation regimes is presented. For the laser mode-locking a NOLM is used. The durations of the output pulses are 640 ps and 85 ps, and they can be tuned by pump power changing. Subnanosecond regime delivers high energy pulses of 5 nJ, and picosecond regime delivers pulses of 0.7 nJ at repetition rate of 5 MHz. The ring cavity is designed using PM fibers only, which ensures low sensitivity of the laser to the environmental conditions and a stable polarization at the output. Numerical simulation of the laser shows a good agreement of experimental and theoretical output characteristics. In the picosecond regime pulses are linearly chirped and can be compressed by an external compressor to duration of 2 ps.

## References

[1] Ozgoren K, Oktem B, Yilmaz S, Ilday F and Eken K 2011 *Opt. Express* **19** 17647.
[2] Muhammad N,Whitehead D,Boor A,Oppenlander W, Liu Z and Li L 2012 *Appl.Phys.A* **106** 607.
[3] Liao K, Cheng M, Flecher E, Smirnov V and Glebov L 2007 *Opt. Express* **15** 4876.
[4] Zhao L, Tang D, Cheng T and Lu C 2007 *Opt. Commun* **272** 431.

[5] Borodkin A, Khudyakov D and Vartapetov S 2015 *Quantum Electron* **45** 98.

Journal of Physics: Conference Series **747** (2016) 012054 doi:10.1088/1742-6596/747/1/012054

# Writing of 3D optical integrated circuits with ultrashort laser pulses in the presence of strong spherical aberration

**M A Bukharin[1,2,4], N N Skryabin[1,2], D V Khudyakov[2,3], and S K Vartapetov[2,3]**
[1]Moscow Institute of Physics and Technology, Institutskiy per.9, Dolgoprudny, Moscow Region, 141700, Russia
[2]Optosystems Ltd., Troitsk, Moscow, 142190, Russia
[3]Physics Instrumentation Center of Prokhorov General Physics Institute RAS, Russian Academy of Sciences, Troitsk, Moscow, 142190, Russia

E-mail: mikhail.bukharin@phystech.edu

**Abstract**. A novel technique was proposed for 3D femtosecond writing of waveguides and optical integrated circuits in the presence of strong spherical aberration, caused by inscription at significantly different depth under the surface of optical glasses and crystals. Strong negative effect of spherical aberration and related asymmetry of created structures was reduced due to transition to the cumulative thermal regime of femtosecond interaction with the material. The differences in the influence of spherical aberration effect in a broad depth range (larger than 200 μm) was compensated by dynamic adjustment of laser pulse energy during the process of waveguides recording. The presented approach has been experimentally implemented in fused silica. Obtained results can be used in production of a broad class of femtosecond written three-dimensional integrated optical systems, inscripted at non-optimal (for focusing lens) optical depth or in significantly extended range of depths.

## 1. Introduction

Femtosecond writing technology provides single-step maskless inscription of waveguides in optical glasses and crystals [1-3]. Due to point-by-point formation process the technique allows easy adjustment of the waveguide's geometry and parameters of the optical integrated circuits based on it. Moreover, thanks to nonlinear nature of the phenomenon, the effect of permanent alteration of the refractive index occurs only in the focal region, providing an opportunity to inscribe three-dimensional structures on optically transparent glasses and crystals [4]. Unfortunately, operational depth range (under the surface of crystal) of 3D optical integrated circuits is essentially narrowed by spherical aberrations of a focusing lens [5] to the width of ±20 Rayleigh length from the depth of optimal aberration correction (±50 μm for NA = 0.45 or ±17 μm for NA = 0.8) [6, 7]. For a number of applications such as large mode area waveguides, multimode waveguides, and optical chips that are easy to couple with standard telecom fibers (described in details below), it is necessary to use direct femtosecond writing technique in a depth range extended to, at least, ±70 Rayleigh length (±180 μm for NA = 0.45 or ±62.5 μm for NA = 0.8). In the investigation we proposed a solution of the challenge and described modified femtosecond inscription technique with an extended operational depth range.

Up to date one of the most widespread techniques of femtosecond written waveguides has consisted in formation of waveguide's cladding with decreased refractive index [4]. Typical geometry

[4] To whom any correspondence should be addressed.

of an induced waveguide is shown in figure 1(a, b). Two neighboring tracks with high aspect ratio are inscribed at optimal depth under the surface of a crystal (depth of optimal spherical aberrations compensation for the lens and ±20 Rayleigh length range, as described above). In such a case, refractive index in the central region between tracks increases due to mechanical stress produced by the written cladding. Such tracks could be written with weak focusing of fs-laser emission (NA = 0.2-0.4), low pulse repetition rate (< 0.5 MHz), and relatively high pulse energy (> 0.1 μJ), which induces strong self-focusing and formation of tracks with high aspect ratio (> 3) in the cross section. The waveguides produced by the described technique exhibit high propagation losses (0.9-3 dB/cm) and only maintain a single linear polarization state (horizontal) due to low containment of light in vertical direction, which sufficiently restrains practical applications.

Typical solution of the problem lies in a full circle writing of cladding, as shown in figure 1(c). The solution increases numerical aperture of the waveguide and reduces optical losses, but unfortunately it fails to resolve such problems as maintenance of all polarization states (propagation conditions for the orthogonal polarization sufficiently differ due to inhomogeneous structure at the top and bottom of the induced tracks, as shown in figure 1(c)) and writing of waveguides at a non-optimal depth under the surface of crystals (due to negative influence of spherical aberration that causes alteration of the track's geometry and leads to the optical breakdown).

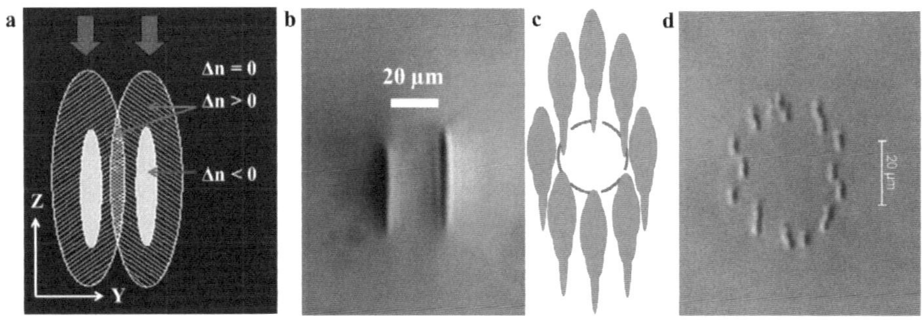


**Figure 1.** (a) Schematic and (b) microphotography of typical anisotropic 2-tracks stressed-induced cladding of waveguide, written with fs-laser; (c) schematic of inhomogeneous full-circle stress-induced cladding of waveguide; (d) microphotography of highly symmetric depressed cladding waveguide, written in thermal cumulative regime.

To resolve the problem and to write highly symmetrical waveguides with both polarization states maintenance even at non-optimal writing depth (in the presence of strong spherical aberration) we have proposed an advanced technique of femtosecond writing. It consists in writing at relatively high NA (0.6), but in a thermal cumulative regime [8], which offers up to twofold increase of the induced refractive index contrast, aspect ratio decrease down to 1.5 (see figure 1(d)), and greater than threefold reduction of the optical losses, down to 0.3-0.5 dB/cm, for both polarizations. Thanks to significant reduction of the aspect ratio and compensation of the spherical aberration influence on the track's geometry due to symmetric thermal effects, the proposed technique provides an opportunity to write waveguides in extended depths range (more than ±70 Rayleigh length, as will be experimentally shown below).

## 2. Experimental setup

For realization of thermal regime, we used experimental setup described in detail in [8] . Femtosecond laser pulses (340 fs, 1040 nm) from the oscillator (HighQ femtoTRAIN, 0.1-10 MHz) were focused with aspheric lens (NA=0.6) into fused silica sample, which was mounted on a 3D-translational stage and moved with velocity 10-800 μm/s to form extended tracks through the total length of the sample (3 cm).

Journal of Physics: Conference Series **747** (2016) 012054 doi:10.1088/1742-6596/747/1/012054

### *2.1. Effect of increased numerical aperture*

Tight focusing with NA = 0.6 instead of widespread 0.3-0.4 provides low size of focusing area (0.9-1.2 μm instead of 2-4 μm). Thus, to maintain induced refractive index effect a relatively low peak power is required (0.1-0.3 MW from low $M^2$ oscillator instead of > 1 MW MOPA-laser), that causes reduced influence of self-focusing and consequently the absence of filamentation, providing formation of nearly-circle cross section tracks with aspect ratio < 2.5 instead of > 3-5 in the case of low NA writing technique.

On the other hand, in the case of tight focusing, even a small alteration of writing depth from the optimal one could sufficiently increase the spherical aberrations effect, that in turn will increase treated volume, required peak power and, as a result, aspect ratio of the induced tracks due to self-focusing. To prevent formation of highly asymmetric tracks, we additionally propose to compensate the effect of spherical aberrations and suppress self-focusing by writing in thermal cumulative regime described below.

### *2.2. Effect of thermal cumulative regime*

Thermal cumulative regime could be realized at high pulse repetition rates (more than 1 MHz), when characteristic time of thermal diffusion from the focal spot exceeds the period between subsequent laser pulses (thermal cumulative parameter α from [8]). In that case, temperature in the focal spot rapidly rises higher than the meting point, and at high temperature the photochemical processes proceed easily even at lower operational energies of the laser pulse. As shown in figure 2(a), this allows to decrease operational pulse energy by more than 2 times, and to decrease self-focusing simultaneously. Moreover, transition to thermal cumulative regime offers to decrease the track's cross section aspect ratio by more than 1.5 times down to 1.5 value due to the isotropic thermal melting effect, as shown in figure 2(b).

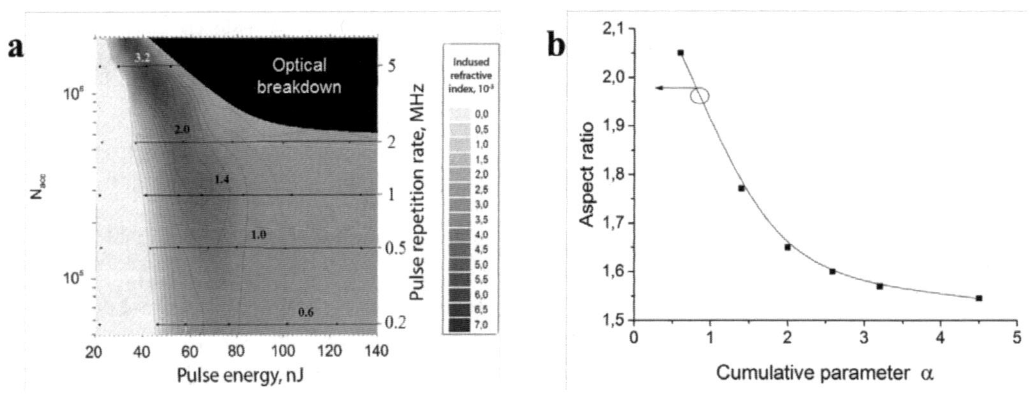


**Figure 2.** (a) Operating range diagram of the induced refractive index effect in fused silica (translation velocity 10 μm/s, pulse repetition rate varies between 0.1 and 10 MHz). Values above horizontal lines represent cumulative regime parameter α [8]; (b) decreasing of the track's cross section aspect ratio with transition to the thermal cumulative regime of femtosecond writing.

## 3. Results

It was experimentally shown that developed technique of femtosecond writing allows to write optical waveguides with low propagation losses (< 0.3 dB/cm) for both polarisation states of propagating light even in case of writing at non-optimal depth (in presence of strong spherical aberrations). Thanks to inscription in thermal cumulative regime, we achieved the operating depth range as broad as ±70 Rayleigh length (±180 μm for NA = 0.45 or ±62.5 μm for NA = 0.8), which is 2 times broader than previously published data[6, 7]. As shown in figure 2(a), induced refractive index contrast was as high as $7\cdot10^{-3}$, which is 2 times larger than the value for non-thermal regime of writing. Higher refractive index contrast provides higher numerical aperture of written waveguides resulting in the values up to 0.08. Achieved characteristics of the femtosecond waveguide writing technique open new opportunities in formation of integrated photonic elements.

## 4. Analysis and discussion

One of the most important challenges in design and creation of optical integrated devices is the optimization of interconnections between bulk and fiber optical elements. For the connection with standard telecom fibers (diameter 125 μm), the optimal depth of waveguides formation equals to 62.5 μm. In this case femtosecond written optical integrated chip and fiber could be placed at mutual alignment base (as shown in figure 3), that offers to adjust only 1 coordinate and 1 angle instead of 2 coordinates and 2 angles in the case of non-optimal depth of waveguide. Common focusing lenses and objectives do not provide correction for spherical aberrations at 62.5 μm depth, thus in order to write waveguides at the depth, one should apply additional techniques for its correction, as described in the paper above. Proposed approach has been experimentally carried out in fused silica for writing of waveguides at depth 62.5 μm (+70 Rayleigh length depth) with high NA (100x, NA = 0.8) micro-objective without spherical aberration compensation, as shown in figure 3(c-d).

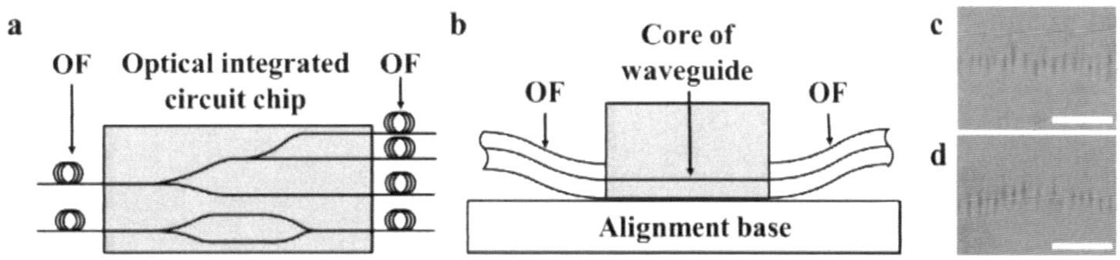


**Figure 3.** Schematic of typical optical integrated circuit chip (a) and alignment process with waveguide written at optimal (for interconnection) depth (b). Microphotographs of the induced tracks written at depth with low (c) and high (+70 Rayleigh length depth) (d) spherical aberrations.

Proposed technique allows to write three-dimensional waveguides at sufficiently different depths. To take into account the uncompensated effect of spherical aberrations, one should dynamically adjust laser pulse energy at each writing depth to maintain constant value of induced refractive index.

## 5. Conclusion

In the paper we have proposed an improved technique for femtosecond waveguide writing that provides an opportunity to reduce self-focusing effect and to compensate negative influence of spherical aberrations. It was experimentally shown that operational depth range of the proposed technique broadens 2 times as compared with low-NA non-thermal one and reaches the value of ±70 Rayleigh length, which opens new opportunities for design and creation of advanced photonic devices that are easily interconnectable with telecom fibers and three dimensional optical integrated circuits.

## Acknowledgements

The work was partially supported by Foundation for Assistance to Small Innovative Enterprises in Science and Technology (4806ГУ1/2014 №0008340 and 8611ГУ/2015 №0018824).

## References

[1] Okhrimchuk A G, ShestakovAV, Khrushchev I and Mitchell J 2005 *Opt. Lett.***30**(17), 2248
[2] Streltsov A M and Borrelli N F 2001 *Opt. Lett.***26** (1), 42–3
[3] Ma Xiao-Song 2014*Nature Photonics***8** 749
[4] Osellame R, Cerullo G, Ramponi R 2012 *Femtosecond Laser Micromachining Photonic and Microfluidic Devices in Transparent Materials*(London: Springer)

[5] Huot N, Stoian R, Mermillod-Blondin A, Mauclair C and Audouard E 2007 *Opt. Exp.* **15** (9) 12395-408
[6] Nnolte S, Will M, Burghoff J and Tuennermann A 2003 *Appl. Phys. A* **77** 109–11
[7] Kowalevicz A M, Sharma V, Ippen E P and Fujimoto J G 2005 *Opt. Lett.* **30** (9) 1060-2
[8] Bukharin M A, Khudyakov D V, Vartapetov S K 2015 *App. Phys. A* **119** (1) 397-403

Journal of Physics: Conference Series **747** (2016) 012055 doi:10.1088/1742-6596/747/1/012055

# The spectra of stimulated concentration scattering (Mie scattering) on nanoparticles latex suspension in the presence of convection

**I S Burkhanov**[1,3], **S V Krivokhizha**[1], **L L Chaikov**[1,2]
[1] P.N. Lebedev Physical Institute of the Russian Academy of Sciences, Leninskiy Prospekt 53, Moscow 119991, Russia
[2] National Research Nuclear University MEPhI, Kashirskoe sh. 31, Moscow 115409, Russia.

E-mail: ilyaburkhanov@gmail.com

**Abstract.** Spectral shifts of the stimulated concentration light scattering (SCLS, stimulated Mie scattering) in suspensions of various sized latex nanoparticles in water were measured by the light guide scheme in conditions of the backscattering in the presence of convection. It is shown that the spectral shift can be either negative or positive depending on the particle size.

## 1. Introduction

Each type of spontaneous light scattering (SpS) has its corresponding type of stimulated scattering (SS) [1], and frequency shifts of most types of SS have been measured. They are $3–9 \cdot 10^{13}$ Hz for SRS [2], 10–1000 MHz for SBS [3], and 10–30 GHz for a newly discovered stimulated globular scattering (SGS, the low-frequency Raman scattering on own vibrations of particles) [4, 5], which are close to the frequency shifts of the respective types of SpS.

Spectral lines of the spontaneous depolarized Rayleigh scattering and the scattering by entropy fluctuations are not shifted relatively the excitation light, but only broadened. At the same time, however, SRWS line [6-8] and STLS line [9-10] are shifted by half of the width of the corresponding spontaneous scattering line. For example, SRWS shift is $\sim 10^{11}$ Hz, and STLS shift is $\sim 3 \cdot 10^{8}$ Hz [11]. For those kinds of scattering, where SpS lines are not shifted with respect to the exciting radiation, the main features of SS are the frequency shift of the scattered light equal to the half-width of SpS lines and a non-linear increase in the scattered light intensity..

However, the SCLS frequency shift was measured only in gas mixtures [12-14]. Authors of recently undertaken investigations of scattering on nanocrystal particles of CdSe / Cds / ZnS in chloroform and on the Au, Au / Ag, Ag, and Pt 10 nm sized nanoparticles in toluene [15-17] discovered the intensity growth and the appearance of a specific peak in the time dependence of the scattering intensity, just as the authors of work [18] have found for the scattering in lutidine-water solution. On this basis, they logically concluded that the observed phenomenon was the SS on variations of particle concentration, and this effect was called stimulated Mie scattering (Mie SS). But neither in liquid solutions nor in suspensions of particles in liquid SS frequency shift has not been measured or even detected, because the half-width of the corresponding spontaneous backscattering lines (and the value of the spectral SS line shift) is 0.3–3 MHz for liquid solutions, and the shift is

[3] To whom any correspondence should be addressed.

even smaller, ~ 30–12000 Hz, for particles in a fluid. Authors of [14–17] used an ordinary for the SS spectra studies method of the spectral measurements by means of a Fabry-Perot interferometer, whose resolution is not enough for the frequency shifts measurements.

For the above mentioned SCLS shifts frequency measurement, the method of correlation spectroscopy with a continuous laser as a source of excitation radiation is appropriate. We first attempted such measurements in suspensions of diamond nanoparticles and latex in water [19, 20, 21].

The purpose of this work was to determine the value and direction of the SCLS (stimulated Mie scattering) spectral shift in the presence of convection in suspensions of latex particles of different sizes.

## 2. Mie SS spectral shift measurement by correlation spectroscopy

In the process of stimulated concentration light scattering, the interference of exciting and scattered waves creates the interference grating of intensity. The particles of suspension are drawn to maxima (or minima) of the grating [22, 23]. This results in formation of the grating of particle concentration and refractive index according to:

$$N(z,t) = N_0 + \delta N(t)\exp(-i\Omega_{st} + i|k_L - k_S|z). \tag{1}$$

Here $N(z,t)$, $N_0$ are local and median concentrations of particles, respectively, $\delta N(t)$ – the amplitude of the lattice concentration, $z$ – the coordinate of the direction of excitation beam, $\boldsymbol{k}_L$ and $\boldsymbol{k}_S$ – wave vector of the exciting and scattered light.

This grating of concentrations (1) moves along the optical axis with a speed $V_{gr} = \Omega_{st}/q$ with respect to the fluid.

It is easy to obtain an expression for the intensity correlation function of the scattered light when a beat takes place between the highlight of the exciting light ($I_h$) and the light of the stimulated scattering ($I_{st}$), in the presence of the spontaneous scattering ($I_{sp}$) when the suspension flows in the direction of the light beam with velocity $V_c$, as it is done in [23] for the particles in the field. The frequency shift of the SS line $\Omega_{st}$ is superimposed by the Doppler shift $\Omega_{Vc}$ resulting from the motion of the particles along the beam with the convective suspension flow, and the intensity correlation function of the light backscattering can be written as:

$$\begin{aligned} G^{(I)}(\tau) = \langle I(t)I(t+\tau)\rangle = I_{sp}^2\exp(-2\Gamma_{sp}\tau) + I_{st}^2\exp(-2\Gamma_{st}\tau) + 2I_{sp}I_h\exp(-\Gamma_{sp}\tau)\cos(\Omega_{Vc}\tau) + \\ +2I_{st}I_h\exp(-\Gamma_{st}\tau)\cos(\Omega_{exp}\tau) + \langle I_{sp} + I_{st} + I_h\rangle^2 \end{aligned} \tag{2}$$

where $G^{(I)}(\tau)$ is the correlation function of the scattered light intensity, $\Gamma_{st}$ – width of the SS line, $\Gamma_{sp} = Dq^2$ – width of the SpS line, $D$ – diffusion coefficient, $\boldsymbol{q} = \boldsymbol{k}_L - \boldsymbol{k}_S$ – scattering vector, $\Omega_{exp} = \Omega_{st} + \Omega_{Vc}$, $\Omega_{exp}$ is total measured SS line shift, $\Omega_{st}$ – shift caused by SS, $\Omega_{Vc} = 4\pi nV_c/\lambda$ – shift due to convective flow, the $n$ – refractive index, $\lambda$ – light wavelength.

Fortunately, the effect of the third item in (2) is small since $\Gamma_{sp} > \Gamma_{st}$, and the total frequency SS shift can be determined by the period $T$ of the cosine $G^{(I)}(\tau)$ component: $\Omega_{exp} = 2\pi/T$.

## 3. Experiment

For the measurements we used spherical monodisperse latex particles suspended in dust free bidistilled water. The particles' radii were 375, 480 and 750 nm. Volume polystyrene concentrations were $C_v = 7\times10^{-4}$% for 480- and 375-nm latex and $C_v = 1.4\times10^{-3}$% for 750-nm latex.

SCLS (Mie SS) was investigated in the fiber probe scheme [20] shown in figure 1. Light from the continuous solid-state laser with $\lambda = 532$ nm with output power up to 50 mW was inputted through the polarizer $P_1$ to the illuminating fiber $IF$ of the optical fiber probe $PR$ by the lens $L_1$. The output end of the $IF$ was embedded in the thin probe cylinder block, together with the input end of the collecting fiber $CF$. These fibers were parallel, and their axes were at a distance of 0.3 mm. The light from the scattering volume placed near the end of the probe was collected by *CF* and arrived to the cathode of a photomultiplier through the polarizer $P_2$ and the spatial coherence providing system *SCS*.

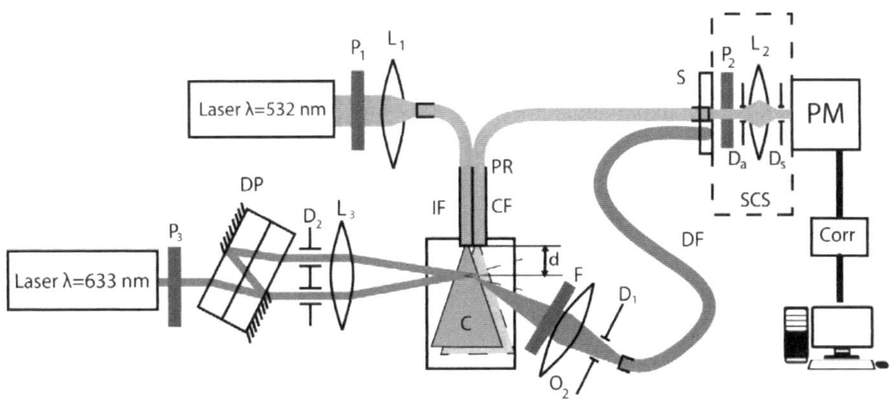


**Figure 1.** Fiber probe optical scheme. Laser – continuous lasers with $\lambda$ = 532 and 633 nm; $P_i$ – polarizers; $L_i$ – lenses, *PR* – fiber optic probe, *IF* – illuminating fiber, *C* – cuvette, *CF* – collecting fiber, *S* – fibers switch, *SCS* – system providing spatial coherence, $D_a$ – aperture diaphragm, $D_s$ – cathode diaphragm (in front of *PM*), $D_2$ – double diaphragm, *PM* – photomultiplier, *Corr* – Correlator, *DP* – beam doubler 50/50%, *F* – red filter, *DF* – optical fiber transmitting the Doppler signal, *d* – distance between the end of the optical fiber probe and the intersection of two red beams, $d \sim 2.5$ mm.

*He-Ne* laser, vertical beam doubler *DP* and a focusing lens $L_3$, being typically two-beam Doppler scheme, were introduced in the optical system to measure the velocity of convection. Light scattered at a small angle relative to the axis of the *He-Ne* laser, through a red filter *F* was focused into the fiber *DF* and through it arrived to *SCS* system. It was easy to switch from the SS measurements to the vertical velocity measurements by switching *CF* and *DF* fibers in the *SCS* system input. Convection velocity was measured at a distance of ~ 2.5 mm from the end of the fiber. This value was estimated to be the most effective for SS observation from the distance dependency of the overlapping areas of illumination cone and the cone of probe viewing field, $I \sim 1/r^2$.

## 4. Results

At a wavelength $\lambda$ = 532 nm the latex particles turned out to have the absorption coefficient magnitude enough to create convective flow in the suspension. Simultaneous measurements of SS concentration grating velocity and convection were conducted for all of the above mentioned suspensions of latex. In the steady state of the light guide scheme, SS correlation functions were measured through *CF*, and immediately after it, with a red filter, the correlation function of the scattered light of two *He-Ne* laser beams was as well measured through the *DF* optical waveguide (Fig. 1). The period *T* of the cosine of the latter correlation function gave us the values of convective flow velocity $V_c$:

$$V_c = \frac{\Lambda}{T} = \frac{\lambda}{2T\sin(\alpha/2)} \tag{3}$$

where $\Lambda$ is the wavelength of grating interference pattern of crossing red laser beams, $\lambda$ – wavelength of the *He-Ne* laser, $\alpha$ – the angle of convergence of the beams in the air. The period of the cosine of the correlation functions of green scattered light received through the optical fiber *CF* gave us the total

SS frequency shift $\Omega_{exp} = \Omega_{st} + \Omega_{Vc}$, including convection shift, and accordingly, the total speed of concentration grating $V_{exp} = |V_c + V_{gr}| = \Omega_{exp}\cdot\lambda/4\pi n$.

These measurements have been carried out in cases when the optical fiber was directed both downwards and upwards.

The measurement results for convective flow velocity $V_c$ and $V_{exp}$ determination depending on the laser power for two light directions are shown in figure 2 a) down, b) up.

## 5. Discussion

Figure 2. shows that if the exciting beam is directed down then for the particles with size 480 nm $V_c > V_{exp}$, and for those with sizes 375 nm and 750 nm $V_c < V_{exp}$. When the light beam is directed up, the situation is opposite. Since the convection velocity is always directed upwards, this means that for the latex particle with a radius of 480 nm the concentration grating velocity (1) with respect to the liquid $V_{gr} = \Omega_{st}\lambda/4\pi n$ is directed along the excitation beam from the radiation source, while for 375- and 750-nm particles, it is directed to the source. That is, SS frequency shift $\Omega_{st} = 4\pi n V_{gr}/\lambda$ is anti-Stokes for 375- and 750-nm particles and it is Stokes for 480-nm particles. This is presumably due to the fact that the direction of the gradient force (to maxima or to minima of the interference grating intensity) depends on the particle size [23]. In [22], the expression is obtained for the Mie SS gain $g$, which includes alternating Bessel function $J_{3/2}(qR)$ and the SS shift $\Omega_{st}$. In order to always satisfy the condition $g > 0$, with alternating $J_{3/2}(qR)$, SS shift must also be alternating in dependence on $R$.

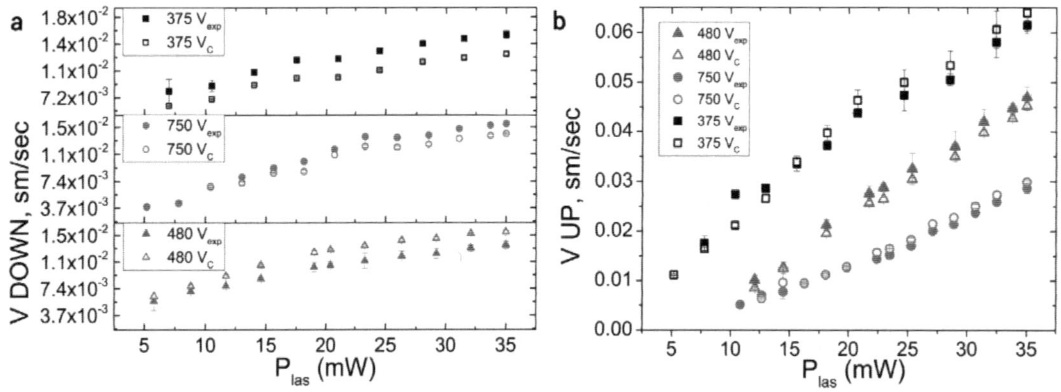


**Figure 2.** Convective flow velocity $V_c$ and total velocity of the concentration wave $V_{exp}$ versus the excitation beam power. The beam is directed (a) down, (b) up.

Figure 3. shows the modulus of the difference between Doppler shift due to the convective flow velocity $\Omega_{Vc}$, and the total measured SS shift $\Omega_{Vexp}$, $\Omega_{st} = |\Omega_{Vexp} - \Omega_{Vc}|$, depending on the power of the excitation beam. It can be seen that the experimental values of SS shift $\Omega_{st}$ are $\Omega_{st}^{E(480)} = 642 \pm 55$ s$^{-1}$, $\Omega_{st}^{E(375)} = 800 \pm 46$ s$^{-1}$, $\Omega_{st}^{E(750)} = 419 \pm 18$ s$^{-1}$. These are close to the theoretical values $\Omega_{st}^{T} = Dq^2$: $\Omega_{st}^{T(480)} = 478 \pm 44$ s$^{-1}$, $\Omega_{st}^{T(375)} = 616 \pm 56$ s$^{-1}$, $\Omega_{st}^{T(750)} = 308 \pm 28$ s$^{-1}$, although experimental values exceed theoretical ones by about 16%. Similar result was obtained for the suspension of the diamond nanoparticles' aggregates in the water in a traditional backscattering scheme [20].

## 6. Conclusion

The value and direction of the spectral Mie SS (SCLC) shift are measured in the light guide scheme in the presence of convective flow. SCLC shift in suspension of latex with $R$ = 480 nm was experimentally found to be Stokes with $\Omega_{st}^{E(480)} = 642 \pm 55$ s$^{-1}$, and SCLC shifts in the suspensions of latex with $R$ = 375 nm and 750 nm were experimentally found to be anti-Stokes with $\Omega_{st}^{E(375)} = 800 \pm 46$ s$^{-1}$ and $\Omega_{st}^{E(750)} = 419 \pm 18$ s$^{-1}$, correspondingly. The difference between the experimental values of the spectral shifts of SCLS and calculated theoretical ones is less than 16%. The frequency shift is sign-alternative that is, apparently, related to the direction of the gradient force (to maxima or to minima of the interference grating intensity) depending on the particle size [23].

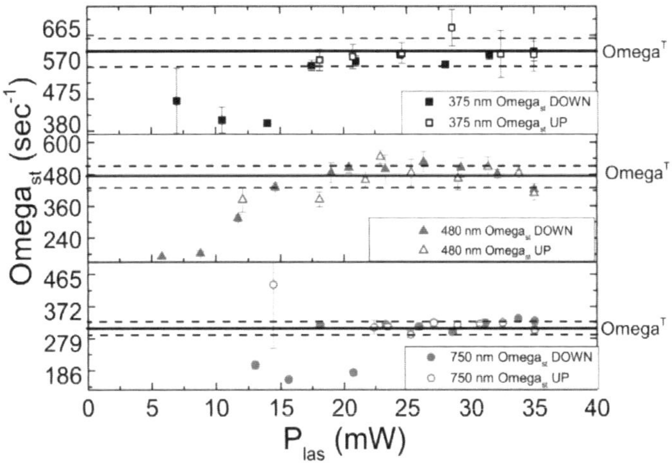


**Figure 3.** The dependence of the modulus of the difference $\Omega_{st} = |\Omega_m - \Omega_{Vc}|$ on the excitation beam power. Theoretical limits are denoted by dashed lines $\Omega_{st}^T = Dq^2$.

**Acknowledgements**
This work was supported by Competitiveness Growth Program of the Federal Autonomous Educational Institution of Higher Professional Education National Research Nuclear University MEPhI (Moscow Engineering Physics Institute).

**References**
[1] Zel'dovich B and Ya Sobel'man I I 1970 *Soviet Physics Uspekhi* **13** 307–317
[2] Sushchinskii M M 1985 *Stimulated Raman scattering* (Moscow: Nauka) 49
[3] Fabelinskii I L 1965 *Molecular Scattering of Light* (Moscow: Nauka)
[4] Gorelik V S, Kudryavtseva V S and Chernega N V 2006 *Bulletin of the Lebedev Physics Institute* **33** 39
[5] Tareeva M V, Gorelik V S, Kudryavtseva A D and Chernega N V 2010 *Bulletin of the Lebedev Physics Institute* **37** 331
[6] Mash D I, Morozov V V, Starunov V S and Fabelinskii I L 1965 *JETP Letters* **2** 25
[7] Sogomonian S, Barille R and Rivoire G 1998 *Optics Communications* **157** 182
[8] Zaitsev G I, Kyzylasov Yu I, Starunov V S and Fabelinskii I L 1967 *JETP Letters* **6** 505
[9] Starunov V S, Fabelinskii I L 1970 *Soviet Physics Uspekhi* **13** 428
[10] Jirauschek C, Jeffrey E M and Faris G W 2001 *Physical Review Letters* **87** 233902
[11] Kyzylasov Yu P, Starunov V S and Fabelinskii I L 1970 *JETP Letters* **11** 66
[12] Aref'ev I M and Morozov V 1969 JETP Letters **9** 260
[13] Lowdermilk W H and Bloembergen N 1972 *Physical Review A* **5** 1423
[14] Bloembergen N, Lowdermilk W H, Matsuoka M and Wang C S 1971 *Physical Review A* **3** 404
[15] Guang S He, Yong K-T, Jing Zhu and Prasad P N 2012 *Physical Review A* **85** 043839
[16] Guang S He, Wing-Cheung Law, Liwei Liu, Xihe Zhang and Prasad Paras N 2012 *Applied Physics Letters* **101** 011110
[17] Guang S He, Wing-Cheung Law, Baev A, Sha Liu, Mark T Swihart and Paras N Prasad 2013 *The Journal of Chemical Physics* **138** 024202
[18] Davydov M A and Shipilov K F 1990 *JETP Letters* **52** 153
[19] Burkhanov I S and Chaikov L L 2012 *Bulletin of the Lebedev Physics Institute* **39** 77

Journal of Physics: Conference Series **747** (2016) 012055 doi:10.1088/1742-6596/747/1/012055

[20] Burkhanov I S and Chaikov L L 2013 *SPIE Proceedings* **9066** 906610
[21] Burkhanov I S, Krivokhizha S V and Chaikov L L 2016 *Quantum Electronics* **7** in press
[22] Afanas'ev A A, Rubinov A N, Mikhnevich S Yu et al. 2007 *Optics and Spectroscopy* **102** 106
[23] Zemanek P, Jonas A 2002 *Optical Society of America* **19** 1025

Journal of Physics: Conference Series **747** (2016) 012056 doi:10.1088/1742-6596/747/1/012056

# Development of the recovery technology for nickel superalloy blades of the aircraft engine by laser cladding

**D P Bykovskiy[1], V N Petrovskiy, V I Polskiy, S N Chirikov and P S Dzhumaev**
National Research Nuclear University MEPhI, Kashirskoe sh. 31, Moscow, Russia, 115409

E-mail: d.bykofsky@gmail.com

**Abstract**. Development of cladding modes was performed with a superalloy nickel based powder on a flat substrate from material identical to compressor and turbine blades. Cross sections were made, and a visual inspection of the shape and the quality of the clad track as well as themetallographic analysis were performed. Microhardness of the deposition zone, chemical composition of the base, cladded metals, and the heat affected zone were determined.

## 1. Introduction

Turbine abrasion occurs during exploitation due to vibrations of the top edge of turbine blades, which leads to deterioration of its working efficiency [1]. Foreign particles inflow in the turbine is also possible, causing damage to the input edges of the blades, appearance of dents and change of their geometries. A turbine may comprise several hundreds of blades, which are subject to strict requirements on the shape stability. Production of a single blade is a long-term and resource-intensive process, which is the cause of its high cost and, therefore, leads to increase of the turbine cost. Replacement of damaged blades with new ones is not economically profitable, so different methods are used to restore turbine blades that reduce maintenance costs, improve reliability and increase service life. To achieve these objectives, it is necessary to repair the turbine periodically.

Studies have shown that the solution of this problem is possible by using one of the many additive technologies – laser gas-powder cladding [2, 3]. Blades form recovery process consists of sequential deposition of the tracks at each other on the edge of the damaged blades. Metal powder is supplied to the repaired surface coaxially with the continuous laser radiation. After melting and cooling of the metal powder, a track is formed on the blade's surface. In this way, it is possible to deposit one track on top of the other, creating a multi-layer structure with required thickness [4].

The first stage in the development of blade recovery technology is to determine the process windowwhere good formation of single tracks is performed [5], which is the topic of this study.

## 2. Experimental equipment

Experimental equipment for laser cladding of metal powder is a system consisting of a module with Huffman HC-205 control unit, a powder feeder Sulzer Metco Twin 10-C (figure 1), and the fiber laser IPG-Photonics LS-3.5 and a Riedel chiller. Huffman HC-205 module with the control unit was originally designed for the repair of aircraft turbine blades yet it is also used for volume shaping and restoration of surfaces of various parts after small software revisions.

[1] To whom any correspondence should be addressed.

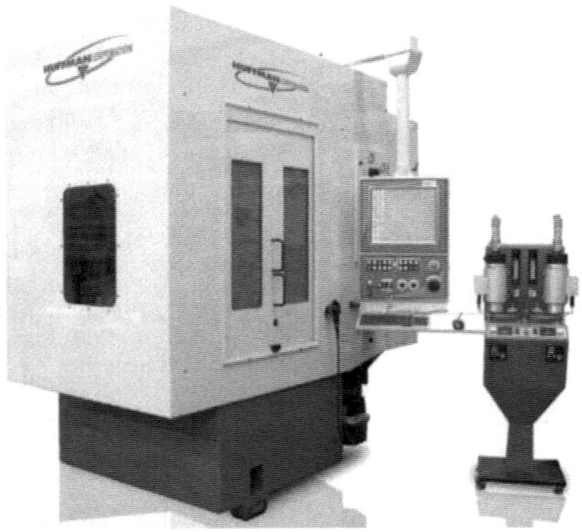

**Figure 1.** Cladding module with powder feeder.

### 3. Experimental results of single track cladding on flat samples

We used the powder of heat-resistant nickel-iron alloy EP718 as a filler material in this study. Cladding was carried out on a plate of the same material. We used three technological windows differing by powder feed rate: the first corresponds to the feed rate of 0.9 g/min, the second – 1.35 g/min, and the third – 1.8 g/min. In each process window the tracks were deposited with different combinations of scanning speed from 6.3 mm/s to 10.5 mm/s and a laser power from 150 to 250 W. Transverse sections of tracks were made after cladding. Studies have shown that an increase in powder feed rate for raising productivity of the process does not lead to good track shaping. Tracks obtained from powder at a small feed rate 0.9 g/min (figure 2) had better adhesion than the tracks obtained at a feed rate of 1.35 and 1.8 g / min (figures 3, 4), which also lacked fusion with the substrate.

**Figure 2.** General view of the cladding microstructure no. 1.

Further detailed analysis of the tracks cladded in the first technological window with a powder flow rate of 0.9 g/min was carried out. Microhardness measurements for these tracks were made in the transverse direction (figure 5) by the Vickers method using HVS-1000 device with automatic indenter loading at $F = 1$ N. Time of loading was 20 s. Coordinates 0–600 μm correspond to the cladded track, 600–1200 μm – to the transition region, and larger than 1200 μm –to the area of the substrate. Studies showed that there is a decrease in microhardness of cladding. However, taking into account the measurement error, it can be concluded that microhardness was not affected by changing the technological modes.

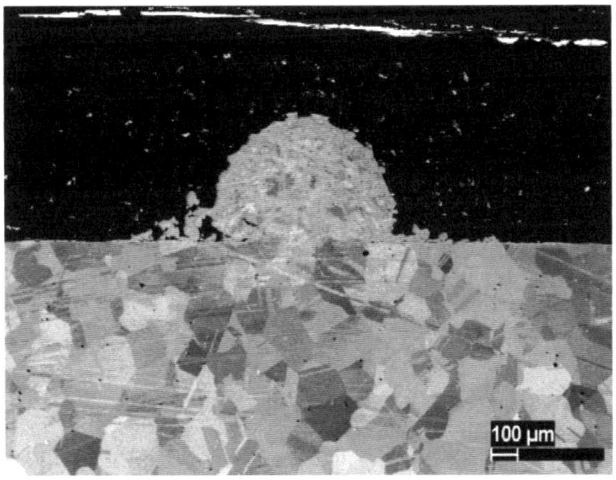


**Figure 3.** General view of the cladding microstructure no. 10.

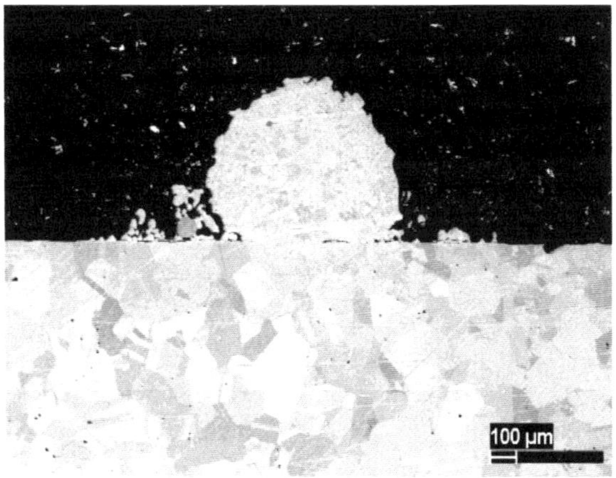


**Figure 4.** General view of the cladding microstructure no. 19.

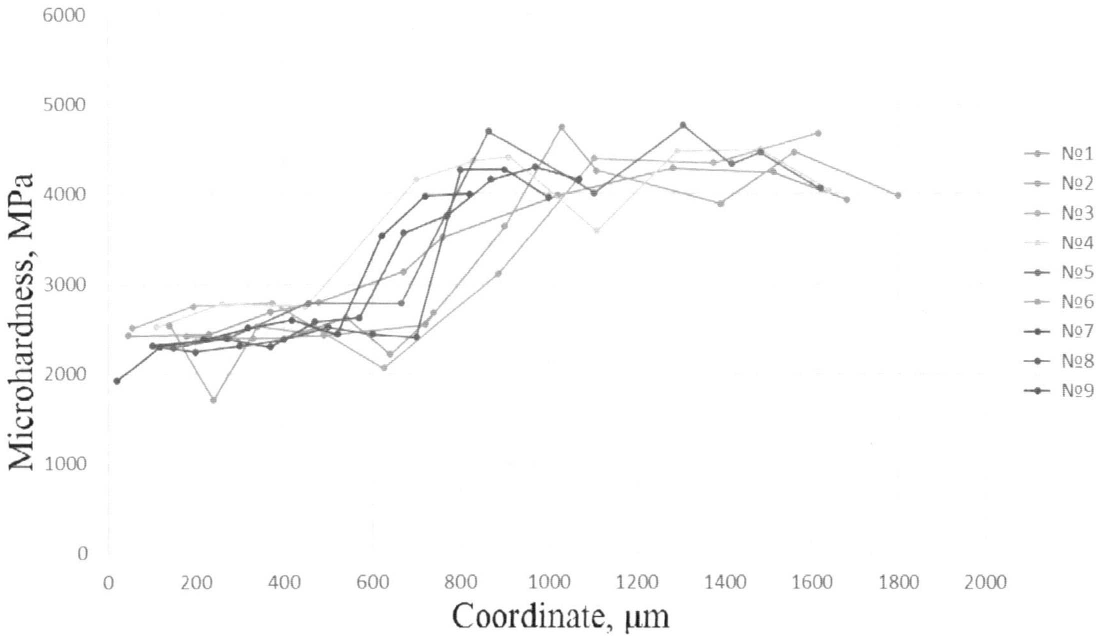


**Figure 5.** Change of the cladded tracks' microhardness over the cross section.

The element composition showed that the cladded material and the substrate are nearly identical, and changes are insignificant in depth. The substrate and the cladded material are 45Ni-28Fe-15Cr-4Mo-3W-2Ti-Al. Microstructure of cladding is a mesh with polyhedral (equiaxed) grain size of 4 ... 5 µm. Microstructure of the substrate is a typical austenite with the traces of plastic deformation at the boundary of substrate and surfacing. Clearly defined heat-affected zones are absent in the structure. Penetration of surfacing material into the base material at the grain boundaries to a depth of 0.02–0.09 mm was observed for all samples and for all technological modes. Taking into consideration the requirements of good fusing track with the substrate, as the absence of defects like cracks, pores, and cavities, an optimum mode can be selected from the first technological window for further work on creating multilayer cladding tracks. Optimal track deposited in this mode (powder flow rate 0.9 g/min, laser radiation power 250 W, surface scanning speed 8.4 mm/s, the gap between the nozzle and the substrate 4 mm) is shown in figure 6.

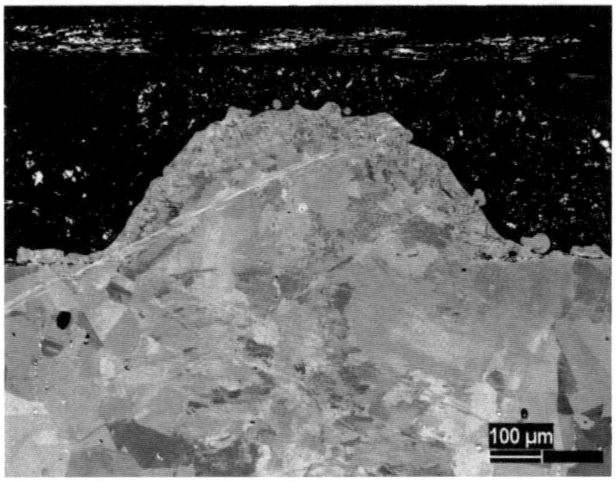


**Figure 6.** General view of the cladding microstructure no. 6.

## 4. Conclusion

Technological modes were obtained for optimal formation of tracks that meet all the requirements of the welding structure, chemical composition, and microhardness. It was revealed that large powder feed rate increases productivity, but it leads to a lack of fusion at the edges of the tracks. The chemical composition of the deposited coating is similar to that of the substrate material. The microhardness decreases near the top of the cladded tracks. Presumably, this is due to the fine grain structure caused be rapid cooling.

**Acknowledgments**

This work was supported by Competitiveness Growth Program of the Federal Autonomous Educational Institution of Higher Professional Education National Research Nuclear University MEPhI (Moscow Engineering Physics Institute).

**References**

[1] Vilar R, Santos E, Ferreira P, Franco N, and da Silva R 2009 *Acta Materialia* **57** 5292–5302
[2] Bendeich P, Alam N, Brandt M, Carr D, Short K, Blevins R, Curfs C, Kirstein O, Atkinson G, Holden T, and Rogge R 2006 *Materials Science and Engineering: A* **437** 70–74
[3] Shepeleva L, Medres B, Kaplan W, Bamberger M and Weisheit A 2000 *Surface and Coatings Technology* **125** 45–48
[4] Bi G and Gasser A 2011 *Physics Procedia* **12 Part A** 402–409
[5] Rottwinkel B, Nlke C, Kaierle S, and Wesling V 2014 *Procedia CIRP* **22** 263–267

Journal of Physics: Conference Series **747** (2016) 012057 doi:10.1088/1742-6596/747/1/012057

# Mechanical properties of the samples produced by volume powder cladding of stainless steel using a continuous fiber laser

**D P Bykovskiy[1], V N Petrovskiy, V D Mironov, A V Osintsev and K Yu Ochkov**
National Research Nuclear University MEPhI, Kashirskoe sh. 31, Moscow, 115409, Russia

E-mail: d.bykofsky@gmail.com

**Abstract.** Samples for tensile tests were manufactured by using one of the additive technologies – direct laser material deposition. Investigations were carried out at the facility Huffman HC-205 equipped with a fiber laser with a power up to 3.5 kW. Various strategies of layering metallic powder of stainless steel 316L were considered to optimize the modes of constructing the samples. We measured the stress-strain state of the produced samples by the method of digital image correlation. It is found that the nominal tensile strength of the samples produced by the direct growing using laser powder of 316L steel is of high level - 767 MPa.

## 1. Introduction

The technology of direct laser deposition of metal powder allows to produce parts layer-by-layer on a pre-programmed path. It is possible to make prototypes and operating samples for further industrial applications [1–3]. However, widespread implementation of this technology is difficult, because dependences of the obtained product properties on the technological mode of their preparation as well as the creating strategy are yet poorly understood. Primarily, this process is characterized by a large number of different parameters. However, the following basic ones should be emphasized since they have the greatest impact on the final quality of the product microstructure. They are power of the laser source, powder feed rate, the scanning speed of the laser beam and the strategy of processing the alloyed product layer by laser beam [4, 5].

## 2. Materials and equipment

In our experiments we used Höganäs 316L powder with a particle size of 20–53 μm and its chemical composition shown in table 1.

**Table 1.** The chemical composition of steel grade 316L powder.

| The content of alloying elements, masses, % | | | | | | |
|---|---|---|---|---|---|---|
| Cr | N | Mo | Mn | Si | C | Fe |
| 17.0 | 12.0 | 2.5 | 1.5 | 0.8 | <0.03 | Balance |

Experiments were carried out in the industrial apparatus Huffman HC-205, equipped with a high-power fiber laser LS-3.5, the coordinate system with five degrees of freedom, and a powder feeding

[1] To whom any correspondence should be addressed.

system. The mixture powder was transported by inert –argon gas and focused into the treatment zone via a coaxial nozzle. Protection of the cladding area was simultaneously produced by the same gas.

## 3. Results of experiments

The samples were created in the form of double-sided blades for tensile testing. The following process parameters were varied during operation in this process: laser power, scanning speed, powder flow rate, the gap between the nozzle and the deposited layer, as well as the direction of the powder coating and subsequent melting. Two directions were chosen for double-sided cladding of blades: parallel to the direction of tension (figure 1) and perpendicular one (figure 2). If the cladding direction is parallel, the contour of the blade 1 would be deposited first, and then it would be the inner part, consisting of several regions, numbered 2–5. Perpendicular deposition consists in drawing the contour, and then filling it with consistent cladding tracks. Also, the start peaks of cladding were alternated to improve thickness uniformity of the sample. After the cladding of two layers with different deposition directions of rolls starting at point A, the nozzle moves to point B, from where the cladding of two layers is made as well. Next, the operation is repeated starting from points C and D. Table 2 summarizes the process modes to obtain the samples. It should be noted that deposited layer for some of the samples was remelted to reduce the roughness and improve layer stability.

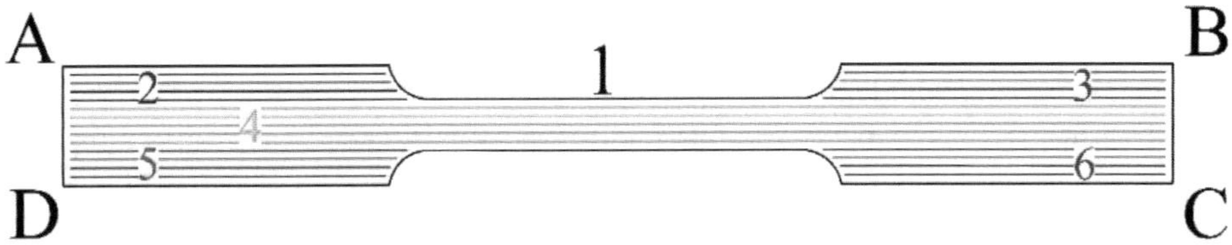


**Figure 1.** Scheme of cladding of rollers in a longitudinal direction.

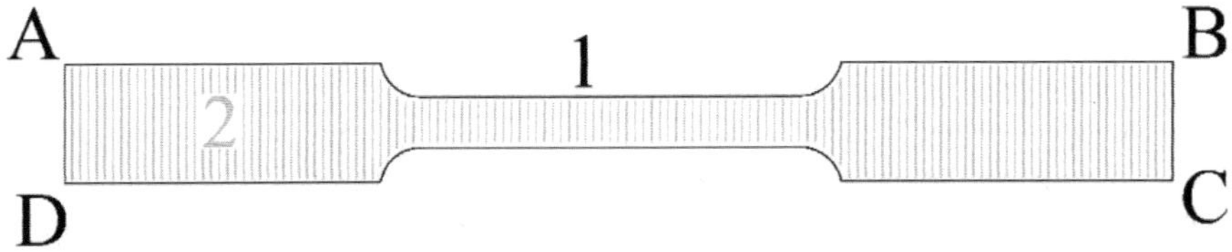


**Figure 2.** Scheme of cladding of rollers in a transverse direction.

**Table 2**. Modes of volume formation samples laser cladding method for tensile testing.

| | Cladding | | | | | Remelting | |
|---|---|---|---|---|---|---|---|
| № | Power, W | Speed, cm/min | Powder federate, g/min | Gap, mm | Gap change after each layer, mm | Power, W | Speed, cm/min |
| 1 | 200 | 100 | 0.8 | 3 | 0.13 | 100 | 127 |
| 2 | 200 | 100 | 1.1 | 3 | 0.2 | 100 | 127 |
| 3 | 300 | 100 | 1.1 | 4 | 0.2 | - | - |
| 4 | 250 | 100 | 0.8 | 3 | 0.2 | - | - |
| 5 | 350 | 100 | 0.8 | 3 | 0.2 | - | - |
| 6 | 250 | 100 | 0.8 | 3 | 0.2 | - | - |

Samples no. 1–6 using EDM wire cutting were divided into separate 1 mm thick bilateral blades. Tests to determine the mechanical properties of the samples were carried out on a tensile testing machine Instron. The results are shown in figure 3.

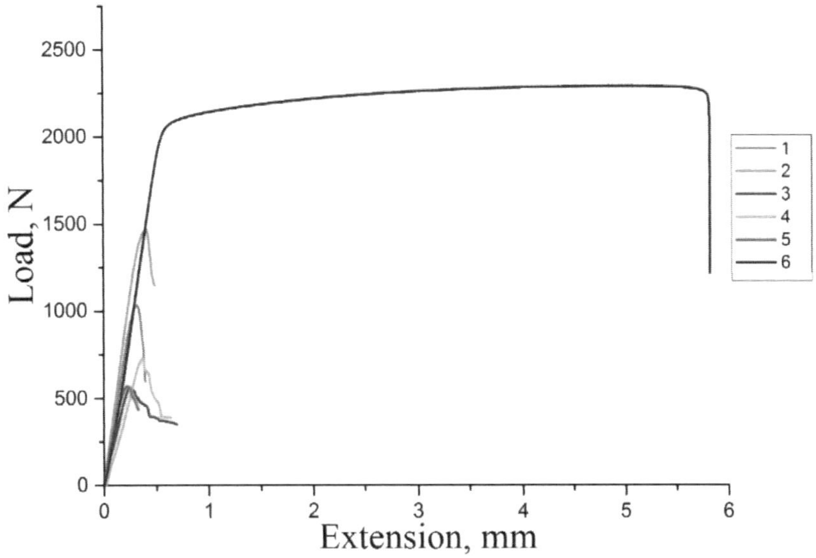


**Figure 3.** Load-displacement diagram for samples no. 1–6.

These results demonstrate that the sample no. 6, which had the largest height stability, and most smooth side surface, weathered larger load compared to samples no. 1–5. Tensile strength of the obtained sample is equal to 767 MPa. It should be noted that for the used steel grade 316L tensile strength after quenching and tempering is equal to 780 MPa.

Three samples in a parallelepiped form were created in the regime no. 6 with different deposition directions of layers: sample no. 7 was created with the direction of the track along the extension line, sample no. 8 – with the track across the extension line, and sample no. 9 was formed in the crossed directions.

Samples no. 7–9 were also cut into 1 mm thick plates, and then were laser cut in a pattern similar in size tono. 6 blades. This was done to study the mechanical properties depending on the direction of sample cladding and the defects on the boundary. Samples no. 7–9 were also tensile tested, and the results are shown in figure 4.

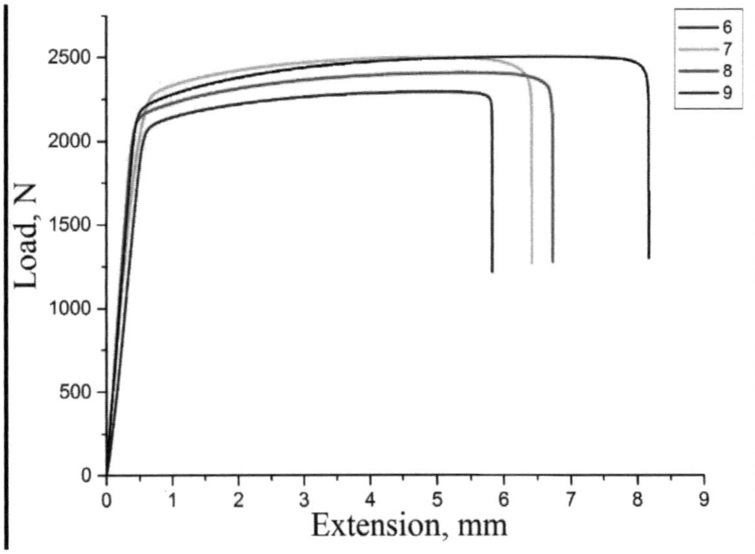


**Figure 4.** Load-displacement diagram for samples no. 6–9.

Results show that the samples no. 7–9 obtained by direct laser deposition of a blank followed by laser cutting withstand higher loads compared with the sample no. 6 obtained by direct laser deposition without post processing. Sample no. 9 with the tracks cladded in crossed directions withstands higher load than the samples cladded both in parallel (no. 7) and perpendicular (no. 8) directions. More uniform distribution of the internal stress after the deposition can be a possible reason for this. It is necessary to conduct more detailed studies of the microstructure of the samples.

## 4. Conclusion

The paper demonstrates the strategy of the optimal mode selection for obtaining products by direct laser deposition. The optimal mode is characterized by a small height of single layer providing its uniformity, and not using the layer re-melting to reduce the heating of the sample. The laser power was 250 W, the gap between the nozzle and the deposited layer was 3 mm, the scanning speed was 100 cm/min, and the powder flow was 0.8 g/min. It is also shown, that defects at the edges, for example, splashes and bumps on the sidewalls, strongly affect the mechanical properties of the sample, reducing the maximum load a sample can withstand.

## Acknowledgments

This work was supported by the Ministry of Education and Science of Russian Federation (unique identifier PNIER RFMEFI58214X0004).

This work was also supported by Competitiveness Growth Program of the Federal Autonomous Educational Institution of Higher Professional Education National Research Nuclear University MEPhI (Moscow Engineering Physics Institute).

## References

[1] Shishkovsky I, Missemer F, and Smurov I 2012 *Phys. Procedia* **39** 382 – 391

[2] Rombouts M, Maes G, Hendrix W, Delarbre E, and Motmans F 2013 *Phys. Procedia* **41** 810 – 814

[3] Tabernero I, Lamikiz A, Martnez S, Ukar E, and Figueras J 2011 *International J. of Machine Tools and Manufacture* **51** 465–470

[4] Bartkowiak K 2010 *Phys. Procedia* **5** Part B 623 – 629

[5] Liu F, Lin X, Song M, Yang H, Zhang Y, Wang L, and Huang W 2015 *J. of Alloys and Compounds* **621** 35–41

Journal of Physics: Conference Series **747** (2016) 012058 doi:10.1088/1742-6596/747/1/012058

# MCP PMT with high time response and linear output current for neutron time-of-flight detectors

**A S Dolotov, P I Konovalov[1], R I Nurtdinov**
Dukhov Research Institute of Automatics (VNIIA), Sushchevskaya str. 22, Moscow, 127055, Russia

e-mail: p_konovalov@mail.ru

**Abstract**. A microchannel plate (MCP) photomultiplier tube (PMT) with a subnanosecond time response and a high linear output current has been developed. PMT is designed for detection of weak pulses of radiation in UV-, visible and nearer-IR ranges and can be used in neutron time-of-flight (nTOF) detectors in experiments on laser compression of thermonuclear fuel. The results of measurements of MCP PMT main parameters are presented: photocathode spectral sensitivity, gain, maximum linear output current, and time response.

## 1. Introduction

In performance of experiments with laser compression of thermonuclear fuel on high-power laser systems, like NIF and OMEGA, the neutron time-of-flight technique is one of the important elements of the system for the plasma nuclear content diagnostics[1, 2]. The main objectives of the technique are to measure the ion temperature of plasma, determine the total neutron yield and the time between the beginning of laser irradiation of a target and the attainment of the maximum neutron yield [3]. The equipment to support the technique at low and medium neutron yields includes detectors based on fast organic scintillators and MCP PMT or a vacuum photocell [3, 4]. The time response of these detectors ranges from 0.2 to 2.5 ns.

This paper deals with VNIIA-developed TNFT25-01 PMT with one MCP and TNFT25-02 PMT with two MCPs that are developed for gamma-ray and neutron detectors to be used in the neutron time-of-flight system. The main peculiarities of these PMTs are narrow FWHM of the output pulse in the linear range of an output current (0.34 ns for PMT with two MCPs and 0.28 ns for PMT with one MCP) at illumination of a photocathode by a light pulse of ~ 0.1 ns, as well as a high linear output current in the pulse mode (more than 1.0 A) at illumination of a photocathode by a light pulse with FWHM of 20 ns.

## 2. MCP PMT design

MCP PMT vacuum unit comprises an entry window made of quartz glass, on the inner surface of which a 25-mm-diam photocathode is formed. Next to the cathode window, the input surface of a microchannel plate is located. One or two MCPs of MKPO25-8 type manufactured by LLC VTC Baspik to meet specific requirements can be used depending on the necessary gain. Electrons arrive from MCP output at the anode that is connected to a coaxial output with wave resistance 50 Ohm.

MCP PMT vacuum unit is placed in a housing which also incorporates a resistive divider that supplies required potentials to the electrodes. On the end-face of the housing there are 50-Ohm

[1] To whom any correspondence should be addressed.

connector of N-type or SMA for recording an output signal, SHV-connector for supplying feed voltage, and a terminal for connecting a grounding wire (figure 1).

**Figure 1.** MCP PMT outward appearance.

## 3. MCP PMT main parameters

### *3.1. Photocathode spectral sensitivity*

nTOF detector sensitivity is determined by the properties of the used scintillator, as well as the spectral sensitivity of a photocathode and PMT gain.

Spectral sensitivity of a photocathode of S20UV type was measured in the range of wavelengths from 200 to 800 nm. Figure 2 shows the obtained spectral characteristic in relative units. PMT attains the maximum spectral sensitivity at a wavelength of 370 nm (a typical value ranges from 70 to 90 mA/W). The maximum light yield of BC-422Q fast scintillator used in nTOF detectors corresponds to a wavelength of 370–380 nm, which perfectly matches PMT maximum spectral sensitivity.

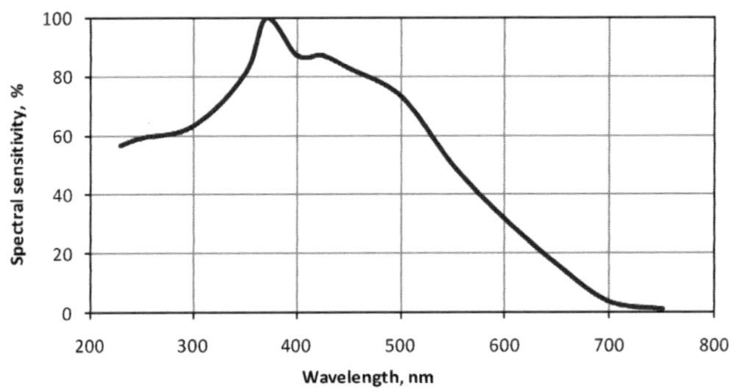


**Figure 2.** Spectral characteristic of MCP PMT with S20UV photocathode.

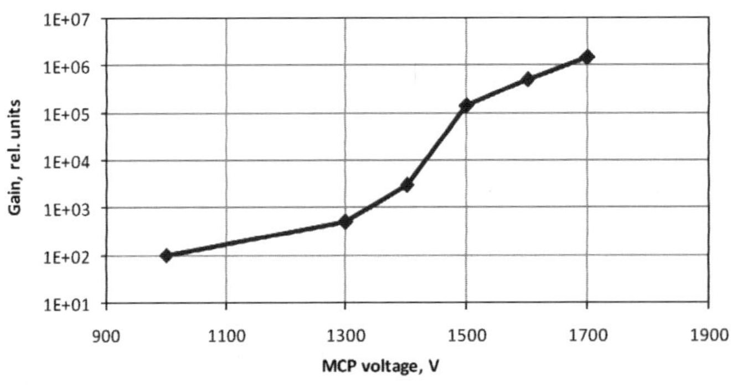


**Figure 3.** Gain factor of PMT with two MCPs as a function of MCP voltage.

Figure 3 shows the dependence of a gain factor of PMT with two MCPs on the voltage applied to MCPs. At MCPs' operating voltages, a gain factor of TNFT25-01 PMT is no less than $5 \cdot 10^3$, and TNFT25-02 PMT – no less than $1 \cdot 10^6$. Detectors based on PMTs with such a gain allow measuring the ion temperature at yields from $1 \cdot 10^9$ to $5 \cdot 10^{13}$ neutron/pulse [1].

### *3.2. Maximum linear output current*

At high neutron yields the performance of the neutron time-of-flight system is determined by the threshold of PMT saturation. At significant luminous fluxes PMT starts operating in a non-linear mode, which can distort the data on the measured ion temperature. The maximum linear output current in the pulse mode is one of the main PMT parameters, being the value of PMT output current, at which a deviation from the direct proportionality of current to a luminous flux impinging on a photocathode does not exceed the specified value. With respect to the effect on the maximum linear output current of MCP PMT caused by the duration of light emission pulse due to low conductivity of MCP channel walls and large capacitance, the measurement were made for a light pulse with a FWHM of no more than 20 ns.

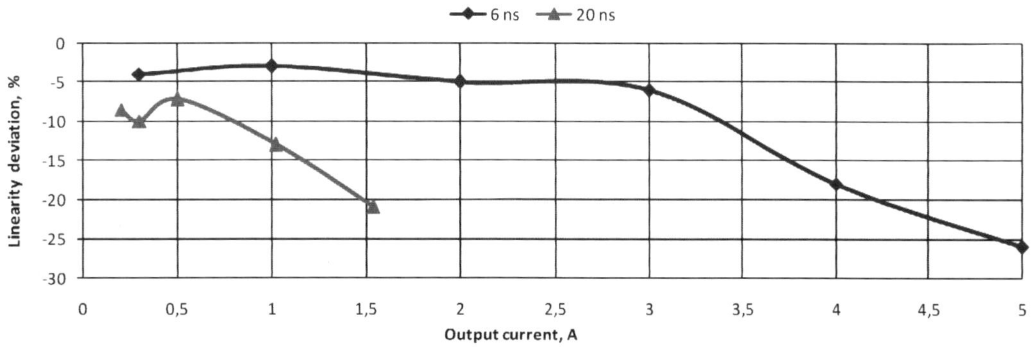


**Figure 4.** Measurement results of MCP PMT maximum linear output current.

The measurement method of the maximum linear output current consists in simultaneous recording of the output pulses from PMT and the photodetector while illuminating their photocathodes by a laser pulse. Monitoring of voltage amplitudes is carried out using an oscilloscope incorporated in the anode circuits of PMT and the photodetector. Linearity deviation was determined mathematically.

Figure 4 shows the measurement results of the maximum linear output current in the pulse mode of PMT with two MCPs at illumination of the cathode by a single laser pulse with a FWHM of 6 and 20 ns. It can be seen that at -15% linearity deviation the amplitude of PMT output current attains 3.7 A in the former case and 1.15 A in the latter case.

### *3.3. Output pulse FWHM and time response*

In the ion temperature measurements of plasma using the time-of-flight technique, the time spread $\Delta t$ of neutrons arriving at nTOF detectors is recorded [5]. The measured parameter $\Delta t$ is summed up with the neutron time of flight through scintillation material and with the values of time characteristics of the detector, cables and digitizers used. To minimize measurement errors, the detector time response should be the minimum, and thus, it should be much less than the time spread of neutrons. For scintillator-based detectors these requirements lead to the use of fast MCP PMT with time resolution of a few hundred picoseconds.

For measurement of an output pulse FWHM ($T_{0.5}$) MCP PMT photocathode was completely illuminated by laser pulses of ~ 0.1 ns FWHM. The pulses were recorded using an oscilloscope. Figure 5 shows oscillograms of PMT with one MCP and PMT with two MCPs.

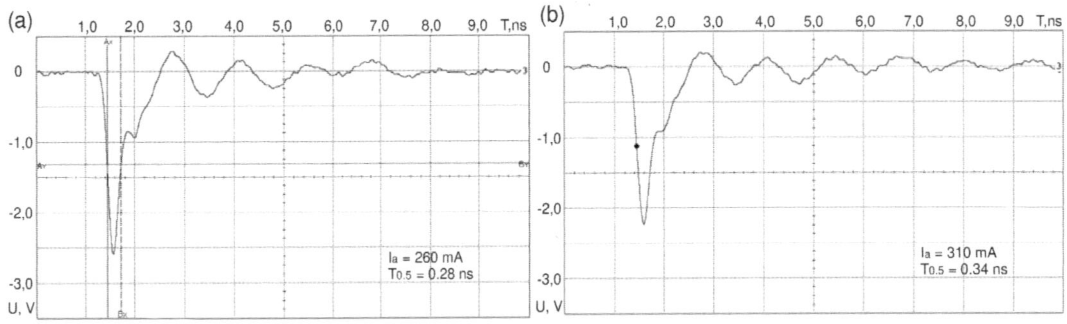


**Figure 5.** Oscillograms of output signals of PMT with one MCP (a) and two MCPs (b) at illumination by a luminous flux with a FWHM of 0.1 ns.

The width of a PMT output pulse depends on the time response of a measuring channel, namely, on the width of a light pulse, time response of a recording channel and an oscilloscope. Therefore, own time response $\tau$ of a PMT is calculated from the formula

$$\tau = \sqrt{T_{0,5}^2 - (T_{ch}^2 + T_{osc}^2 + T_r^2)}, \qquad (1)$$

where $T_{0,5}$ – output pulse FWHM of PMT,
$T_{ch}$ – FWHM of a pulse characteristic of electric signal recording channel,
$T_r$ – radiation pulse FWHM,
$T_{osc}$ – oscilloscope pulse characteristic FWHM.

MCP PMT's own time response calculated from the formula (1):
- for TNFT25-01 – 0.16 ns at output current of 260 mA;
- for TNFT25-02 – 0.25 ns at output current of 310 mA.

## 4. Conclusion

Application of original decisions in designing PMT's individual components allowed VNIIA specialists to create a microchannel plate photomultiplier tube with high time and current characteristics. The obtained results demonstrate that the photomultiplier tube parameters are on a par with those of the best international analogues, and even better in some cases [6, 7]. PMT can be used in nTOF detectors during the experiments on laser compression of thermonuclear fuel.

## References

[1] Ali Z A et al 2008 *Rev. Sci. Instrum.* **79** 10E527.
[2] Glebov V Y et al 2010 *Rev. Sci. Instrum.* **81** 10D325.
[3] Lerche R A et al 2010 *Rev. Sci. Instrum.* **81** 10D319.
[4] Stoeckl C et al 2010 *Rev. Sci. Instrum.* **81** 10D302.
[5] Glebov V Y et al 2004 *Rev. Sci. Instrum.* **75**
[6] Milnes J S and Howorth JR 2005 *Proc. SPIE* **5580** 730–40.
[7] Howorth J R et al 2006 *HSPP Conference.*

Journal of Physics: Conference Series **747** (2016) 012059 doi:10.1088/1742-6596/747/1/012059

# Metrological characterization of nm-range dynamic etalons using a heterodyne interferometer

**T V Kazieva[1,3], A P Kuznetsov[1], M V Ponarina[1], K L Gubskiy[1], and V N Reshetov[2]**

[1]National Research Nuclear University MEPhI, Kashirskoye sh. 31, Moscow, 115409, Russia

[2]Technological Institute for Superhard and Novel Carbon Materials, Tsentralnaya str. 7a, Troitsk, Moscow, 142190, Russia

E-mail: glizerogen@gmail.com

**Abstract.** Test structures used for calibration of scanning probe microscopes have certain limitations. They are short-lived, their work surface becomes coated with microparticles over time, and it gradually wears out by contact with the measuring probe, resulting in the etalon geometry deformation. Dynamic etalons allow calibrating SPMs in ranges from pm to nm. In this article we present the results of dynamic etalon metrological characteristics research using an SPM equipped with tree-coordinate heterodyne laser interferometer. Obtained data indicates stability of piezoelectric modulus and absence of piezoelectric hysteresis phenomena in the etalon samples used.

## 1. Introduction

Reaching the subnanometer resolution and accuracy of the dimension measurements during topography characterization of micro- and nanostructures is provided by using scanning probe (tunneling, atomic force) and scanning electron microscopies [1]. The main problem regarding linear dimensions measurement in the nanometer range is associated with intricate and often ambiguous relationships between the measured object and its image in the scanning microscope [2]. The recorded image does not definitely agree with the real profile of the measured surface topography element of the object. At present, various topography etalons of the geometric dimensions (height, period) remain the only possible means of calibration of scanning probe microscopes (SPMs). They represent an array of periodic structures of specific geometrical shapes mounted on the surface of a silicon single crystal square plate, its surface oriented parallel to the crystallographic plane (100) [3]. The transfer of the length unit from etalon to SPM is carried out by comparing the results of etalon surface profile measurements using the device being calibrated, with the etalon technical data as measured by the metrological instrument (device equipped with a laser interferometer) [4]. However, such etalons are short-lived. Their work surface becomes coated with microparticles over time, and it gradually wears out by contact with the measuring probe, causing changes in the etalon's geometric dimensions. Besides, the measurement results of surface profile always depend on the shape of the probe, since the image obtained using a scanning probe microscope (SPM) represents convolution of the profile surface of the probe and the surface of the topography etalon.

[3] To whom any correspondence should be addressed.

Journal of Physics: Conference Series **747** (2016) 012059 doi:10.1088/1742-6596/747/1/012059

Instead of the static ones, the so-called dynamic action etalons may be used that are free from these drawbacks. Dynamic etalons of length unit allow to simplify the calibration procedure of SPMs and greatly enhance the accuracy of the results. The principle of dynamic action is based on the controlled movement of the surface of material with inverse piezoelectric effect, under the influence of a regulated voltage. This etalon allows us to replace a set of static measures, producing the specified required displacements in the range from tens of microns to tens of picometers. They are durable, and their operation does not require any special shape of the probe tip. Due to high resonance frequencies (over 10 kHz), dynamic etalons can measure the SPM response time and the dynamic parameters of its nanopositioning system. However, just as any static etalon, the dynamic one requires calibration using a metrological SPM. This paper presents the results of metrological characteristics study of dynamic action etalons carried out by means of a heterodyne laser interferometer.

## 2. Dynamic etalon of geometric dimensions for scanning probe microscopes calibration

At present, two types of dynamic (tunable) etalons have been well-studied. They differ in the active piezoelectric material: piezo-ceramic etalons [5] and single crystal piezomaterial ones (typically, lithium niobate) [6]. Both these materials change their dimensions when the control voltage is applied. The advantage of ceramics is that the piezoelectric modulus (factor of voltage-displacement conversion) is tenfold that of the single crystal, i. e. for the same magnitude of displacement of the etalon's surface, the applied voltage for ceramics would be less. Disadvantages of the ceramic etalon are the instability of the piezoelectric modulus and the dependence of deformation on the electric field frequency. Piezo-ceramics always has polarizing noise due to fluctuations in the electrical domain dipoles. Ambient temperature, heating of ceramics during operation, and mechanical stresses also affect the voltage-displacement conversion. Etalons based on a single crystal have more stable piezoelectric modulus, while the voltage-limiting restrictions can be eliminated through design modification, in particular, by using the multilayer structures that significantly extend the range of attainable displacements. In the present study we investigated the characteristics of the etalons made of single-crystal piezo-ceramics.

Figure 1 presents a schematic design of dynamic etalons of vertical and horizontal displacement. Type of displacement depends on the relative orientation of the crystallographic axes and the external electric field. A single crystal is mounted between two platforms, one of which serves as a carrier, the second being the reference.

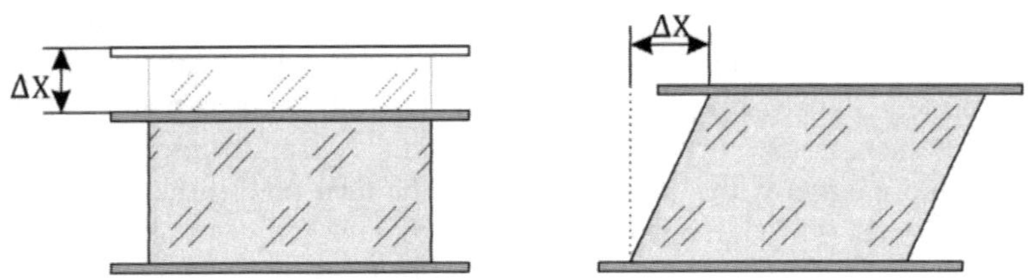


**Figure 1.** Schematic design of vertical and horizontal displacement dynamic etalons

When calibrating the SPM etalon of vertical type, it is located on a microscope table, and SPM probe is brought to its reference surface. Control voltage is applied to the etalon with a certain magnitude, and the etalon's surface is displaced by a known amount, which is measured using SPM servosystem. Etalon of horizontal displacement has an array of reference marks (typically a series of parallel submicron grooves) on its surface. During operation, the etalon surface moves and SPM scanning system records the movement of the reference marks with respect to their initial position. Values obtained from SPM are compared to the corresponding calibration data of etalons. In this way SPM measurement channels are calibrated using dynamic etalons.

## 3. Heterodyne laser interferometer NanoScan-3Di

The research of characteristics of the single-crystal piezoelectric dynamic etalon was conducted in the measuring SPM NanoScan-3Di [7]. This device ensures the accuracy of the data through the use of hard piezo-resonant probes and accommodation of a three-channel heterodyne interferometer that controls movement of the piezoelectric table. The radiation source in the interferometer is a frequency-stabilized He-Ne laser with a power of 1mW, and a relative instability of wavelength $10^{-9}$ for 8 hours. Dimensions of the optical interferometer unit are 260 × 260 × 60 mm, which allows embedding it inside the commercially available SPMs, including SPM NanoScan-3D [8].

Special design features including placing the triple-prisms on the moving axes of table in three orthogonal planes allowed minimizing Abbe error that is typical for measuring interferometers [2]. Reducing the influence of temperature fluctuations is achieved by removal of the laser radiation source outside the optical scheme [9]. Interferometer allows measurements in the range of 100 × 100 × 10 mm along the axes X, Y, Z, respectively, with a resolution of 0.01 nm.

## 4. The results of dynamic etalon characterisation

This work deals with a study of the metrological characteristics of dynamic etalons of vertical and horizontal displacement. Dynamic etalon was placed on the surface of a nanopositioner PI P-733.3 XY(Z) [10] of metrological probe microscope NanoScan-3Di. Scanning probe of SPM was brought to the surface of dynamic etalon, and the automatic control system was activated to maintain the contact of the probe tip with the etalon surface. After that, the pulsed control voltage of rectangular or trapezoidal shape with adjustable magnitude and period was applied to the etalon. Several series of measurements were performed for the applied control voltage in the range 1–1000 V. The etalon surface was measured using the interferometer. The precautions of acoustic and thermal insulation of measuring SPM greatly reduced phase noise in the interferometer that was determined by the temperature drift of the structure and by the fluctuations of the refractive index of air in the measuring arms. Measurements of dynamic etalons were carried out in a special enclosure box after continuous thermostating for 3–6 hours. In this manner, monotonic drift of the SPM piezoelectric table positioning was eliminated, and phase noise associated with the convective air flow inside the box was less than 2 nm.

For example, figure 2 demonstrates the time-resolved movement of a dynamic etalon at a voltage of 100 V when the displacement range was 74.56–75.30 nm.

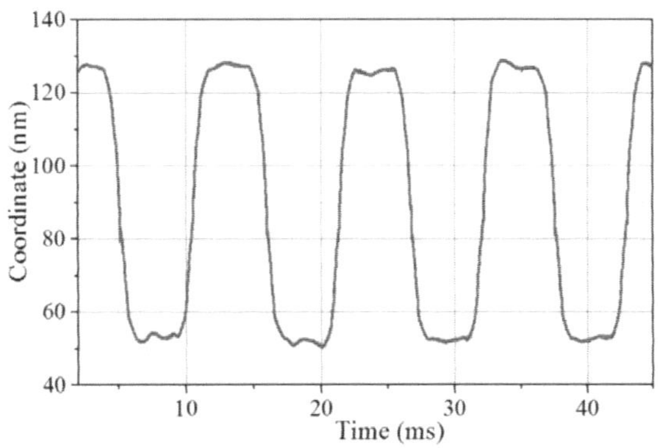


**Figure 2.** Time-resolved movement of a dynamic etalon at a voltage of 100 V

Figure 3 demonstrates the displacement of the etalon's surface when a voltage of 10 V was applied. Here, the displacement range was 0.95–1.23 nm.

Journal of Physics: Conference Series **747** (2016) 012059 doi:10.1088/1742-6596/747/1/012059

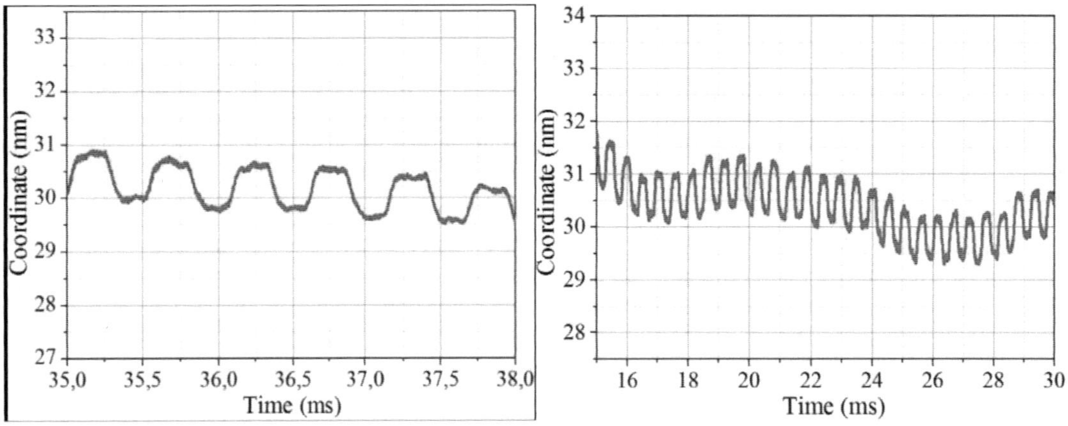


**Figure 3.** Time-resolved movement of a dynamic etalon at a voltage of 10 V

Data obtained by the interferometer was processed using an algorithm involving gradient filtering based on the convolution with the second derivative of the Gaussian function, and the extraction of areas corresponding to the maximum and minimum coordinates of the etalon's surface position from the original data set. Figure 4 shows the resulting data set obtained after applying the processing algorithm.

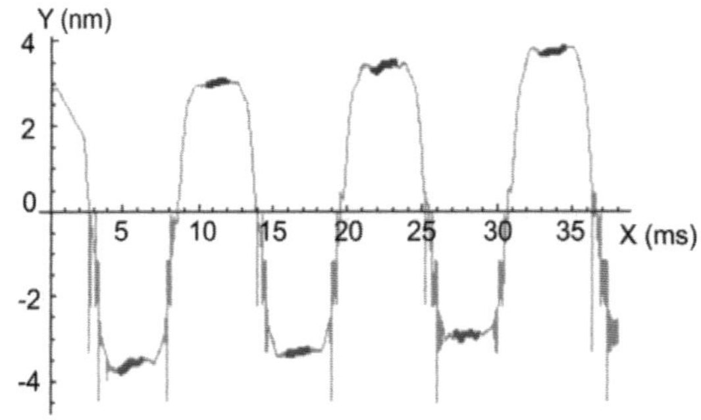


**Figure 4.** The data set obtained after selecting maximum and minimum positions of the etalon

The histogram distribution of maxima and minima of the etalon's positions was then built on the basis of acquired data. The displacement of the etalon's surface is defined as the algebraic difference between the values of the histogram peaks corresponding to the maximum and minimum position of the etalon (see figure 5).

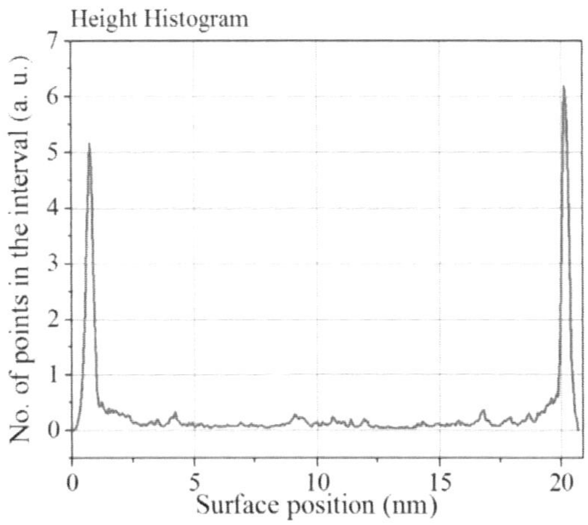


**Figure 5.** Histogram of measured step height distribution

Using the above mentioned algorithm, we obtained the displacements of etalons for different control voltages applied. For voltage 100 V the displacement was 73.91 ± 0.41 nm (see figure 3), and for 10 V it was 0.97 ± 0.13 nm (see figure 4). The displacement plots of two etalons' surfaces as functions of applied voltage in the range 100–1000 V are shown in figure 6. These measurements were performed both for increasing (black curves) and decreasing (red curves) voltage.

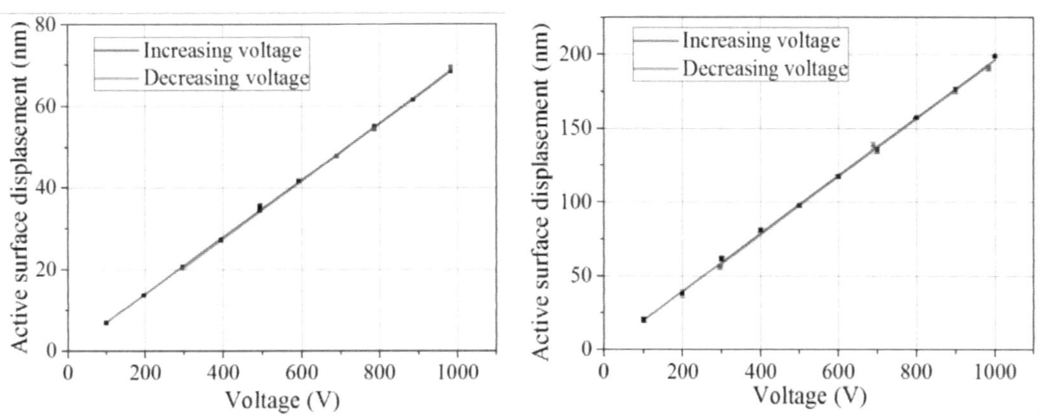


**Figure 6.** The displacement of two etalons' surfaces as a function of applied voltage in the range 100–1000 V: sample no. 1—vertical type and sample no. 2—horizontal type

Analysis of the curves shows that the dependence of displacement on the applied voltage for samples no. 1 and no. 2 can be approximated with a function of $y = kx + b$. Factor $k$ stands for the sensitivity of the etalons to the control voltage and is expressed in pm/V. Sensitivity of etalon no. 1 was 196.2 ± 8.5 pm/V, and of the etalon no. 2 — 70.2 ± 1.3 pm/V. Sample no. 1 examined with the increasing voltage had $k_i$ = 196.4 ± 8.5 pm/V, and with the decreasing voltage—$k_d$ = 195.8 ± 2.1 pm/V. Meanwhile, for sample no. 2 these values were $k_i$ = 70.8 ± 1.3 pm/V and $k_d$ = 69.5 ± 0.4 pm/V, respectively. The relative discrepancy of the factor $k$ may be defined as $K = (k_i–k_d)/k_d$ and used to estimate the nonlinearity of piezoelectric material. For sample no. 1 it is 0.4%, and 1.8% — for the sample no. 2. These values do not exceed the random error in determining the slope coefficient. Hence the results show that the piezoelectric material of the samples does not exhibit hysteresis effects.

## 5. Conclusion

The metrological SPM NanoScan-3Di equipped with three-axis heterodyne interferometer was used to study the characteristics of two single-crystal dynamic etalons of vertical displacement. The use of a heterodyne laser interferometer allowed us to calibrate etalons in the nanometer range with an accuracy of 0.1 nm. The dynamic range of the test etalon samples was 200 nm and 80 nm. Dependencies of the etalon's surface displacement on the control voltage were obtained for increasing and decreasing voltage magnitude. The obtained trends were approximated with functions $y = kx + b$, resulting in sensitivity of the etalons 196.2 ± 8.4 pm/V and 70.2 ± 1.3 pm/V, and non-linearity of the etalons of 0.4% and 1.8%, respectively, indicating the stability of piezoelectric modulus and the absence of piezoelectric hysteresis phenomena.

## 6. Acknowledgments

The present work was supported by a research grant from Russian Ministry of Education and Science 14.577.21.0088 (unique identification number RFMEFI57714X0088).

This work was also supported by Competitiveness Growth Program of the Federal Autonomous Educational Institution of Higher Professional Education National Research Nuclear University MEPhI (Moscow Engineering Physics Institute).

## References

[1] Wilkening G and Koenders L 2005 *Nanoscale Calibration Standards and Methods: Dimensional and Related Measurements in the Micro and Nanometer Range* (Germany: Wiley-VCH Verlag GmbH & Co. KGaA)
[2] Harding K 2013 *Handbook of Optical Dimensional Metrology* (New York: CRC Press Taylor & Francis Group)
[3] Bykov V and A Gologanov 1998 *Appl. Phys. A* **66** 499–502
[4] Leach R 2010 *Fundamental Principles of Engineering Nanometrology* (New York: Elsevier)
[5] Golubev S and Zaharinet V et al 2012 *Nano industry* **38** 42–46 (in Russian)
[6] Luskinovich P and Zhabotinskiy V et al 2015 *Nano industry* **59** 74–80 (in Russian)
[7] Gogolinskii K V and Kuznetsov A P et al 2012 *Measurement Techniques* **554** 400–405.
[8] http://nanoscan.info/eng/
[9] Kazieva T V and Kuznetsov A P et al 2014 *Proc. of SPIE* **9442** 94420K.
[10] www.physikinstrumente.com

Journal of Physics: Conference Series **747** (2016) 012060 doi:10.1088/1742-6596/747/1/012060

# Research of energy characteristics of power amplifier containing KNFS Nd:phosphate glass slabs and MIRO Silver foil reflectors at the "Luch" facility

**I A Belov, S A Bel'kov, I N Voronich, S G Garanin, V N Derkach, S V Koshechkin[1], M I Lysov, S S Markov and S V Savkin**

Russian Federal Nuclear Center "All-Russian Research Institute of Experimental Physics", pr. Mira 37, 607190 Sarov, Nizhnii Novgorod region, Russian Federation

oefimova@otd13.vniief.ru

**Abstract**. The amplifier elements upgrade at the "Luch" laser facility was carried out. Measurements showed that the upgrade of the amplifier elements resulted in the amplifier's small signal gain coefficient $K_0$ increase from 12.9% to 14.3% depending on the capacitor charging voltage; the linear gain coefficient increase was about $g_0 \approx$ (6-8)%. Full-scale laser experiments at the facility showed the power amplifier gain coefficient increase consistent with active medium gain coefficient measurement results.

## 1. Introduction

Upgrade operations of amplifiers at the "Luch" facility [1, 2] were started in 2013. Elements upgrade operations included the replacement of amplifiers' blastshields made of FLOAT glass by K-8 glasses with antireflection coatings, and the replacement of flat MIRO [3] foil reflectors by MIRO Silver [4] foil reflectors. Measurements of the gain coefficient showed that amplifier elements upgrade resulted in stored energy increase from 6.4% to 7.0%. The regular gain coefficient level of power amplifier under reduced flashlamp explosion fraction $f_x = 0.2$ (the capacitor charging voltage was equal to $U = 20$ kV) instead of 0.25 ($U = 22$ kV) was achieved in the full-scale experiments at the facility due to stored energy increase.

In the course of above mentioned work, the reequipment of the amplifiers with KNFS (KNFS-1 and KNFS-3) laser glass slabs [5] was carried out in 2015 at the "Luch" facility in the process of planned refit of the amplifiers. The reequipment with new slabs and other slabs rearrangement in the amplifiers were fulfilled in channel 1 (bottom channel) and channel 2 (top channel) of the facility. The main principles of slab placing into amplifiers were as follows. Firstly, all available laser slabs made of new KNFS-1 and KNFS-3 glass were placed into the channel 1. All other sites were equipped with KGSS-0180/35 glass slabs chosen among those that had been in use before. Secondly, among the slabs selected for channel 1 and channel 2, higher optical quality slabs were placed in the A1 amplifier, where the laser energy density was maximal. Overall, eight KNFS-1 slabs, eight KNFS-3 slabs and twenty KGSS-0180/35 slabs were placed in channel 1 and channel 2 of the amplifiers.

The active medium in the channel 1 of the A1 amplifier was formed with KNFS glass laser slabs, mainly KNFS-3 (seven KNFS-3 slabs and two KNFS-1 slabs) as a result of slabs rearrangement. The

[1] To whom any correspondence should be addressed.

active medium in the channel 1 of the A2 amplifier consisted of six KNFS-1 slabs, one KNFS-3 slab, and two KGSS-0180/35 slabs. In the channel 2 both A1 and A2 amplifiers were equipped with KGSS-0180/35 slabs.

Besides the reequipment of amplifiers with laser slabs made of KNFS laser glass, the total replacement of MIRO foil rhomb-shaped reflectors by MIRO Silver foil rhomb-shaped reflectors in flashlamp cassettes was carried out in the course of the amplifiers refit.

Experimental research of the active medium gain of the amplifiers was performed after the above described operations. The results are presented below.

## 2. The gain coefficient measurement experimental scheme

Gain coefficient measurements were performed by laser radiation amplification of a regular alignment laser in the amplifier channel. The amplifier channel contained 9 slabs. The scheme for the gain coefficient measurements was similar to the one described in [2, 3]. The principle scheme for the gain coefficient measurements is shown in Figure 1.

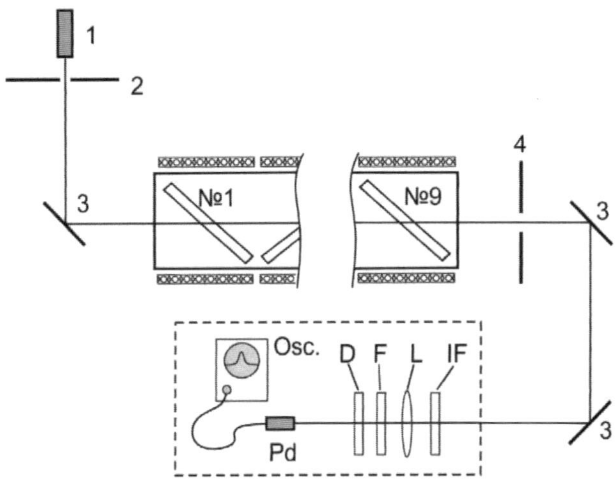


**Figure 1.** Principal optical scheme for the gain measurement. (1 – test laser, 2 – aperture diaphragm, 3 – dielectric mirrors,4 – probing diaphragm, IF –interference filter, L – lens, F – optical filter, D – diffuser, Pd – photodiode FD256, Osc – oscilloscope TDS-3052).

The regular alignment laser (1) radiation with the wavelength of 1053 nm and beam size of $\approx 205\times 205$ mm$^2$ was passed through the amplifier. The probing diaphragm (4) (probing diaphragm diameter was equaled to 15 mm) was located in the optical path at the amplifier output to select the probe area on the amplifier aperture. The diaphragm (4) had also restricted pump light ingress into the registration scheme and acted as the selective diaphragm. After passing the diaphragm (4), laser beam with diameter 15 mm and power 1 mW (that corresponded to the oscilloscope signal value $U \approx 20$ mV) was transported to the registration scheme by dielectric mirrors (3). The registration scheme was located in a separate room. The aperture diaphragm (2) with diameter 20 mm was placed in line with the diaphragm (4) at the amplifier input. The laser radiation was focused on the photodiode (Pd) by means of the lens (L) in the registration scheme. The interference filter (IF) ($\lambda = 1053$ nm, $\Delta\lambda_{0.5}$=0.05 μm) was placed before the lens to provide pump radiation interception. The IKS-7 optical filter (F) attenuated laser radiation power up to the desired level. The depolished glass diffuser (D) was placed before the photodiode. Registration scheme elements were protected from ambient illumination by means of a non-transparent tube. Measurements were performed with the time resolution ~ 30 ns using an FD256 silicone photodiode and a TDS-3052B oscilloscope.

Journal of Physics: Conference Series **747** (2016) 012060 doi:10.1088/1742-6596/747/1/012060

Small signal gain coefficient measurements were carried out depending on the pump value at capacitor charging voltages 18, 20, 22, and 24 kV in channel 1 and channel 2 of the A1 amplifier at central aperture points and in the channel 1 of A2 amplifier at central aperture point. The geometric center of the fifth laser slab was taken as a central point of the amplifier optical aperture. Slabs pump was performed only in the A1 amplifier while performing the measurements on the A1 amplifier, and in the A2 amplifier while performing the measurements on the A2 amplifier.

The characteristic oscillogram of the amplified alignment radiation $U(t)$ and the oscilloscope zero readout taken 1 minute before the experiment without alignment irradiation and flashlamp pumping $U_0(t)$ are shown in Figure 2.

The amplified radiation pulse shape $U'(t)$ was calculated using the following formula:

$$U'(t) = U(t) - U_0(t) . \tag{1}$$

To obtain the time dependence of the gain coefficient $K(t)$, the region before the moment of pump start was selected out of the distribution $U'(t)$, and the level of reference signal $U_{ref}$ was evaluated under the average value in this region. The time dependence of the gain coefficient is presented in Figure 3. It was calculated according to the formula:

$$K(t) = \frac{U'(t)}{U_{ref}} . \tag{2}$$

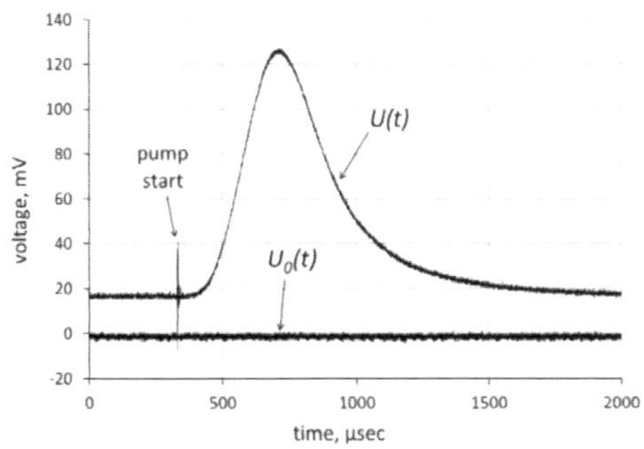


**Figure 2.** The characteristic oscillogram of temporal pulse shape of amplified radiation of the alignment laser $U(t)$, and the oscilloscope zero signal $U_0(t)$.

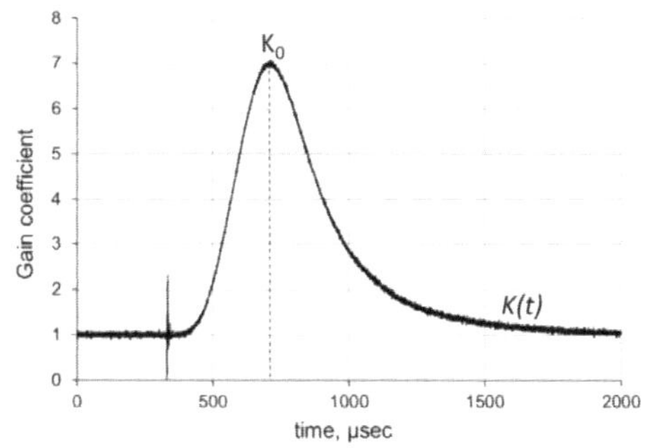


**Figure 3.** Gain coefficient temporal dependence.

The small signal gain coefficient value $K_0$ corresponds to the distribution maximum of $K(t)$ and it was achieved in about 370 µs after the pump start (that approximately corresponds to monopulse arrival time in full-scale experiments). The measurement error of $K_0$ was about ±5%. The small signal gain coefficient $g_0$ was calculated as:

$$g_0 = \frac{1}{L} \times \ln(K_0) , \tag{3}$$

where $L$ = 43.2 cm – active medium length (9 slabs at Brewster's angle). The relative measurement error of the linear gain coefficient $g_0$ was estimated by the formula:

$$\frac{\Delta g_0}{g_0} = \frac{\Delta K_0}{K_0 \times \ln(K_0)} \approx \pm 2.5\% . \tag{4}$$

## 3. Small signal gain experimental studies

Gain measurements in the channel 1 of the A1 amplifier (the A1 amplifier contained KNFS Nd-phosphate glass slabs) at capacitor charging voltages $U$ = 18, 20, 22, and 24 kV were carried out.

Obtained time dependences of the gain coefficient at capacitor charging voltages $U = 18$, 20, 22, and 24 kV are presented in Figure 4. Figure 5 illustrates characteristic of the pump current pulse, the light pump pulse oscillogram registered in red spectral region in one of the experiments (behind the KS-13 filter), and the time dependence of the gain coefficient.

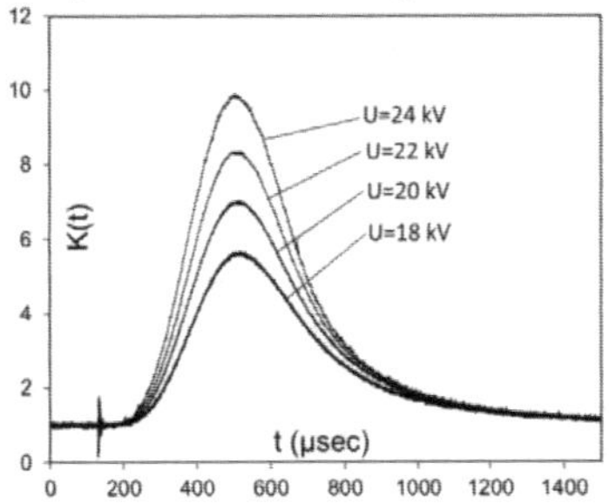


**Figure 4.** The time dependences of the gain coefficient at various capacitor charging voltages.

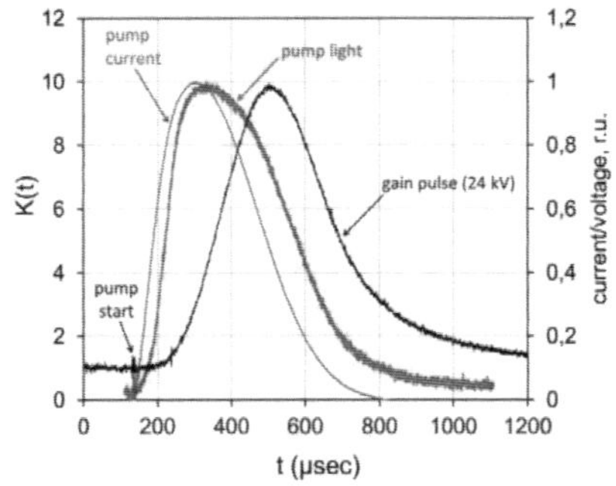


**Figure 5.** The time dependence of the gain coefficient; the light pump pulse oscillogram ($\lambda \geq 600$ nm) and the pump current oscillogram.

Gain coefficient measurement data at the central aperture point in the amplifier channel containing KNFS slabs are presented in Table 1. Defined capacitor charging voltages, values of electric energy delivered to a flashlamp $E_{del}$, values of small signal gain coefficients $K_0$, and values of linear gain coefficient $g_0$ calculated according to small signal gain coefficients $K_0$ are presented in Table 1. The electric energy delivered to the flashlamp was calculated using formula:

$$E_{del} = \frac{T}{N} \cdot \sum_{i=1}^{n} \left[ \frac{\left( \sum_{j=1}^{m} C_{ji} \right) \cdot U_i^2}{2} \right], \tag{5}$$

where N=144 – the number of flashlamps in 9 amplifier modules, n=9 – the number of amplifier modules, m=4 – the number of capacitors supplying pumping lamps of one of the amplifier module, C – value of capacitance supplying 4 flashlamp series chain (certified value), U – capacitor charging voltage at the moment of flashlamps firing, $T \approx 0.8$ [2] – the bank-to-lamp energy transfer efficiency.

**Table 1.** Gain data obtained in the amplifier channel containing KNFS glass slabs.

| U, kV | $E_{del}$, kJ | $K_0$ | $g_0 \times 10^2$, cm$^{-1}$ |
|---|---|---|---|
| 18 | 5.4 | 5.7±0.3 | 4.0±0.1 |
| 20 | 6.6 | 7.1±0.4 | 4.5±0.1 |
| 22 | 8.0 | 8.3±0.4 | 4.9±0.1 |
| 24 | 9.5 | 9.7±0.5 | 5.3±0.1 |

Figure 6 shows the comparison of gain coefficient $g_0$ dependences on the delivered energy $E_{del}$, obtained after blastshields and flat reflectors replacement in 2013, and after equipping the amplifier with KNFS slabs and upgrade of the rhomb-shaped reflectors.

As shown in Figure 6, significant increase of linear gain coefficient was achieved as a result of the reequipment of the channel 1 with slabs made of KNFS laser glass and the replacement of rhomb-shaped reflectors in flashlamp cassettes. The increment of the linear gain coefficient $g_0$ value varied from ≈6.2% to ≈8.4% depending on the electric energy delivered to the flashlamps, while the electric energy $E_{del}$ varied from ≈5.4 kJ to ≈9.6 kJ (charging voltage equalled to 18-24 kV).

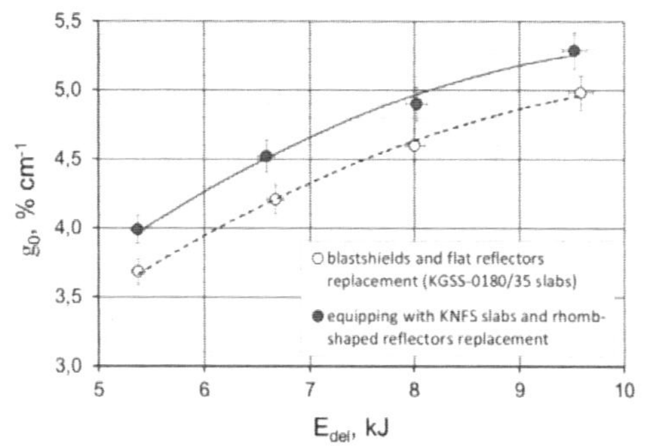


**Figure 6.** The gain coefficient dependence on the delivered energy before and after the amplifier's equipment with the KNFS slabs.

Significant increment of the gain coefficient value is unclear because technical requirements of the KGSS-0180/35 and KNFS laser slabs material are identical. The increase of the pump radiation flux after replacement of the MIRO foil rhomb-shaped reflectors by MIRO Silver foil rhomb-shaped reflectors was expected to be less than 3% (reflection coefficients of foils were equal to 95% and 98% respectively [4]).

Therefore, it was decided to carry out the study of the gain coefficient more thoroughly in order to allow to directly compare the values of the gain coefficients obtained on KNFS and KGSS-0180/35 glass laser slabs, eliminating the influence of the rhomb-shaped reflectors material.

### *3.1. Active medium gain coefficient research of different laser glass types*

To exclude the impact of reflectors material, the gain coefficient measurements were carried out in A1 amplifier channel 2 containing KGSS-0180/35 slabs only. Since the gain studies of A1 amplifier channel 1 described above had been finished, a series of over 20 full-scale experiments run on the facility. Therefore, to correctly compare the KGSS-0180/35 and KNFS gain characteristics, after the measurements in A1 amplifier channel 2, the gain measurements were made in A1 amplifier channel 1 again.

It is obvious that both channels experience equal pump terms due to the amplifier symmetry.

Gain measurement results for channels 1 and 2 of A1 amplifier are given in Tables 2 and 3, respectively.

**Table 2.** Gains measured in the amplifier channel containing KGSS-0180/35 glass slabs.

| U, kV | $E_{del}$, kJ | $K_0$ | $g_0\times10^2$, cm$^{-1}$ |
|---|---|---|---|
| 18 | 5.4 | 5.0±0.3 | 3.7±0.1 |
| 20 | 6.6 | 6.0±0.3 | 4.2±0.1 |
| 22 | 8.0 | 7.3±0.4 | 4.6±0.1 |
| 24 | 9.5 | 8.4±0.4 | 4.9±0.1 |

**Table 3.** Gains measured in the amplifier channel containing KNFS glass slabs.

| U, kV | $E_{del}$, kJ | $K_0$ | $g_0\times10^2$, cm$^{-1}$ |
|---|---|---|---|
| 18 | 5.4 | 5.6±0.3 | 4.0±0.1 |
| 20 | 6.6 | 6.8±0.3 | 4.5±0.1 |
| 24 | 9.5 | 9.4±0.5 | 5.2±0.1 |

Figure 7 illustrates dependences of the gain coefficient on the energy delivered to the flashlamps for the KNFS glass amplifier channel and for the KGSS-0180/35 glass amplifier channel.

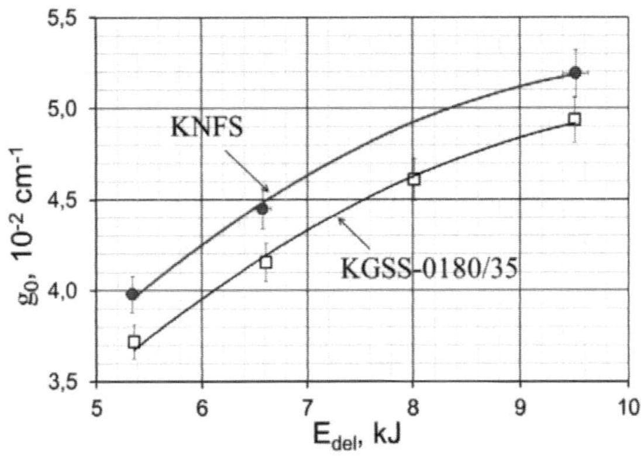


**Figure 7.** Dependences of the gain coefficient on the electric energy delivered to the flashlamps for the KNFS glass amplifier channel and for the KGSS-0180/35 glass amplifier channel.

As illustrated in Figure 7, values of linear gain coefficients obtained in the amplifier channel with KGSS-0180/35 glass active medium varied from $\approx 3.7 \cdot 10^{-2}$ cm$^{-1}$ to $\approx 4.9 \cdot 10^{-2}$ cm$^{-1}$ in the range of charging voltages from 18 to 24 kV ($E_{del}$ from 5.4 to 9.6 kJ). Values of linear gain coefficients obtained in the amplifier channel with KNFS (mainly KNFS-3) glass active medium under the same pumping conditions varied from $\approx 4.0 \cdot 10^{-2}$ cm$^{-1}$ to $\approx 5.2 \cdot 10^{-2}$ cm$^{-1}$. Thus, linear gain coefficient values obtained for the KNFS glass active medium were by about 7% higher than that those for the KGSS-0180/35 glass slabs.

The gain coefficient value of the A2 amplifier channel 1 based on KNFS-1, KNFS-3, and KGSS-0180/35 slabs was also of interest.

Gain coefficient measurements in the A2 amplifier channel 1 with KNFS and KGSS-0180/35 slabs were carried out with the use of a similar scheme as for the A1 amplifier. Measurement data are presented in Table 4 and Figure 8.

**Table 4.** Gains measured in the amplifier channel containing KGSS-0180/35 and KNFS laser glass slabs.

| U, kV | $E_{del}$, kJ | $K_0$ | $g_0 \times 10^2$, cm$^{-1}$ |
|---|---|---|---|
| 18 | 5.4 | 5.3±0.3 | 3.9±0.1 |
| 20 | 6.6 | 6.5±0.3 | 4.3±0.1 |
| 22 | 8.0 | 7.8±0.4 | 4.8±0.1 |
| 24 | 9.5 | 9.0±0.5 | 5.1±0.1 |

Solid curve in Figure 8 shows $g_0$ gain coefficient dependence on the delivered energy for mixed type active media, dashed curves show $g_0(E_{del})$ dependences for KNFS only and for KGSS-0180/35 only containing amplifier channels. As shown in Figure 8, mixed media $g_0$ values are predictably situated in an intermediate position between special KNFS and KGSS gain results.

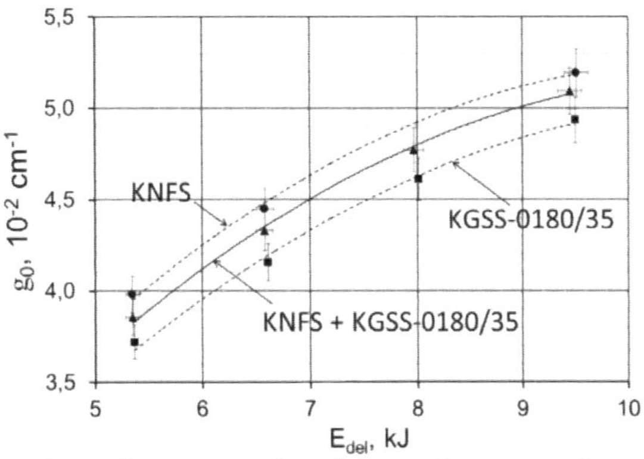


**Figure 8.** Gain coefficient dependence on the delivered energy for three amplifier channel configurations.

As shown in Figure 8, obtained values of linear gain coefficients predictably fall between the results for KNFS glass and KGSS-0180/35 glass.

*3.2. Active media gain coefficient studies under different pumping conditions*

Comparison of $g_0$ values in the A1 amplifier channel 2 obtained before rhomb-shaped reflectors upgrade in 2013 and after the reflectors replacement and laser slabs rearrangement in 2015 was provided to reveal the influence of the rhomb-shaped reflectors replacement on the value of the linear gain coefficient $g_0$. The A1 amplifier channel 2 before and after the rearrangement of laser slabs (the reequipment of amplifiers by KNFS slabs) included only KGSS-0180/35 slabs.

Dependences of the gain coefficient on the delivered energy obtained before and after the replacement of rhomb-shaped reflectors in the A1 amplifier channel 2 are presented in Figure 9 (2013 and 2015). In the first case rhomb-shaped reflectors were made of MIRO foil, and in the second case, they were made of MIRO Silver foil.

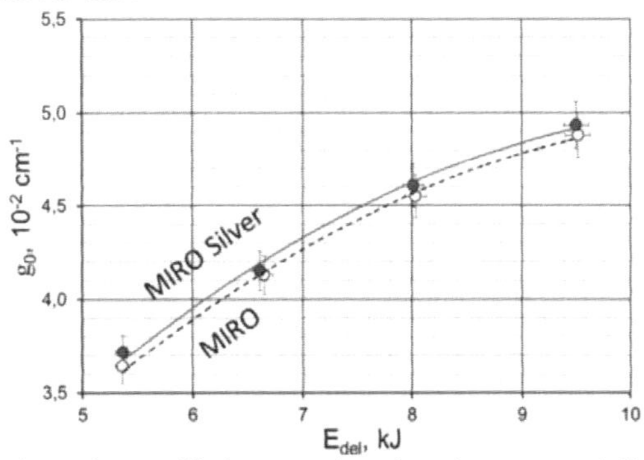


**Figure 9.** Dependences of the gain coefficient on the electric energy delivered to the flashlamps in the amplifier channel with MIRO and MIRO Silver foil rhomb-shaped reflectors.

As one can see from the graph presented in Figure 9, values of linear gain coefficient obtained after the replacement of the rhomb-shaped reflectors are situated higher than those obtained with formerly used reflectors. However, the difference of linear gain coefficient values is from about 1.1% to 1.6% over the whole pumping energy range. It is less than measurement error and processing error of experimental data.

## 4. The power amplifier efficiency increase at the "Luch" facility

Figure 10 shows dependences of the linear gain coefficient on the electric energy delivered to the flashlamps. These dependences were obtained before the amplifiers' elements upgrade, after the blastshields and flat reflectors replacement, and after the reequipping with KNFS slabs and the replacement of the rhomb-shaped reflectors.

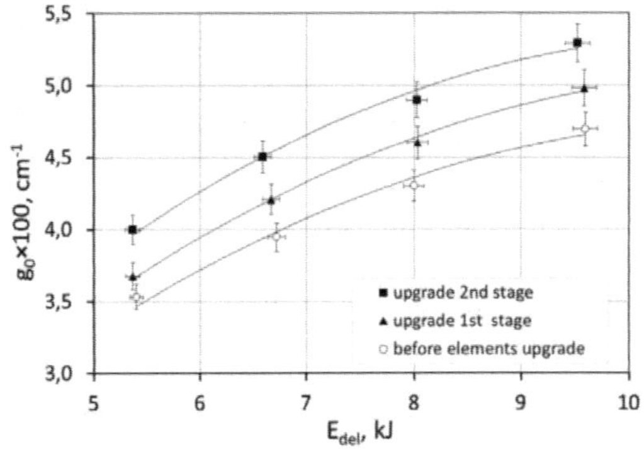


**Figure 10.** The dependence of the gain coefficient on the electric energy delivered to the flashlamps.

As can be seen from the graph presented in Figure 10, upgrade operations of amplifier elements performed at the first stage and at the second stage both allowed to significantly increase the value of the amplifier active medium gain coefficient at the "Luch" facility. The linear gain coefficient was increased from ≈12% to ≈14% after two upgrade stages over the delivered electric energy range from 5.3 kJ to 9.4 kJ. The value of the linear gain coefficient achieved about $5.3{\cdot}10^{-2}$ $cm^{-1}$ at the capacitor charging voltage 24 kV. Obtained gain coefficient increase (up to 14%) enabled receiving the regular output energy level while conducting the four-channel experiments at the facility, since it is known that operating under the four-channel regime results in the average linear gain coefficient decrease by about 15%.

There is no opportunity to measure the amplifiers active medium small signal gain coefficient value during each full-scale experiment carried out at the facility. Therefore, the power amplifier (implying both A1 and A2 amplifiers divided by cavity spatial filter) efficiency is most conveniently characterized by using the channel double-pass gain coefficient $K_{12}$ for the purpose of control and analysis of the gain coefficient changing dynamics. The double-pass gain coefficient is determined from the formula:

$$K_{12}=\frac{E_{rev}}{E_{input}}, \tag{6}$$

where $E_{rev}$ – the energy at the output of the second pass (or at the reverser input); $E_{input}$ – the energy at the input of the laser channel.

Since the laser radiation amplification regime in the first 2 passes is linear, the double-pass gain coefficient $K_{12}$ is related to small signal gain coefficients of A1 and A2 amplifiers according to the equation:

$$K_{12}=K_{A1}^2\times K_{A2}^2\times T \tag{7}$$

where $K_{A1}=\exp(g_{0\,A1}{\cdot}l)$ and $K_{A2}=\exp(g_{0\,A2}{\cdot}l)$ – small signal gain in A1 and A2 amplifiers, respectively, l – active medium length, T – laser channel double-pass transmission.

The dependence of the gain coefficient $K_{12}$ on the experiment number in the full-scale experiments at the facility from 2011 till 2015 is presented in Figure 11. Periods before the upgrade of the amplifiers' elements, after the first and second upgrade stages are also demonstrated in Figure 11.

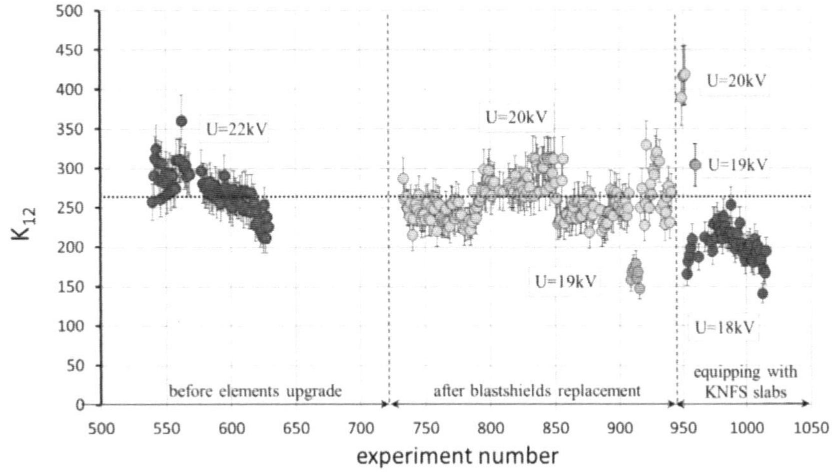


**Figure 11.** The dependence of the gain coefficient $K_{12}$ on the experiment number.

As shown in Figure 11, the amplifier blastshields replacement made it possible to achieve regular level of the power amplifier gain coefficient ($K_{12}\approx270$) in the full-scale experiments, i.e. regular output energy level at the same input signal, at a reduced value of the capacitor charging voltage U=20 kV, instead of 22 kV (under reduced flashlamps explosion fraction $f_x$=0.2, instead of 0.25). The realization of the second upgrade stage (i.e. the reequipping of amplifiers with KNFS slabs) resulted in the gain coefficient $K_{12}$ increase up to ≈415 under U=20 kV. Obtained values of the gain coefficient $K_{12}$ at the capacitor charging voltage 19 and 18 kV were $K_{12}\approx305$ and 220, respectively. Thus, the second upgrade stage made it possible to achieve regular level of the laser channel output energy at the capacitor charging voltage U=18 kV ($f_x\approx0.17$) with insignificant increase of the input signal from SRFS (supporting radiation forming system).

The facility power amplifier gain coefficient evolution after performing of two amplifiers upgrade stages is illustrated in Figure 12 as the dependence of the power amplifier gain coefficient $K_{12}$ value on the squared capacitor charging voltage. Experimental and calculated results in Figure 12 are shown as filled and opened markers, respectively. Experimental dependences of the reverser input energy on the amplifier channel input energy in the full-scale experiments in the channel 1 at the facility are presented in Figure 13. The slopes of lines represented in Figure 13 correspond to the gain coefficient $K_{12}$.

Dependences in Figure 13 also indicate that the blastshields replacement allowed to reduce the charging voltage from 22 kV to 20 kV obtaining regular level of the power amplifier gain coefficient (dashed lines in Figure 13). This gain level occupies an intermediate position between $E_{rev}(E_{input})$ lines corresponding to data obtained after equipping amplifier with KNFS slabs at U=18 kV and 19 kV.

## 5. Conclusion

The amplifiers' elements upgrade was carried out. It included the equipment of the channel 1 with the KNFS laser glass slabs, the slabs rearrangement in the amplifiers, and the replacement of MIRO foil rhomb-shaped reflectors by MIRO Silver foil rhomb-shaped reflectors.

Active medium gain coefficient measurements in the channel 1 of the A1 amplifier were performed. The gain coefficient increase of 8.4% was obtained. The maximum value of the linear gain coefficient at the capacitor charging voltage 24 kV was about $g_0\approx5.3\cdot10^{-2}$ cm$^{-1}$.

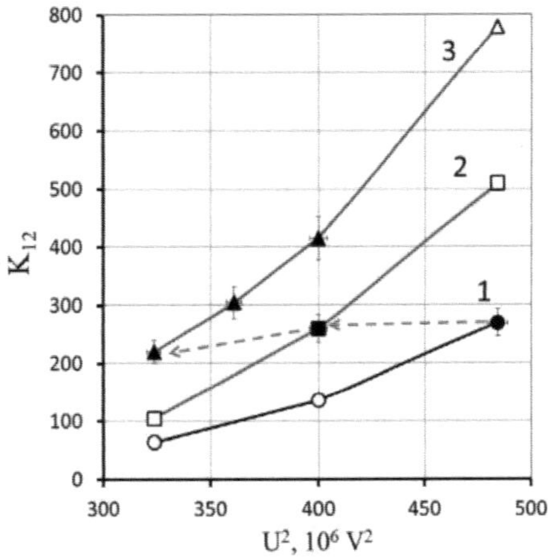


**Figure 12.** Dependence of $K_{12}$ coefficient versus squared capacitor charging voltage (filled markers - experimental data; opened markers - calculated data). 1 – before the amplifiers elements upgrade, 2 – after the blastshields replacement, 3 – after the equipping of amplifiers with KNFS slabs.

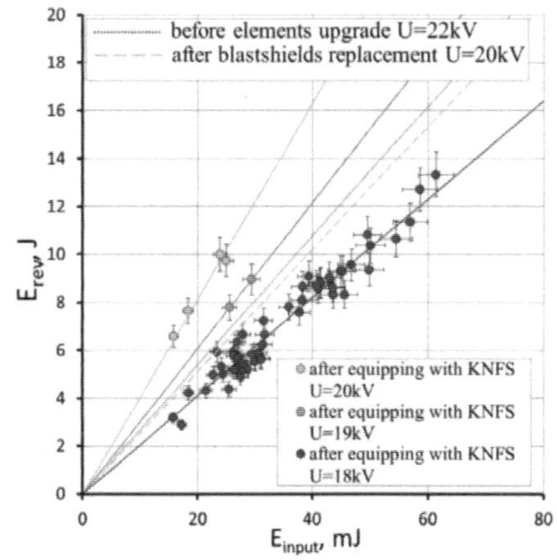


**Figure 13.** Experimental dependences of the energy at the reverser input on the energy at the channel input: red dashed line corresponds to the case before the amplifiers' upgrade, 22 kV; yellow dashed line corresponds to the case after the blastshields replacement, 20 kV; blue, green, and yellow solid lines correspond to the case after equipping with KNFS slabs, 18, 19, and 20 kV, respectively.

Gain coefficient studies in the amplifier channels based on KNFS and KGSS-0180/35 laser glass slabs were carried out. It was shown that the linear gain coefficient values of KNFS glass active medium were about 7% higher than the linear gain coefficient values for KGSS-0180/35 slabs. Gain coefficient measurements in the channel of the amplifier containing both KNFS glass laser slabs and KGSS-0180/35 laser glass slabs were carried out. Obtained linear gain coefficients values fall between the results for KNFS glass and KGSS-0180/35 glass.

Active media gain coefficient measurements data under different pumping conditions (different rhomb-shaped reflectors) were presented. It was shown that the rhomb-shaped reflectors replacement resulted in the about 1.5% increase of the gain coefficient.

It was demonstrated that the equipment of the amplifier modules with KNFS glass slabs, the replacement of the reflectors, and previous work with the blastshields replacement resulted in about 14% increase of the small signal linear gain $g_0$.

## References

[1] Garanin S G, Zaretski A I, et al 2005 *Quantum Electron* **35** (4), 299-301.
[2] Voronich I N, Galakhov I V, et al 2003 *Quantum Electron* **33** (6), 485-488.
[3] Voronich I N, Garanin S G, et al 2004 *Quantum Electron* **34** (6), 509-510.
[4] http://www.alanod.com
[5] Avakyants L I, Ignatov A N, et al 2014 *Opt. Technol.* **81** (12) 719-722.

# Fiber Lasers Application for Welding of Titanium Alloys With 16 mm Thickness

**N N Evtihiev[1,2,3], N V Grezev[2], Y V Markushov[1] and M A Murzakov[1]**
[1] National Research Nuclear University MEPhI, Kashirskoe sh. 31, Moscow, 115409, Russia
[2]LLC NTO "IRE—POLUS", Vvedenskogo Sq. 1, Fryazino, Moscow Region, 141120, Russia

E-mail: SinNerRus@mail.ru

**Abstract.** This article illustrates the use of fiber laser welding of a titanium alloy with 16 mm thickness. The basic advantages of the laser welding process over the traditional methods of arc welding of titanium are demonstrated. Destructive testing of welds was performed to confirm the quality of the welding. The results of the static tensile tests, static bending and toughness at room temperature are presented. All tests confirmed the high quality of the welded joint.

## 1. Introduction

Titanium and its alloys are successfully used in many spheres of industry because they have lots of advantages and unique properties, such as: low specific weight, high toughness, excellent corrosion resistance, high melting point. Titanium and titanium alloys are widely used, for example, in aircraft, space, shipbuilding, chemical industries, nuclear power, and in the medicine.

Welding of pure titanium and its alloys is in most cases complicated because of its high reactivity with atmospheric gases. Titanium is an extremely active chemical element and at a high temperature it easily absorbs harmful gases from ambient air (oxygen, nitrogen, hydrogen), making the weld much worse. The activity of titanium takes place provided the temperature of titanium alloys is above 300° C. And in case of poor protection, heated titanium surface absorbs harmful gases from ambient air resulting in the formation of brittle phases of titanium: carbides, oxides, nitrides. Brittle phases lead to increase in hardness and decrease in ductility. As a result, strength and resilience of the fusion zone and the heat affect zone decrease [1].

Another feature of titanium welding is the tendency to grain growth. Large volume of weld pool, occurring during arc welding, leads to grain growth in the fusion zone metal and in the heat affected zone. As a result, brittle phases are formed in cooling phase, which leads to reduction in weld strength [2].

Laser welding is one of the modern methods of welding, and it has a number of advantages relevant to the titanium alloys treatment: low heat input, low residual stress, minimum temperature limits, and high physical and mechanical properties of weld.

## 2. Equipment and materials

[3] To whom any correspondence should be addressed.

A complex for laser welding on the basis of the fiber laser production company OOO NTO "IRE-Polus" was used. Technical titanium, with a titanium content of at least 99 percent, was selected as a material for laser welding, and the wire feed had similar chemical composition. Inert gas argon was used for protection of welding pool and liquid metal during laser welding.

## 3. Results

A 16 mm thick titanium weld joint was obtained by selecting an optimal mode for laser welding process. Sample back-to-back weld was performed with a special edge preparation for multi-pass laser welding technology. Principle of a multi-pass technology of laser welding is described in [3]. Titanium weld joint 16 mm thick plates are shown in Figure 1.

**Figure 1.** Right side of the weld joint of titanium alloy.

**Figure 2.** Reverse side of the weld joint of titanium alloy.

## 4. Macro- and microstructure of the 16 mm thick titanium alloy weld

Figure 3 shows a macrosection of the welded titanium. The width of the weld is about 13–14 mm from the surface, and 6 mm in the central part of the seam. Heat affected zone (HAZ) is about 1.0–1.5 mm. Defects in form of scores on the front and back sides do not exceed 0.4 mm.

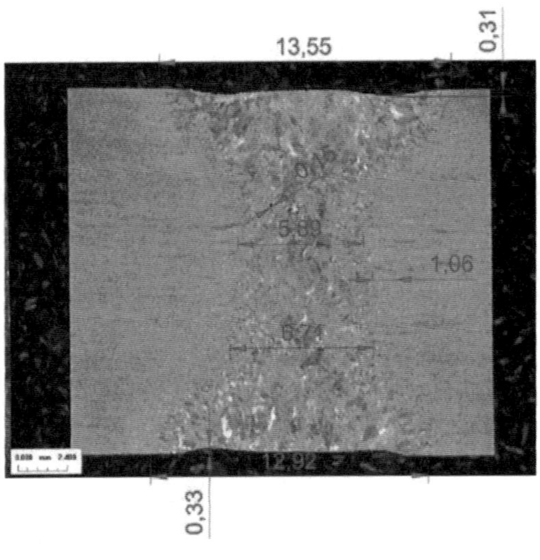


**Figure 3.** Macrosection of welded joint.

Microstructure is shown in Figure 4. Welded joint has a fine grain structure. Basic metal (BM) comprises α phase, the HAZ zone – α, α', β phases, and the fusion zone (FZ) consists of α' and β phases. A single pore has a transverse dimension of 150 mm, which is 0.94% of thickness of the welded joint shown in Figure 3.

Single-phase α alloys show no plasticity and they are not embrittled by the heat treatment. Phase of fusion zone and phase of HAZ are more durable, well forged, stamped, and amenable to thermal

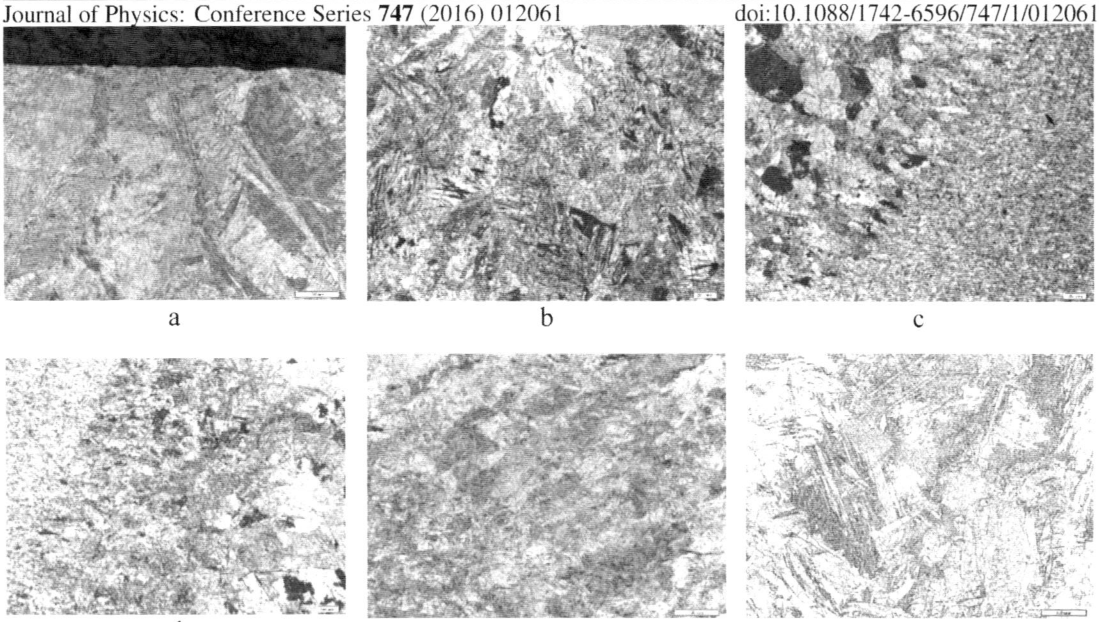


**Figure 4.** Photographs of the microstructure of different areas: a) Near-surface structure, 200× magnification; b) Center of fusion zone, 50× magnification; c) Area of transition zone FZ-HAZ-BM, 50× magnification; d) Area of transition zone BM-HAZ-FZ, 50× magnification; e) Heat affected zone, 200× magnification; f) Center of fusion zone, 200× magnification.

treatment than alloys with single α phase. Bar in the designation "α" indicates the formation of martensitic structure of alpha phase.

It should be noted that α-titanium alloys in the martensite do not lead to embrittlement of the welded joint. [4].

## 5. Microhardness measurement of the weld

Microhardness measurements were performed on 70 DuraScan devices. Measurements were made across the weld at a 300 μm distance between dots. Figure 5 below shows the results of the microhardness measurements: for the "up" case at a depth of 2 mm and for the "down" case to a depth of 14 mm.

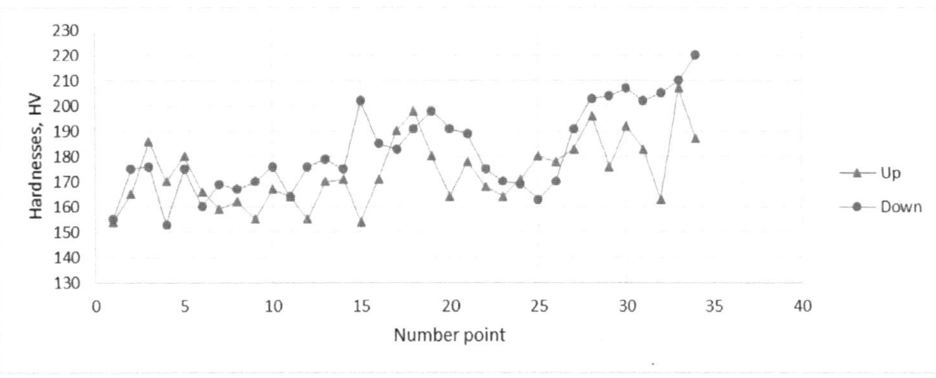


**Figure 5.** Measurement results of microhardness no. 1.

Measurements showed that microhardness of the weld metal and HAZ zone is in the range of 150 HV – 225 HV. It indicates the absence of oxides, nitrides, carbides and other harmful inclusions having microhardness higher than 400 HV [2]. Based on these results, the quality of shielding gas may be argued.

## 6. Destructive testing of welded joints

### *6.1. Results of mechanical tests on the static breaking.*

Tests were conducted in accordance with GOST 6996-66. Results are presented in Table 1.

**Table 1.** Test results on static breaking.

| Specimen № | Width (mm) | *Thickness (mm)* | Yield (0,2%) (MPa) | *Strength limit (MPa)* | Relative extension (%) |
|---|---|---|---|---|---|
| 1 | 24,6 | 16,4 | 358,15 | 413,24 | 32,22 |
| 2 | 24,6 | 16,4 | 375,74 | 413,23 | 33,89 |
| 3 | 24,6 | 16,4 | 365,28 | 410,01 | 33,33 |
| 4 | 24,6 | 16,4 | 376,41 | 409,70 | 33,33 |
| **Average** | **24,6** | **16,4** | **368,90** | **411,55** | **33,19** |
| Standard deviation | | | 1,85 | 0,37 | 1,63 |

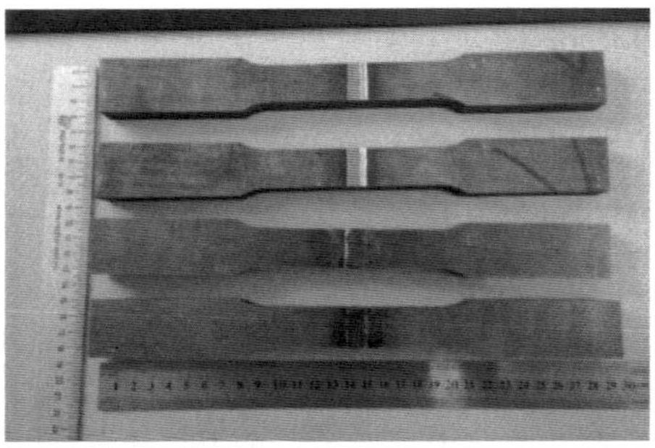

**Figure 6.** Before tests.

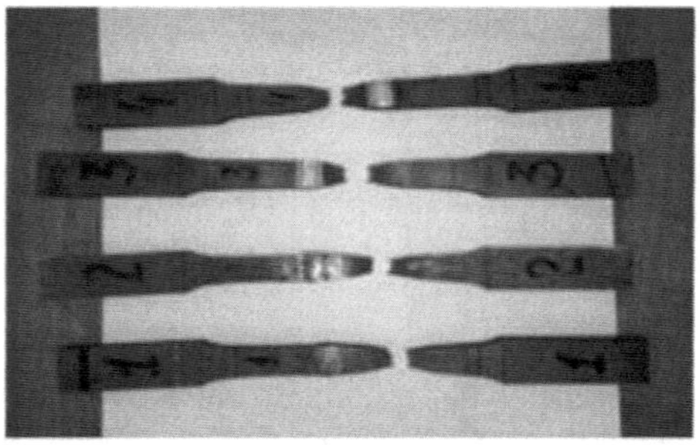

**Figure 7.** After tests.

### *6.2. The results of mechanical tests on toughness.*

Tests were conducted in accordance with GOST 9454-78. Results are presented in Table 2.

**Table 2.** Test results on toughness.

| Specimen № | Place of cut | *Impact energy K (J)* | Toughness KCV (J/cm$^2$) |
|---|---|---|---|
| 1 | Base metal | 110,21 | 138 |
| 2 | Base metal | 112,10 | 140 |
| 3 | Base metal | 111,99 | 140 |
| 4 | Base metal | 109,33 | 137 |
| **Base metal average value** | | **110,91** | **139** |
| 5 | Weld metal (mid) | 153,29 | 192 |
| 6 | Weld metal (mid) | 126,28 | 158 |
| 7 | Weld metal (mid) | 144,49 | 181 |
| 8 | Weld metal (mid) | 132,47 | 166 |
| 9 | Weld metal (mid) | 141,99 | 177 |

| | | | |
|---|---|---|---|
| 10 | Weld metal (mid) | 145,59 | 182 |
| **Weld metal average value** | | **140,69** | **176** |

Average value of toughness of the welded metal was 127% of the toughness of base metal.

*6.3. Results of mechanical tests on static bending.*

Figures 8 and 9 show that the samples have passed the tests without formation of defects. The maximum bending angle was 70 degrees, indicating high plastic properties of weld.

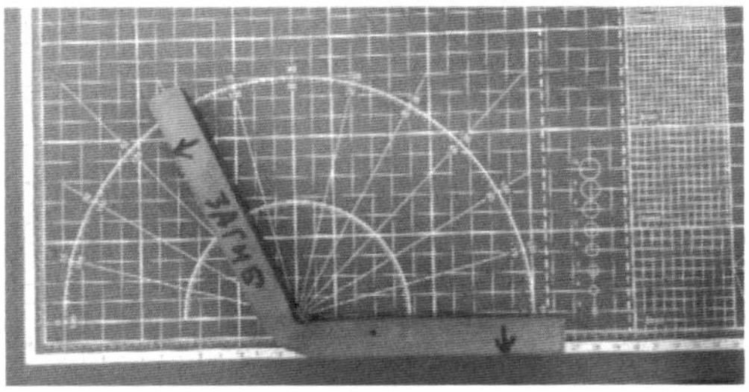

**Figure 8.** Sample after bending at 70°.

**Figure 9.** Welded joint after bending. Top sample - face, lower sample - flip side.

**7. Conclusion**

• The possibility of using fiber lasers for welding of titanium alloy with 16 mm thickness was demonstrated.

• Microhardness measurements revealed no region of increased hardness that indicates good protection of molten bath during solidification of weld.

• Studies of physical and mechanical properties of welded joints on static gap in accordance with GOST 6996 showed that all the samples submitted for testing torn in base metal, which indicates high quality of weld joint.

• Technological bend at an angle of 70° without cracking was shown.

• Studies of the physical and mechanical properties of the welded joints showed increase of the weld metal toughness up to 127% compared to the base metal.

**References**

[1] Lahtin Y M 1983 *Metallurgy and heat treatment of metals (Moscow: "Metallurgy")*

[2] Tretyakov F E 1968 *Fusion welding of titanium and its alloys (Moscow: "Engineering")*

[3] Evtikhiev N N, Markushov Y V, Grezev N V, Murzakov M A 2015 *Multipass narrow gap of heavy gauge steel with filler wire,* Physics Procedia 71 267–271

[4] Gulyaev A P 1986 *Metallurgy (Moscow: "Metallurgy")*

Journal of Physics: Conference Series **747** (2016) 012062 doi:10.1088/1742-6596/747/1/012062

# Research on microstructure and wear resistance of coatings obtained by adding nanoparticles of refractory compounds in laser cladding

**M A Murzakov[1], S N Chirikov, Y V Markushov**

National Research Nuclear University MEPhI, Moscow, Russia

E-mail: clericfull@mail.ru

**Abstract.** The paper is aimed at research of coatings, which are achieved by means of laser cladding with additives of nanoparticles of high-melting compounds in form of tungsten carbide and tantalum (WC and TaC). In the course of experiment, various ceramic powder concentrations were tested. Main technological characteristics were determined. Power density amounted to 0.68–0.98 MW/cm$^2$. During the coating wear resistance measurement, it was discovered that increase in nanopowder concentration extended wear resistance of coating 2–6 times. Wear resistance measurement and wear coefficient calculation were performed using Brinell-Howarth method. The load was 15 N, load time was 10 minutes. Optical metallographic microscope Neophot-30 was used to study microstructure of the deposited coatings. To reveal microstructure of the deposited coatings, the samples were exposed to chemical etching. Elemental composition of the samples was determined by the methods of X-ray microanalysis in testing solution using electron microscope EVO-50 under acceleration voltage 10–20 kV (probe current 5–50 nA) using energy- and wavelength-dispersive spectrometers.

## 1. Introduction
Laser cladding of powder is widely used to recover and produce protective coatings in various fields of mechanical engineering, mining, oil and gas industries. In recent years, scientists work actively to produce coatings with nanopowder additives [1–5]. The work involves such studies as microstructural analysis and wear resistance measuring. It is noted that with the addition of nanoparticles, microstructure is improved and wearing is reduced.

This work represents a comparative analysis of coating microstructures obtained with the addition of tungsten and tantalum carbide nanoparticles. Brinell-Howarth method was used to measure wear resistance. It was found that with increasing nanoceramics concentration, wear resistance of coatings with nano-additives is growing 2–6 times compared to conventional steel substrate.

## 2. Experiment methodology
Experiments were conducted on a C5140 steel substrate. Its chemical composition is shown in Table 1.

[1] To whom any correspondence should be addressed.

**Table 1.** C5140 chemical composition.

| | Ni | C | Fe | Cr | Si |
|---|---|---|---|---|---|
| Substrate | 0.3 | 0.40 | 97 | 0.9 | 0.20 |

As the main powder, 1559-40 by Höganäs was chosen. Different concentrations of TaC and WC nanopowders (5, 10, 15, and 20 wt.%, respectively) were added to this powder. Powders were premixed to obtain homogeneous distribution of particles, and then the binder, cellulose aqueous solution, was added. This mixture was placed on steel plates with dimensions 60×20×10 mm, and then well dried.

Selection of parameters was carried out with ytterbium 5 kW fiber laser as follows: in the first part of the experiment, process parameters on the substrate without the powder were selected. The following parameters were varied: processing speed, blooming, power of the laser radiation. Substrate surface temperature was measured using an infrared FLIR camera. According to [6], temperature significantly influences content of tungsten carbide in the coating, thus the parameters ensuring surface temperature of 1100°C were selected.

After the first part of the experiment was completed, these parameters were adjusted on already powdered substrate with and without nanoparticles additives. Power densities used in these experiments were 0.68–0.98 $MW/cm^2$. Heat input was 7.5 kJ /m.

## 3. Results and discussion

Abrasion testing was carried out using Brinell-Howarth method. The test procedure is shown in figure 1: a rotating rubber disk 1 is pressed against the sample 2, and abrasive (river sand) pours through the feed magazine 4. The load (number 3) is 15 N, exposure time is 10 minutes.

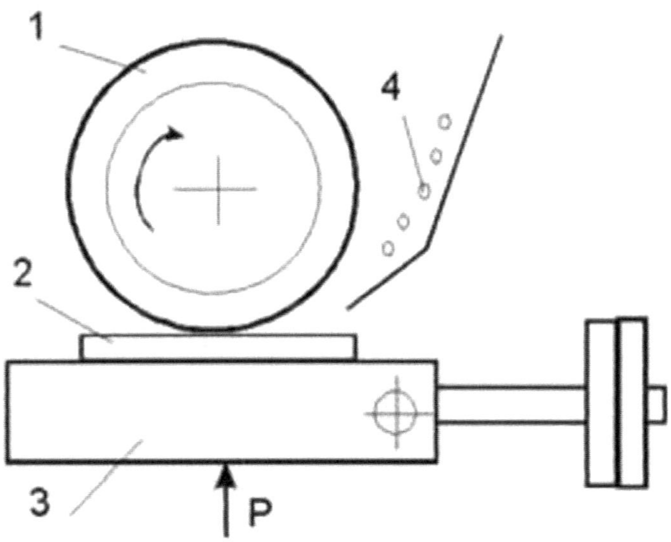


**Figure 1.** Overall schematic of the wearing test according to Brinell-Howarth method.

Figure 2 shows the results of calculation of the wear resistance coefficient *K*, obtained at different power densities and different nanopowder concentrations of tantalum carbide and tungsten carbide.

It can be seen in Figure 2 that wear resistance coefficient increases with increasing concentration of ceramic nanoparticles. It was also observed that with an increase of power density, wear resistance decreases, which is because WC and TaC decompose and form secondary carbides or other compounds.

Journal of Physics: Conference Series **747** (2016) 012062 doi:10.1088/1742-6596/747/1/012062

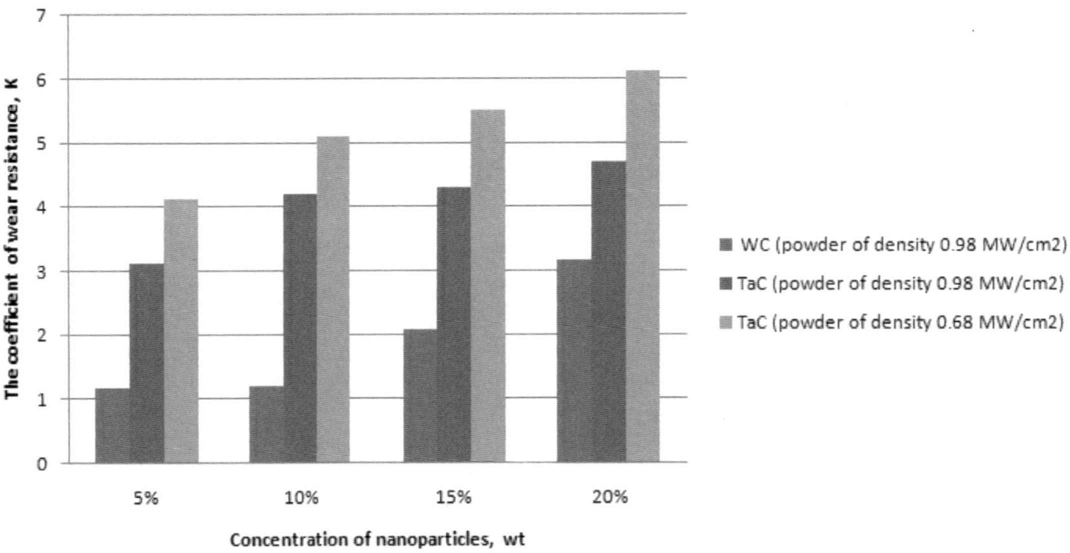


**Figure 2.** Wear resistance coefficient at different concentrations of nanoparticles and different power densities.

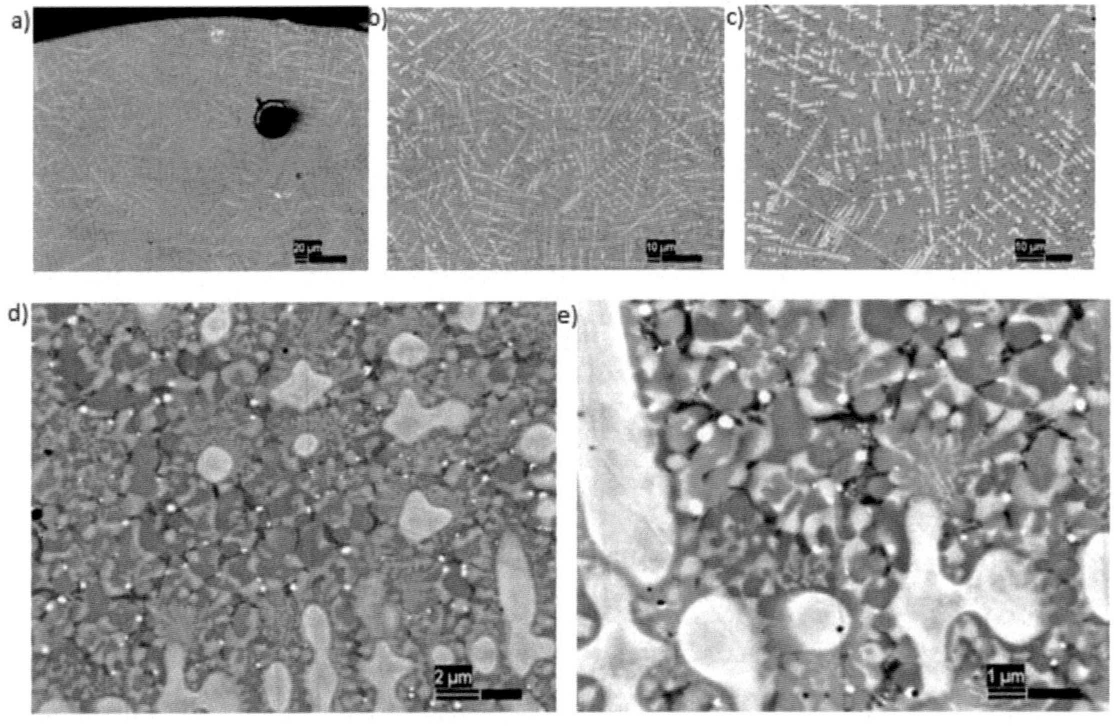


**Figure 3.** Microstructure of the coatings with nano-WC additives (concentration 10%): a) 250x magnification; b) 500x magnification; c) 1000x magnification; d) 2500x magnification; e) 5000x magnification.

Figure 3 shows coating microstructure produced by laser cladding with additives of tungsten carbide nanopowder at different magnifications. There is an acicular structure. In figure 3 (d, e) white patches are clearly seen, these are nanoparticles of tungsten carbide. The microstructure is fine dispersed, fine-grained because of the content of the nanoparticles in the coating.

**Table 2.** Measurement results of the reflection spectrum of backscattered electrons and chemical composition at each point by laser cladding with additives of nano-W (concentration 7.5%).

| Spectrum | C | Si | Cr | Fe | Ni | W |
|---|---|---|---|---|---|---|
| Top clad | 1.97 | 2.74 | 0.44 | 1.48 | 87.52 | 5.85 |
| Boundary clad/substrate | 1.54 | 2.60 | 0.45 | 3.89 | 85.55 | 5.96 |
| Substrate | - | - | 0.12 | 99.02 | - | - |

As we can see from the Table 2, tungsten is present in a single track, which is consistent with the spectrum of the backscattered electrons in the top clad, the clad/substrate boundary and the substrate. Exactly at the "coating-substrate" transition boundary tungsten disappears, therefore iron, which is present in the substrate, partially diffuses into the coating body. Cladding boundary is sharp, and the mixing is minimal.

## 4. Conclusion

Laser cladding deposit was made with powder additives of TaC and WC ceramic nanoparticles. It was revealed that nanopowder additives affect both microstructure and mechanical properties of the coatings, particularly, their wear resistance. The fine-grained, acicular microstructure containing WC nanoparticles between grains was observed. No defects, such as cracks, were detected in the process of cladding deposition.

Wear resistance of the coating is improved 2–6 times compared to the ordinary steel, which also indicates the effect that nanoparticles produce on its physical and mechanical properties. During the process of cladding deposition, temperature did not exceed 1100°C, ensuring the WC and TaC nanoparticles presence in the coating.

The power density ranged from 0.68 to 0.98 $MW/cm^2$, and the heat input was 7.5 kJ /m.

**Acknowledgements**

The study was financially supported by the Ministry of the Education and Science (unique identifier PNIER RFMEFI58214X0004).

This work was supported by Competitiveness Growth Program of the Federal Autonomous Educational Institution of Higher Professional Education National Research Nuclear University MEPhI (Moscow Engineering Physics Institute).

The reported study was funded by RFBR according to the research project No.16-32-00083 мол_а.

**References**

[1] Zhang Wei, Yao Jian Hua, Zhang Qun Li, Kong Fan Zhi 2006 *Solid State Fenomena* **118**, p. 579-584

[2] Yong-Chwang Chen, Kwo-An Chiang 2006 *Journal of the Chines Institute of Engineers* **29 3**, p. 423-431

[3] Hu Cheng, Jian Yi, Zhigang Fang, Sheng Dai, Xianrui Zhao 2013*Material Transactions* **1**, p. 50-55

[4] Qianlin Wu, Wenge Li 2011 *Material Transactions* 52 3, p. 560-563

[5] Parisa Farahmand, Shuang Liu, Zhe Zhang, Radovan Kovacevic2014*Ceramics International* **40**, p. 15421-15438

Journal of Physics: Conference Series **747** (2016) 012063 doi:10.1088/1742-6596/747/1/012063

# Analysis of a multi-loop temperature stabilization system for compact rubidium standard of frequency and time

**S V Norkin[1,3] and V N Reshetov[2]**

[1]National Research Nuclear University MEPhI, Kashirskoe sh. 31, Moscow, 115409, Russia
[2]Technological Institute for Superhard and Novel Carbon Materials, Centralnaya str. 7a, Troitsk, Moscow, 142190, Russia

E-mail: power.knowleg@yandex.ru

**Abstract.** A possible design of a thermal stabilization system for compact atomic clock based on coherent population trapping is analyzed. The feasibility of independent control of the semiconductor laser's characteristics, such as wavelength and optical power, is considered. An algorithm is suggested for choosing the performance parameters of various control circuits, eliminating the occurrence of self-oscillation mode.

## 1. Introduction

Development of methods for precise frequency and time measurements was crucial for progress in science and technology. The researches devoted to standards of time are closely connected with achievements in fundamental physics and global telecommunication networks. Improvement of frequency standards is important for understanding many physical phenomena and searching for new approaches in technological development.

At the same time, simultaneously with the creation of extremely accurate and stable quantum clock, the work on designing the compact ultra-precise atomic clocks is progressing. To meet the needs of technology, Miniature Atomic Clock (MAC) on $^{87}$Rb atoms with relative long-term stability better than $10^{-11}$ are being developed [1-5] and are already commercially available [6,7].

## 2. A method for stabilizing the probe laser

Miniature Atomic Clock based on the effect of coherent population trapping (CPT) contains a number of elements requiring precise temperature maintenance. A generalized scheme of the MAC including the adjustment elements is shown in Figure 1, taken from [8]. An atomic cell with rubidium vapor is thermally stabilized and used as a quantum discriminator for RF generator. The wavelength of the semiconductor laser can be adjusted by heating. A separate loop is used to stabilize the reference crystal oscillator. The external thermostat regulates the temperature of the MAC case.

Advances in miniaturization of the rubidium frequency and time standard are associated with the use of semiconductor lasers for generating optical emission used to stabilize the radio-frequency signal with CPT effect in rubidium vapor. The appearance of high speed single-mode semiconductor lasers with a vertical structure of VCSEL allowed creating a compact version of MAC in the form factor of

[3] To whom any correspondence should be addressed.

conventional high stable crystal oscillators. The long-term stability of this atomic clock is several orders higher than the highest achieved for quartz oscillators.

The basic problem concerning thermal stabilization of the separate units of MAC is taking into account the mutual influence of the different control systems on each other. In this case, only part of the systems operates using the thermal sensors data, while the others rely on the optical effects and receive the control signal from the quantum discriminator.

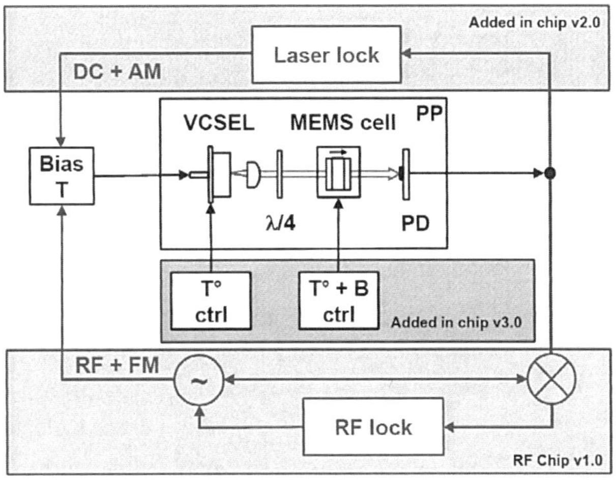


**Figure 1.** The generalized scheme of one of MAC options

For example, to adjust the laser wavelength, an effect of the optical absorption of light in the rubidium vapor cell is used. To fine tune to the absorption line, the frequency of the laser emission is changed by several hundred MHz with a frequency of tens of kHz, by modulating the current through the VCSEL. The signal received by the photodetector is detected by synchronous detector and enters the control circuit of the Peltier element that changes the average temperature of the laser diode. The dependence of the wavelength λ on the current $I$ and temperature $T$ in the first-order approximation is linear and can be described by equation (1) with typical values $k_i = 0.6\ ^{\mathrm{nm}}/_{\mathrm{mA}}$, $k_t = 0.06\ ^{\mathrm{nm}}/_{\mathrm{K}}$:

$$\lambda = \lambda_0 + k_i(I - I_0) + k_t(T - T_0) \tag{1}$$

This type of dependence holds for modulation frequency of the laser current much lower than 3.4 GHz, at which the laser light is modulated to bind to the CPT resonance. Linear relationship between the wavelength of the laser and the current and temperature is due to thermal expansion of the laser cavity caused by its heating.

Thus, laser power in a first-order approximation is linearly dependent on the current and temperature. This relationship is shown in equation (2) where the coefficients have the following values: $\eta_s = 0.3\ ^{\mathrm{mW}}/_{\mathrm{mA}}$, $\eta_t = 0.01\ ^{\mathrm{mW}}/_{\mathrm{K}}$.

$$P = P_0 + \eta_s(I - I_0) - \eta_t(T - T_0) \tag{2}$$

Solving equations (1) and (2) can yield the conditions under which the simultaneous variation of temperature and current leaves unchanged the wavelength or power of the laser. To stabilize the wavelength when the current changes, it is necessary to have $\Delta T = (-k_i \cdot k_t^{-1})\,\Delta I$, and to stabilize the power when the temperature changes, the correction current is required: $\Delta I = (\eta_t \cdot \eta_s^{-1})\,\Delta T$. Both these conditions, apparently, are not equivalent since $(\eta_t \cdot \eta_s^{-1})(k_i \cdot k_t^{-1}) \neq 1$, and it is generally impossible to stabilize the power and the wavelength of the laser diode simultaneously. However, the very possibility of such modulation of solely amplitude or frequency is important and can be used to optimize the algorithm of the MAC.

## 3. Analysis of a multi-loop system of thermal stabilization

When creating MAC, different variants of thermal stabilization of the cell and laser are used. Examples of such circuits, leading to long-term stability of various MACs, are shown in Figure 2, taken from [9].

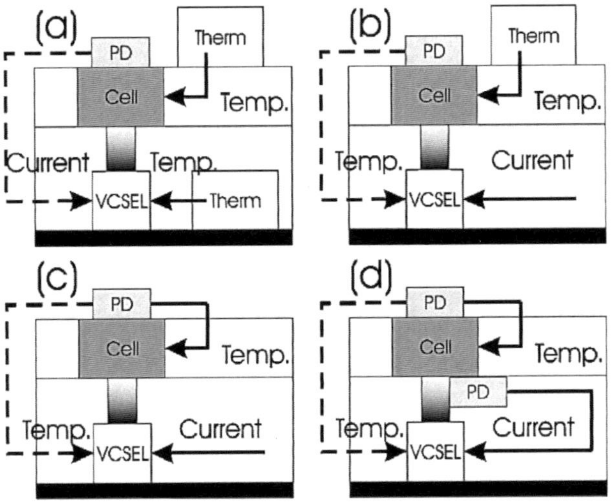


**Figure 2.** Experimental setup for chip scale atomic clock parameter control. Therm – thermistor; Temp. – temperature control; Current – current control; PD – photodiode; VCSEL – vertical-cavity surface-emitting laser, solid lines – dc signals; dashed lines – phase detection signals. (a) Conventional setup. (b) Laser frequency is controlled by temperature feedback. (c) In addition to (b), cell temperature is controlled by atomic absorption. (d) In addition to (c), laser intensity is controlled by use of a photodetector.

A fairly full review of the basic approaches to the design of MAC is given in [10-12]. As temperature sensors in such devices thermistors are traditionally used, having a high sensitivity to temperature change. The exponential resistance versus temperature dependence makes these sensors greatly nonlinear and complicates creating analog thermoregulation systems with optimal speed and accuracy. The resistance of the thermistor and hence its sensitivity changes by several times (usually falls from 10 kOhm to 2 kOhm) while MAC reaches its operating conditions, that is becomes heated from 20°C to 60°Ctimes. Such a significant change of the thermal control system parameters requires separate modes of operation for the first turning on and then for consequent holding the operating parameters. The most advisable in such cases is to make a digital temperature control system capable of adapting to the conditions of work and algorithmically suppressing the oscillations emerging in the system.

The fastest channel is the current control channel of the laser and its time constant in terms of heating the crystal is typically less than a few microseconds. Next in terms of speed of response is the Peltier element that regulates average temperature of the laser. The time constant of this element integrated in the VCSEL structure is a few tenths of a second. Then there are two systems with approximately the same operating speed – the base of heating, where a Peltier element with a laser are located, and an optical cell with rubidium vapor. Performance of these heaters varies from tens to hundreds of seconds, depending on the dimensions of the elements used. The most inertial system of the MAC is a general thermal stabilization circuit of casing as a whole. Here, using a thermal insulation system of the MAC from the environment, characteristic time of the transition process can reach several thousand seconds. Crystal oscillator, essential in any MAC, has its own, sometimes double-loop, thermal stabilization system, and in this case operates as independent system, disturbing the thermal equilibrium of the MAC stabilization system. Actually, it is a simple source of heat, whose power depends on the ambient temperature of the crystal oscillator.

Using existing data, following approach to the implementation of sustainable multi-loop temperature control system for MAC can be proposed. Based on a clear hierarchy of the response times of different sub-systems, the method of frequency division of signals can be used, and the whole system can be designed so that the effect of low-frequency control systems on the higher-speed ones occurs in the plateau region of the amplitude-frequency response characteristic of the higher-speed systems. In general, the system of equations describing the operation of a multi-loop interconnected temperature control system can be determined by the equation (3):

$$\dot{T}_i = b_i P_i + q_i T_i + \sum_1^k a_{ij}(T_j - T_i), \quad (3)$$

where $T_i$ – deviation of temperature of the $i$-th element from the case temperature, $P_i$ – power of the $i$-th heater, $q_i$ – coefficient characterizing the thermal conductivity of the $i$-th element to the casing of the device, $b_i$ – coefficient characterizing specific heat of the $i$-th element, $a_{ij}$ – coefficients describing mutual influence of various elements on each other's system.

This system of equations must be supplemented with the equations describing the behavior of specific thermal regulators which in the simplest cases are represented by the PID regulators used to control the power of the heater [13]. Purpose of the PID controller is to maintain the setpoint $T_0$ of a certain quantity $T$ by changing another quantity $u$. The value of $T_0$ is a predetermined value, and the difference $e = (T_0 - T)$ is a regulation error. Controller output is given by equation (4):

$$u(t) = P + I + D = K_p e(t) + K_i \int_0^t e(\tau)d\tau + K_d \frac{de}{dt} \quad (4)$$

where $K_p$, $K_i$, $K_d$ – respectively, the proportional, integral and derivative gain components of the controller.

Introducing the PID controllers into consideration allows optimizing the transition process for each of the control channels and ensuring its sustainability. In this case the behavior of the individual control loop is described by the differential equation of the second order, and the analysis of its stability does not represent any particular problems. In simple systems, each PID controller operates independently of the other elements of the system, and the mutual influence is manifested in the physical layer through the terms of the form $\sum_1^k a_{ij}(T_j - T_i)$. However, in the case of discussed multi-loop control system, this approach is not the productive one. When creating digital temperature control system, it is more rational to use all information available to the controller to manage the system as a whole. However, to properly configure such a system one needs detailed information about all the relationships, and the only opportunity to gather it is during the testing of the finished product. Considering the strict requirements on Allan dispersion and fluctuations range of the reference frequency signal generated by the MAC, it is possible to propose the following temperature control circuit for basic operational components of atomic clock.

Using LED current (and thus its temperature) modulation, the optical wavelength tuning is carried out, and the signal of absorption line capture is formed. The data from this very fast system of regulation is used to govern the Peltier element controlling the average temperature of the laser. The time constant of the optical wavelength capture system may be of the order of 1–10 ms, which is significantly less than the Peltier element response time of 0.1–1 s. Tenfold margin of the cutoff frequency corresponding to the frequency response ensures the stability of the control loop.

The temperature of the substrate, where the Peltier element regulating the wavelength of the laser is situated, and the temperature of the optical cell with rubidium vapor can be the same in the working MAC version, but in any case, the information about the required temperature data values is taken from the thermistors and used in the control circuit with a characteristic speed 10–100 seconds. The proximity of time constants of these control objects indicates the inexpediency of organizing a parallel operation of these circuits. Therefore, based on the data of both thermistors (when entering the operating mode) and using only the temperature of the cells for the fine tuning, this unit can be made single, and a distributed heater can be used to reduce the possibility of temperature fluctuations in the quantum discriminator. This control loop based on the Peltier element and the loop of cells are well spaced in their typical response times (more than 10 times) and do not interfere with each other.

The circuit involved in the maintenance of the working temperature of the MAC casing in this case performs a purely subsidiary function, accelerating the system entering the operating mode and facilitating the work of the internal systems of fine thermal stabilization. Its main purpose is to reduce the external temperature drops, reaching tens of degrees, down to a value not exceeding a degree Celsius.

## 4. Conclusion

The analysis shows the importance of performance of the control channels included in the MAC and indicates the need to measure the response of the system as a whole and its individual components, in particular, to the work of each individual regulatory system. Since the performance between parts of a multi-loop control system differs by tens and hundreds of times, the sustainability of the whole system can be ensured by the control of the specific parameters of the PID controllers.

## 5. Acknowledgments

The present work was supported by a research grant from Russian Ministry of Education and Science 14.577.21.0088 (unique identification number RFMEFI57714X0088).
This work was also supported by Competitiveness Growth Program of the Federal Autonomous Educational Institution of Higher Professional Education National Research Nuclear University MEPhI (Moscow Engineering Physics Institute).

## References

[1] Lutwak R, Vlitas P, Varghes M, Mescher M, Serkland D K, Peake GM 2005 *Freq. Cont. Symp.* IEEE 752-57
[2] Knappe S, Shah V, Schwindt P D D, Hollberg L, Kitching J, Liew L A, Moreland J 2004 *Appl.Phys. Lett.* **85 (9)** 1460-62
[3] Guo T *et al.* 2009 *Applied Physics letters* 94
[4] Boudot R *et al.* 2011 *Journal of Applied Physics* 109
[5] Zibrov S *et al.* 2010 *Physical Review* A 81
[6] http://www.microsemi.com/products/timing-synchronization-systems/embedded-timing-solutions/components/sa-45s-chip-scale-atomic-clock#overview
[7] Deng J et al. 2008 A commercial CPT rubidium clock *European Frequency and Time Forum*
[8] Haesler J, Bennès J, Overstolz T, Pierer J, James RJ, Ruffieux D , Lecomte S 2012 *European Frequency and Time Forum*
[9] Gerginov V, Shah V, Knappe S, Hollberg L and Kitching J 2006 *Optics letters* **31(12)**
[10] https://physique.cuso.ch/fileadmin/physique/document/2014_Mileti_atomic_4.pdf
[11] http://www.ion.org
[12] http://www.eftf.org
[13] Poles K Yu 2008 Theory of automatic control for the "Dummies"

Journal of Physics: Conference Series **747** (2016) 012064 doi:10.1088/1742-6596/747/1/012064

# Properties of multilayer coatings produced by coaxial laser cladding

**V N Petrovskiy[1], D P Bykovskiy, P S Dzhumaev, V I Polskiy, N M Prokopova and S N Chirikov**

National Research Nuclear University MEPhI, Department of Laser Physics, Kashirskoe highway 31, Moscow, Russia, 115409

E-mail: VNPetrovskij@mephi.ru

**Abstract.** This article contains results of the study of multilayer coatings produced by laser cladding on the substrate steel 34HMA using iron based powder PR-10R6M5 as the filler material. The coatings were produced with consistent application of the tracks with fixed overlapping. The dependencies between the characteristics of tracks and the technological mode of deposition were revealed. Properties of coatings were determined for various overlapping of tracks and directions of the cladding layers.

## 1. Introduction

Laser cladding is an additive technology for materials processing. It is an actively developing processing technique, which is penetrating to the industry. This technology allows to create coating on the surface of the workpiece to restore the damaged parts whose manufacturing is time-consuming and expensive. In addition, claddings are used to impart desirable properties to the workpiece surface, e.g. corrosion or wear resistance. The destruction of parts often starts with surface defects [1], thereby laser cladding technology allows to improve part characteristics by creating layer with the desired properties on the surface.

Apart from a laser, the energy source for cladding may be an arc, electron beam, plasma and others. The laser cladding features the possibility of creating coatings with minimal heat affected zone (HAZ) and minimal mixing of filler material with the substrate [1]. Wires, ribbons, or powders may be used as a filler material. Cladding using a metal powder allows to achieve maximum uniformity of coating, enables local surface treatment and producing any existing coating by mixing the powders. Other advantages of laser cladding of powders were analyzed in more detail in [2–6].

## 2. Experimental equipment

Experimental equipment is shown in figure 1. Cladding processing cell (1) is equipped with a control unit (2). Figure 2 shows the inside of the cell (1) consisting of a 5-axis positioning system, optical head with water cooling and supply of protective gas and powder from the feeder (3) SulzerMetco Twin 10-C, as well as an industrial digital camera for positioning parts and monitoring the process. As an energy source we used fiber laser (4) LS-3.5 with power up to 3.5 kW, manufactured by IPG.

The metal powder is fed into the treatment zone via a coaxial nozzle. Under the impact of laser radiation it is heating and melting. Simultaneously, the substrate surface is melting. The track is produced during the motion of the nozzle relative to the substrate after cooling. New layers are formed

[1] To whom any correspondence should be addressed.

by sequential application of tracks with some overlapping. Workpiece of the substrate was prepared for cladding before the experiment. First, the cleaning of the working surface of samples with a wire brush was performed. Then the work surface was treated by acetone wiping.

**Figure 1.** Experimental equipment.

**Figure 2.** View of optical head and positioning system inside the cell.

## 3. Results of the cladding study

### *3.1. One track*

Initially, we produced cladding of tracks with different technological modes and determined their geometric characteristics: width and height of the track, and the depth of penetration of filler material in the substrate. The measurements showed that the width of the cladding track and the filler material penetration depth in the substrate are mainly influenced by a change of the laser power, while the height of the tracks is mostly affected by the powder feed rate. We selected the most appropriate mode for further research based on the following parameters: uniformity of the track form, the absence of pores and cracks, the minimum level of mixing between substrate materials and powder, and a minimum depth of penetration of filler material in the substrate combined with good adhesion of the coating without delamination from the substrate.

Optimal mode for a single track cladding was selected in accordance with all the above mentioned requirements: laser power 400 W, substrate movement speed 37.5 cm/min, powder feed rate 1 g/min, and the height of the nozzle above the surface of 5 mm. We carried out a study of topography and microstructure of the surface as well as the elemental composition of the samples using the electron microscope (SEM) Carl Zeiss EVO 50 XVP. The microhardness was determined by the Vickers method in the HVS-1000 setup with automated loading of the indenter to the load of 1 N. The dwell time under the load has been fixed at 20 s. The results showed (figure 3) that the HAZ is 300 μm, cladding structure is the equiaxed mesh, dendritic structure is observed near the substrate boundary, and the grain boundaries are enriched in heavy elements (Mo, W). The structure of the HAZ is a milled grain ferrite-pearlite with traces of partial martensitic transformation. The structure of the substrate is the polyhedral (equiaxed) grains of ferrite and pearlite. The substrate-cladding boundary is distinct, there is no peeling, cracking, or porosity. Microhardness (figure 4) changes at the boundary of the substrate and cladding, from 5700 MPa (cladding) to 2000 MPa (substrate material).

Figure 3. Single track produced in optimal regime.

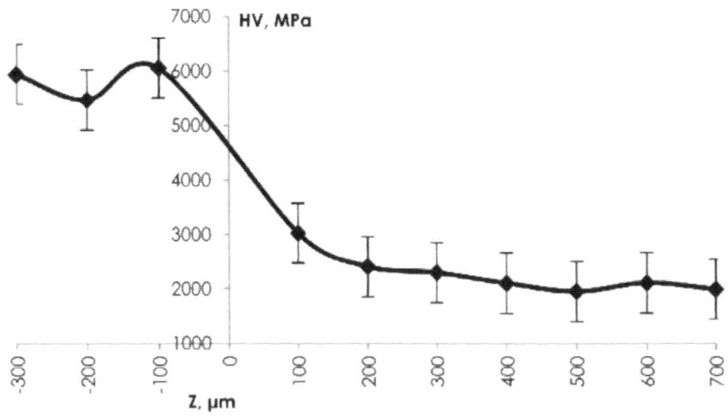


**Figure 4.** Change of microhardness over the cross section.

### *3.2. Single layer coatings*
The layers were created under selected track mode with different overlapping: 0.33, 0.5, 0.66. We conducted a similar study of the microstructure and microhardness of coatings. For all coatings with different surfacing overlapping, the structure is equiaxed mesh, near the boundary with the substrate the dendritic structure is observed, grain boundaries are enriched in heavy elements (Mo, W). The structure of the HAZ is a milled grain ferrite pearlite with traces of partial martensitic transformation. The boundary between the substrate and cladding is distinct, there is no peeling, cracking, or porosity. Microhardness changes at the boundary with the cladding and the substrate from 5000 MPa (cladding) to 2000 MPa (substrate material). The Fe content in the powder is 80%, and about 98% in the substrate. The measurements of the Fe content in the cladding, depending on the coefficient of overlapping shows that a layer with overlapping 0.33 has 86% content of Fe, with overlapping 0.5 – 85%, and with overlapping 0.66 – 83% . It can be seen that increasing overlap reduces the content of Fe in the coating, thus decreasing the mixing of materials and improving the quality of the cladding. Iron content in a layer depends on the strategy of its preparation.

### *3.3. Two layer coatings*
The layers were created with the same mode of overlapping the tracks, but with a different direction of the second layer cladding: 0°, 45°, 90°. Structure of coatings is presented in figures 5–7. It is a non-equiaxed mesh, with the dendritic structure near the boundary with the substrate -, and the grain boundaries enriched in heavy elements (Mo, W). The width of the HAZ is 300 µm, its structure is the crushed grain ferrite-pearlite with traces of partial martensitic transformation. The iron content in the three cases was 85%. The microhardness increased to 6000 MPa (figure 8), its dependence is more uniform in the case with the direction of the clad layers of 0°. Since the remaining measured coating characteristics are the same, one can conclude that the best quality of cladding is obtained when depositing layers at 0° angle.

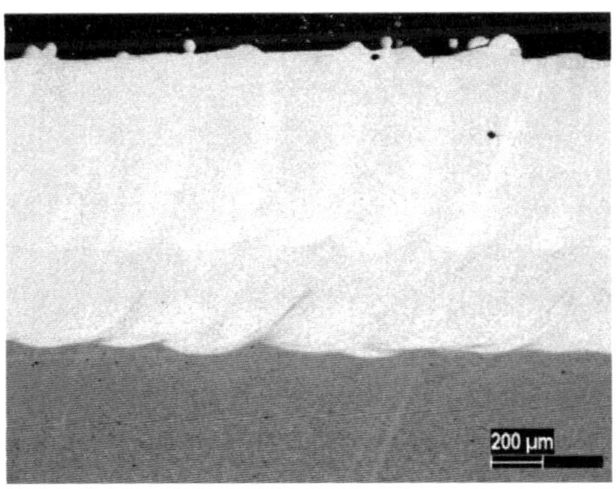

**Figure 5.** Bilayer with direction of second layer 0°.

**Figure 6.** Bilayer with direction of second layer 45°.

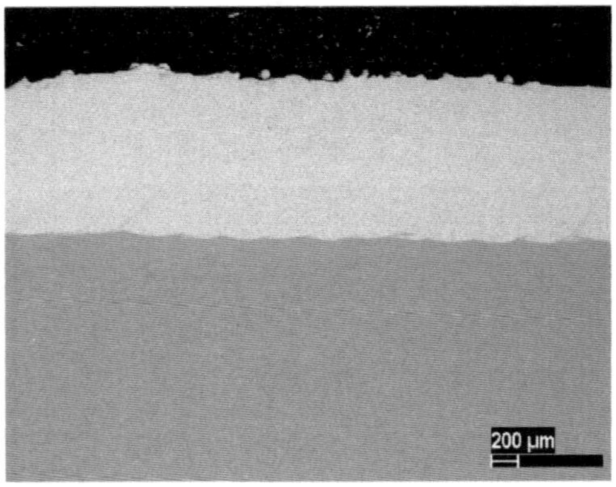


**Figure 7.** Bilayer with direction of second layer 90°.

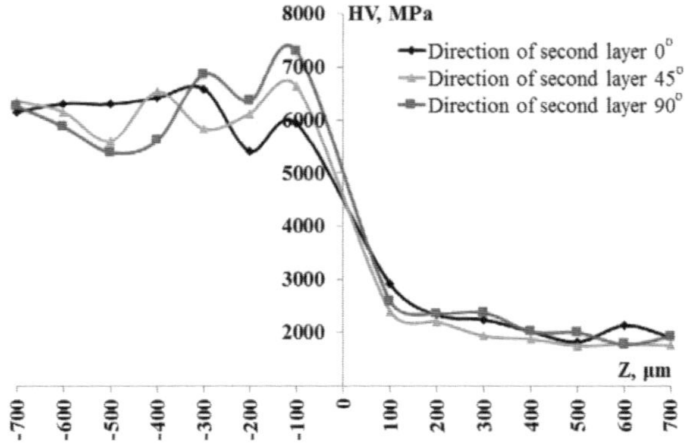


**Figure 8.** Change of microhardness over the cross section for bilayer coatings.

## 4. Conclusion

Results of the study showed that the width of the cladding track and the penetration depth of filler material in the substrate are mainly influenced by a change of the laser power, while the height of the tracks is mostly affected by the powder feed rate. Monolayer coatings with a coefficient of tracks overlapping 0.66 exhibit lower level of mixing between powder material and the substrate than the coatings with coefficients 0.33 and 0.5. This is a consequence of smaller area of thermal effects on the substrate. Bilayer coatings with the direction of the clad layers 0° are more uniform than those with a direction angle of 45° and 90°. Differences in the ratio of mixing for the three different directions are not observed.

**Acknowledgments**

This work was financial supported by the Ministry of Education and Science of Russian Federation (unique identier PNIER RFMEFI58214X0004).

This work was also supported by Competitiveness Growth Program of the Federal Autonomous Educational Institution of Higher Professional Education National Research Nuclear University MEPhI (Moscow Engineering Physics Institute).

## 5. References

[1] Grigoryants G, Shiganov I, and Misyurov I 2006 *Technological processes of laser processing* (Moscow: MGTU) (In Russian)

[2] Murzakov M, Petrovskiy V, Birukov V, Dzhumaev P, Polski V, Markushov Y, and Bykovskiy D 2015 *Phys. Procedia* **71** 202–206

[3] De Damborenea J and Vzquez A 1993 *J. of Materials Science* **28** 4775–4780

[4] De Oliveira U, Ocelk V, and Hosson J D 2005 *Surface and Coatings Technology* **197** 127–136

[5] Nenadl O, Ocelk V, Palavra A, and Hosson J T D 2014 *Phys. Procedia* **56** 220–227

[6] Hasui A and Morigaki O 1985 *Surfacing and spraying* (Moscow: Mashinostroenie) (In Russian)

Journal of Physics: Conference Series **747** (2016) 012065 doi:10.1088/1742-6596/747/1/012065

# NMR Express-analyser for quality monitoring of motor fuel

**E A Protasov[1], D E Protasov**
National Research Nuclear University MEPhI, Kashirskoe sh. 31, Moscow, 115409, Russia

E-mail: protasovea@mail.ru

**Abstract**. A method for the rapid analysis of motor fuel quality was developed by artificial increase of the octane number through dissolving ferrocene in a low-octane gasoline ($C_{10}H_{10}Fe$). Measurements of the spin-lattice relaxation time of nuclear magnetic resonance is used for determination of ferrocene presence in standardized and real fuel from gas stations. The results of measurements of the relaxation characteristics among certain grades of motor fuel with dissolving ferrocene therein are presented.

## 1. Introduction

Problem of motor fuel quality at gas stations becomes very important since the presence of heavy metal salts and iron in gasoline leads to failure of the expensive machinery and environmental deterioration. Standard laboratory techniques, which are used in oil refining, usually give results not earlier than 2 to 3 hours. In this paper it is proposed to conduct a rapid analysis of the fuel quality in 3 to 4 minutes with an instrument, based on the measurement of the spin-lattice relaxation time of protons ($T_1$) of nuclear magnetic resonance (NMR), which has high precision and reproducibility of the obtained data and small measurement duration.

It is known that by mixing two substances with different viscosity or between molecules which have a chemical interaction, the spin-lattice relaxation time $T_1$ decreases versus time $T_1$ in the pure liquid. Note that this time also depends on other factors, such as type of nucleus, the magnetic field strength, mobility of spins (microviscosity), and temperature. Furthermore, due to the presence of the magnetic moments of atoms and molecules, real substances have paramagnetic properties. A particularly strong influence on the time $T_1$ have paramagnetic ions that may be contained in materials, through which the octane number of gasoline can be increased. As it is known, a magnetic moment of a paramagnetic ion is of the order of the Bohr magneton, and almost $10^3$ times greater than nuclear magnetic moment. Paramagnetic ions ($Mn^{2+}$, $Cu^{2+}$, $Fe^{2+}$, $Gd^{3+}$) contain unpaired electrons whose reorientation creates a very strong fluctuating magnetic field that leads to a substantial decrease of the relaxation time. In this case, even a small concentration of paramagnetic ions in solutions containing such ions has a significant impact on the value of the spin-lattice relaxation time $T_1$ NMR. For example, the content of 0.01 mol of iron salt in a solution reduces the magnitude of $T_1$ compared with deoxygenated water from 3.6 to 0.01 s [2]. Thus, the effect of the paramagnetic ions in a liquid sample may significantly reduce the spin-lattice relaxation time.

In practice, ferrocene ($C_{10}H_{10}Fe$) is commonly used to increase the octane number of fuel by dissolving it into the fuel with a lower octane number. Ferrocene is the cheapest and most effective means for the octane number magnification. In order to increase the octane number by 4 to 5 units, it is enough to add 1 t of gasoline to 170 g of ferrocene.

[1] To whom any correspondence should be addressed.

In this paper, studies on influence of the dissolved ferrocene concentration in pure gasoline Ai-92 and Ai-95 on the spin-lattice relaxation time were carried out with the help of NMR relaxometer [3, 4], which was described in [2]. Ferrocene was used for studies in which the content of free iron was about 0.03%, and the contents of other substances insoluble in gasoline did not exceed 0.5%.

## 2. Experimental procedure

In this paper measurement of the spin-lattice relaxation time $T_1$ is performed by magnetic resonance absorption at the double modulation of the magnetic field by sawtooth pulses of low frequencies and sinusoidal vibrations of sound frequencies. The use of dual modulation can significantly increase the signal to noise ratio and thereby reduce error in determining the time $T_1$.

Signal amplitudes arising from the implementation of the nuclear resonance condition are measured in the device. This condition is described by equation of Larmor:

$$f = \gamma \cdot H_0 , \quad (1)$$

where $f$ – frequency of the HF generator, $\gamma$ – gyromagnetic ratio (for a proton $\gamma$ is equal to 42.58 MHz/T), $H_0$ – the constant magnetic field strength, in this case amounted to 0.37 T.

Information about the spin-lattice relaxation time $T_1$ is retrieved from the formula:

$$T_1 = \tau \left( -\ln \left(1 - \frac{A(\tau)}{A_1}\right)\right)^{-1} \quad (2)$$

where $A_1$ – amplitude of the first signal (saturation), which is always greater than the second signal and remains the same during the measurement, $A(\tau)$ – amplitude of the second signal, $\tau$ – time between a pair of signals (see figure 1).

The range of values of the first signal determines the measurement error $T_1$. In contrast to the signal processing method described in [3, 4], in this research the period between a pair of absorption pulses was set from 0.9 to 1.0 s.

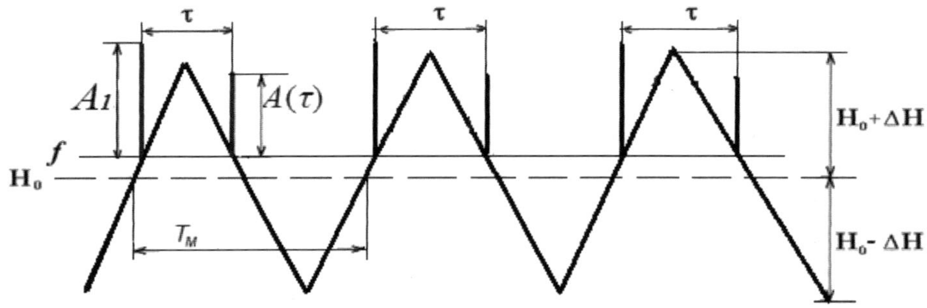


**Figure 1**. Form of absorption signals at a fixed frequency of HF generator. At this frequency, period between absorption signals was set at about 1.0 s. $T_M$ is a period of triangular modulation.

One of the conditions for optimum operation of the device is that triangular modulation period should be from 3 to 5 times larger than the spin-lattice relaxation time $T_1$, i.e. $T_M = (3–5)\ T_1$. Inside the couple of absorption signals, the time delay must be less or comparable to $T_1$. Then all the information about changing of the second signal amplitude goes to a computer where a specially designed program starts processing of this data array. The relaxation time is determined by a number of a predetermined amount of measurements (up to 30), averaged and displayed at the measurement results panel (see figure 2).

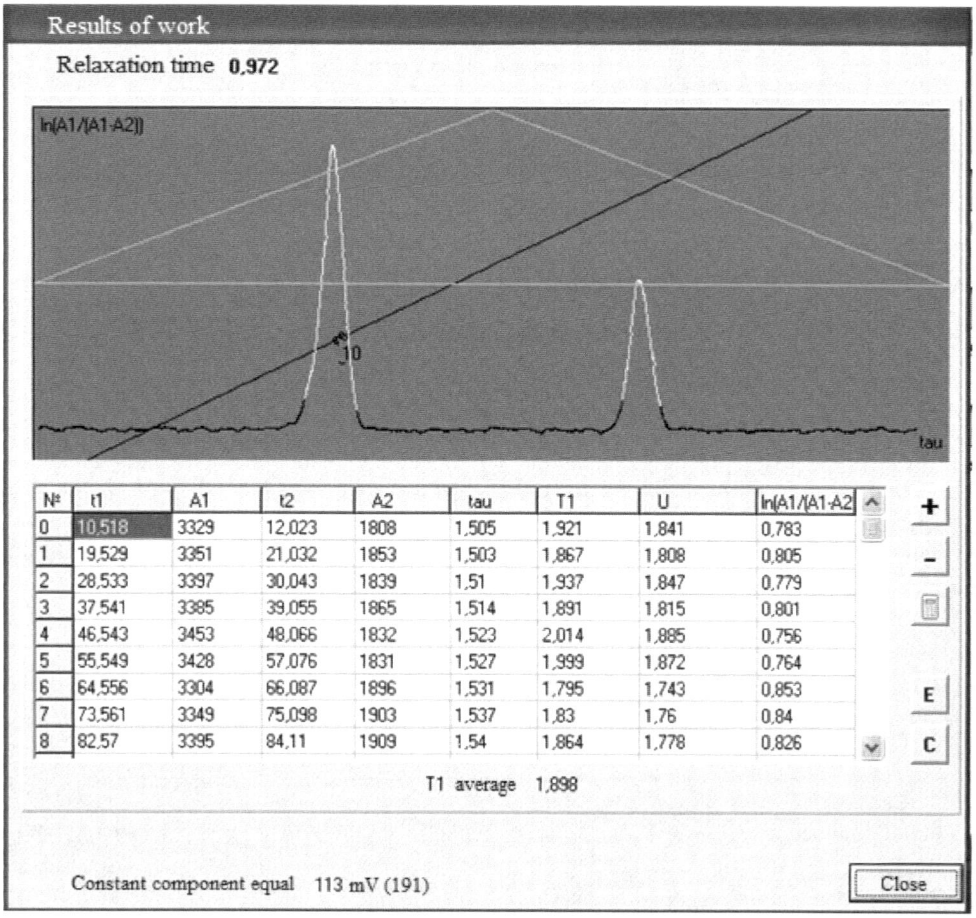


| № | t1 | A1 | t2 | A2 | tau | T1 | U | ln(A1/(A1-A2 |
|---|---|---|---|---|---|---|---|---|
| 0 | 10,518 | 3329 | 12,023 | 1808 | 1,505 | 1,921 | 1,841 | 0,783 |
| 1 | 19,529 | 3351 | 21,032 | 1853 | 1,503 | 1,867 | 1,808 | 0,805 |
| 2 | 28,533 | 3397 | 30,043 | 1839 | 1,51 | 1,937 | 1,847 | 0,779 |
| 3 | 37,541 | 3385 | 39,055 | 1865 | 1,514 | 1,891 | 1,815 | 0,801 |
| 4 | 46,543 | 3453 | 48,066 | 1832 | 1,523 | 2,014 | 1,885 | 0,756 |
| 5 | 55,549 | 3428 | 57,076 | 1831 | 1,527 | 1,999 | 1,872 | 0,764 |
| 6 | 64,556 | 3304 | 66,087 | 1896 | 1,531 | 1,795 | 1,743 | 0,853 |
| 7 | 73,561 | 3349 | 75,098 | 1903 | 1,537 | 1,83 | 1,76 | 0,84 |
| 8 | 82,57 | 3395 | 84,11 | 1909 | 1,54 | 1,864 | 1,778 | 0,826 |

**Figure 2.** The measurement results panel.

In the case of such device realization in the automotive industry and energy sector, it is supposed to adhere to the following algorithm of device operation. Amplitude data and information about the spin-lattice relaxation time ($T_1$) of fuels, which correspond to their standardized quality at a suitable temperature, are programmatically entered in a computer.

The device automatically measures the specified parameters of the real fuel, compares them with the ideal fuel parameters, which are set in the program, and if they do not match the standard, a warning signal is triggered. Differences of the characteristics of the fuel from the standard can be fixed with an accuracy of ± 5%. Device control and signal processing is performed using the built-in processor, excluding the manual adjustments.

## 3. Measurement results

To check the correct operation of relaxometer, test measurements of time $T_1$ of distilled water were conducted. This time period was equal to 2.42 s. The actions have been taken to exclude the fuel evaporation during conduction of the measurement. About 3.0 ml of the fuel was required for this analysis. The measurements were performed three times by 30 points for each fuel grade, and then the average $T_1$ was calculated. Table 1 shows the measurement data regarding time $T_1$ at the temperature of 21°C for various grades of fuel produced at different gas stations in Moscow and Moscow area.

**Table 1.** The spin-lattice relaxation time dependence on fuel type

| Fuel type | $T_1$, s |
|---|---|
| Diesel fuel | 0.78±0.05 |
| KO brand Kerosene | 0.93±0.05 |
| Kerosene “Ekspohimtreyd” | 1.15±0.05 |
| Gasoline “The galosh” | 2.45±0.05 |
| Gasoline Ai-95 (2015 year) | 3.16±0.07 |
| Gasoline Ai-95 (ten years old) | 2.54±0.07 |
| Gasoline Ai-92 (2015 year) | 3.62±0.07 |
| Gasoline Ai-92 (ten years old) | 2.40±0.05 |

As it can be seen from the table, the relaxation time decreases significantly with the increase of fuel storage period. Presumably, it can be associated with the fact that during long storage period various photochemical processes can occur, and also evaporation of the fuel's light fraction leads to the increase of viscosity which consequently decreases the relaxation time.

To measure the influence of ferrocene concentration on the spin-lattice relaxation time $T_1$, the appropriate amount of ferrocene was dissolved in 20 g of gasoline Ai-92.

Figure 3 shows a graph of time $T_1$ dependence on the concentration of ferrocene in gasoline Ai-92 and Ai-95.

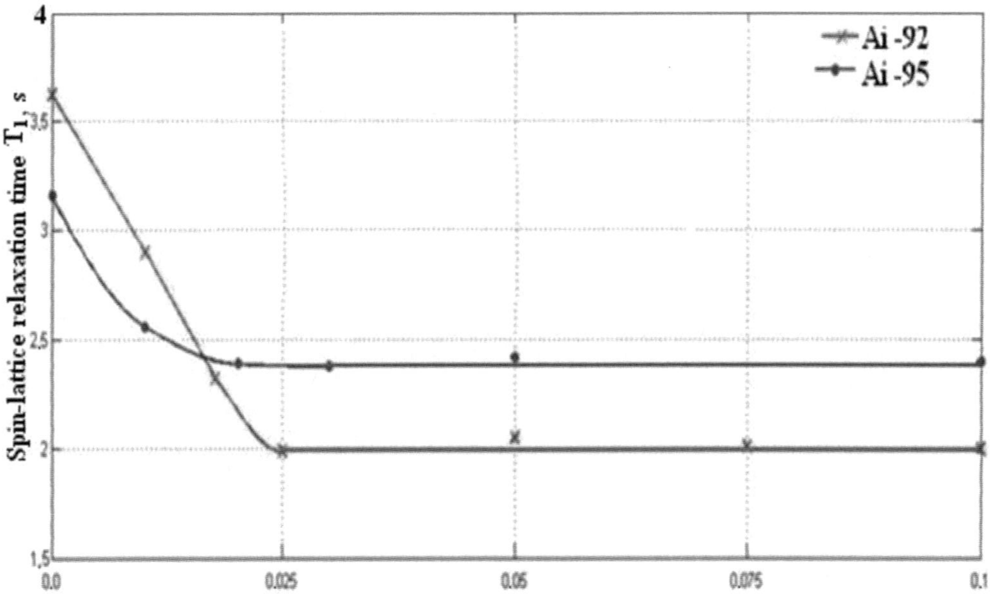


**Figure 3.** Spin-lattice relaxation time $T_1$ dependence on the concentration of ferrocene dissolved in gasoline Ai-92 and Ai-95.

Comparing the data for gasoline Ai-92 and Ai-95, it can be concluded that in measured gasoline Ai-95, ferrocene probably was already present.

As it can be seen from the graph, at ferrocene concentration of 0.025% and higher, the relaxation time $T_1$ within the margin of error ceases to depend on the amount of ferrocene, i.e. there is a saturation effect. Obtained error is mainly associated with the inaccuracy of components' weighing. Nature of the saturation effect is unknown yet and requires further research. Presumably, in this case the presence of free atoms of iron in ferrocene matters, which in its turn can prevent heat exchange between nuclear moments and the lattice.

Thus, the time difference of the spin-lattice relaxation time between pure Ai-92 gasoline and gasoline dissolved therein with ferrocene concentration even less than 170 grams per ton of gasoline is 1.5 s. This fact allows to implement a reliable detection of this substance presence in gasoline.

## Acknowledgments

In conclusion, the authors thank the commercial director of "Company Fuel Region" Yanchenkov A S for provided ferrocene.
This work was supported by Competitiveness Growth Program of the Federal Autonomous Educational Institution of Higher Professional Education National Research Nuclear University MEPhI (Moscow Engineering Physics Institute).

## References

[1] Vashman A A, Pronin A S 1986 *Nuclear magnetic relaxation spectroscopy* (Moscow: Energoizdat)
[2] Lesha A 1963 *Nuclear induction* Edited by P.M.Borodina.
[3] Bondarchuk E Y, Komarova S E, Protasov E A 2010 *Medical Physics* **4 (48)** 65-70
[4] Dudova D, Protasov D E, Protasov E A 2013 *Applied Magnetic Resonance* **44 (7)** 781-790

Journal of Physics: Conference Series **747** (2016) 012066 doi:10.1088/1742-6596/747/1/012066

# Effects of laser interaction with living human tissues

**O E Molchanova, E A Protasov[1], D E Protasov and A V Smirnova**
National Research Nuclear University MEPhI, Kashirskoe sh. 31, Moscow, 115409, Russia

E-mail: protasovea@mail.ru

**Abstract.** With the help of a highly sensitive laser device with the wavelength $\lambda$ = 0.808 μm, which is optimal for deep penetration of the radiation into biological tissues, the effects associated with the appearance of uncontrolled human infrasonic vibrations of different frequencies were investigated. It was established that the observed fluctuations are associated with the vascular system which is characterized by its own respiratory movements, occurring synchronously with the movements of the respiratory muscles, the operation of the heart muscle, and the effect of compression ischemia. The effect of "enlightenment" of a tissue is observed with stopping of blood flow in vessels by applying a tourniquet on the wrist.

## Introduction

It is known that during the passage of optical radiation through living human biological tissues, which are penetrated by blood vessels, there are signals related to heart operation. These periodic vibrations of volume of blood vessels are associated with heartbeat functioning due to the dynamics of blood filling during one cardiac cycle. Furthermore, in the human body mechanical vibrations of the human body surface can be generated (infrasonic waves) [1]. Own acoustic radiation is limited in the short-wave area by ultrasonic radiation. Low frequency radiation (frequencies below $10^3$ Hz) is generated by physiological processes like heartbeat, breathing movements, blood flow in blood vessels and other processes, which illustrate variations of the human body surface within the range of about 0.01–$10^3$ Hz.

Underlying methods of registration of the human body radiation and its own physical fields are important for medical diagnostics because they are non-invasive. Therefore, the task of measuring and research of human infrasonic fields may be of a great interest.

## Experimental technique

Most often, the pulse is measured using a heart rate monitor, which includes an optical radiation source and the photodetector. It is natural that we should observe the pulse signals [2] in the chosen scheme of the measuring device (figure 1).

A semiconductor laser module KLM-H808-120-5 was used as a radiation source with wavelength $\lambda$ = 0.808 μm (1). It was powered by a DC supply (2). The output power of the laser was 130 mW. It is possible to avoid significant absorption of the laser radiation by bones (below 80%) in some places in the palm (3). Radiation transmitted through the palm goes to the photodetector (4), and then to the amplifier (5) and digital voltmeter Tektronix TDS 1001C (6).

[1] To whom any correspondence should be addressed.

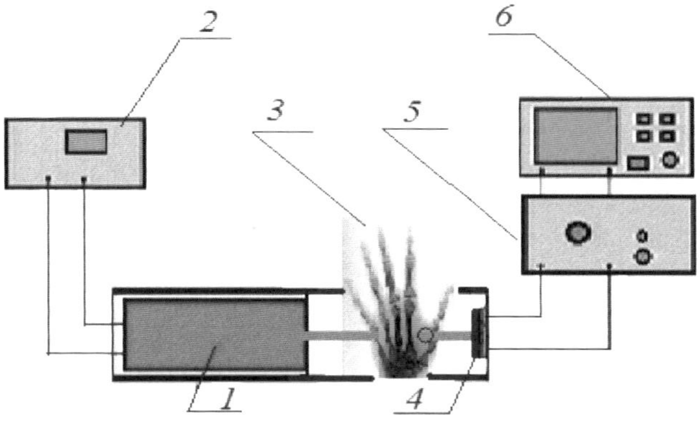


**Figure 1.** A block diagram of the device: 1 – semiconductor laser, 2 – power supply, 3 – location of the object, 4 – photodetector, 5 – amplifier U7-1, 6 – TektronixTDS 1001C.

**Measurement results and discussion**

Owing to a detailed study of pulse oscillation at a higher time sweep of the oscilloscope (~ 5 s/segment), modulation of the pulse oscillations with a frequency of 0.01–0.1 Hz was observed. The results of these studies are presented in figure 2.

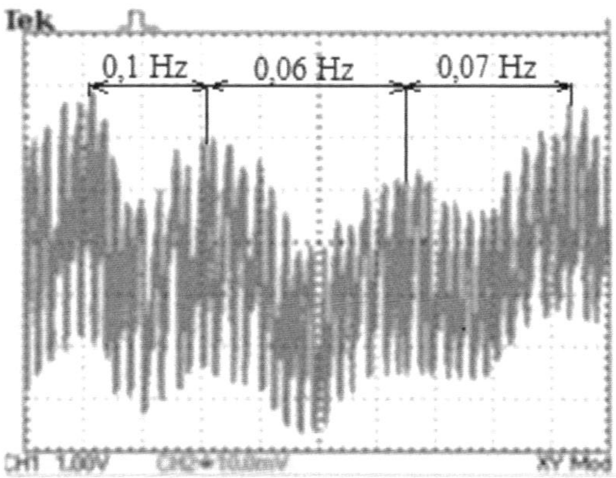


**Figure 2.** Photodetector signal of the laser radiation transmitted through the palm (axis reamer "Y" - 10 mV/ division, "X"- 5 s/ division).

From these figures it can be seen that the infrasonic vibrations have different frequencies 0.06–0.1 Hz and have the form of sin($x$). Presumably, this is due to the uncontrolled behavior of the nerve impulses, which are generated by the human brain. Research in this direction will be continued.

It is known [3], that any human body has different physical fields around itself – acoustic and electromagnetic, – which are characterized by the intrinsic processes, i.e. fields which are generated by the organism during its operation. Low-frequency radiation (frequencies below $10^3$ Hz) is formed by physiological processes: heartbeat, breathing movements, blood flow in blood vessels and other processes, illustrating variations of vibrations of the human body surface within the range of about 0.01–$10^3$ Hz.

To check the connection between the observed features of the signals' behavior and the respiratory movements, experiments with breath holding were conducted. These experiments had 3 stages: sigh for about 2 s (point A in figure 3), breath-holding for approximately 7 s, and exhalation for 2 s (point B). After this cycle, the process was repeated. The result of this experiment is shown in figure 3.

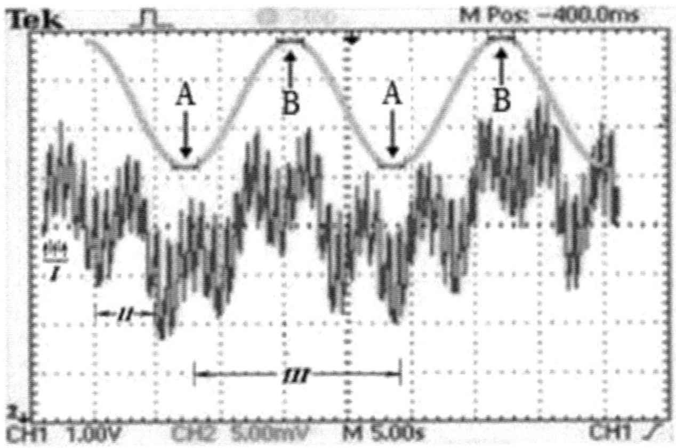


**Figure 3.** Signal of the laser radiation transmitted through the palm with periodic breath-holding. Point A corresponds to the breathing in, point B corresponds to the exhalation; ***I, II, III*** – periods of different wave types; upper curve is an envelope curve of waves (reamer of axis "Y" - 1V/division, "X" - 5 s/division).

As it can be seen, there are several types of waves in this case. Waves of the first type (***I***) are associated with the reduction of the heart (pulse), the second type (***II***) waves are linked to the rhythm of human breathing, and the third type waves (upper curve) are called waves of Traube-Hering (***III***) in the literature [4]. With the help of the "acute" experiments on the dog, Goering concluded that the vascular system is characterized by its own breathing movements, occurring synchronously with the movements of the respiratory muscles. Both of these types arise because of nerve impulses coming from the respiratory center, which is located in the medulla oblongata. Normally, these waves are practically not observed, but in a state of emotional tension, occurrence of these waves is very common. Figure 4 shows the waveform at long breath hold. Breathing stopped at the point A, and after 8 s the signal began to decrease to a point B within 25 s. As it is known [5], hemoglobin concentration increases during breath-holding, hence the absorption of laser radiation grows, and the decrease of the signal intensity is observed. After breathing resumed, the signal slowly comes back to its initial value. The figure shows that the oscillations observed in a breath-hold have higher frequency than the the oscillations during an exhalation. Oxygen does not enter the blood in the circulatory system during cessation of breathing, so the concentration of $CO_2$ increases leading to the fact that breathing becomes more frequent and heart rate increases.

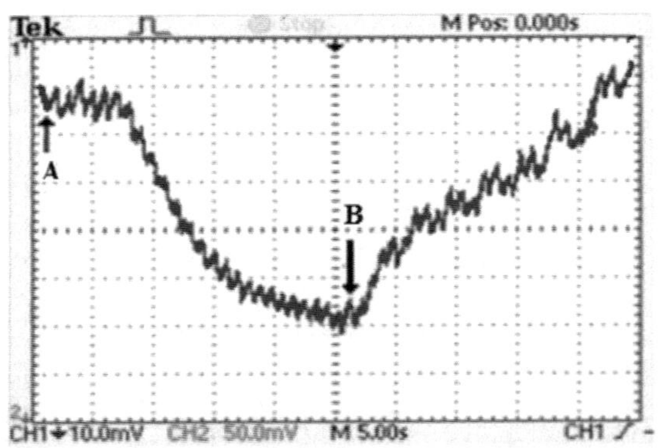


**Figure 4.** The nature of the signal changes at a long breath-holding. Reamer of "X" axis: 5 s/division.

In order to clarify the influence of pulse oscillation on the effects described above, experiments were carried out to study the processes associated with blood flow in the vessels of the heart beats using the tourniquet which was superimposed on the wrist for 12.5 s. The result of this experiment is shown in figure 5.

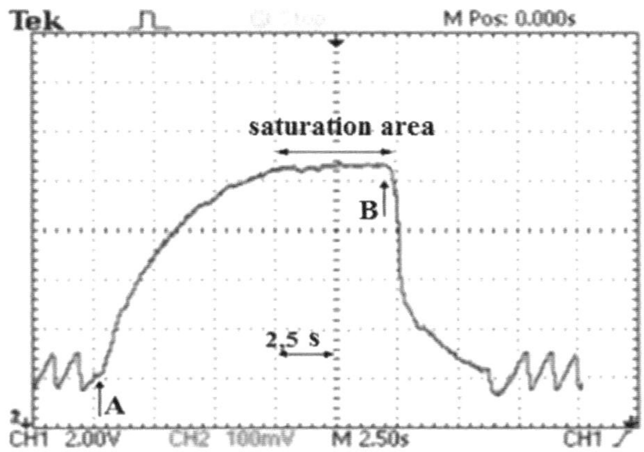


**Figure 5.** Laser radiation signals from the detector. This radiation passed through the palm which was tied at the wrist. Point A corresponds to the tourniquet application, and B – to its removal (reamer "Y" axis - 2 V /division, "X" - 2.5 s /division).

At the moment of the termination of the blood flow, an increase of signal at the photodetector was observed, which corresponds to the effect of "enlightenment" of tissues. In this case enlightenment was approximately 30%. The saturation occurred over the next few seconds, i.e. signal did not change in magnitude. After removing the tourniquet, the signal decreased by 2 times in 0.5 s and then came to its normal value in about 3.5 s. As it can be seen from the waveform in figure 5, at the termination of the blood movement, the infrasonic vibrations are absent, i.e. observed waves are directly related to circulation.

Generally speaking, the effect of tissue "enlightenment" was earlier observed in article [6], where this effect was caused by the compression and puncture at the tissue area which was illuminated by light. This phenomenon was attributed to the departure of blood and squeezing the muscles of the area of compression. It was also noted that the compression layer reduces light scattering and absorption of the stain, due to the blood displacement from the area of pressing. By eliminating the pressing, enlightenment does not disappear immediately – footprint of a pressing continues to transmit the light for 1-3 s. It is noted that the thicker part of the palm, such as flesh from the edges and at the region of phalanges, tends to be more permeable and prone to light illumination by compression than the center of the palm. In our case, this effect is related to the cessation of blood flow into vessels located in the palm. Thus, besides reducing the absorption and scattering of the radiation, reduction in tissue oxygenation can also be substantial, since light is mainly absorbed by protein molecules and oxygen in the near-infrared region.

**Acknowledgements**

This work was also supported by Competitiveness Growth Program of the Federal Autonomous Educational Institution of Higher Professional Education National Research Nuclear University MEPhI (Moscow Engineering Physics Institute).

Journal of Physics: Conference Series **747** (2016) 012066 doi:10.1088/1742-6596/747/1/012066

**References**

[1] Godik E E, Gulyaev Y 1990 Physical fields of human and animal *"In the world of science"*
[2] Protasov E A, Protasov D E, Ryzhkova A V 2015 *Physics Procedia* **71** 261-266.
[3] Antonov V F, Chernysh A M, Voznesensky S A, Kozlova E K 2000 *Biophysics* (Moscow: Humanity ed. Centre VLADOS)
[4] 1985 *The Great Medical Encyclopedia*. ed. by acad. Petrovsky B V, Soviet Encyclopedia
[5] Richardson M. X., Engan H. K., Lodin-Sundström A, Schagatay E 2012 *National Center for Biotechnology Information* **42(1)** 4-9,
[6] Askaryan G A 1982 *Quantum Electronics* **9** 7 1370-1383

Journal of Physics: Conference Series **747** (2016) 012067 doi:10.1088/1742-6596/747/1/012067

# Streak Tubes for Diagnostics of Lasers and Plasmas

**A Yu Sokolov[1], P I Konovalov, R I Nurtdinov, M P Vikulin, I G Pryanishnikov, A S Dolotov, P S Krapiva**

Research & Production Center for Pulse Technique, Dukhov Research Institute of Automatics (VNIIA), Sushchevskaya str. 22, Moscow, 127055, Russia

E-mail: SokolovAYu@gmail.com

**Abstract**. Designing a facility for laser fusion research requires sufficient advancement in diagnostics techniques for lasers and plasmas, including those involving streak camera imaging. Maximum specifications of streak cameras depend on the parameters of streak tubes. The paper illustrates how these devices function, and which of their parameters are limiting. The paper presents a novel technological platform designed at VNIIA, which was used to develop a new generation of streak tubes. Using these streak tubes in streak cameras, the efficiency of streak camera imaging techniques can be improved by several orders of magnitude, and new techniques can be designed.

## 1. Introduction

Work at laser fusion research facilities, such as NIF, OMEGA, LMJ, ISKRA-5, involves a wide range of complex scientific problems, one of them being the measurement of the laser and plasma pulses parameters lasting ~ $10^{-12}$ ... $10^{-9}$ s [1–10]. Streak cameras with streak tubes as the key components are used to address these issues. Information capacity of a streak camera image depends on its dynamic range and spatial resolution, which are limited by the similar parameters of streak tube. Improving these characteristics in new designs of streak tubes is one of the main objectives of VNIIA Research & Production Center for Pulsed Technique, which has over 30 years of experience in developing and producing streak tubes, successfully used by VNIIEF in their methods [1, 3].

## 2. Effect of streak tube parameters on the effectiveness of diagnostics methods

A typical streak camera image, which provides power measurement over time for ten different channels of igniting laser radiation, is shown in Figure 1. The number of channels in a single streak camera image depends on two parameters: the number of resolution elements and dynamic range of streak camera, and the dynamic range and resolution of streak tube, respectively. Increase of the number of channels will allow to reduce the necessary amount of streak cameras used in the detection methods of the laser time structure [3, 5, 8] and to avoid non-simultaneous arrival of laser beams to the thermonuclear target from different channels [4].

[1] To whom any correspondence should be addressed.

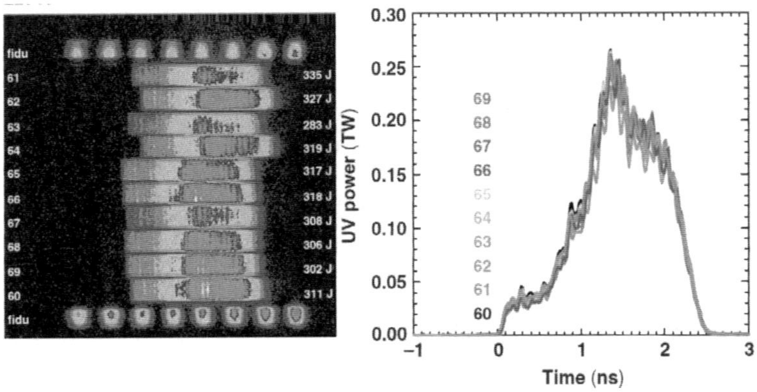


**Figure 1.** Ten-channel streak cameras provide pulse shape and power measurements at OMEGA [8].

Dynamic range is also the key parameter in the measurement method of contrast and profile of the laser pulse front, which define the fusion target implosion [2, 3]. At up-to-date facilities, the ratio of the maximum value of emission power vs. the value at the beginning of a pulse is of seven orders of magnitude or larger [2, 3], which forces to use ranging principle for such measurements (see Figure 2). The number of ranges is defined by the dynamic range of streak tube.

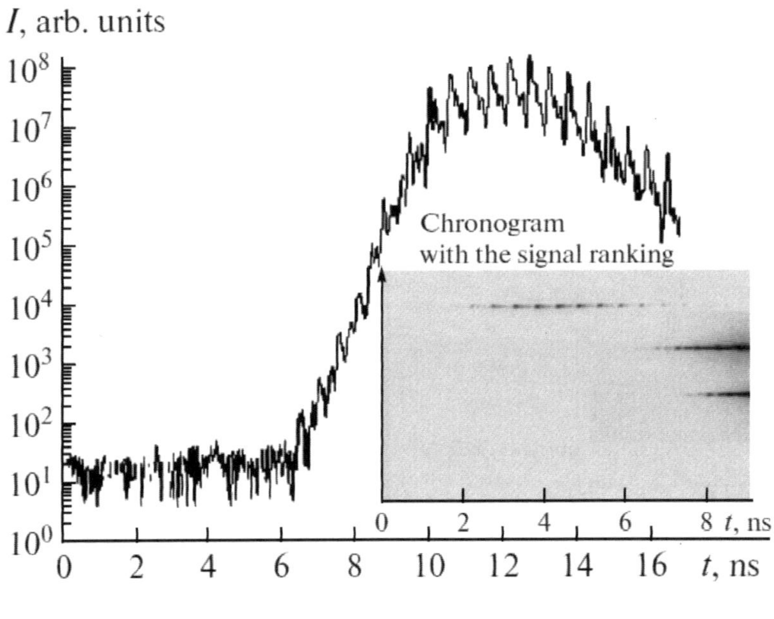


**Figure 2.** Result of the laser pulse deep front reconstruction [3].

The spatial resolution is the key characteristic for methods involving variation analysis over time of the laser spectral intensity scattered from the hohlraum plasma [6, 9]. The streak camera resolution determines the spectral resolution of such methods and measurement precision of the major plasma parameters (composition, temperature, and density).

## 3. Factors affecting the performance of streak tubes

As it was demonstrated above, the main streak tube parameters affecting the efficiency of diagnostic methods are the dynamic range and the limiting spatial resolution.

The streak tube dynamic range depends on the maximum photocurrent, which can be removed from the photocathode without image distortion on the screen, as well as on the signal-to-noise ratio of a streak tube [10]. Theoretically, the maximum current is limited by Child-Langmuir law, so its value is directly proportional to the electric field strength that accelerates and focuses photoelectrons in streak tube. Actually, streak tube with semitransparent photocathode rarely operates in the Child-Langmuir space-charge-limited current mode because image distortion on the streak tube screen appears much earlier due to electron trajectory disturbance as a result of the space-charge repulsion in crossover or violation of the electric field equipotentiality close to photocathode. The higher the resistance of the photocathode, the smaller the maximum photocurrent, which can be removed from it without violating the equipotentiality.

The signal-to-noise ratio is determined by the streak tube photocathode sensitivity and its dark current noise, which depends on the number of parasitic electrons sources in the streak tube envelope: free cesium, parasitic photocathodes on the electrodes of the cathode chamber, field emission points, etc.

Regardless of the streak tube maximum spatial resolution being mainly determined by its electron-optical system [11, 12], the calculated values are rarely achievable in practice due to significant deterioration introduced by the device assembly error (apart from considering the influence of other technological factors). Inaccuracy of the assembly in the previous generation devices produced by VNIIA Research & Production Center for Pulsed Technique up to 2015 was due to the quality of metal-glass body and specific limitation of tubulation technology of photocathode production.

## 4. New generation streak tubes

In 2015, VNIIA Research & Production Center for Pulsed Technique implemented a new technological platform for manufacturing of vacuum electronic devices that enabled producing streak tubes with improved key parameters. The platform is based on the finish processing of tubes using transfer technology, which involves producing a photocathode within an individual vacuum volume followed by its transfer to the case and sealing. This technology has significantly increased the photocathode sensitivity and reduced the dark current noise thanks to the absence of parasitic photocathodes on the electrodes of the streak tube cathode chamber and reduction of the alkali metal residual vapor in the volume of the device.

Using the transfer technology allowed to avoid direct contact of alkali metals with the streak tube internal parts during photocathode formation, which in turn, together with the introduction of ceramic-and-metal envelope, allowed significantly increasing the dielectric strength of insulators. As a result, the new generation of streak tubes operates at high electric field strength modes and allows to obtain the maximum time resolution up to 0.7 ps. Transition to ceramic-and-metal envelope increased the accuracy and reproducibility of its dimension. High-precision assembly of streak tubes has become possible using a special equipment.

Thanks to introduction of a number of special technologies for photocathode substrate preparation, its effective resistance was reduced and its sensitivity was increased manyfold.

Based on the previous developments [3, 11, 12], the introduction of technological improvements mentioned above allowed to create a new generation of streak tubes (see Figure 3) with improved capabilities. High-precision assembly and increase of the field strength made it possible to increase the streak tube resolution from 20 lines/mm to 30 lines/mm. Increase of the field strength, reduction of the dark current noise, lower photocathode resistance, and increase of its sensitivity led to an increase of the dynamic range of a streak tube by an order of magnitude from 1300 to 10000 (measurement method is described in [3]). The new generation streak tubes have smaller weight and dimensions and outperform their predecessors in almost all features, including time resolution and life time.

**Figure 3.** New generation streak tubes: TPO29 series (left) and TPO30 series (right).

## 5. Conclusion

According to our estimates, using streak cameras based on the new generation of streak tubes will increase the effectiveness of the methods described above in the following way: using one streak camera image it will be possible to record more than 1.5 times the number of laser channels at the same time, 2 times as low number of ranges will have to be introduced in the pulse shape measurements, while the spectral resolution of the stimulated Raman scattering recording method will increase by a factor of 1.5. The developed new generation of streak tubes will no longer limit the parameters of the existing streak cameras [3]: other components of these systems will require improvement.

## References

[1] Vatulin V V et al 2010 *Plasma Physics Reports* **36** 413–19
[2] Kirdyashkin M Yu et al 1992 *Soviet Journal of Quantum Electronics* **22** 961
[3] Kornienko D S et al 2014 *Instruments and Experimental Techniques* **57** 165–75
[4] Vorontsov E N et al 1994 *Quantum Electronics* **24** 927
[5] Clément Chollet et al 2009 *Proc. of SPIE* **7126** 712610-1
[6] Kirkwood R K et al 2004 *Rev. Sci. Instrum.* **75** 4174
[7] A Self-Calibrating, Multichannel Streak Camera for Inertial Confinement Fusion Applications, *Laboratory for Laser Energetics LLE Review* **63** 109–21
[8] Jaanimagi P A et al 2004 *26th International Congress on High Speed Photography and Photonics*
[9] Hinkel D E et al 2011 *Phys. Plasmas* **18** 056312
[10] Lerche R A et al 2004 *Lawrence Livermore National Laboratory* UCRL-TR-208078
[11] Badin L V et al 2009 Cathode assembly of time-analysing electronic optical image converter *Patent* RU2374719
[12] Badin L V et al 2010 Time-analysing electron-optical image converter *Patent* RU2378734

# Modeling of micrometeoric streams under the action of the high-power laser pulse on multicomponent polycrystal rocks

**I S Timofeev[1], I N Burdonskiy, A Yu Goltsov, A G Leonov, K N Makarov, V N Yufa**
Moscow Institute of physics and technology, Moscow, Russia

E-mail: i.s.timofeev@gmail.com

**Abstract.** The paper presents the analysis of interaction of the high-power laser pulse with multicomponent polycrystal rocks. Experiments were completed on a laser facility "Saturn" with intensity of $10^{10}$–$10^{13}$ W/cm$^2$. Structural analysis of the materials from the spall crater and from the plasma flame show significant differences. The article demonstrates experimental results of the spall formation moment depending on the thickness of the radiated target. The scale of damage to aluminium 6 μm thick foil at the rare side of the target is illustrated when it is hit by the andesite fragments from a spall crater.

## 1. Introduction

Today, due to LTS program research, laser radiation interaction with metals and dielectrics is well studied in a wide range of intensities. However, the mechanism of laser radiation interaction with multicomponent polycrystalline rocks at $I \sim 10^{10}$–$10^{13}$ W/cm$^2$ is studied insufficiently and poorly described in literature.

It is of great interest for a variety of technological applications, in particular, for laser simulation of micrometeorite impact. There were a number of articles in scientific literature on laser simulation of micrometeorite impact with the help of metallic foils acceleration [1]. Though it was known that most of micrometeorites are particles of stony dust [2], the information about laser simulation of micrometeorite impact was poorly described in literature.

The problem of laser simulation of corresponding micrometeorite impacts is therefore considered to be studied thoroughly. Andesite can be used as the main material. Its composition is close to that of meteorites and Martian dust [3].

Taking these facts into consideration, the aim of this research was the detailed study of the high-power laser radiation interaction with the multicomponent polycrystalline rocks and the development of a laser-plasma source of μm-sized particles with solid density with the speed close to that in the micrometeorite flows.

## 2. Experimental results

The experiments were carried out at the high-power facility "Saturn" described in detail in [4]. Laser pulse was formed at the output of the final amplifier with the energy $E_L \sim 20$–$50$ J, FWHM duration $\tau_{FWHM} = 30$ ns and divergence $\theta \sim 1.5 \cdot 10^{-4}$ rad. Focal spot diameter $D_L$ varied from 100 μm to 300 μm.

The research of the crater formation process requires accurate parameters of the plasma flame and the magnitude of the ablation pressure, generating a shock wave in the unevaporated part of the target.

[1] To whom any correspondence should be addressed.

A simplified analytical model of a stationary spherical expansion of plasma, proposed by P. Mora [5], was used to calculate the flame parameters. It is supposed that the main mechanism of laser radiation is the inverse bremsstrahlung.

It should be pointed out that the plasma flame temperature can be used as a parameter for model verification. The method of absorbers (filters) was used to determine the temperature of the resulting dense plasma. As the test objects for verification method, Al targets were used.

The data obtained for aluminium and andesite targets [6] agree within the experimental error with the results calculated on the basis of Mora's plasma expansion model [5]. It is shown that Mora's model can be used to obtain reliable data on the plasma ablation pressure and temperature. Theoretical calculations according to Mora's model show ablation pressure of about 4 Mbar for laser radiation intensity $I = 10^{13}$ W/cm$^2$.

Crater parameters in the studied range of intensity can be described by means of an analytical model suggested by S. Yu. Guskov [7]. Crater formation during the laser interaction with the target is the result of phase transformations of pressed and heated substance behind the shock wave front.

The calculation results [4] have demonstrated that the estimated crater's diameter and depth for aluminum and andesite targets agree with the experimental data. The model allows to estimate the shock wave speed. According to the experiments with 2 cm thick andesite targets, the shock wave speed is about 5 km/s and it decreases before coming to free surface [4].

To study the shock wave exit onto the rear side, series of shootings at thin andesite [8] targets were made. As a result of all our experiments with thin targets, the spall crater was formed from rear side of the targets. In some cases, a through channel was formed inside the spall crater. As a rule, it was formed on the targets with thickness less than 350 μm.

To study the material ejected from the plasma flame and the spall crater, targets material spraying on special silicon plates was researched [4]. The analysis of spraying structure is carried out with the help of the electron microscope, whereas chemical analysis is studied using the X-ray spectrometer.

There is a large number of liquid fragments in the material ejected from the crater into the front space (figure 1). At the same time, there are only solid fragments (figure 2) in the material ejected from the spall. This fact proves that the temperature of the material from the spall crater is lower than the melting temperature. Sizes of the fragments vary from 0.5 μm to 50 μm [8].

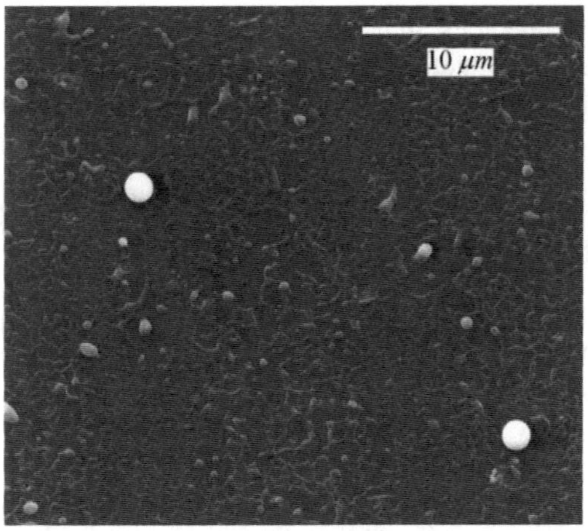


**Figure 1.** Electron photographs of the andesite condensate film in the front space of the target $D_L$ = 100 μm, $I = 5 \cdot 10^{12}$ W/cm$^2$.

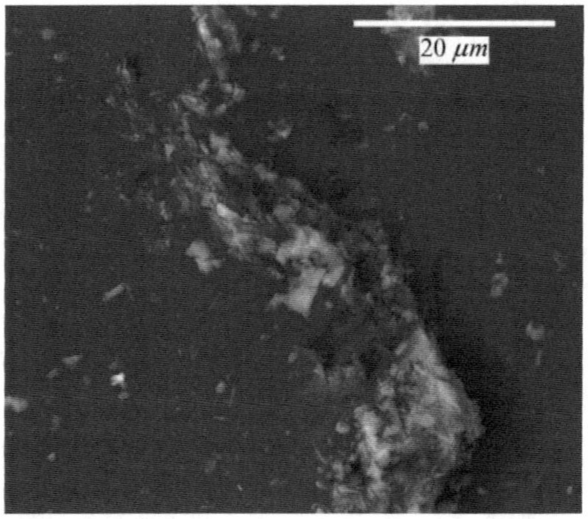


**Figure 2.** Electron photographs of the andesite condensate film from the rare side of the target $D_L$ = 100 μm, $I = 5 \cdot 10^{12}$ W/cm$^2$.

To study the moment of spall formation in thin targets, the scheme presented in figure 3 was developed. A thin stripe of 6 μm thick Al foil is placed into the base holes and is adjusted by contacts C1 and C2. The target is fixed on the base close to the stripe.

When the foil is being torn, the oscilloscope records potential shift. The moment of diode $D_1$ switching on is synchronized with the laser impulse input on the front surface of the target.

Figure 4 shows signals registered at Al foil tearing during irradiation of 400 μm thick andesite targets. The signal of the foil tearing on one of the targets is shown with negative magnitude for better visualization. One should note satisfactory repeatability of the results obtained at the same thickness of the target and at the same laser intensity.

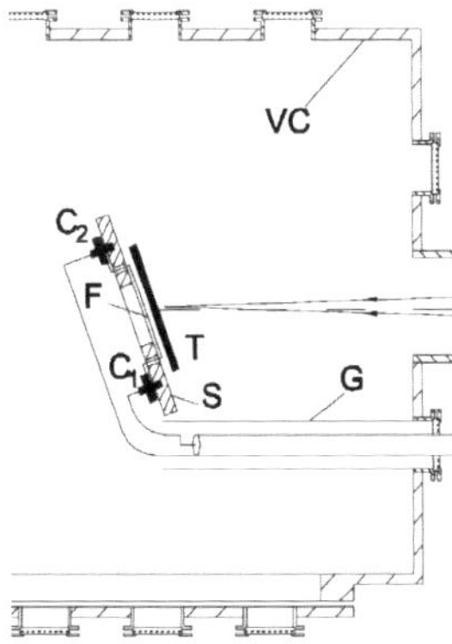


**Figure 3.** The scheme of the research of spall formation moment. VC – vacuum chamber, $L_1$ – focus lens, $D_1$ – photodiodes, $F_1$ – filters, $P_1$ – dividing plates, T – thin target, F – aluminum foil, S – organic glass substrate, $C_1$, $C_2$ – contacts to fix foil, G – grounded copper braid, MS – measuring system.

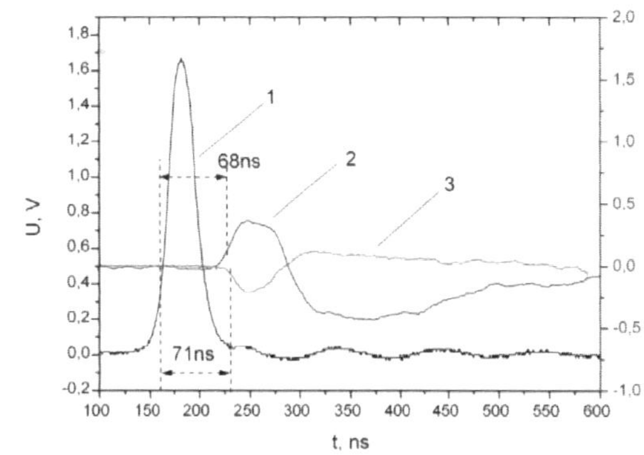


**Figure 4.** Signals registered when laser impulse with energy of 8 J interacts with andesite targets (thickness 400 μm): 1 – impulse from diode $D_1$, 2 – dependence of voltage on time from the foil (target no. 1), 3 – dependence of voltage on time from the foil (target no. 2).

Figure 5 shows the dependence of the spall formation moment on the thickness of andesite targets. The spall is formed at shock wave ejection on the free surface of the target and depression wave generation. Prior to the moment of spall formation, the shock wave passes a distance of about a target thickness. The speed of the shock wave for targets with thicknesses from 400 to 700 μm is $W \approx 7$ km/s. This data interpretation is coherent with the calculation data [4]. Taking in consideration the material speed doubling at shock wave ejection on the free surface and approximation of a strong shock wave, expansion velocity of fragments from the spall crater is proved to be around $4W/(\gamma+1) \sim$ 10 km/s.

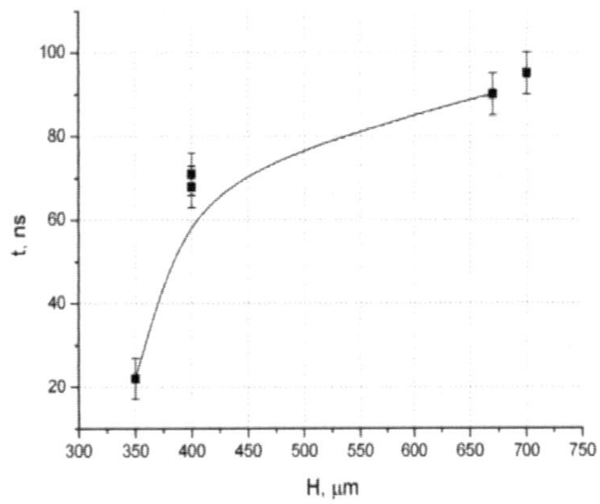


**Figure 5.** Dependence of the spall formation moment on the thickness of an andesite target.

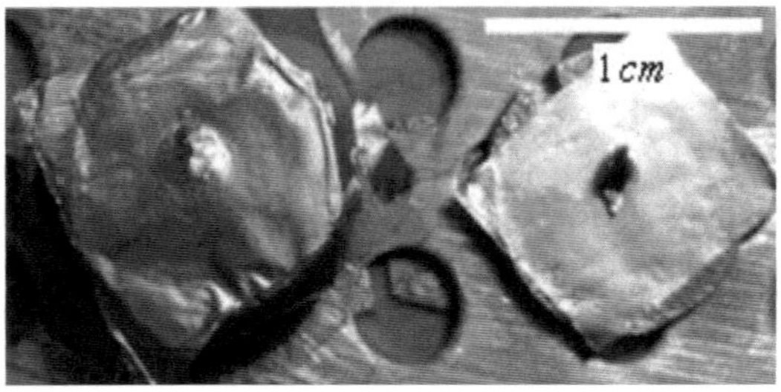


**Figure 6.** Diagnostic Al foil fixed at the rare side of the target.

If the thickness of the target is getting less than 400 µm, a through channel can be formed in the spall crater. The glow at the rare side of the target was registered during the experiments with a through channel formation [8].

To show the scale of surface damage when andesite fragments from the spall crater hit it, a piece of 6 µm thick aluminum foil was fixed 2 mm and 6 mm off the rare side of the target. In figure 6 one can see the traces of the damage to foil induced by the fragments. The sizes of the craters in foil vary from 1–20 µm (left in figure 6) to 2 mm (right in figure 6).

## 3. Conclusion

Thus, it has been demonstrated that under the action of the laser pulse of moderate intensity on the multicomponent polycrystal rocks, µm-sized fragments are formed at the rare sides of targets. These fragments are equivalent in composition, temperature, dimension, and speed to the micrometeorite flows in space.

## References

[1] Ripin B, Decoste R, Obenschain S 1980 *Phys. Fluids* vol 23, pp1012–1026
[2] Zook D, Flaherty R, Kessler D 1970 *Planet Space Sci.*vol 18, pp 953–964
[3] Rieder R, Economou T, Waeenke H 1997 *Science* vol 278, pp 1771–1774
[4] Burdonsky I, Goltsov A, Leonov A, Makarov K, Timofeev I, Yufa V 2013 *Problems of Atomic Science and Technology* vol 36, pp 8-19
[5] Mora P 1982 *Phys. Fluids* vol 25, pp 1051–1056
[6] Timofeev I, Aleksandrov N, Burdonskiy I, Goltsov A, Leonov A, Makarov K, Putilin M, Yufa V, 2014 *Laser Phys* vol 24, 126002
[7] Guskov S, Borodzyuk S, Kalal M, 2004 *Quantum electronics* vol 34, pp 989-1003
[8] Timofeev I, Burdonskiy I, Goltsov A, Leonov A, Makarov K, Yufa V, 2014 *Physics of Extreme States of Matter* pp 123–125

Journal of Physics: Conference Series **747** (2016) 012069 doi:10.1088/1742-6596/747/1/012069

# Simulation of Spin-orbit Dynamics in Storage Rings

**A. Ivanov[1,3], S. Andrianov[1] and Yu. Senichev[2]**
[1] Saint Petersburg State University, St. Petersburg, Russia
[2] Institute for Nuclear Physics, Forschungszentrum Jülich, Germany

E-mail: 05x.andrey@gmail.com, sandrianov@yandex.ru

**Abstract.** In the article a mapping approach based on nonlinear matrix integration for long-term spin-orbit dynamics simulation is briefly described. Using this technique the nonlinear effects of spin dynamics in an electrostatics storage ring are investigated. Namely, the fringe fields, the energy conservation law and the random field errors are considered. The necessity of examination of such effects arises, for example, in the storage ring design for search the Electrical Dipole Moment of proton and deuteron. The EDM ring is proposed to measure EDM using the spin transformation of polarized particle in the magneto-electrostatic elements of the ring. The article consists of short description of the spin-orbit simulation results based on the nonlinear model.

## 1. Introduction

Simulation of spin-orbit dynamics plays a crucial role in a storage ring design for search of EDM [1]. Such nonlinear effects as the fringe fields, the energy conservation law and the random field errors should be carefully investigated. The main problem is that one has to distinguish the fake EDM signal from presumably existing EDM signal. For example there were fears that fringe fields may change reference orbit dramatically in sense of spin dynamics and lead to unexpected spin behaviour. The electrostatic fields which are going to be used induce additional problems. In particular, consideration of the energy conservation law means we should take into account accordance between the particle energy and its coordinates in electrostatic fields. To investigate such effects on spin dynamics the numerical analysis can be used only. In this section we will briefly describe our mapping approach for nonlinear spin-orbit dynamics simulation. Then we will present results of simulation and indicate further directions of numerical method analysis.

For numerical integration of differential equation of spin-orbit dynamics the nonlinear matrix integration is used. This mapping approach allows describing particle beam dynamics for whole ring (see [2])

$$X = R_0 + R_1 X_0 + R_2 X_0^{[2]} + \ldots + R_k X_0^{[k]}, \tag{1}$$

where $X_0$ means initial state of a particle (space coordinates, energy and spin) and $R_i$ are 2 dimensional matrices that describe nonlinear map of particle dynamics for a ring. In general case this map depends on lattice of storage ring and can also include effects of fringe fields and random errors.

### *1.1. Simulation*

For building the map (1) we have developed the MODE software that allows creating the parametric nonlinear matrix map for arbitrary field distribution in indicated lattice (see [3]). Aside from the

[3] To whom any correspondence should be addressed.

standard magnets and lenses we have implemented additional elements, such as fringe fields and fields with random errors. The MODE allows building the parametric map, and the simulation can be performed just varying set of parameters (such as field distribution characteristic, strength and length of fringe field, etc.)Spin-orbit dynamics investigation

*1.2. Fringe fields*

Modeling of fringe fields shows that there are no dramatic influences on spin dynamics at all. Fringe fields change reference orbit and energy. Nevertheless one can optimize parameters of a lens or magnet to achieve the same spin behaviour as in case without fringe field.

*1.3. Energy conservation low*

The spin-orbit dynamics of charged particles in electrostatic fields has more complicated behaviour than in magnetic ones. If in the magnetostatic field the force acting on the particle is always perpendicular to the motion direction and the kinetic energy remains constant, in the electrostatic field the kinetic energy is continuously transferred into the potential energy and vice versa. This means that the velocity of a particle v is not more a constant value, but a function of coordinates. More precisely one can write v = v(u(x, y, z)), where u is a scalar potential of an electrostatic field and x, y, z are the space coordinates of a particle.

As shown in [4] the low-order electrostatic fields introduce the high orders in actual force that effect on particle. For instance, in the electrostatic quadrupoles (first order field) there are the second order aberrations in spin dynamics.

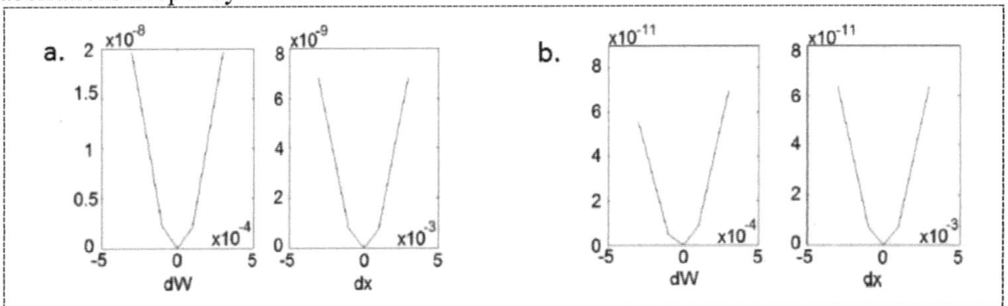


**Figure 1.** (a) spin aberrations without lattice optimization; (b) with quadrupoles optimization.

*1.4. Random fields errors*

Spin dynamics due to the beam dynamics nonlinearity has complex behaviour with respect to the random errors. Various types of error fields in the elements lead to systematic errors, limiting the possibility of the EDM experiment. One way is to decrease the systematic errors it is the method of direct and reverse direction of the beam in the storage ring, the so-called clockwise and contra-clockwise method.

Image that we have error field Eer in electrostatic storage ring that belongs to normal distribution. Then the frequency of spin precision

$$w(E_{er}) \notin N, \langle w(E_{er}) \rangle \neq 0, w_{cw} \neq w_{ccw},$$

where <a> denotes mean value of random variable a. So it is not possible to easy use clockwise ($w_{cw}$) and contra-clockwise ($w_{ccw}$) beams to measure EDM. Moreover spin tune now depends on beam position and it is not clear how to correct spin aberration without impact on EDM.

## 2. Conclusion

Thus we have developed a mathematical method of the spin-orbital dynamics simulation, which is necessary for the design of the storage ring to search for EDM. This method allows taking into account all the necessary properties of the system, affecting the spin. The method became the basis for writing the program MODE, which was tested by other codes.

## References

[1] Senichev Y., Lehrach A., Maier R., Zyuzin D., Andrianov S., Ivanov A, Berz M., Makino K. 2012. Storage Ring EDM Simulation: Methods and Result. Proc. of ICAP2012, 99–103

[2] Ivanov A., Andrianov S. 2012. Matrix Formalism for Long-term Evolution of Charged Particle and Spin Dynamics in Electrostatic Fields. Proc. of ICAP2012, 187–189

[3] Ivanov A., Senichev Y. 2015. Matrix Integration of ODEs for Spin-orbit Dynamics Simulation. Proc. of IPAC2015, 400–402.

[4] Doroshko A., Ivanov A. 2014. Inuence of Electrostatic Multipoles on Spin-orbit Dynamics via Energy Conservation Consideration. Proc. of BDO2014.

# Multicriterial approach to beam dynamics optimization problem

**L V Vladimirova**[1]

St. Petersburg State University, 7/9 Universitetskaya nab., St. Petersburg, 199034, Russia

E-mail: l.vladimirova@spbu.ru

**Abstract.** The problem of optimization of particle beam dynamics in accelerating system is considered in the case when control process quality is estimated by several functionals. Multicriterial approach is used. When there are two criteria, compromise curve may be obtained. If the number of criteria is three or more, one can select some criteria to be main and impose the constraints on the remaining criteria. The optimization result is the set of efficient controls; a user has an opportunity to select the most appropriate control among them. The paper presents the results of multicriteria optimization of beam dynamics in linear accelerator LEA-15-M.

## 1. Introduction

The paper is devoted to multicriteria optimization of charged particle beam dynamics. The problem of beam dynamics optimization in accelerating and focusing systems is urgent and widely studied by many researchers. Saint-Petersburg State University takes part in the research in this area; in particular, the books and papers [1-9] should be mentioned.

Many beam dynamics optimization problems are conceptually multicriterial because there are several beam characteristics to be improved or restricted, and one has to introduce multiple quality criteria in order to find the best solution. Such an approach is considered in [10-11].

When several criteria are conflicting, the improvement of one of them leads to the worsening of the others. In this case multicriteria optimization is preferable. The optimization result is the set of efficient controls providing quality criteria vectors that can't be improved. The optimization problems with conflicting criteria are, in particular, treated in [10], [12-15].

## 2. Trajectory ensemble control problem

Let particle beam evolution be described by the following differential equations presented in [1]:

$$\frac{dx}{dt} = f(t,x,u),$$

$$\frac{\partial \rho}{\partial t} + \frac{\partial \rho}{\partial x} f(t,x,u) + \rho(t,x) div_x f(t,x,u) = 0,$$

with initial conditions:

[1] To whom any correspondence should be addressed.

$$x(0) = x_0 \in M_0,\ \rho(0,x) = \rho_0(x).$$

Here $t \in [0,T]$ is independent variable; $T$ is a constant; $x \in \Omega \subset E^n$ is $n$-dimensional vector of particle phase coordinates; $E^n$ is $n$-dimensional Euclidean space; $u = u(t)$ is $r$-dimensional control function; $f(t,x,u)$ is $n$- dimensional vector function; $M_0$ is an open bounded set of non-zero measure; $\rho_0(x)$ is continuously differentiable nonnegative function and $\int_{M_0} \rho_0(x_0)dx_0 = 1$.

Let us assume the initial particle phase state $x_0$ to be the value of random variable $X_0$ with probability density $\rho_0(x_0)$. Consider particle phase state $x(t,x_0,u)$ to be the value of an $n$-dimensional random variable $X_t$ with probability density $\rho(t,x)$. Let us assume $f(t,x,u)$ to be sufficiently smooth function and control $u = u(t)$ to be piecewise continuous.

So for every admissible control $u(t)$ the ensemble of trajectories $x(t,x_0,u)$ emanating from $M_0$ is introduced. The set $M_{t,u} = \{x_t = x(t,x_0,u) : x_0 \in M_0\}$ is called trajectory ensemble cross-section.

Let us introduce beam dynamics quality criterion to be a functional of the form

$$I(u) = \int_0^T \int_{M_{t,u}} \Phi(t,x_t)\rho(t,x_t)dx_t dt + \int_{M_{T,u}} g(x_T)\rho(T,x_T)dx_T. \quad (1)$$

where $\Phi(t,x)$, $g(x)$ are smooth nonnegative functions. The integrands mentioned are chosen to correspond the physical meaning of quality functional. For example, they may be penalty functions providing the restrictions account [2], [15-18].

The problem of functional (1) minimization with respect to control vector $u(t)$ is defined to be trajectory ensemble programmed control problem. The function $u^{(0)}(t)$ providing functional (1) minimum is determined to be optimal control.

## 3. Formulation of multicriteria optimization problem

When treating beam dynamics optimization problems, we often use several criteria, sometimes conflicting. Let us consider the example, specifically, the problem of longitudinal beam dynamics control in travelling-wave linear accelerator. Particle phase state is described by the vector $x(t) = (\gamma(t), \varphi(t))$, where $\gamma$ is reduced energy and $\varphi$ is the phase. The optimization objectives are as follows: 1) to maximize bunching coefficient; 2) to provide the required energy $\hat{\gamma}$ at device exit; 3) to maximize capture coefficient $k_c$ in the acceleration mode. The analogous problems are investigated in [1-2], [18].

We can formulate trajectory ensemble control problem to be multicriteria optimization problem following [9], [14]. Let us introduce the criteria $K_i(u)$, $i = \overline{1,3}$ corresponding to the objectives mentioned above. The particular criteria $K_i(u)$, $i = \overline{1,2}$, may be suggested in the form

$$K_1(u) = \int_0^T \int_{M_{t,u}} \Phi(\varphi_t)\rho(t,\gamma_t,\varphi_t)d\gamma_t d\varphi_t dt,\ K_2(u) = \int_{M_{T,u}} G(\gamma_T)\rho(T,\gamma_T,\varphi_T)dx_T$$

The integrand $\Phi(\varphi)$ is penalty function defined as follows:

$$\Phi(\varphi) = g(\varphi,\varphi_{\min}(t),\varphi_{\max}(t)),\ g(\omega,\omega_1,\omega_2) = \begin{cases} (\omega-\omega_1)^2, & \omega < \omega_1 \\ 0, & \omega \in [\omega_1,\omega_2] \\ (\omega-\omega_2)^2, & \omega > \omega_2 \end{cases}.$$

Here $\varphi_{\min}(t), \varphi_{\max}(t)$ are given functions describing the admissible phase boundaries. The integrand for the second particular criterion is $G(\gamma) = (\gamma - \hat{\gamma})^2$. The third criterion $K_3(u) = k_c$ is not represented in the integral form and is calculated numerically.

So we consider the criteria vector $K(u) = (K_1(u), K_2(u), K_3(u))$ instead of the functional (1). The problem is formulated to minimize the criteria $K_i(u)$, $i = \overline{1,2}$ and to maximize the criterion $K_3(u)$.

Now we will transform the problem formulation to obtain the convenient visual representation of the Pareto set (i.e. the set of unimprovable criteria vectors). Let us choose the first two criteria to be main and impose the restriction on the third criterion:

$$K_3(u) \geq k_c^*,$$

where $k_c^*$ is capture coefficient lower boundary. Using this approach we can obtain the Pareto set for two main criteria (it is the compromise curve).

## 4. Optimization algorithm

The algorithm of approximate compromise curve constructing is described in [9-10] in detail. In short form it can be presented as follows. Let us assume the control vector $u$ to depend on the vector of parameters $\theta = (\theta_1, \ldots, \theta_p)$ taking values in a given set $\Theta \subset E^p$, i.e. $u = u(\theta)$.

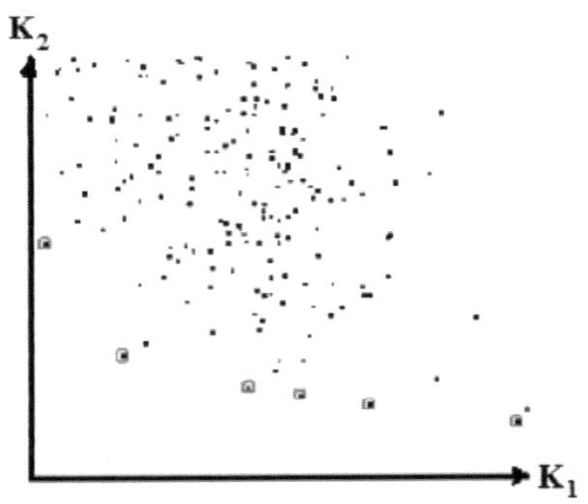


**Figure 1.** The points of the set V in main criteria space.

1). Simulation of $N$ random vectors $\theta^{(i)}, i = \overline{1,N}$ uniformly distributed in $\Theta$. The simulation methods are described in [14,19]. Control functions $u^{(i)} = u(\theta^{(i)})$, $i = \overline{1,N}$ obtaining.

2). Approximate calculation of trajectories of $J$ model particles for every control $u^{(i)}$, $i = \overline{1,N}$. Calculation of criteria values $K_1(u^{(i)})$, $K_2(u^{(i)})$, $K_3(u^{(i)})$, $i = \overline{1,N}$ with the use of Monte Carlo method expounded in [15,19].

3). Constructing of the set $V = \{(K_1(u^{(i)}), K_2(u^{(i)})) : K_3(u^{(i)}) \geq k_c^*, i = \overline{1,N}\}$ in main criteria space.

**Table 1.** The values of main criteria and capture coefficient.

| $K_1$ | $K_2$ | $K_3$ |
|---|---|---|
| 0.0027 | 1.4563 | 0.94 |
| 0.0066 | 1.1284 | 0.90 |
| 0.0080 | 0.9685 | 0.92 |
| 0.0014 | 1.6221 | 0.90 |
| 0.0120 | 0.6424 | 0.91 |
| 0.0065 | 1.2440 | 0.90 |

4). Selection of unimprovable points in the set $V$ according to the method given in [9-10], [14]. The points selected constitute the approximate compromise curve. The corresponding controls are called the *approximately efficient controls*.

## 5. Numerical results

The device under consideration is travelling-wave accelerator LEA-15-M [9] with initial energy $W_0 = 40$ keV, structure length $L = 0.78$ m and accelerating wavelength $\lambda = 0.1$m. The numerical experiment was carried out for $N = 150$ random vectors of parameters and $J = 50$ model particles.

The points of the set $V$ in main criteria space are presented at Fig. 1. The points of the approximate compromise curve are marked.

Table 1 presents the values of main criteria for the points of the approximate compromise curve and corresponding values of capture coefficient.

The controls obtained as a result of multicriteria optimization provide sufficiently high beam quality at accelerator exit. A user can select the proper control among approximately efficient controls taking into account beam and device characteristics obtained.

## 6. Conclusion

When treating beam dynamics optimization problems, we often use several criteria, sometimes
In this paper multicriteria approach is applied to beam dynamics optimization problem. Multicriteria optimization of longitudinal beam dynamics in linear accelerator is carried out. The approximate compromise curve is obtained.

Multicriteria optimization method is especially effective in combination with directed methods based on the analytical representation of quality functional variation obtained by Ovsyannikov [1]. Any approximately efficient control may be used as initial control for the directed method.

**Acknowledgments**

The author is grateful to Professor D A Ovsyannikov for valuable advices.

**References**

[1] Ovsyannikov D A 1990 *Modeling and Optimization of Charged Particle Beam Dynamics* (Leningrad: Leningrad State University Press)

[2] Vladimirova L V 1995 Optimization of interacting particle beam dynamics in linear accelerator *Journal of Computer and Systems Sciences Int.* **6** 178-83

[3] Bondarev B I, Durkin A P et al 2001 The LIDOS.RFQ.DESIGNER development. *Proc. IEEE Particle Accelerator Conf. (Chicago, IL)* sponsors: US Department of Energy Office of Naval Research ed P Lucas and S Webber (IEEE) pp 2947-49

[4] Ovsyannikov D A, Ovsyannikov A D, Antropov I V and Kozynchenko V A 2015 Software complex BDO-RFQ *Proc. III Int. Conf. "Stability and Control Processes" in Memory of V.I.Zubov (SCP) (5-9 October 2015 St. Petersbirg*) ed L A Petrosyan and A P Zhabko (St. Petersburg: IEEE) pp 335-37 DOI: 10.1109/SCP.2015. 7342132

[5] Ovsyannikov D A, Ovsyannikov A D, Antropov I V and Kozynchenko V A 2005 BDO-RFQ code and optimization models *2$^{nd}$ Int. Conf. on Physics and Control (PhysCon 2005) (24-26 August 2005 St. Petersburg)* 1513994 pp 282-88

[6] Ovsyannikov A D, Ovsyannikov D A, Balabanov M Yu and Chung S-L 2009 On the beam dynamics optimization problem *Int. Journal of Modern Physics* A **24** (5) 941–51

[7] Svistunov Yu A, Zuev Yu V, Ovsyannikov A D and Ovsyannikov D A 2011 The elaboration small size deuteron Accelerator for neutron generator 1 MeV energy *Vestnik St. Petersburg University, Ser. 10: Applied Mathematics, Informatics, Control Processes* **1** 49-59

[8] Ovsyannikov D A 2012 Mathematical modeling and optimization of beam dynamics in accelerators *Proc. XXIII Russian Particle Accelerator Conf. RuPAC-2012 (24-28 September 2012 St. Petersburg)*(Geneva: JACoW http://www.JACoW.org) pp 68-72

[9] Vladimirova L V 2014 Multicriteria optimization of beam dynamics *The Bulletin of Irkutsk State University. Series "Mathematics"* **7** pp 3-18

[10] Sobol I M and Statnikov R B 2006 *Optimal Parameters Choice in the Problems with Multiple Criteria* (Moscow: Nauka)

[11] Vladimirova L V Global extremum search on the basis of density and its mode estimation *Proc. 20th Int. Workshop: Beam Dynamics & Optimization (BDO) (30 June-4 July 2014 St. Petersburg)* (St. Petersburg: IEEE) p 186

[12] Gubanov V A, Zaharov V V and Kovalenko A N 1988 *Introduction to System Analysis* (Leningrad: Leningrad University Publishing)

[13] Podinovskiy V V and Nogin V D 1982 *Pareto-optimal Solutions of Multicriteria Problems*

Journal of Physics: Conference Series **747** (2016) 012070 doi:10.1088/1742-6596/747/1/012070

(Moscow: Nauka)

[14] Vladimirova L V 2001 On application of multicriterial optimization in partical beam control problem *Proc. 11th Int. IFAC Workshop: Control Applications of Optimization* ed V Zakharov vol 1 (Oxford: Pergamon Press) pp 363-67

[15] Vladimirova L V 2014 Monte Carlo methods in beam dynamics optimization problem *Vestnik St. Petersburg University, Ser. 10: Applied Mathematics, Informatics, Control Processes* **1** 30-39

[16] Drivotin O I and Ovsyannikov D A 2006 Modeling of self-consistent distributions for longitudinally non-uniform beams. *Nuclear Instruments and Methods in Physics Research* A **558** (1) 112-118

[17] Vladimirova L V and Fatyanova I A 2015 Construction of regression experiment optimal plan using parallel computing. *Proc. III Int. Conf. "Stability and Control Processes" in Memory of V.I.Zubov (SCP) (5-9 October 2015 St. Petersbirg*) ed L A Petrosyan and A P Zhabko (St. Petersburg: IEEE) pp 361-363 DOI: 10.1109/SCP.2015.7342140

[18] Rubtsova I D 2015 Optimization of iterative beam dynamic process *Proc. III Int. Conf. "Stability and Control Processes" in Memory of V.I.Zubov (SCP) (5-9 October 2015 St. Petersbirg*) ed L A Petrosyan and A P Zhabko (St. Petersburg: IEEE) pp 198-200 DOI: 10.1109/SCP.2015.7342092

[19] Ermakov S M 1975 *Monte Carlo Methods and Related Questions* (Moscow: Nauka)

Journal of Physics: Conference Series **747** (2016) 012071 doi:10.1088/1742-6596/747/1/012071

# Results of using the axisymmetric RF focusing by means of field spatial harmonics at 7 MeV proton linac

**V S Dyubkov**[1]

Electrophysical facilities Department, National Research Nuclear University MEPhI, Kashirskoe shosse 31, Moscow 115409, Russian Federation

E-mail: VSDyubkov@mephi.ru

**Abstract**. For several decades, axially symmetric channels with RF focusing by means of nonsynchronous spatial harmonics of the accelerating field are offered as an attractive alternative to proven and reliable RFQ linacs. In a number of works an effectiveness of channels with axially symmetric RF focusing by means of the nonsynchronous harmonics of the field was demonstrated in the proton energy range up to 2 MeV. An effectiveness of discussed channels for protons at energies up to 7 MeV is considered in this paper. Numerical simulation results of proton self-consistent dynamics in a channel with axisymmetric RF focusing are presented and discussed in this article.

## 1. Introduction

The emergence of particle accelerators stimulated a development of various fields of science. In turn there has been a rapid growth of the industry and its applications. Today modern high-tech medicine cannot exist without accelerators. Proton and ion accelerators are used for medical radionuclide production, as well as for radiation therapy. Generally radionuclide production is performed by means of cyclotrons, to be used for medical diagnostics ($^{18}$F, $^{13}$N, $^{11}$C, $^{15}$O, $^{64}$Cu, $^{67}$Ga, $^{68}$Ga, $^{111}$In, $^{123}$I, $^{124}$I, $^{81m}$Kr, $^{99}$Mo, $^{82}$Rb, $^{201}$Tl) or therapies ($^{67}$Cu, $^{57}$Co, $^{82}$Sr, $^{68}$Ge) in nuclear medicine [1]. There are several commercial companies with a number of basic cyclotrons (Advanced Cyclotron Systems Inc., Advanced Biomarkers Technology, Best Cyclotrons Systems Inc., GE Healthcare, IBA, Siemens, JSC "Efremov Institute of Electrophysical Apparatus" and other). The cyclotrons are well recommended and reliable machines. Nevertheless there are some aspects of cyclotrons related to biological shield, to high weight, as well as to its rather high power supply [2]. Systems based on linacs are free from these weak points. It makes sense to enlarge the use of linacs for positron emission tomography (PET) isotopes production. The widely used radioisotope for non-invasive PET is the short-lived positron emitter $^{18}$F. It can be produced due to reactions $^{18}$O(p,n)$^{18}$F, $^{20}$Ne(d,α)$^{18}$F, $^{17}$O(d,n)$^{18}$F. Portable complexes PULSAR™ 7 (AccSys Technology Inc.) for $^{18}$F production, based on linac, exist and it is commercially realized nowadays [3]. The PULSAR™ 7 linac consists of 3.5 MeV RFQ part and 3.5 MeV DTL, which provides protons with energy is equal to 7 MeV. There is a project of $^{3}He^{2+}$ RFQ accelerator for the production of PET isotopes [4, 5]. This 10.5 MeV linac was planned to be based on RFQ sections only. Possible PET isotope production using linear deuteron accelerators is discussed in [6]. There are several proposals to use initial parts of the existing middle energy linacs (up to 200 MeV) for PET radionuclide production in scientific centers [7–9]. A possibility of using the RF

[1] To whom any correspondence should be addressed.

focusing by means of the nonsynchronous spatial electromagnetic (EM) field harmonic at proton linac with energy up to 7 MeV is considered in this paper and suggested as a part of PET isotope production complex. A structure with RF focusing by means of the nonsynchronous spatial EM field harmonic is regarded as alternative to structure with RFQ for protons energies up to 3 MeV traditionally [10]. This structure allows one to increase beam current in compare with RFQ structure. An effectiveness of applying the RF focusing by means of the nonsynchronous harmonic for low energy protons was presented in the earlier papers [11, 12]. The main goal of current paper is to investigate applying the RF focusing by means of the nonsynchronous harmonic for protons with energies in the range from 2 MeV to 7 MeV.

## 2. Basic relations

The analytical investigation of the beam dynamics in a polyharmonical field is a difficult mathematical problem. Rapid longitudinal and transverse oscillations as well as a strong dependence of field components on transverse coordinates do not allow one to use the linear approximation in the paraxial region for a field series. Nevertheless, the analytical beam dynamics investigation in the oscillating fields can be carried out by means of the smooth approximation. RF field is expressed as the Fourier expansion by the standing wave spatial harmonics in an axisymmetric periodic resonant structure as it was done in [11], assuming that the structure period is a slowly varying function of the longitudinal coordinate.

$$E_{\|} = \sum_{n=0}^{\infty} E_n I_0(k_n r)\cos\left(\int k_n dz\right)\cos\omega t;\; E_{\perp} = \sum_{n=0}^{\infty} E_n I_1(k_n r)\sin\left(\int k_n dz\right)\cos\omega t,$$

$$k_n = (\theta + 2\pi n)/D$$

where $E_n$ is the $n$th harmonic amplitude of RF field on the axis; $k_n$ is the propagation wave number for the $n$th RF field spatial harmonic; $D$ is the resonant structure geometric period; $\theta$ is the phase advance per $D$ period; $\omega$ is the circular frequency; $I_0$, $I_1$ are modified Bessel functions of the first kind.

It is considered here that there are two spatial harmonics at the linac. One of it is the synchronous harmonic with $s = 0$, and another one is the nonsynchronous (focusing) with $n = 1$. In order to estimate basic linac parameters one can use results presented in [11] for the mentioned conditions and $\theta = \pi$. Thus, one can write the following expressions for squares of eigenfrequencies of small linear vibrations:

$$\omega_{\|}^2 = -\frac{1}{2} e_0 \sin\varphi_s - \frac{1}{8} e_0 e_1 \cos 2\varphi_s, \tag{1}$$

$$\omega_{\perp}^2 = \frac{1}{4} e_0 \sin\varphi_s + \frac{13}{64} e_0 e_1 \cos 2\varphi_s + \frac{3}{128} e_0^2 + \frac{135}{512} e_1^2, \tag{2}$$

where

$$e_i = \frac{eE_i\lambda}{2\pi m\beta_s c^2}$$

$e$ and $m$ are a charge and a mass of particle, $\lambda$ is a free-space RF wavelength, $c$ is the speed of light, $\beta_s$ is the normalized velocity of the synchronous (equilibrium) particle and $\varphi_s$ is the synchronous particle phase. It is necessary that the parameters of the channel will be chosen in terms of the conditions $\omega_{\|}^2 > 0$ and $\omega_{\perp}^2 > 0$ (for the simultaneous transverse and longitudinal focusing). In terms of the equation (2) one can obtain a set of curves defining threshold value of $e_0/e_1$ versus $\varphi_s$ under different $e_1$ value (for instance see figure 1).

On the basis of the eigenfrequencies analysis the following fixed basic parameters of linac sections are chosen and presented in the table 1.

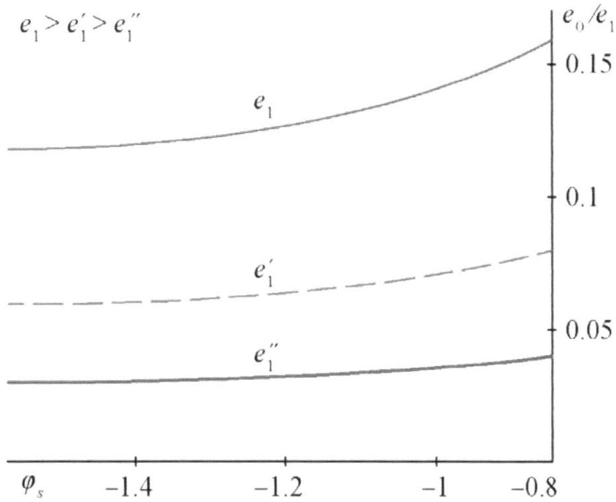


**Figure 1.** Harmonics amplitudes ratio vs equilibrium particle phase.

**Table 1.** Basic linac parameters.

| Parameter | Section 1 | Section2 |
|---|---|---|
| Operational frequency, MHz | 162 | 162 |
| Section length, m | 2 | 2 |
| Linac half-aperture, mm | 5 | 5 |
| Equilibrium particle phase | −60° | −45° |
| Eynchronous harmonic amplitude, kV/cm | 36 | 40 |
| Harmonics amplitudes ratio, % | 11 | 12 |

## 3. Numerical simulation results

Self-consistent beam dynamics simulations were conducted by means of a modified version of the specialized computer code BEAMDULAC–ARF3 based on CIC technique to calculate beam self-space-charge field [10].

The following input beam parameters were used for a simulation in the first linac section: injection energy was equal to 2 MeV, energy spread – 30%, phase width – 190°, radius – 2.5 mm, current – 100 mA, transversal rms emittance – 1.5π mm·mrad. Particle distributions in the longitudinal and transversal phase spaces were uniform and "water-bag" respectively.

Beam emittances at the first section output are presented in figure 2 (line 1 is rms emittance and 2 is Floquet ellipse at the part b of figure). One can see that the beam radius increased slightly under chosen parameters of the first linac section. Output beam energy was equal to 4 MeV and output transversal rms emittance – 8π mm·mrad under 99% current transmission.

Journal of Physics: Conference Series **747** (2016) 012071 doi:10.1088/1742-6596/747/1/012071

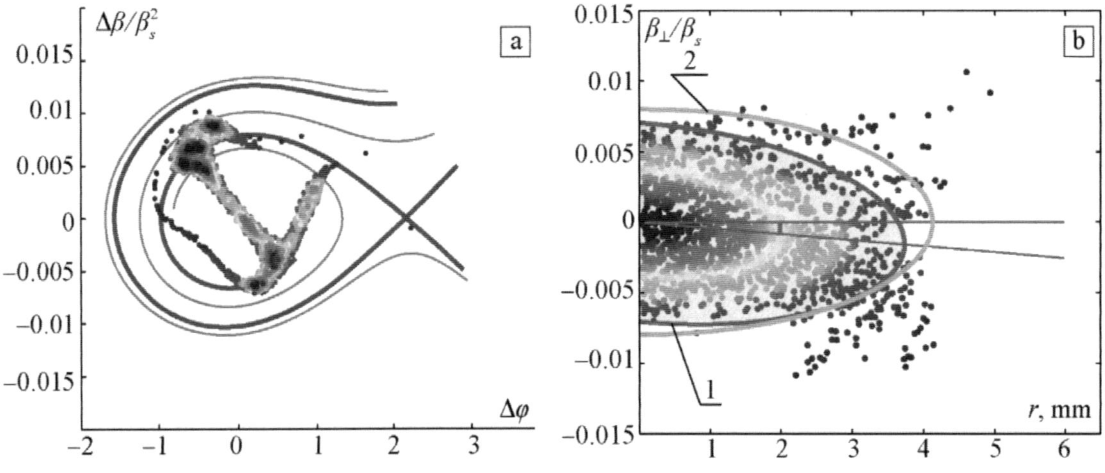


**Figure 2.** Output longitudinal (a) and transversal (b) beam particles distributions.

Further, computer simulation of self-consistent beam dynamics was carried out for the second section. In order to evaluate the effectiveness of RF focusing by means of the nonsynchronous harmonic in the proton energy range from 4 MeV to 7 MeV the following beam parameters were used: injection energy was equal to 4 MeV, energy spread – 30%, phase width – 130°, radius – 2.5 mm, current – 100 mA, transversal rms emittance – $\pi$ mm·mrad. Particle distributions in the longitudinal and transversal phase spaces were the same as for the first section. Note, that synchronous phase value was changed along this section unlike the first section where it was the constant.

Beam emittances at the second section output are presented in figure 3. Output beam energy was equal to 7 MeV, output transversal rms emittance – 5$\pi$ mm·mrad, current transmission – 96%.

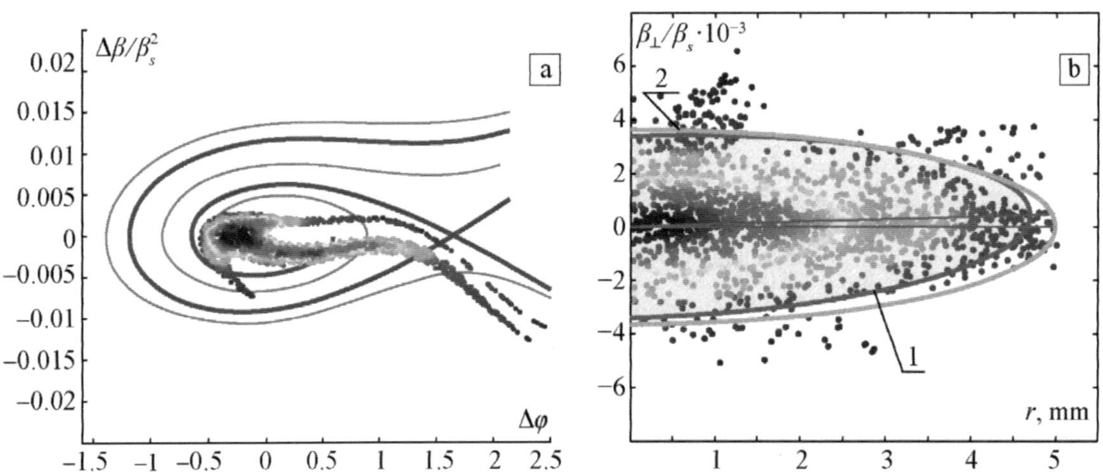


**Figure 3.** Output longitudinal (a) and transversal (b) beam particles distributions.

Numerical simulations showed that a rigidity of RF focusing by means of the first spatial harmonic decreased in the energy range from 5 to 7 MeV under chosen parameters. The particle loss is observed in the transversal direction. For the rigidity of RF focusing to be increased maximal amplitude value of electric field must be ramped up and/or harmonics amplitudes ratio must be decreased for the fixed value of equilibrium particle phase. Note, that maximal amplitude value of electric field is limited by field value at which discharges appear (Kilpatrick limit). In turn, small value of harmonics amplitudes ratio leads to energy gain decrease.

## 4. Conclusion

Using the $\pi$-mode structures with the focusing by means of the first spatial harmonic of accelerating field in proton energy range from 2 MeV to 7 MeV is considered. Self-consistent proton beam dynamics simulations were carried out. It is shown that the focusing technique considered can be used at 7 MeV proton linacs, which can be used for PET isotope production and neutron generation.

## References

[1] Reina A and Jimenez V 2011 *12 Chapters on Nuclear Medicine* ed A Gholamrezanezhad (Rijeka: InTech) chapter 3 pp 47–72
[2] Mandelkern M A 1995 *Annu. Rev. Nucl. Part. Sci.* **45** 205
[3] Pramudita A 2011 *Atom Indonesia* **37**(1) 1
[4] Pasquinelli R J 1997 *Proc. Particle Accelerator Conf. (Vancouver)* vol 3 (IEEE) p 3804
[5] Link J M, Krohn K A, Bida J T, DeHaas R, Larson D J, Johanning J, Young P E and Pasquinelli R J 1999 *AIP Conf. Proc. (Texas)* **475** 1010
[6] Volkovitsky P and Gilliam D M 2005 *NIM* A **548** 571
[7] Picardi L *et al* 1997 Progetto del TOP Linac *ENEA-CR (Frascati)* RT/INN/97-17
[8] Akulinichev S V, Baklanov S I, Gavrilov Yu K, Matushko V L and Sorokin V M 2008 *Problems of atomic science and technology (Nuclear Physics)* **49**(3) 184
[9] Courtney W, Sowder K, McGyver M, Stevenson N and Brown D 2007 *NIM* B **261** 739
[10] Masunov E S and Vinogradov N E 2001 *Zhurnal Tekhnicheskoi Fiziki* **71**(9) 79
[11] Dyubkov V S and Masunov E S 2009 *Int. J. Modern Phys.* A **24**(5) 843
[12] Dyubkov V S and Masunov E S 2010 *Problems of atomic science and technology (Nuclear Phys. Investigations)* **54**(3) 94

Journal of Physics: Conference Series **747** (2016) 012072 doi:10.1088/1742-6596/747/1/012072

# BDO-RFQ Program Complex of Modelling and Optimization of Charged Particle Dynamics

**D A Ovsyannikov, A D Ovsyannikov, I V Antropov[1] and V A Kozynchenko**

Saint-Petersburg State University, 7/9 Universitetskaya nab., Saint-Petersburg, 199034, Russia

Email: i.antropov@spbu.ru

**Abstract.** The article is dedicated to BDO Code program complex used for modelling and optimization of charged particle dynamics with consideration of interaction in RFQ accelerating structures. The structure of the program complex and its functionality are described; mathematical models of charged particle dynamics, interaction models and methods of optimization are given.

## 1. Introduction

When an accelerator of the charged particles is being developed an important task is an optimization of charged particle dynamics in order to improve the quality of the output beam. Different mathematical models of beams dynamic, models of particles interaction and methods of optimization are used for optimization problem solving in accelerator structures. BDO Code is a program complex, which has a developed graphic interface and different sets of instruments, methods and algorithms and allows researcher to concentrate time on implementation of the model. BDO RFQ project is included into this complex and it's general description is given in this article.

## 2. Program Complex

BDO Code Program Complex consists of managing program, models and libraries (Fig. 1):

- BDO Shell – managing program, which has developed graphic interface, libraries of the parameters of the model and system functions. The specialties of this product are a possibility to divide into the stages of the calculation process with setting of your own group of input and output data files; a possibility of automatization of the modelling process for the given parameters ranges; usage of the scenarios with setting of the determined parameters, input and output data files, setting up of the calculation stages.
- BDO RFQ – project for modelling and optimization of charged particle dynamics in RFQ structure [1]. The project contains several models, which allow to perform a step-by-step optimization on "from easy to hard" basis, and library of the consideration of the particle interaction and library of optimization methods.

## 3. Models

Consider in details the models realized in BDO-RFQ project [2-7]:

[1] To whom any correspondence should be addressed.

*3.1. BDO RFQ TW (Travelling wave)*

This model was proposed in works [5,6] and successfully used for modelling and optimization of the longitudinal motion of the beam of charged particles in the field of equivalent travelling wave in RFQ structure [5-8]. The functional realized in this model allow modelling the beam dynamic with interaction of particles, to perform the optimization with usage of the gradient method. Main window BDO RFQ TW is presented on Figure 2.

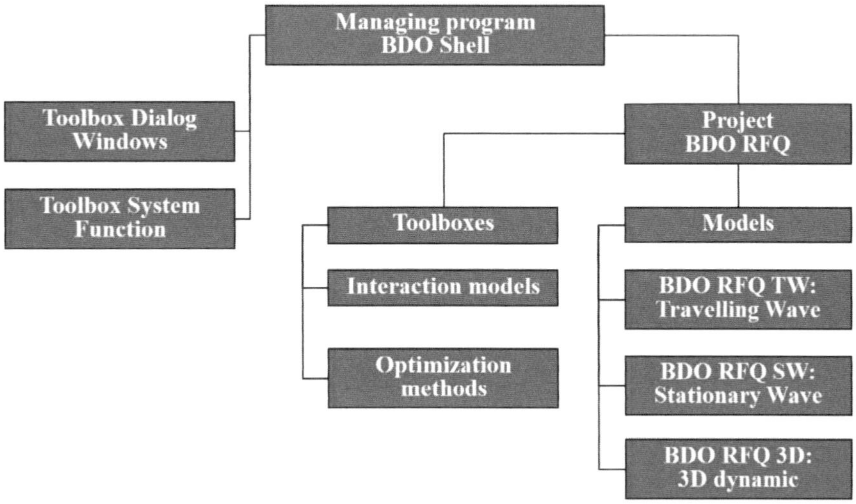


**Figure 1.** Structure of BDO Code Program Complex

The advantages of this model are the following:

- The consideration of the particle interaction is performed by the method of macro particles for conventional and periodic cases;
- The joint optimization for program and perturbed motions is performed;
- Different quality functionals are used for characteristics at the exit of the accelerator and also for along the structure;
- Recalculation of the managing parameters of the model after optimization for the model of the dynamics in stationary wave.

*3.2. BDO RFQ SW (Stationary wave)*

This model is used for modelling and optimization of the longitudinal and transversal motion of the beam of charged particles in the field of stationary wave in RFQ structure. The realized functional of this model allows to calculate acceptance of the channel, to model and optimize the matcher, to model the dynamic of the beam for the envelopes with the interaction of particles, to optimize by different methods.

The advantages of this model are the following:

- The consideration of the particle interaction is performed with cylindrical models;
- The joint optimization for longitudinal and transversal motion of the beam is performed;
- Optimization of a radial matching section.

*3.3. BDO RFQ 3D*

The results of the optimization are checked with the current 3D dynamic model. The consideration of the particles interaction is performed by the method of the pair interaction or by the grid method.
There is also a possibility of comparison of the coulomb field on the axis of accelerator calculated with the different models of the particle interaction.

## 4. Interaction models toolbox

In modelling of the dynamic of the charged particles beams with particle interaction in BDO Code program complex the following calculation methods of the coulomb field of the beam are realized:

- Method of thin discs – for calculation of the coulomb field, the beam is a set of infinitely thin and proportionally charged discs with the same radius.
- Method of thick discs – for calculation of the coulomb field, the beam is a set of proportionally charged discs with the finite thickness of the same radius.

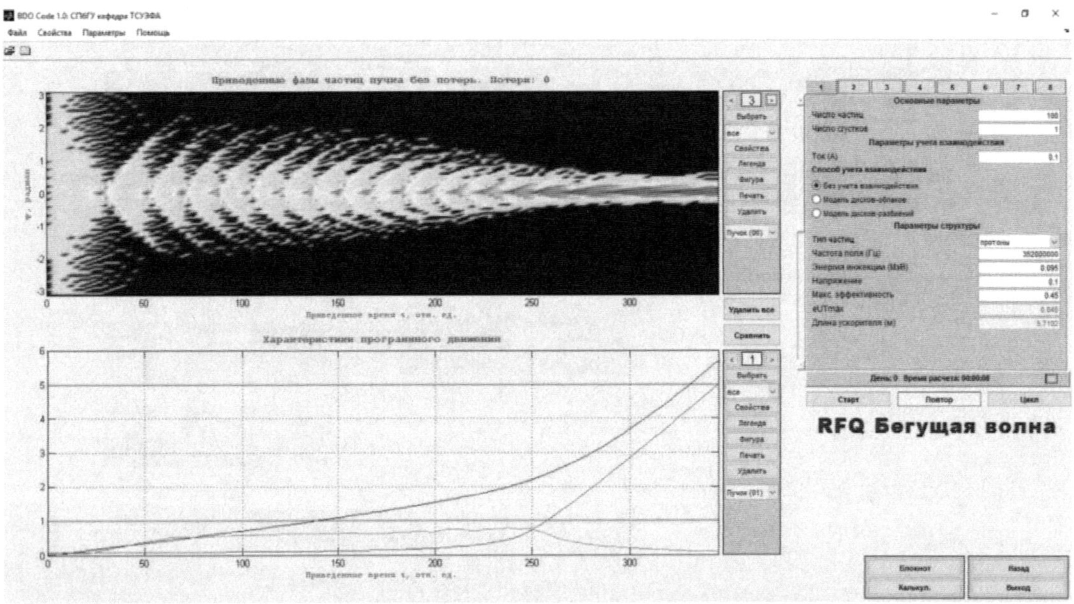


**Figure 2.** Main window BDO RFQ TW

These methods allow to calculate approximately only longitudinal component of the coulomb field of the beam and are used for modelling of the longitudinal beam dynamic and also for comparison of the calculation results of longitudinal component of the coulomb field on the axis of accelerator by other methods.

- Cylindrical model – for calculation of the coulomb field the beam is a cylinder with a constant radius, proportionally charged in each transversal section with heterogeneous distribution of the charge along the axis of accelerator.
- Method of nested circular cylinders [9] – for calculation of the coulomb field the beam is a set of embed cylinders with constant radius. Each cylinder is proportionally charged in each transversal section with heterogeneous distribution of the charge along the axis of accelerator.
- Pair interaction method - for calculation of the coulomb field the beam is a set of proportionally charged balls with the same radius. The coulomb field of the beam is calculated as a vector sum of the fields, made by each ball.
- Grid method - the coulomb field the beam is calculated by the grid method, boundary value problem for Poisson equation is solved.

## 5. Optimization toolbox

For optimization of charged particle dynamics in BDO-RFQ system, various optimization methods have been developed [7,10-14]. In directed optimization methods of charged particle dynamics external accelerating and focusing fields in the accelerator and internal coulomb field of the beam are calculated with analytic approximate expressions. Analytic models of the external and internal fields allow using analytic expressions for the gradient of the optimized functional. In optimization with analytic formulas for functional gradient in BDO-RFQ system there are two optimization methods realized: gradient descent method (method of steepest descent) and ravine method. Along with directed optimization techniques and other methods of numerical optimization are used. There are two numerical optimization methods used: genetic algorithm and Box-Wilson method. When genetic algorithm is used the system user defines the parameters of the crossing, mutation and population size.

In conclusion we would like to mention, that there is a possibility to exchange files of BDO-RFQ program complex with files of LIDOS.RFQ program complex [6].

**References**

[1] Kapchinsky I M and Teplyakov V A 1970 Linear Ion Accelerator with Spartially Homogeneous Strong Focusing *Prib. Tekh. Eksp.* no 2 pp 19-22

[2] Ovsyannikov D A, Ovsyannikov A D, Antropov I V and Kozynchenko V A 2015 Software complex BDO-RFQ *Proc. Int. Conf. Stability and Control Processes in Memory of V.I. Zubov (SCP)* (Saint-Petersburg: Saint-Petersburg State University) pp 335-37 DOI: 10.1109/SCP.2015.7342132

[3] Ovsyannikov D A 2012 Mathematical modeling and optimization of beam dynamics in accelerators *RuPAC 2012 Contributions to the Proc. – 23rd Russian Particle Acclerator Conference* pp 68-72

[4] Ovsyannikov D A, Ovsyannikov A D, Antropov I V and Kozynchenko V A 2005 BDO-RFQ code and optimization models *Proc. Int. Conf. on Physics and Control* pp 282-88 DOI: 10.1109/PHYCON.2005.1513994

[5] Bondarev B I, Durkin A P and Ovsyannikov A D 1999 New mathematical optimization models for RFQ structures *Proc. of the IEEE Particle Acclerator Conf. The 18th Biennial Particle Acclerator Conf.* (New York, NY, USA) pp 2808-10

[6] Bondarev B, Durkin A, Ivanov Y, Shumakov I, Vinogradov S, Ovsyannikov A and Ovsyannikov D 2001 The Lidos RFQ designer development *Proc. of the IEEE Particle Acclerator Conf. 2001 Particle Acclerator Conf.* (Chicago, IL, USA) pp 2947-49

[7] Ovsyannikov D A, Ovsyannikov A D, Vorogushin M F , Svistunov Yu A and Durkin A P 2006 Beam dynamics optimization: models, methods and applications *Nuclear Instruments and Methods in Physics Research. Section A: Accelerators, Spectrometers, Detectors and Associated Equipment* vol 558 no 1 pp 11-19

[8] Svistunov Y, Durkin A and Ovsyannikov A D 2012 Beam dynamics investigations for 433 MHz RFQ accelerator *RuPAC 2012 Contributions to the Proc. – 23rd Russian Particle Acclerator Conference* pp 82-84

[9] Kozynchenko V A and Boyko A A Modeling and simulation of beam dynamics in linear accelerator with RFQ *20th International Workshop on Beam Dynamics and Optimization (BDO 2014)* pp 96-97 DOI: 10.1109/BDO.2014.6890041

[10] Ovsyannikov A D, Ovsyannikov D A, Balabanov M Yu and Chung S L 2009 On the beam dynamics optimization problem *Int. Journal of Modern Physics A* vol 24 no 5 pp 941-51

[11] Ovsyannikov D A 1997 Modelling and optimization problems of charged particle beams dynamics *ECC 1997 – European Control Conf. 4* pp 1463-67

[12] Antropov I V and Ovsyannikov A D 2013 Modelling and optimization radial motion in charged particles accelerator with Radio Frequency Quadrupole *Vestnik of St. Petersburg State University of Technology and Design (Series 1 Natural and Engineering Sciences)* no 4 pp 19-23

[13] Antropov I V and Ovsyannikov A D 2009 Modelling and optimization in Radio Frequence Quadrupole structure *Vestnik of Saint-Petersburg State University (Series 10: Applied mathematics. Computer science. Control processes)* no 4 pp 12-24

[14] Ovsyannikov A D, Ovsyannikov D A and Chung S L 2009 Optimization of a radial matching section *Int. Journal of Modern Physics A* vol 24 no 5 pp 952-58

# Comparison of accelerating structures for the first cavity of the main part of the INR linac

**I V Rybakov[1], Y Z Kalinin, V N Leontev, A N Naboka, V V Paramonov, V L Serov and A V Feschenko**

Institute for Nuclear Research of the RAS, 117312, 60-th October Anniversary Prospect, 7a, Moscow, Russia

E-mail: irybakov@inr.ru

**Abstract**. For the beam power improvement of the hydrogen-ion INR linac replacement of the first four-section cavity in the main part of linac is required. Existent cavity is realized using DAW structure on 991 MHz operating frequency. The new cavity should at least not lose in parameters to the current structure and essential changes in other linac systems are not wish able. Parameters of accelerating structures possible for such application are compared.

## 1. Introduction

The main objective of our work is to perform a comparison between different accelerating structures which could be used in the first cavity of the INR linac, and choose the optimal option for practical realization of a single cavity. We consider both time-proven in existent intense linacs structures –Disk and Washer (DAW) structure [1], Side Coupled Structure (SCS) [2], Annular Coupled Structure [3], and promising development Cut Disk Structure (CDS) [4]. These structures are shown schematically at figure 1.

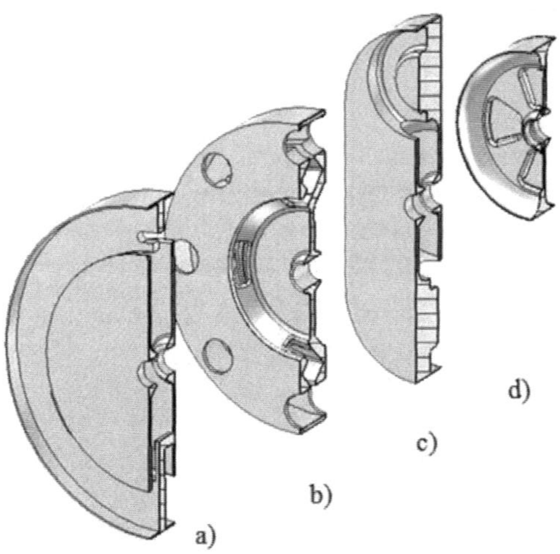


**Figure 1**. Artistic view of the structures: a) DAW, b) ACS, c) SCS, d) CDS

[1] To whom any correspondence should be addressed.

All structures are compared for parameters of the first INR linac cavity, operating frequency $f_a$=991 MHz, relative particles velocity β=0.4313, aperture radius 17 mm, acceleration rate $E_0 T cos\varphi_s$=2.5 MV/m with synchronous phase $\varphi_s$=-33 degrees, RF pulse length 200 μs and repetition rate 100 Hz.

## 2. Electro dynamical characteristics

The reference DAW1 design [1], currently in use, has the maximum electric field at the drift tube surface $E_{s.max}$=0.5 $E_k$ where $E_k$ is the Kilpatrick limit. Drift tube optimization has been performed to improve the value of effective shunt impedance $Z_e$. Drift tube with two rounding radiuses and optimized gap ration was applied. With $E_{s.max}$ increasing to safe value of 1$E_k$ it results in approximately 10% $Z_e$ increase. The optimized drift tube shape was applied for all other structures under consideration. Calculated electro dynamical parameters of the structures are listed in table 1.

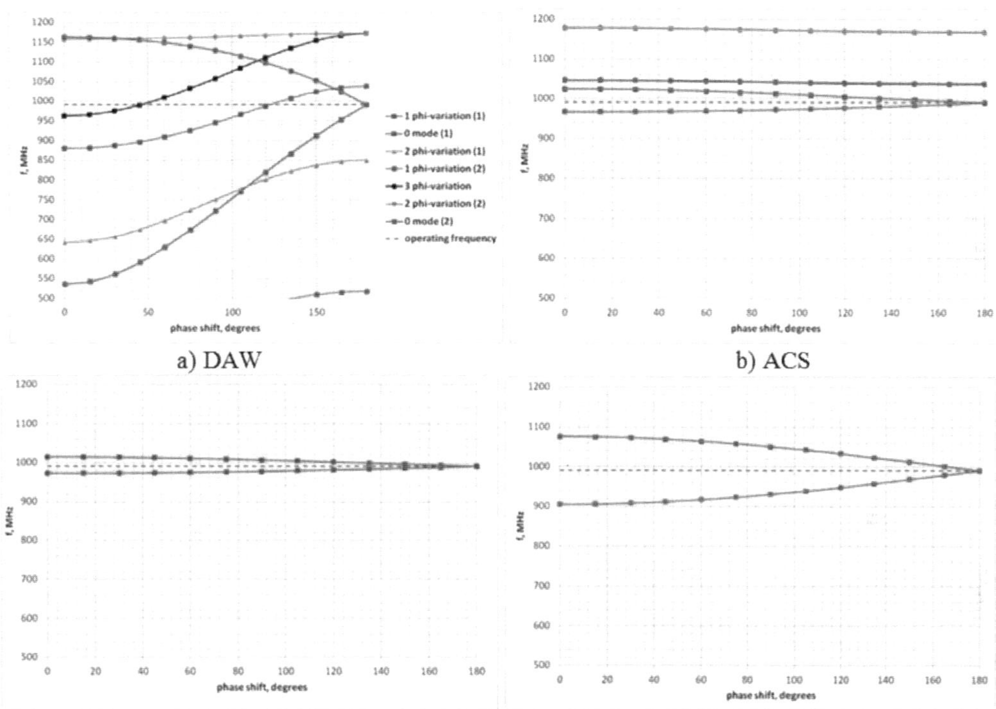


a) DAW b) ACS

c) SCS d) CDS

**Figure 2.** Brillouine diagrams for the structures

To compare dispersion properties the Brillouine diagrams for each structure at the same scale are presented at figure 2. The DAW structure has the maximal coupling coefficient $K_c$ but it has a set of parasitic modes near the operating frequency. The displacemet of these modes requires introduction of additional components into the structure which complicates the cavity production and RF tuning. The value $Z_e$ for the DAW1 option takes into account stems for washer support and the slots for parasitic modes shift in the exsisting cavity. It is a reference $Z_e$ value which cannot be lowered in the new cavity. The analysis for DAW, ACS, SCS and CDS structures was performed in 3-D approximation with all geometry features taken into account. As we see from table 1, with the optimized dtrift tube shape, ACS, SCS and CDS structures have $Z_e$ higher than reference value of DAW1 and satisfy requiremt of RF efficiency.

## 3. Structures cooling

In INR intense linac the first cavity should operate with rather high average heat load ≥7 kW/m. Two options of cooling circuits were considered for each structure, both with external cooling channels only and external and internal cooling channels together (see figure 3). Preliminary study has shown that cooling with external channels only results in not tolerable temperature rise $\Delta T$ and related frequency shift $\Delta f_a$ in each structure (see for example CDS1 in table 1). Internal cooling channels are required. CDS1 option has the highest $Z_e$ value due to thin walls. In this case internal channels are not possible. The second option CDS2 with increased walls thickness was considered instead of lower $Z_e$.

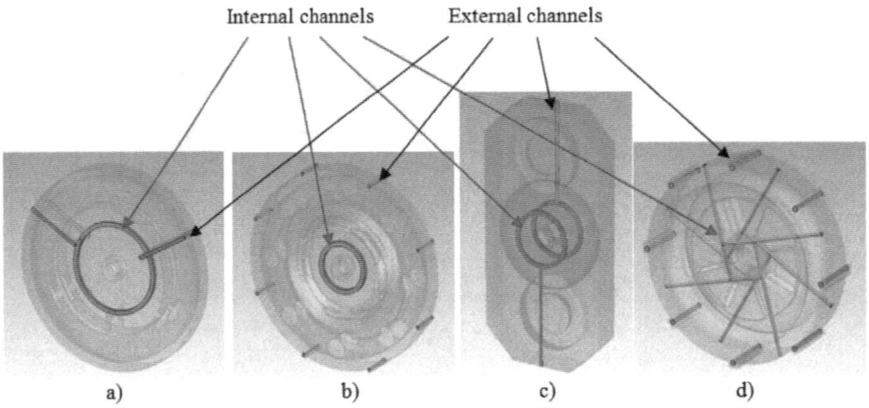


**Figure 3**. Cooling circuit: a) DAW, b) ACS, c) SCS, d) CDS2

The applied cooling circuits for every structure are presented at figure 3. The complete coupled analysis was performed with procedure described in [5] and based on ANSYS software [6]. The cooling conditions were considered for the safe value of the flow velocity 2 m/s at the turbulent regime with Reynolds number $Re \geq 10^4$. Simulated temperature distributions for structures are shown at figure 4 and the main results of this thermal-structural analysis are listed in table 1.

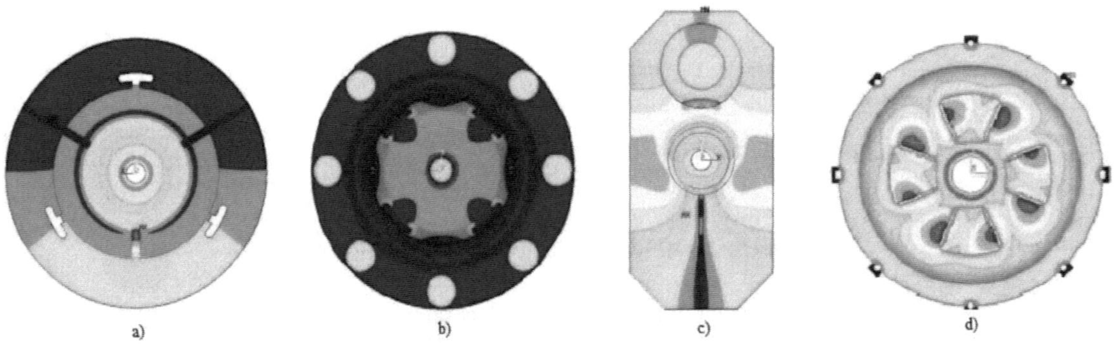


**Figure 4.** Thermal map: a) DAW, b) ACS, c) SCS, d) CDS2

In DAW, SCS and ACS internal cooling channels could be only produced with brazed joints water-vacuum. It reduces reliability of long term operation. In CDS2 design we can avoid such brazed joints [4].

In the table 1 $R_c$ is the structure outer radius, $\lambda$ is the operating wavelength, $Q$ is the quality factor, T is the transit time factor, $\Delta T$ is the temperature rise with respect of cooling water temperature and $\sigma_{max}$ is the maximum stress value in the cavity, assuming the Young module for fully annealed OFE copper ~ 100 GPa.

**Table 1.** Parameters of the structures

| Structure\ Parameter | $R_c/\lambda$ | $K_c$ | $Q*10^4$ | $Z_e$, MOm/m | $T$ | $E_{s.max}/E_0T$ | $\Delta T$, °C | $\sigma_{max}$, MPa | $\Delta f_a$, kHz |
|---|---|---|---|---|---|---|---|---|---|
| DAW1 | 0.71 | 0.52 | 2.08 | 23.65 | 0.82 | 5.83 | - | - | - |
| DAW2* | 0.71 | 0.56 | 2.34 | 30.76 | 0.85 | 9.29 | 14.98 | 44.50 | -86.19 |
| ACS | 0.67 | 0.05 | 1.56 | 28.66 | 0.84 | 8.62 | 15.55 | 5.00 | -61.36 |
| SCS | 0.77 | 0.04 | 1.70 | 30.04 | 0.84 | 8.81 | 8.53 | 6.09 | -101.74 |
| CDS1 | 0.37 | 0.17 | 1.74 | 31.16 | 0.83 | 7.94 | 13.49 | 31.00 | -365.75 |
| CDS2 | 0.35 | 0.16 | 1.47 | 25.64 | 0.86 | 8.40 | 3.10 | 2.74 | -46.83 |

*without stems for washer support

## 4. Results analysis

As we see from table 1 with the optimized drift tube shape all options exceed in $Z_e$ the reference structure DAW1 which is now in use. DAW structures has the highest $K_c$ value, but with a lot of HOM parasitic modes near the operating frequency. Removal of these modes requires additional efforts during structure construction and RF tuning. ACS and SCS have too low $K_c$ providing extra requirements for precision of structure manufacturing and tuning. The CDS1, 2 options realize acceptable intermediate $K_c$ value, which allow to avoid individual cell tuning [4].

For reliable operation internal cooling is necessary for all structures. With internal cooling expected internal stress values for all structures are below the limit of elastic deformations. In DAW, SCS and ACS internal cooling channels could be only produced with brazed joints water-vacuum. Avoiding such joints in CDS2 option will have more reliable structure operation.

As we see from figure 1, DAW, ACS and SCS structures have large outer diameter ~450 mm. Accelerating structures normally are produced using expensive material - OFE copper. Large outer diameter of DAW, ACS and SCS causes additional difficulties starting with raw material to additional requirements for processing equipment. The outer diameter of CDS structure is approximately two times lower and we can expect essential cost reduction in cavity construction of single cavity unit.

## 5. Conclusion

Consideration shows that in case of accepted DAW, ACS and SCS in construction of single cavity unit we will get similar problems in mechanical structure processing. Taking into account the existing experience in DAW construction and estimated parameters set, application of ACS and SCS looks not reasonable. Until now CDS is proven in electron ($\beta$=1) linacs [4]. The analysis shows that for proton acceleration with $\beta$=0.43 using optimized drift tube shape, CDS satisfies the requirements for applying into the first cavity of the main part of the INR linac, even with increased wall thickness. As we see from table 1 CDS has counter-balanced set of parameters. Avoiding brazed water-vacuum joints improves cavity reliability. Smaller transverse dimensions results at least in cost reduction for cavity manufacturing. Analysis performed shows that CDS structure is the most promising for structure replacement of the first cavity of the INR linac.

## References

[1] Murin B P 1981 *Linear ion accelerators* (Moscow M.-Atomizdat)
[2] Henderson S et. al. 2014 The SNS accelerator design. *NIM* **A763** p 610-673
[3] *Accelerator Technical Design Report for J-PARC* 2003 KEK Report 2002-13 Japan
[4] Paramonov V et. al 2010 The PITZ CDS Booster Cavity RF Tuning and Start of Conditioning. *Proc. 2010 Linac Conference* p 241
[5] Paramonov V, Skassyrskaya A, Hayashizaki N et. al. 2002 The complete 3-D coupled rf-thermal-structural-rf analysis *Proc. 2002 Linac Conference* p 216
[6] www.ansoft.com

# Analytical approach to quasiperiodic beam Coulomb field modeling

**I D Rubtsova**[1]

St. Petersburg State University, 7/9 Universitetskaya nab., St. Petersburg, 199034, Russia

E-mail: i.ribtsova@spbu.ru

**Abstract.** The paper is devoted to modeling of space charge field of quasiperiodic axial-symmetric beam. Particle beam is simulated by charged disks. Two analytical Coulomb field expressions are presented, namely, Fourier-Bessel series and trigonometric polynomial. Both expressions permit the integral representation. It provides the possibility of integro-differential beam dynamics description. Consequently, when beam dynamics optimization problem is considered, it is possible to derive the analytical formula for quality functional gradient and to apply directed optimization methods. In addition, the paper presents the method of testing of space charge simulation code.

## 1. Introduction

The problems of modeling of self-consistent charged particle distributions are urgent and widely discussed by the researchers [1-6]. This paper deals with analytical representation of space charge field. Such an approach is successfully used by many authors e.g. [1-4,7-13], when treating different problems of beam dynamics investigation and optimization.

The paper is devoted to modeling of Coulomb field of quasiperiodic axial-symmetric beam. Particle beam is represented by the set of charged disks. Two Coulomb field models considered are presented in terms of Fourier-Bessel series and trigonometric polynomial. The analytical expressions make it possible to analyze particle interaction and to test space charge field simulation program.

Both models allow integral representation of Coulomb force and provide the possibility of integro-differential beam dynamics description. Consequently, beam dynamics optimization problems may be treated on the basis of the approach suggested by D A Ovsyannikov [1]. Analytical expression for quality criterion gradient may be obtained; one can use directed optimization methods. This approach is successfully used in the papers [7,8,14-19].

## 2. Beam dynamics equations

For beam dynamics investigation particle-in-cell method is used. The beam is considered to have constant radius $R$ and to move inside the conducting channel of radius $a$. Every bunch is represented by $N$ disks-clouds (cylinders) with radius $R$ and thickness $2\Delta$. May independent variable be $\tau = ct$, where $c$ is the velocity of light, $t$ is the time. Particle phase state is characterized by the vector $(z, p)$, where $z$ is longitudinal coordinate, $p$ is reduced impulse. Particle dynamics equations are as follows:

[1] To whom any correspondence should be addressed.

$$\frac{dz}{d\tau} = \frac{p}{\sqrt{1+p^2}}, \quad \frac{dp}{d\tau} = \frac{e}{m_0 c^2}\left(E^{(RF)}(\tau,z) + E(\tau,z)\right). \tag{1}$$

Here $e = Q/N$, $m_0 = M_0/N$, $Q$ and $M_0$ are the charge and rest-mass of a bunch correspondingly; $E^{(RF)}$, $E$ are the longitudinal intensity components characterizing respectively the effect of RF field and space charge field on the disk as a whole.

## 3. Model 1 of particle interaction account: Fourier-Bessel series

Let us assume particle beam to be periodic sequence of bunches. May $M_\tau$ be the set of particle phase states of a bunch for $\tau$ value of independent variable. Let us suppose this set to be continuous medium volume. Potential field intensity characterizing the action of the beam on model particle with $z$ longitudinal coordinate is described by the following expression [4,9]:

$$E(\tau,z) = \int_{M_\tau} G_1(z-\hat{z},\hat{p})\rho(\tau,\hat{z},\hat{p})d\hat{z}d\hat{p}, \tag{2}$$

$$G_1(x,y) = \frac{Qa^2}{2\pi\varepsilon_0 R^2\Delta^2}\sum_{k=1}^{\infty}\frac{J_1^2(\mu_k R/a)}{\mu_k^4 J_1^2(\mu_k)\, sh(\mu_k\sqrt{1+y^2}H/a)} V_k\left(\sqrt{1+y^2}x, \sqrt{1+y^2}H\right), \tag{3}$$

$$V_k(\eta_1,\eta_2) = 2g_k(\eta_1,\eta_2) - g_k(\eta_1+2\Delta,\eta_2) - g_k(\eta_1-2\Delta,\eta_2), \tag{4}$$

$$g_k(\xi_1,\xi_2) = sign\,(\xi_1)\left[sh\left(\mu_k\xi_2/a\right) - sh\left(\mu_k\left(\xi_2-|\xi_1|\right)/a\right)\right]. \tag{5}$$

Here $\rho(\tau,z,p)$ is phase density corresponding to dynamic system (1); $\varepsilon_0$ is electric constant; $J_1(\eta)$ is Bessel function of the first order; $\mu_k, k=1,2,\ldots$ are the roots of Bessel function $J_0(\eta)$ of zero order; $2H$ is the length of moving beam spatial quasiperiod. When using the formulae (3)-(5), it is supposed $|x| \le H$ in view of the assumption of beam periodicity. Phase density satisfies the following transfer equation on the trajectories of the system (1) [1]:

$$\frac{\partial\rho}{\partial\tau} + \frac{\partial\rho}{\partial z}\frac{p}{\sqrt{1+p^2}} + \frac{\partial\rho}{\partial p}\frac{e}{m_0c^2}\left(E^{(RF)} + E\right) = 0.$$

For numerical Coulomb field simulation one can calculate the integral (2) using Monte Carlo method. The approximate formula is as follows:

$$E(\tau,z) = \frac{1}{N}\sum_{n=1}^{N} G_1(z-z_n, p_n), \tag{6}$$

where $z_n, p_n$, $n = \overline{1,N}$ are respectively longitudinal coordinates and reduced impulses of model particles.

## 4. Model 2 of particle interaction account: trigonometric polynomial

Consider the bunch with center coordinate $z_c$ and average reduced impulse $p_c$. One can approximate charge density of a bunch by grid function

$$\tilde{S} = \left\{S_j,\ j = \overline{0,\ 2M-1}\right\}, \tag{7}$$

where $2M$ is the number of grid cells. This function may be obtained on the basis of model particles positions and the bunch center state $(z_c, p_c)$ using cloud-in-cell method.

Journal of Physics: Conference Series **747** (2016) 012074 doi:10.1088/1742-6596/747/1/012074

Let us introduce analytical model of charge density in the form of trigonometric polynomial assuming the values (7) at grid points:

$$\eta(r,z,z_c)=\sum_{m=0}^{M}\left[A_m\cos\left(\frac{m\pi}{H}(z-z_c)\right)+B_m\sin\left(\frac{m\pi}{H}(z-z_c)\right)\right]\cdot\begin{cases}1, & r\le R\\ 0, & R<r\le a\end{cases}, \tag{8}$$

$$A_0=\frac{Q}{2\pi R^2H},\quad A_m=\frac{(-1)^m}{M\left[1+m/M\right]}\sum_{j=0}^{2M-1}S_j\cos(m\,j\pi/M),\ m=\overline{1,M}\,, \tag{9}$$

$$B_m=\frac{(-1)^m}{M}\sum_{j=0}^{2M-1}S_j\sin(m\,j\pi/M),\ m=\overline{0,M}\,. \tag{10}$$

In formula (9) $\left[1+m/M\right]$ is integer part of the value $1+m/M$.

When modeling Coulomb field, let us suppose every particle to have reduced impulse $p=p_c$. Potential field intensity characterizing the action of periodic sequence of bunches with charge density (8) on model particle with $z$ longitudinal coordinate is determined by the following expression [4,9]:

$$E(z,z_c,p_c)=\frac{2a}{R\Delta}\sum_{m=1}^{M}\alpha_m\sin\left(\frac{m\pi\Delta}{H\sqrt{1+p_c^2}}\right)\left(A_m\sin\left(\frac{m\pi}{H}(z-z_c)\right)-B_m\cos\left(\frac{m\pi}{H}(z-z_c)\right)\right), \tag{11}$$

$$\alpha_m=\frac{2aRH^2\sqrt{1+p_c^2}}{\varepsilon_0}\sum_{k=1}^{\infty}\frac{J_1^2(\mu_k R/a)}{\mu_k^2J_1^2(\mu_k)\left(m^2\pi^2a^2+\mu_k^2H^2(1+p_c^2)\right)},\ m=\overline{1,M}\,. \tag{12}$$

It should be remarked that beam evolution mathematical model in this case includes bunch center dynamics equations.

To obtain Coulomb field integral representation, the dependence of coefficients (9), (10) on particle positions and bunch center state $(z_c,p_c)$ should be expressed analytically. Using the approach suggested in [20] one can derive the formula

$$E(\tau,z,z_c,p_c)=\int_{M_\tau}G_2(z,\hat{z},z_c,p_c)\rho(\tau,\hat{z},\hat{p})\mathrm{d}\hat{z}\mathrm{d}\hat{p},$$

where the function $G_2(z,\hat{z},z_c,p_c)$ is determined by the mode of grid function (7) and polynomial (8) constructing. To obtain smooth integrand (that is necessary for mathematical optimization methods developing), the proper form-factor and interpolation function should be chosen; in this case model particles may be charged nonuniformly.

## 5. Method of testing of space charge simulation code

We have two models of particle interaction account, but it is not sufficient for code testing. Indeed, the models are too different because model 2 includes the constructing of grid function (7) and trigonometric polynomial (8).

To overcome this difficulty the following idea is suggested in [4,9]. We can introduce the special beam charge density given by trigonometric polynomial (8), where $A_0=Q/(2\pi R^2H)$, and the remaining coefficients $A_m,B_m$, $m=\overline{1,M}$ are arbitrary. This density may be simulated with the use of special particle distribution. Coulomb field generated may be calculated using two ways: model 1 and formulae (11)-(12). Due to special charge distribution we avoid constructing of grid function representing charge density as well as trigonometric polynomial (8). To test the code realizing model 1, one can compare the results obtained.

To implement this idea let us prescribe the number $M$ and coefficients $A_m, B_m$, $m=\overline{1,M}$ and introduce the uniform grid at spatial quasiperiod. May $J >> M$ be the number of grid cells. The number $J$ is to be chosen rather large to provide the proper accuracy of special charge density simulation. To simulate the density (8) we locate $N = J$ model particles in the centers $z_j$, $j=\overline{1,J}$ of grid cells. Let particle charge values be

$$q_j = \frac{2\pi R^2 H}{J}\sum_{m=0}^{M}\left[A_m \cos\left(\frac{m\pi}{H}(z_j - z_c)\right) + B_m \sin\left(\frac{m\pi}{H}(z_j - z_c)\right)\right],\ j=\overline{1,J} \tag{13}$$

and particle reduced impulses be $p_c$. Assume the cloud size not to exceed the grid cell size.

It is easy to modify the formulae (3)-(6) for model particles with reduced impulse $p_c$ and different charges (13). For particle distribution presented we have two ways to calculate Coulomb field: the analogue of formulae (3)-(6) and the formulae (11)-(12). For the testing of the code realizing model 1, one can compare the space charge field values in the points $z_j$, $j=\overline{1,J}$ calculated by two ways mentioned. If the code is correct we can provide (by increasing $J$) the required matching accuracy $\varepsilon$.

## 6. Numerical experiments

Beam dynamics was investigated for klystron-type buncher with following main characteristics: initial energy of electrons $W_0 = 0.5$ MeV, average beam current $I = 15$ A. Beam dynamics simulation code was developed in cooperation with B.S. Zhuravlev. Two interaction account modes (discussed in this paper) are realized. The numerical results obtained support the conclusion that model 2 is preferred to use when electron bunches are mostly formed.

The testing of code realizing model 1 was successive. The coefficients of trigonometric polynomial (8) were chosen as follows: $A_0 = Q/(2\pi R^2 H)$, $A_m = 0.000003m$, $B_m = 0.00007\cos(m+4)$, $m=\overline{1,M}$. The testing was carried out repeatedly. In particular, when $M = 3$, the accuracy $\varepsilon = 0.05$ was achieved for $J \geq 150$.

## 7. Conclusion

We discussed two models of particle interaction account in axial-symmetric quasiperiodic beam. Coulomb field representations in the form of Fourier-Bessel series and trigonometric polynomial may be used both for simulation and analysis. The method of testing of space charge simulation code is worked out on the basis of analytical approach.

In addition, both analytical expressions allow the integral representation of space charge field. It provides the possibility of beam dynamics description by integro-differential equations. Consequently, beam dynamics optimization problems may be treated on the basis of the approach suggested by D A Ovsyannikov [1]: analytical expression of quality functional variation may be obtained to enable the use of directed optimization methods.

## References

[1] Ovsyannikov D A 1990 *Modeling and Optimization of Charged Particle Beam Dynamics* (Leningrad: Leningrad State University Press)

[2] Ovsyannikov D A and Edamenko N S 2013 Modeling of charged particle beam dynamics *Vestnik St. Petersburg University, Ser. 10: Applied Mathematics, Informatics, Control Processes* **2** 61–66

[3] Kozynchenko V A 2014 The modeling of charged particle interactions in the elliptic beam *Proc. 20th Int. Workshop: Beam Dynamics & Optimization (BDO) (30 June-4 July 2014 St. Petersburg)* (St. Petersburg: IEEE) pp 94-95 DOI: 10.1109/BDO.2014.6890040

[4] Rubtsova I D 2014 On quasiperiodic beam of interacting particles dynamics modeling *Vestnik*

Journal of Physics: Conference Series **747** (2016) 012074 doi:10.1088/1742-6596/747/1/012074

*St. Petersburg University, Ser. 10: Applied Mathematics, Informatics, Control Processes* **1** 104–19

[5] Drivotin O I and Ovsyannikov D A 2006 Modeling of self-consistent distributions for longitudinally nonuniform beam *Nucl. Instr. Meth. Phys. Res.* A **558** pp 112-18

[6] Drivotin O I and Ovsyannikov D A 2009 Self-consistent distributions for charged particle beam in magnetic field *Int. J. Modern Phys.* A **24** (5) pp 816-42

[7] Ovsyannikov D A, Ovsyannikov A D, Vorogushin M F, Svistunov Yu A and Durkin A P 2006 Beam dynamics optimization: models, methods and applications *Nuclear Instruments and Methods in Physics Research* A **558** (1) 11-19

[8] Ovsyannikov D A 2012 Mathematical modeling and optimization of beam dynamics in accelerators *Proc. XXIII Russian Particle Accelerator Conf. RuPAC-2012 (24-28 September 2012 St. Petersburg)*(Geneva: JACoW http://www.JACoW.org) pp 68-72

[9] Ovsyannikov D A, Rubtsova I D and Kozynchenko V A 2013 *Some Problems of Intense Charged Particle Beams Modeling in Linear Accelerators* (St. Petersburg: VVM publishing office)

[10] Rubtsova I D and Suddenko E N 2012 Investigation of program and perturbed motions of particles in linear accelerator *Proc. XXIII Russian Particle Accelerator Conf. RuPAC-2012 (24-28 September 2012 St. Petersburg)*(Geneva: JACoW http://www.JACoW.org) pp 367-69

[11] Ovsyannikov D A, Ovsyannikov A D, Antropov I V and Kozynchenko V A 2015 Software complex BDO-RFQ *Proc. III Int. Conf. "Stability and Control Processes" in Memory of V.I.Zubov (SCP) (5-9 October 2015 St. Petersbirg)* ed L A Petrosyan and A P Zhabko (St. Petersburg: IEEE) pp 335-37 DOI: 10.1109/SCP.2015. 7342132

[12] Kozynchenko V.A. and Kozynchenko S.A. 2014 Parallel beam dynamics simulation in injection systems taking into account particle interactions *Proc. 20th Int. Workshop: Beam Dynamics & Optimization (BDO) (30 June-4 July 2014 St. Petersburg)* (St. Petersburg: IEEE) pp 98-99 DOI:10.1109/BDO.2014.6890042

[13] Kozynchenko V.A. and Boyko A.A. 2014 Modeling and simulation of beam dynamics in linear accelerator with RFQ *Proc. 20th Int. Workshop: Beam Dynamics & Optimization (BDO) (30 June-4 July 2014 St. Petersburg)* (St. Petersburg: IEEE) pp 96-97 DOI:10.1109/BDO.2014.6890041

[14] Ovsyannikov D A, Ovsyannikov A D, Antropov I V and Kozynchenko V A 2005 BDO-RFQ code and optimization models *2nd Int. Conf. on Physics and Control (PhysCon 2005) (24-26 August 2005 St. Petersburg)* 1513994 pp 282-88

[15] Ovsyannikov A D, Ovsyannikov D A, Balabanov M Yu and Chung S-L 2009 On the beam dynamics optimization problem *International Journal of Modern Physics* A **24** (5) 941–51

[16] Ovsyannikov A D, Ovsyannikov D A, Durkin A P and Chung S-L 2009 Optimization of matching section of an accelerator with a spatially uniform quadrupole focusing *Technical Physics. The Russian Journal of Applied Physics* **54** (11) 1663-66

[17] Rubtsova I D 2014 Mathematical optimization model of longitudinal beam dynamics in klystron-type buncher *Proc. XXIV Russian Particle Accelerator Conf. RuPAC-2014 (6-10 October 2014 Obninsk)(Geneva: JACoW* http://www.JACoW.org) pp 66-68

[18] Rubtsova I D 2015 Optimization of iterative beam dynamic process *Proc. III Int. Conf. "Stability and Control Processes" in Memory of V.I.Zubov (SCP) (5-9 October 2015 St. Petersbirg)* ed L A Petrosyan and A P Zhabko (St. Petersburg: IEEE) pp 198-200 DOI: 10.1109/SCP.2015.7342092

[19] Drivotin O I and Starikov D A 2014 Second order method for beam dynamics optimization *Proc. XXIV Russian Particle Accelerator Conf. RuPAC-2014 (6-10 October 2014 Obninsk)(Geneva: JACoW* http://www.JACoW.org) pp 69-71

[20] Rubtsova I D 2014 Integral-differential model of quasi-periodic beam longitudinal dynamics *Proc. 20th Int. Workshop: Beam Dynamics & Optimization (BDO) (30 June-4 July 2014 St. Petersburg)* (St. Petersburg: IEEE) p 144

Journal of Physics: Conference Series **747** (2016) 012075 doi:10.1088/1742-6596/747/1/012075

# Capturing coefficient increase and energy spread decrease in LPWA

**S M Polozov[1] and V I Rashchikov**

National Research Nuclear University - Moscow Engineering Physics Institute, Moscow 115409, Russia

SMPolozov@mephi.ru

**Abstract**. The acceleration of electrons in laser plasma channels is one of the contemporary ideas on energy frontier of accelerators. Many simulations and experimental studies are now provided. But two main laser plasma acceleration problems are not solved until now: low fraction of particles is capturing into acceleration and accelerated electrons have very broad band. Especial techniques should be proposed to preserve such negative effects. New results of beam dynamics simulations in laser plasma channel having pre-bunching stage are discussed in article. Results are obtained for two pre-bunching techniques which are analogous to waveguide and klystron type buncher used in RF accelerators.

## 1. Introduction

A number of ideas for increasing the rate of the energy gain have been discussed in the last few decades. This rate is limited by discharge for conventional RF accelerators. The idea of electrons acceleration in a modulated plasma channel was proposed by Ya.B. Feinberg in the 1950's [1]. Possible schemes for the plasma wakefield acceleration (PWA) differing in ways of modulating the plasma channel were developed later. The first one uses a high energy (tens of GeV) beam of particles to form a plasma wave and accelerate a fraction of the injected particles or a probe beam [2]. Another method is the laser plasma wakefield acceleration (LPWA) [3], in which a laser pulse is used to create a plasma wave. The problem of electron acceleration in plasma channel with varying density produced by power laser pulse or short electron bunch is under attentive consideration [1, 4]. The accelerating gradient in this channel is limited not by the discharge effects as in usual accelerating structures but the plasma density and laser-plasma transmission efficiency. Depending of laser intensity and plasma channel parameters there are linear and non-linear LPWA modes: in the underdense plasma, in which $\pi^2 r_l^2 / \lambda_p^2 >> a_0^2 / 2\gamma_l$ , (quasi linear regime) and the non-linear regime with $\pi^2 r_l^2 / \lambda_p^2 << a_0^2 / 2\gamma_l$. Here rl is the laser spot size, $a_0 = eA/W_0$ normalized laser intensity, $\gamma_l = (1 + a_0^2 / 2)^{1/2}$. The electron beam dynamics is different in the two regimes. The theory of the laser-plasma interaction and acceleration in the plasma channel are discussed in [5-7].

A few methods for improving the energy spread in the non-linear regime have been proposed. The first is to use two plasma stages with constant but not equal plasma densities and a transient stage with varying density between them for the beam modulation [8]. An energy spectrum better than 3 % for a 1 GeV beam has been numerically and in experiment has demonstrated a low energy spectrum

[1] To whom any correspondence should be addressed.

< ± 3 % [9] for a similar distribution of the plasma density (decreasing in the first stage and constant in the second one).

A ponderomotive injection using two synchronized laser pulses was proposed in [10]. Two lasers can also excite a beat wave in the plasma, which is then used for capturing of the shot bunch [11]. With a third laser pulse this method can produce "cooled" electron beams [12]. The method of controlled electron self-injection in wave breaking regime has been also proposed [13], and an energy spread of ± 3 % has been demonstrated experimentally.

These methods improve the energy spread to about 3 % for a 1 GeV beam. Still, this number is too high for many applications. The electron capturing efficiency also remains problematic. All the methods described above apply to the non-linear or wave breaking regimes. However, the linear LPWA mode is also interesting for practical use. The rate of the energy gain can still be very high, while the laser power requirements are comparatively moderate, meaning that compact, laboratory scale facilities could be designed for accelerating electron beams to hundreds of MeV.

## 2. Pre-bunching for linear mode LPWA

Two ideas of pre-bunching realisation for linear mode were proposed [14, 15]. In the first the bunching scheme similar to waveguide buncher in conventional RF lnac was studied. The plasma channel is divided into two stages. In the first the bunching scheme similar to waveguide buncher in conventional RF lnac was studied. The plasma channel is divided into two stages. The plasma density slowly decreases in the first, pre-modulation stage, and is constant in the second, the main accelerating stage (figure 1a). The following assumptions are made: the beam is injected externally, the amplitude of the electric field does not vary on the scale of the time of flight, the plasma is cold, linear and collisionless, and the space charge field of the injected electrons is much lower than the plasma. This scheme allowed to increase capture ratio to $K_t$=40 %, but now sufficient energy spectrum dissipation $\delta W$=5-6 %. This problem can be solved by using a number of short low density ($10^{18}$ $cm^{-3}$) plasma stages with gaps between of them (figure 1b). This scheme is similar to the multigap klystron buncher of conventional RF linac and based on a number of short plasma sub-stages (several $\lambda_l$ long each) separated by drift gaps. At this case we obtained Kt more than 50 % and energy spectrum reduction to 3 % for energy of 100 MeV [16]. Simulations were done using new BEAMDULAC-LWA2D code version which was designed to study the beam dynamics in LPWA channel.

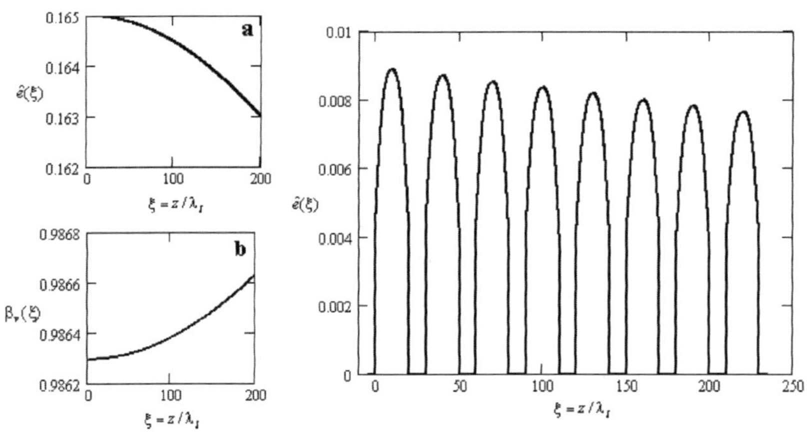


**Figure 1.** Two possible schemes of electrons pre-bunching for linear-mode LPWA.

## 3. Particle-in-cell simulations.

2.5D particle in cell code SUMA [17, 18] was used for electrons dynamics simulations in laser plasma channel taking into account real field distribution in the channel and to confirm previously results of theoretical investigation and simulations. The system of equations used in mathematical model

Journal of Physics: Conference Series **747** (2016) 012075 doi:10.1088/1742-6596/747/1/012075

consists of the Maxwell equations, the equation of the medium, and the equation of motion. Unperturbed plasma oscillations with different density values were simulated for thin code parameters tuning. Longitudinal oscillation frequencies can be obtained analytically and are in a good agreement with simulation results .After that computer simulation of waveguide buncher with 200 $\lambda_l$ length, where $\lambda_l$ = 1.06μm, was conducted. To obtain effective beam bunching we need to choose some plasma and laser pulse parameters, which are linked to each other. For example optimal electron beam characteristic after buncher requires electric field amplitude as less as possible but enough for space charge forces compensation. It leads to plasma density reduction. But plasma density reduction brings to electric field amplitude decreasing and as a result to not enough electron beam mean velocity accession rate with compare to wave phase velocity.

Figure 2a shows longitudinal (red dash line) and transverse (blue dash line) plasma wave electric field distribution in buncher at the mean beam radius. Plasma density is equal to $10^{24}$ m$^3$, laser field 4x$10^{11}$ V/m, laser pulse duration 40 fs, laser spot radius $r$ = 30μm, capillary radius 120 μm. External electron current pulse duration is 0.15 ps. Electron bunch (green dots) is injected 60 fs after laser pulse. Figure 2a corresponds to time 300 fs after laser pulse start. Electrons are injected close to zero of reducing longitudinal electric field of plasma wave. For the time at Figure 2a initial longitudinal size of injected bunch is reduced twice during the bunching. As shown, electrons are at the focusing transverse electric field. Figure 2b shows the combine figure, which is divided in horizontal direction. The upper part shows plasma self field potentials distribution and lower - charge particles distribution. Black and red points correspond to plasma electrons and ions accordingly. Injected electrons are presented at the lower part of the figure by blue points near the potential maximum. The potential wall size is presented as the first longitudinal field osculation. The form and depth of the wall depends on the plasma density and laser pulse parameters that allows us to change the number of electrons to be captured in acceleration. Injected electron initial energy in a sensible range do not effect on the result due to essential potential wall depth.

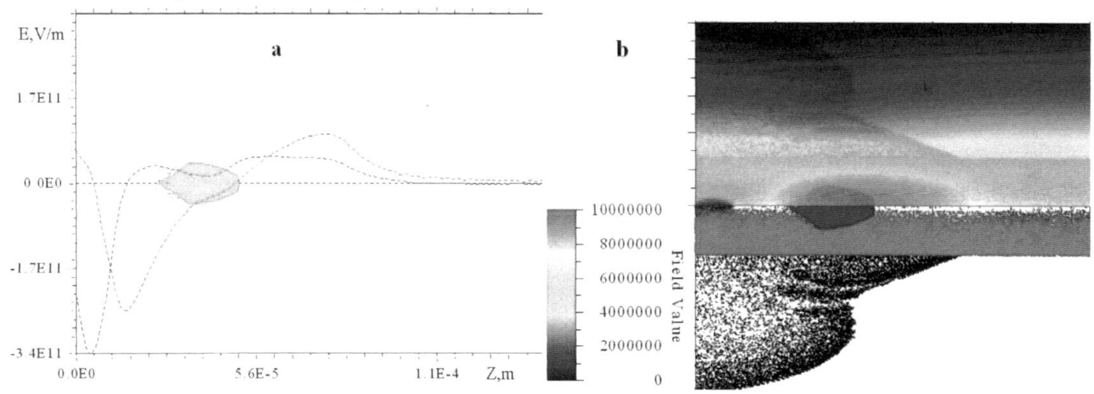


**Figure 2.** Plasma wave electric field (a), potential and charge particle (b) distribution in buncher.

Figure 3a shows time depended (time step 100 fs) potential and charged particles distributions in the channel with constant plasma density $10^{24}$ m$^3$. Electron bunch is tightened in both directions at the front and then get to the end of potential well where electrons are in the defocusing transverse electric field and get off the axis. Longitudinal size of the electron bunch starts to rise. Decreasing plasma density along the channel (see Figure 3b) is used to prevent bunch lag behind the well.

The correct chose of plasma density gradient allows to keep electrons in the potential well and to exclude the defocusing.

## 4. Conclusion

Principal possibility to obtain sufficient energy spectrum of injected electron beam at the channel with low plasma density (< 1018cm3) was shown. The proper chose of plasma density gradient in the buncher may considerably improve capture rate and energy spectrum.

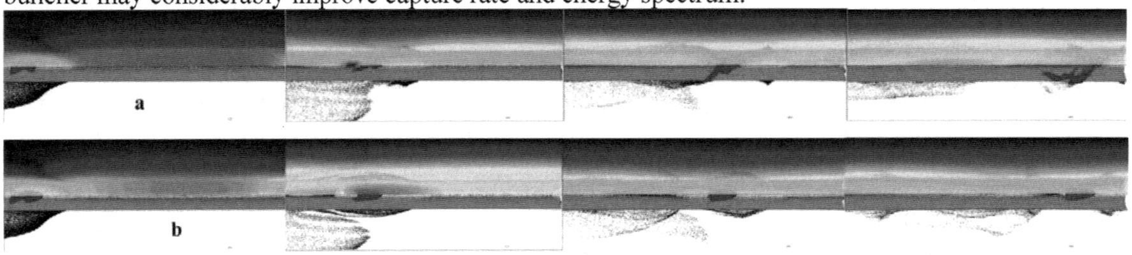

**Figure 3.** Time depended potential and charged particles distributions

## References


[1] Feinberg Ya.B., 1959. Charge particle acceleration in plasma. Sov. Atomic Energy 6, 1084.

[2] Muggli P. et al., 2008. High-Gradient Plasma-Wakefield Acceleration with Two Subpicosecond Electron Bunches. Phys. Rev. Lett. 100, 074802.

[3] Tajima T., Dowson J.M., 1979. Laser Electron Accelerator. Phys. Rev. Lett. 43 (4), 267.

[4] Hogan M.J., Barnes C.D., Clayton C.E. et al., 2005. Multi-GeV Energy Gain in a Plasma-Wakefield Accelerator, Phys. Rev. Lett., 95, 054802-1-4.

[5] Akhiezer A.I., Polovin R.V., 1956. Theory of wave motion of an electron plasma Sov. Phys. JETP 30, 915.

[6] Gorbunov L.M., Kirsanov V.I., 1987. Excitation of plasma waves by an electromagnetic wave packet. Sov. Phys. JETP 66, 290.

[7] Esarey E., Schroeder C.B., Leemans W.P., 2009. Physics of laser-driven plasma-based electron accelerators. Rev. of Modern Phys. 9, 1229.

[8] Bulanov S.V. et al., 2008. Phys. Plasmas15, 073111.

[9] Gonsalves A.J. et al., 2011. Nature Physics 7, 862.

[10] Umstadter D., Kim J.K., Dodd E., 1996. Phys. Rev. Lett76, 2073.

[11] Esarey E. et al., 1997. Phys. Rev. Lett. 79, 2682.

[12] Esarey E., Leemans W.P., 1999. Phys. Rev. E 59, 1082.

[13] Mangles S.P.D. et al., 2004. Nature 431, 535.

[14] Polozov S.M., 2012. Energy Spread Decreasing in Linear Mode Operating Laser Plasma Wakefield Accelerator. In: Proc. of. PuPAC 2012, 251.

[15] Polozov S.M., 2013. A possible scheme of electron beam bunching in laser plasma accelerators. NIM A 729, p.517.

[16] Polozov S.M., 2013. 2D Beam Dynamics Simulation in Linear Mode LPWA Channel with Pre-modulation Stage. Problems of Atomic Science and Technology. Series: Nuclear Physics Investigations 6 (88), p. 29-34.

[17] Rashchikov V.I., 1990. Electromagnetic field simulation in complex geometry structures, Problems of Atomic Science and Technology. Series: Nuclear Physics Investigations 10 (18), 50-53.

[18] Didenko A.N., Rashchikov V.I., Fortov V.E., 2011. Terahertz emission from target under the action of powerful laser pulses, Technical Physics 56, 10, pp. 1535-1538.

# Facility for thermoradiotherapy of deep-seated tumors

**A.M. Fadeev[1,4], S.M. Ivanov[1,2], E.A. Perelstein[3] and S.M. Polozov[1]**

[1] National Research Nuclear University - Moscow Engineering Physics Institute, Moscow 115409, Russia
[1] N.N. Blokhin Russian Cancer Research Center, Moscow 114478, Russia
[3] Joint Institute for Nuclear Research, Dubna 141980, Russia
SMPolozov@mephi.ru

**Abstract**. Therapeutic hyperthermia in combination with radiotherapy which is realized by means of modern linear electron accelerators (called thermoradiotherapy) is one of widely used contemporary cancer treatment methods. Hyperthermia provides increasing of the tumors sensibility to radiation. Thus the radiation dose could be decreased. The experience achieved at N.N. Blokhin Russian Cancer Research Center during 40 years shows that 5-year survival rate increases (15-40%) for a number of cancer types and localization. High dose can be easily delivered to the deep seated tumors but local heating is a big problem so far. One possible way to solve the problem of deep tissues local heating is discussed in this article, the brief review of previous simulations and latest results for experimental prototype setup are presented.

## 1. Introduction

Hyperthermia is an adjuvant methods of cancer treatment in which tumor temperature is increased to high values (40-44 °C). Many researches have shown that high temperature can damage and kill tumor cells, thus reduces tumor size. However the main advantage is that hyperthermia is a promising approach to increase efficiency of chemotherapy or radiation therapy. Under hyperthermia some tumor cells become more sensitive to the radiation and anticancer drugs. The effect on surviving fraction depends both on the temperature increase and on the duration of the expose. The main mechanism for cell death is probably protein denaturation at temperatures above 40 °C, which leads to changes in molecular structures such as cytoskeleton and membranes, and changes in enzyme complex for DNA synthesis and repair [1]. Heat also enhances the cytotoxicity of X-rays. Increased cytotoxicity is maximized when radiation and hyperthermia are given simultaneously. The combined effect decreases with time when the treatments are separated by more than one hour [2]. When cells are exposed to increased temperatures to anticancer drugs, their response is often different from the one at normal temperature. Drugs whose rate-limiting reaction is primarily chemical are expected to be more efficient at higher temperatures. Thus the combination of chemotherapy with hyperthermia has high potential in clinical practice [3].

The thermal therapy combined with the radiation (thermoradiotherapy, TRT) has been applied in N.N. Blokhin Russian Oncological Research Center (RORC) since 1980th [4]. More than 1000 patients have been treated to date. Such program allows to sufficient and authoritative reduce of the regional cancer recrudescence and metastases comparatively to the surgery or independently radiotherapy (RT). As an example, the rectum cancer recrudescence was observed for 0.9±0.6 % (2 form 220 patients) and metastases for 5% (11 from 220 patients) comparatively with 16,2±1.9% (64

[4] To whom any correspondence should be addressed.

from 395) for surgery and 9,6±2,1% (26 from 272) for surgery and before radiotherapy [4]. Frequency of full regression after TRT also increased: for prostate cancer form 69±8.2 % for RT to 94±2.3 % for TRT, for soft tissue sarcoma form 14±5.1 to 45±5.2 %, for regional metastasis of neck epidermoid cancer from 12±7.8 to 57±6 % etc. [5].

## 2. Problem of local heating of a deep seated tissues

Several hyperthermia systems have been proposed, however non-invasive heating of deep seated tumors is still a difficult technical challenge. In this case the electromagnetic (EM) fields have an advantage over other methods such as ultrasound: radio frequency (RF) fields have a greater penetration rate than optic or infrared waves. Single applicators are not effective in treating tumors located deeper than 2-3 cm. First applicators were not able to steer and focus energy without shifting patient position relative to the radiator. In these initial studies 27 MHz ridged waveguide [6], 70 MHz coax TEM-applicators [7] and 82 MHz helix [8] were used. Several applicators based on dielectric filled waveguide or horn antennae have been investigated for many years. In most cases, those devices produce an EM field distribution with maximum field intensity at the surface and in the center of the applicator. The intensity of the EM fields is rapidly falling off with both depth into tissue and transverse distance from the applicator center. To overcome these limitations the feasibility of using microstrip technique was investigated. For instance, spiral microstrip antenna were designed and integrated into applicators suitable for clinical use [9]. Because of their small size these antennas can be arranged into the arrays. They operate over the range of 400 MHz to 1200 MHz and produce sufficient specific absorption rate (SAR) distribution at a depth of few centimeters from the surface [10].

One of the major problems of the devices mentioned above is the limited depth of penetration due to the principle of skin-effect. Only tumors located 2-3 cm (7-8 cm for the contact flexible microstrip applicator [11] from the surface can be heated by these applicators. To increase SAR value in the tumors situated deeper than 10 cm relative to the surface SAR value it is necessary to focus energy of electric fields produced by an array of applicators. It consists of several antennae surrounding the patient and emitting radio-waves. Single antenna or groups of antennae are fed separately. Thus by proper selection of amplitudes and phases the interference patterns of the produced fields can be focused to create desirable temperature distribution [5]. Significant research successes have been obtained for the deep hyperthermia with the phased array. Simulation and measurements results are presented in [12].

RORC clinical studies demonstrate improving results of treatment by combined using of hyperthermia and radiation for the several tumor localizations. But only applicators for superficial hyperthermia were used in RORC. Common RORC-MEPhI-JINR project is pointed to expand the range of utilizing devices, i.e. using of devices for the regional hyperthermia gives more advantages for an oncological disease treatment.

## 3. Prototype of the local termoradiotherapy system

The most evident approach is using an annular array of applicator situated around the patient body [13]. Top view of this structure is shown in the figure 1. Arrays of applicators with variations in frequency, phase, amplitude and orientation in space give more possibilities to control heating pattern during hyperthermia treatment [12]. RF power feeding scheme is presented in [14]. Thus the phased array provides deeper tissue penetration of electromagnetic waves in comparison with single applicator, reduce undesirable heating of healthy tissues situated between applicator and tumor and improve local control for heating area. Also using array of applicators gives ability to control and to plan heating process without changing of patient position. Suggested phased array consists of eight copper dipoles, attached on the inner side of the dielectric cylinder, and surrounds a patient body. The aperture radius is up to 60 cm which can be applied in more cases. Dipoles are fed independently permitting to control phases and amplitudes of waves. Space between dipoles and the patient body is filled by deionized water (conductivity $\sigma \approx 0.001$ S/m). Thus applicators are squeezed from the inner side by lossy medium with high permittivity (deionized water ε≈80), and from the outer side by medium with low permittivity (air ε=1). The conducting elements of antenna are isolated from lossy medium by thin layer of an insulator (thickness h≈1 mm). Because of energy density of electrical fields ($\vec{E}\vec{D}/2$) inside the dielectric tank is higher by a factor of ε (the relative dielectric constant of medium), energy is mainly concentrated inside the array. Thus deionized water not only cools body surface and superficial tissues but is also a matching medium. Electric field lines inside the phased array are parallel to the axis of dipoles. That's why heat absorption in the surface and superficial tissues (such as skin and fatty tissue) which is proportional to tissue conductivity (σ, these values for fat and skin are significant lower than for muscle or tumor tissue) will be substantially lower than in deeper tissues. Thus skin and fatty tissue overheating is reduced in comparison with the using capacitive applicators. So $E_z$ (z is direction along of the patient body) is the only component, which may be able to control by shifting the amplitudes and phases of eight dipoles. Ex and Ey components are not under control. E-field generated by each of the dipoles is given by $E_j = A_j E_{j,0}(x,y)\cdot\exp(-i(\omega t - \Phi_j))$, where $E_{j,0}$ – is the complex E-field for $A_j$=1, $\Phi_j$=0, and $A_k$=1 for $j \neq k$, $A_j$ – is a scaling factor of amplitude, $\Phi_j$ – wave phase, $j$ and $k$ are numbers of dipoles, $i$ is imaginary unit. It is possible to move peak of interference pattern and to focus it into the tumor site with the variation of these two parameters (phase and amplitude). The measure of the rate at which energy is absorbed by the body when exposed to a radio frequency (RF) electromagnetic field is a Specific Absorption Rate (SAR). It is defined as the power absorbed per mass of tissue and has units of watts per kilogram (W/kg): $SAR = (\sigma\vec{E}^2)/\rho$, ρ – is the density of the tissue (kg/m3), $E$ – is the root of mean square electric field.

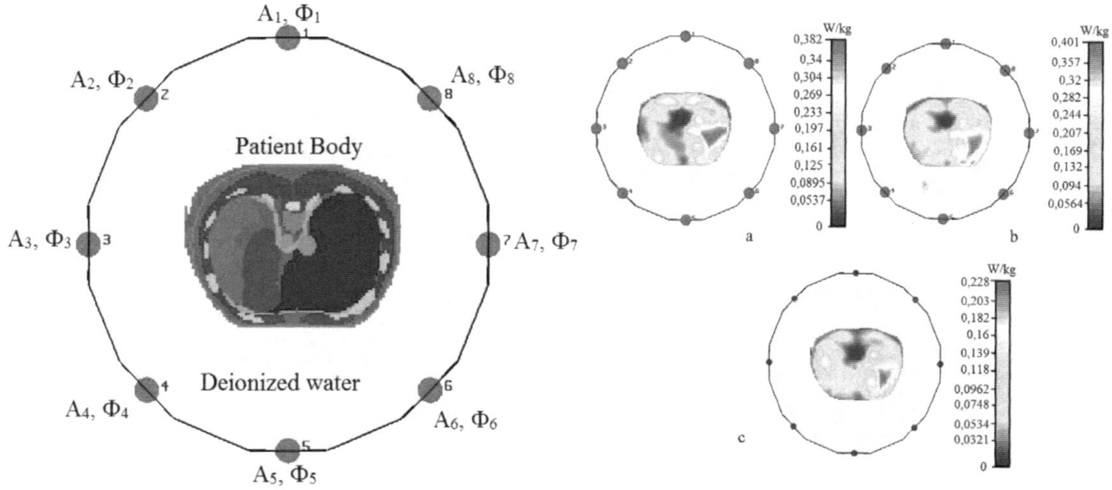


**Figure 1.** Top view of phased array surrounding patient body (left) and cross-section patterns of the SAR distribution with different operation frequencies (a) 150 MHz; (b) 100 MHz; (c) 80 MHz and with input phases of 50°, 50°, 50°, 50°, 0°, -30°, -40°, -10° applied to channels 1, 2, ..., 8 respectively.

The simulations are performed with CST Microwave Studio which uses the Finite Element Method (FEM) for Maxell's equation solving. Their solvers feature curved elements of arbitrary order. These elements enable a conformal representation of the geometry which improves the simulation accuracy. In combination with the unstructured FEM grid, which can resolve small structure details very efficiently, it can increase the simulation performance dramatically. All simulations were performed for input rms power 0.5 W per each channel. To demonstrate that the phased array can produce a maximum SAR distribution inside patient body, electromagnetic simulations were performed at different operating frequencies. The voxel model which was used is presented in [15]. This model doesn't contain tumor tissue. Material properties for water bolus, enclosure and antennae are also included in simulation. The blood flow and the thermal conductivity aren't factored in the simulation, thus the simulated pattern of SAR distribution have local maximum within vessels. The real SAR distribution hasn't these locals.

There are three operating frequencies were used for SAR simulation: a) 150 MHz, b) 100 MHz, c) 80 MHz. Cross-section patterns of SAR distributions with relative phases are depicted on Fig. 1 also. The simulated focus is steered to the liver with input phases of 50°, 50°, 50°, 50°, 0°, -30°, -40°, -10° applied to channels 1, 2, …, 8 respectively. Also the goal was to prevent occurring hot spots in other regions and to maximize the SAR value in relative to the SAR of healthy tissue. The phases were chosen manually by the principle that the peak SAR shifts away from the dipole which has phase delay, i.e. the peak SAR will shift away from the dipole #3 due to the phase delay on this dipole will be 50°. Amplitudes were not varied. Three operating frequencies mentioned above were used. It is clear that variation both amplitudes and phases leads to further localization of the heating region comparatively phases variation only as it was realized in [13].

## 4. Experimental results

Based on previous simulations [16] and RF feeding system design [14] the first experimental prototype was developed and constructed. It consists of two RF dipoles and one feeding system. Series of local heating experiments were done using such prototype. Experimental demonstration of the local energy distribution maximum and its motion versus of phase and amplitudes variation were the main goals of such experiments. Two dipoles were mounted on external side of a dialectical cylinder which was filled by deionized water. The tissue-equivalent phantom was prepared using a number of thin-walled dielectric tubes filled by salt water which permittivity and conductivity are close to human tissues. Tubes were placed on line connects centers of RF dipoles. Such phantom correctly imitates the deep suited tissue and can be easily used in temperature control which was organized by means of a number of thermocouples which were placed into all thin-walled tubes. The initial temperature was constant for all system points. Water starts the energy absorption after RF power on and the temperature in the local heating volume starts growth. Temperature distributions after heating are shown in figure 2 for in-phase dipoles on (a) and for 60 degrees phase shift (b). The heating was provided 20 minutes for in-phase experiments and 12 minutes for experiments with phase-shift. Zero-point on x-axe corresponds to the center of main cylinder. It is clear from figures that 50 mm of localization can be achieved in hyperthermia system proposed. Experimental heating results are compared with CST Studio Suite simulations (dot curves in figure 2) and have very good accuracy.

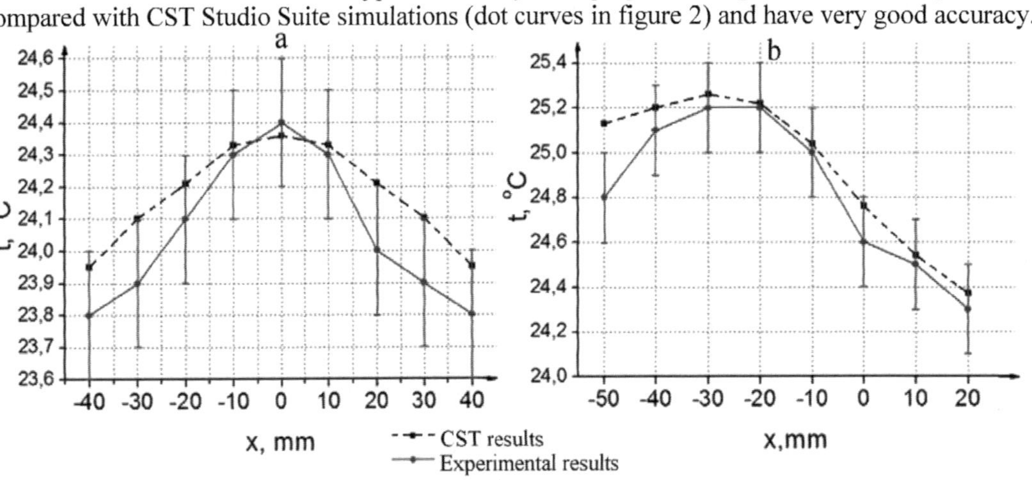


**Figure 2.** Temperature distributions after heating for in-phase dipoles on (a) and for 60 degrees phase shift (b).

## 5. Conclusions

The system for local hyperthermia of deep seated tumours which is planning to use further for combined cancer thermoradiotherapy is presented. It was shown by means of electrodynamics and thermal simulations that system of a number of independently phased dipoles is suitable for local heating of deep suited tissues. The first experimental prototype was constructed and numbers of heating localization experiments were carried out. Comparison of simulation and experiments shows that the phased array solve the local heating problem and deep suited tumorous can be locally heated and treated.

## References

[1] Dewey W C 1994 Arrhenius relationships from the molecule and cell to the clinic *Int. J Hyperthermia* **10** 457–83

[2] Hahn G M 1982. Hyperthermia and cancer (New York: Plenum)

[3] Kampinga H H 2006 Cell biological effects of hyperthermia alone or combined with radiation or drugs: a short introduction to newcomers in the field *Int. J Hyperthermia* **22** 191–96

[4] Tkachev S I, Barsukov Y A and Gordeyev S S 2010 New combined treatment regimen for primary unresectable locally advanced rectal cancer *Eur. Journal of Surgical Oncology* **36**(9) 877

[5] Ivanov S M *et al.* 2012 Facility development for a combined cancer therapy *Problems of Atomic Science and Technology. Series: Nuclear Physics Investigations* **3**(79) 191-194

[6] Paglione R *et al.* 1981 27 MHz ridged waveguide applicators for localized hyperthermia treatment of deep-seated malignant tumors *Microwave J.* **24** 71–80

[7] Van Es C A *et al* 1995 Regional hyperthermia of pelvic tumours using the Utrecgt Coaxial TEM system: a feasibility study. *Int. J. Hyperthermia* **11** 173–186

[8] Croghan M K *et al.* 1993. A phase I study of the toxicity of regional hyperthermia with systemic warming. *Am. J. Clin. Oncol.* **16** 354–58

[9] Tanabe E *et al.* 1985 Microstrip antenna applicators-design and clinical experience. Hyperthermia in cancer therapy (Tokyo: MAG Bros., Inc.) 151-153

[10] Samulski T V *et al.* 1990 Spiral Microstrip Hyperthermia Applicators: technical Design and Clinical Performance *I. J. Radiation Oncology. Biology. Physics* **18**(1) 233-240

[11] Gelvich E, Mazohkin V, Tkachev S and Trochin I 1996 Clinical possibilities of CFMA applicators *Proc. of the 7th International Congress on Hyperthermia Oncology* 449-451

[12] Fadeev A M *et al.* 2012 Cylindrical Phased Dipoles Array for Hyperthermia of Deep-situated Tumors *Proc. of RUPAC'2012* 521-523

[13] Wust P *et al.* 1996 Simulation studies promote technological development of radiofrequency phased array hyperthermia *Int. J. Hyperthermia* **12**(4) 477-494

[14] Polozov S M *et al.* 2012 RF Power and Control Systems for Phased Dipoles Array System for Hyperthermia *Proc. of RUPAC'2012* 524-525

[15] Paulides M M, Bakker J F, Zwamborn A P and Van Rhoon G C 2007 A head and neck hyperthermia applicator: theoretical antenna array design. *Int. J. Hyperthermia* **23**(1) 59-67

[16] Ivanov S M, Perelshtein E A, Polozov S M and Fadeev A M 2013 Facility for the electromagnetic hyperthermia based on the phased array *Problems of Atomic Science and Technology. Series: Nuclear Physics Investigations* **6**(88) 220-224

# The processes of nonequilibrium exchange in rotating plasma flows

**A R Karimov**[1,2,7], **V L Shatokhin**[1], **M Y Yu**[3,4] **and L Stenflo**[5,6]

[1] Department of Electrophysical Facilities, National Research Nuclear University MEPhI, Kashirskoye shosse 31, Moscow, 115409, Russia
[2] Institute for High Temperatures, Russian Academy of Sciences, Izhorskaya 13/19, Moscow, 127412, Russia
[3] Institute for Fusion Theory and Simulation, Hangzhou, 310027, China
[4] Ruhr-Universitat, Bochum, D-44780, Germany
[5] Linkoping University, Linkoping ,SE-58183, Sweden
[6] Umea University, Umea, SE-90187, Sweden

E-mail: ARKarimov@mephi.ru

**Abstract**. The mechanisms of energy/momentum exchange in rotating and compressing plasma flows have been discussed. It has been shown that such flows are capable of transforming the energy of different degrees of freedom into the energy of one degree owing to the interaction of the coupled nonlinear radial, axial and azimuthal electron-ion oscillations. These processes may lead to the additional acceleration of the flow in azimuthal or axial direction so they might be instrumental for the creation of space thrusters employing pulse transformations for propulsion.

## 1. The problem to be analyzed

In the present piece, we shall consider the processes of energy/momentum exchange in rotating two-dimensional, cold plasmas which may come about when the coupled nonlinear electron-ion oscillations are excited in the plasma. Our starting point is a natural statement to the effect that under some conditions, the strong nonequilibrium systems are capable of absorbing and accumulating energy/momentum in one macroscopic degree of freedom from other degrees of freedom (here we call such process as a intermode exchange).

There is some evidence that this assumption stands to reason. For example, everyone knows the phenomenon of energy localization of initial excitations in the chain of one-dimensional nonlinear linked oscillators, so called Fermi-Pasta-Ulam recurrence effect [1]. It was found that under some initial conditions, such a simple dynamic system, instead of manifesting equipartition of energy on all degrees of freedom, shows an unbalanced distribution of energy. Proceeding from this point, one can expect similar behavior for the multidimensional, time-dependent system too [2 - 6]. Owing to different kinds of interactions which may exist in these systems and due to a variety of their initial states, different modes of collective motion are possible. Therefore, unlike one-dimensional systems, time-dependent nonequlibrium states for these systems arising, for example, in the nonlinear stage of a few instabilities will be more versatile in character and then one can expect more favorable conditions

[7] To whom any correspondence should be addressed.

for the energy/momentum exchange. As result, the energy/momentum from different scales may be transformed into one macroscopic degree of freedom [7, 8, 2].

## 2. The coupled nonlinear electron-ion oscillations

Such intermode exchange was established under studying the interaction of radial, axial and azimuthal nonlinear waves in the rotating cold plasmas. We have studied a plasma flow rotating around the z-axis and propagating along this axis. It is assumed that at the initial moment of time the plasma fills the cylinder of some radius and unlimited in z-direction. The displacement current is assumed to be fully compensated by the conduction current, so that the temporal dependence of the magnetic field can be represented by an azimuthal electric field. The dynamics of such system is investigated analytically by solving the cold-fluid equations of the electrons and ions together with the Maxwell's equations.

Depending on the initial state, these flows are capable of transforming the energy of different degrees of freedom into the energy of one degree of freedom. In particular, one can obtain such effect for the coupled azimuthal and radial oscillations when the radial velocity $v_{jr}$ and the azimuthal velocity $v_{j\phi}$ for electron and ion fluids are defined by the relations [9, 10]:

$$v_{jr} = A(t)r\sin(2\phi) + B(t)r\cos(2\phi) + u_{jr}(t), \quad v_{j\phi} = B(t)r\sin(2\phi) - A(t)r\cos(2\phi) + u_{j\phi}(t) \tag{1}$$

Here the electron and ion fluids are marked by the indices $e$ and $i$, respectively. The corresponding electric field has the form

$$\vec{E} = [C(t)r\cos(2\phi) + D(t)r\sin(2\phi) + r\varepsilon_r(t)/2]\vec{e}_r + [D(t)r\cos(2\phi) - C(t)r\sin(2\phi) + r\varepsilon_\phi(t)/2]\vec{e}_\phi \tag{2}$$

In this case one can get the transfer of energy to the rotation mode of the flow. The similar nonlinear plasma waves were studied in [11, 12].

Under exciting the coupled radial, axial and azimuthal oscillations when the velocity field is specified by the dependence

$$v_{jr} = A_j(t)r, \quad v_{j\phi} = C_j(t)r, \quad v_{jz} = B_j(t)z \tag{3}$$

and the wave electric field is

$$\vec{E} = r\varepsilon_r(t)\vec{e}_r + r\varepsilon_\phi(t)\vec{e}_\phi + z\varepsilon_z(t)\vec{e}_z \tag{4}$$

we also observed the intermode exchange [13, 14]. It should be noted that the time-dependent amplitudes in (1)-(4) are associated with the radial, azimuthal and axial components of velocity and electric fields.

Figure 1 and figure 2 present the typical behavior for these cases [13]. For the convenience of presentation, here and in all further calculations, all values are presented in dimensionless form: we take the initial electron density $n_0$, the radius of plasma flow $R$ and the inverse electron frequency $\omega_{pe}$ as the scale of all densities, lengths and times. So all velocities are normalized by $R\omega_{pe}$ and electric field is normalized by $4\pi eRn_0$. Also in all computations we shall use $\mu = m_e/m_i = 10^{-5}$. Comparing figure 1 and figure 2, we see that a small difference in the initial flow leads to very different oscillation patterns. In particular, it is possible to have oscillations occurring only in the azimuthal flow component. We note that the initial disturbance of the basis radial flow has a strong effect on the azimuthal and axial flows. However, initial disturbance of the basis azimuthal flow, even when strong, does not affect the radial and axial flow components. That is, the energy in the azimuthal flow oscillations cannot be converted into the other degrees of freedom, but the energy in the latter can be converted into the azimuthal component.

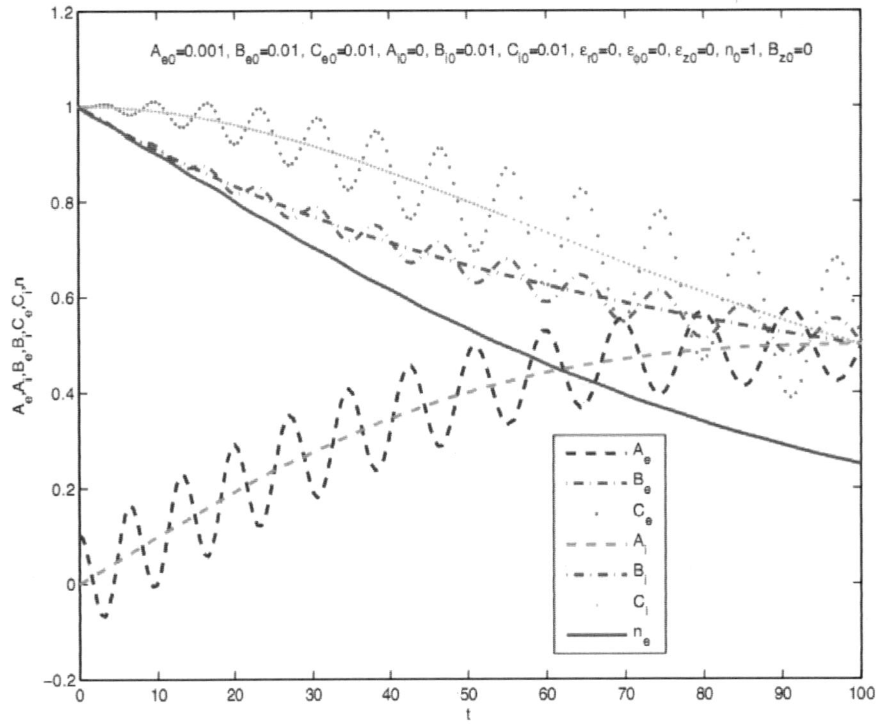


Figure 1. The evolution of $A_j$ ,$B_j$ and $C_j$ given by (3) where $j = e,i$ for $A_{e0} = 10^{-3}$.

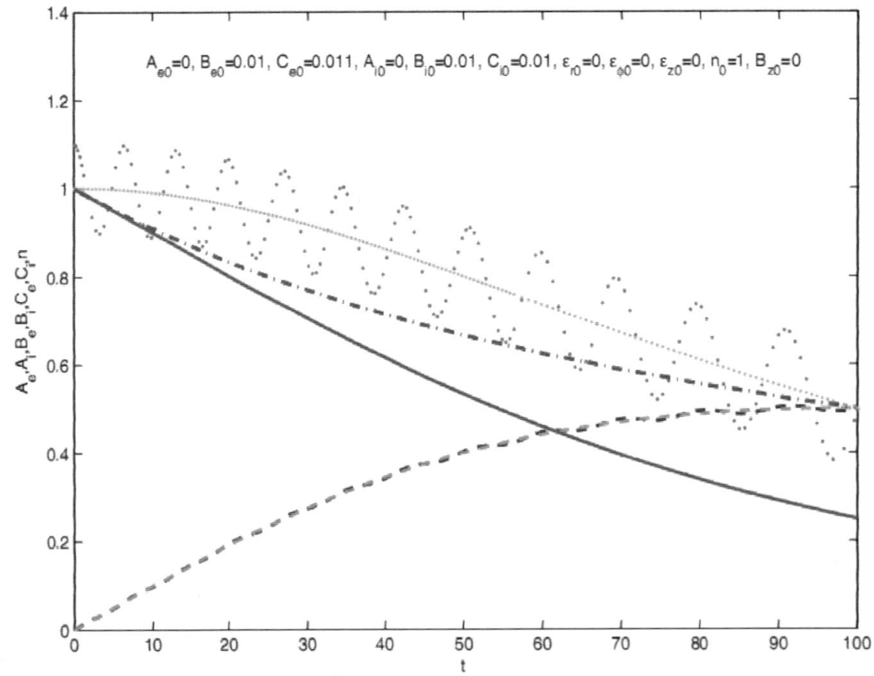


Figure 2. The evolution of $A_j$ ,$B_j$ and $C_j$ given by (3) where $j = e,i$ for $A_{e0} = 0$.

Such behavior is caused by some modulation between the velocity of electron and ion fluids and the electric field (see equations (1) and (2)). Moreover, it was found that the flow asymmetry tend to enhance the axial acceleration of the plasma fluid, and the oscillation energy tend to remain in the azimuthal and axial directions. Such situation is depicted in figure 3 for $V_{rj}(0) = -10^{-2}$, $V_{\phi j}(0) = V_{zi}(0) = 10^{-2}$, $V_{ze}(0) = 1.1\times10^{-2}$ [14].

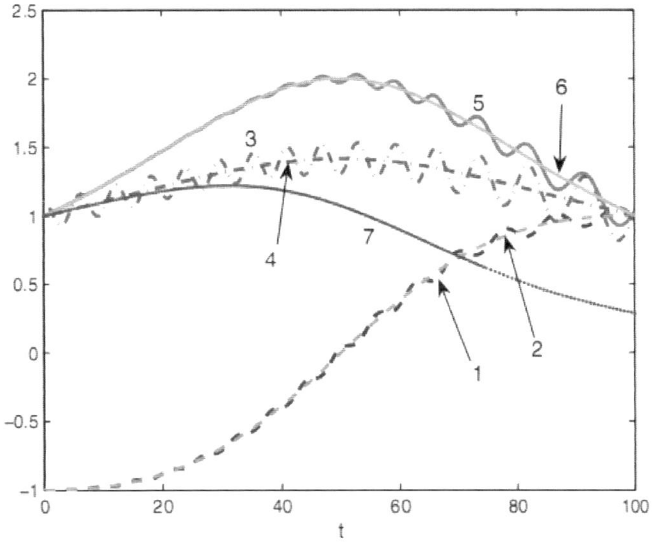


Figure 3. The time dependence of $V_{re}$ (curve 1) ,$V_{ri}$ (curve 2), $V_{ze}$ (curve 3), $V_{zi}$ (curve 4), $V_{\phi e}$ (curve 5), $V_{\phi i}$ (curve 6) and $n_e$ (curve 7).

In fact, the oscillation energy in the radial mode tends to be transferred to the other degrees of freedom.

## 3. Discussion

The present results allow us to speak about the redistribution of energy/momentum between the macroscopic degrees of freedom of the cold plasma flow. They show that the strong nonlinear cylindrical waves can perform nonzero work on the closed curve, as result, the plasma particles will acquire kinetic energy from the waves [15, 9, 10, 13, 14]. In particular, for axisymmetric flows the energy in the radial and axial flow components can be transferred to the rotating component but not vice versa [9, 13]. If we pay no regard to the cylindrical flow symmetry, then there is a connection between radial and axial flows which in the case of the radial compression of the flow leads to the axial acceleration in the asymmetric flow [14, 16]. The symmetrical flows are an exception, often there are asymmetrical flows. So it would be of interest to study how the violation of cylindrical flow symmetry affects the energy transfer among the different flow directions taking into account the plasma inhomogeneity on the nonlinear coupling among the electron and ion flow components and oscillations [17, 18].

Besides, the effect of intermode exchange has been established in the hydrodynamic approach for a cold, neutral flow model of the cylindrical plasma. However, these results suggest that the processes of energy/momentum transfer in the real systems may come about on the same footing. So we may venture a guess that there are some conditions under which the thermal mode might serve as a tank of "internal" energy which can be transformed in one degree of freedom. For example, as a result of this process the energy of chaotic degrees of freedom will be transferred into the rotation mode of the flow. We assume that this peculiarity is the basic property of rotating, strong non-equilibrium plasma flows which differs them from other hydrodynamic system. In order to show this we have to expand our pattern to include the finite electron and ion pressure. In fact, this process is directly opposite to the

Journal of Physics: Conference Series **747** (2016) 012077 doi:10.1088/1742-6596/747/1/012077

Kolmogorov cascade related to the energy transfer from large to small scales where it is dissipated [5]. In this case, unlike the Carnot thermal machine being an evident technical embodiment of the local balance principle, here the mechanism beyond local equilibrium physical paradigm, energy production stems from redistributing the energy in space and time.

In this regard, the connection of present researches with kinetic problem of particle-wave interaction should be noted. The two-dimensional hydrodynamic solutions obtained in [13, 14] are very close in the form to the moments of the distribution function obtained in [19, 13, 20]. Such similarity suggests that these hydrodynamic solutions may also be obtained in the framework of the kinetic formalism extended on the cylindrical geometry to describe the intermode exchange in the rotating plasma flows [19, 18, 20]. This opens up fundamentally new opportunities since more complex problems can be considered further on. In the one-dimensional geometry there is no possibility to get energy and momentum redistribution, one can obtain only phenomena like the Landau damping. However, such opportunities appear in the two-dimensional rotating plasmas. So we should try to explore this partial model and the resulting physical effects in more detail for multi-dimensional cases.

**References**


[1] Fermi E, Pasta J and Ulam S 1955 *Collected Papers of E. Fermi* vol 2 (Chicago: Chicago Univ. Press) p 978

[2] Sreenivasan K R 1999 Fluid turbulence *Reviews of Modern Physics* **71** 383-395

[3] Saffman P G 1992 *Vortex Dynamics* (Camridge: Camridge University Press)

[4] Petviashvili V and Pokhotelov O 1992 *Solitary Waves in Plasma and in the Atmosphere* (Philadelphia: Gordon & Breach Science Publisher)

[5] Eyink G L and Sreenivasan K R 2006 Onsager and the theory of hydrodynamic turbulence *Reviews of Modern Physics* **78** 87-135

[6] Kuznetsov E A 2008 Mixed Lagrangian-Eulerian description of vortical flows for ideal and viscous fluids *Journal of Fluid Mechanics* **600** 167-180

[7] Kadomtsev B B 1994 Dynamics and information *Physics-Uspekhi* **37** 425-499

[8] Bohr T, Jensen M H, Paladin G and Vulpiani A 1998 *Dynamics Systems Approach to Turbulence* (Cambridge: Cambridge University Press)

[9] Karimov A R 2009 Nonlinear waves in twirling plasmas *Journal of Plasma Physics* **75** 817-828

[10] Karimov A R and Godin S M 2009 Coupled Radial–Azimuthal Oscillations in Twirling Cylindrical Plasmas *Physica Scripta* **80** 035503 1-6

[11] Stenflo L 1990 Strongly nonlinear cylindrical surface waves *Physica Scripta* **41** 643-644

[12] Stenflo L and Yu M Y 1995 Nonlinear wave evolution in bounded plasmas *Physics of Plasmas* **2** 1494-96

[13] Karimov A R, Stenflo L and Yu M Y 2009 Coupled azimuthal and radial flows and oscillations in a rotating plasma *Physics of Plasmas* **16** 062313 1-6

[14] Karimov A R, Stenflo L and Yu M Y 2009 Coupled flows and oscillations in asymmetric rotating plasmas *Physics of Plasmas* **16** 102303 1-4

[15] Karimov A R 2002 Influence of inhomogeneity and finite pressure of a plasma on nonlinear electron oscillations *Physica Scripta* **65** 356-359

[16] Karimov A R, Yu M Y and Stenflo L 2013 Asymmetric oscillatory expansion of a cylindrical plasma *J. Plasma Physics* **79** 6 1007-09

[17] Karimov A R, Stenflo L and Yu M Y 2011 Flow oscillations in radial expansion of an inhomogeneous plasma layer *Physical Letters* A **375** 2629-2636

[18] Karimov A R 2001 Nonlinear solutions of a maxwellian type for the Vlasov–Poisson equations *Physics of Plasmas* **5** 1533-1537

[19] Yu M Y, Chen Z Z and Stenflo L 2001 A new class of exact solutions of the Vlasov equation *Physics of Plasmas* **8** 5081-5085

[20] Karimov A R 2013 Coupled electron and ion nonlinear oscillations in a collisionless plasma *Physics of Plasmas* **20** 052305 1-6

Journal of Physics: Conference Series **747** (2016) 012078 doi:10.1088/1742-6596/747/1/012078

# Investigation of the hybrid electron linac with negative group velocity

**E.A. Savin[1], R.Yu. Alekhanov, A.V. Bulanov, V. I. Kaminskiy, S.V. Matsievskiy and N.P. Sobenin**
National Research Nuclear University MEPhI (Moscow Engineering Physics Institute),
Kashirskoye shosse 31, Moscow, 115409, Russia

E-mail: es-abyss@yandex.ru

**Abstract.** Hybrid accelerator, incorporating travelling wave (TW) and standing wave ( SW ) structures is proposed and discussed. Accelerator can sum up an RF focusing in the SW buncher and lower losses in the TW accelerating structure walls. Moreover, the structure without dumping load is proposed. Input power and beam loading are chosen to minimize power reflection from buncher maintain travelling wave regime in the accelerating structure while beam loading is on. In this case SW buncher operates as a dumping load, so all input power either goes to beam or dissipates in the wall losses, it increases structure efficiency. RF characteristics and beam dynamics simulations have been performed.

## 1. Accelerating and bunching cells geometry

Well studied [1, 2] magnetic coupled diaphragm loaded structure (DLS-M) has been chosen as a TW part of the accelerator (Fig. 1). It operates on the $2\pi/3$ mode at 2856 MHz. Magnetic coupling provides a higher group velocity (i.e. structure filling time) compared to the conventional DLS structures and also allows to control it without changing an aperture radius. One of the aspects of using such a structure is it having a so-called negative dispersion curve, which corresponds to power flow and wave phase velocity travelling in opposite directions. For this project DLS-M has been tuned to the 1% group velocity.

[1] To whom any correspondence should be addressed.

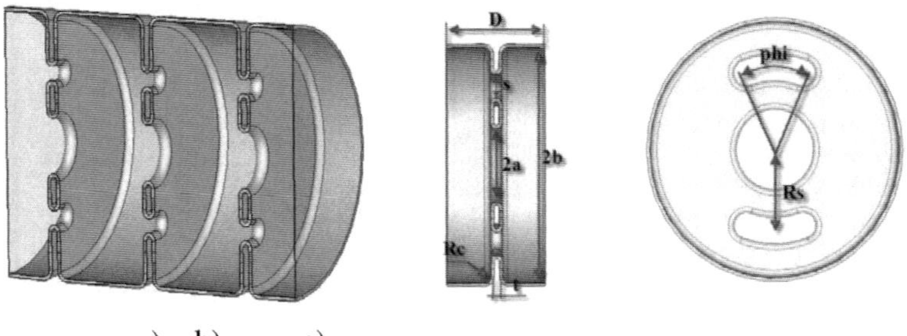


a) b) c)

**Figure 1.** DLS-M geometry

### *1.1. Buncher*

Bi-periodic accelerating structure (BAS) [3] has been chosen to bunch an electron beam, injected from the 30 kV electron gun. Three bunching cells and on-axis accelerating field distribution are presented in Fig. 2c. To obtain maximum capture and minimal energy spread, in according to beam dynamics study, it appears that $E_{z1}/E_{z2}/E_{z3}$ should be as shown in the Fig. 2c. Beam dynamics simulation model with the 1 MW RF input power and current of 0.5 A is presented in Fig. 3.

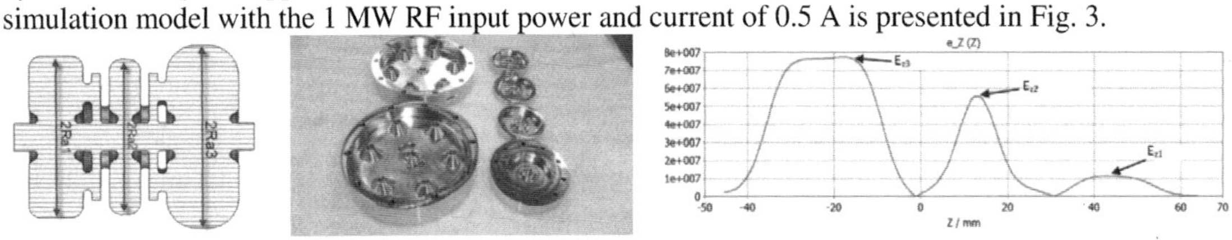


a) b) c)

**Figure 2.** (a) Buncher cells geometry; (b) generator-accelerator buncher cells; (c) on-axis accelerating electric field distribution

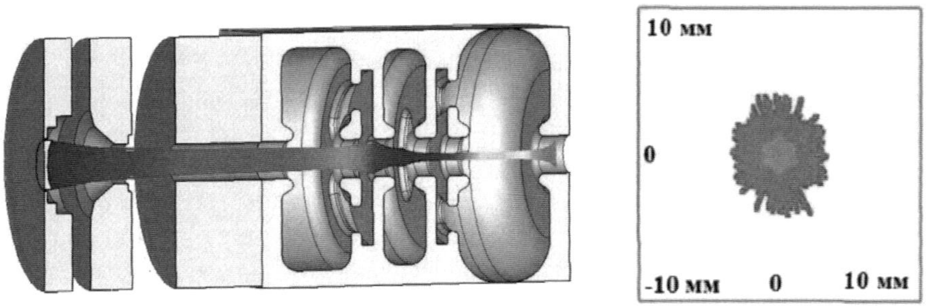


a) b)

**Figure 3.** (a) Beam dynamics in the buncher cells; (b) electron beam cross section at the end of the buncher

## 2. Hybrid structure

DLS-M is connected to the BAS via the rectangular waveguide as shown in Fig. 4. The main difference between this structure and ones, utilizing electrically coupled DLS, is that the RF power

Journal of Physics: Conference Series **747** (2016) 012078 doi:10.1088/1742-6596/747/1/012078

flows through such a system "backwards", deriving the TW part and the SW part of the accelerating structure. With the optimized number of cells in both parts, the excessive power from the TW part if fully absorbed by the SW part, thus eliminating the need for the dumping loads and circulators. In the Fig. 5 transient process of filling of the structure with RF power (without beam) is presented.

## 3. Beam dynamics

In this paper beam dynamics was calculated for the stationary RF fields. Output beam parameters structure, incorporating 3 BAS and 40 DLS-M cells are presented in the Table 1 and in the Fig. 6.

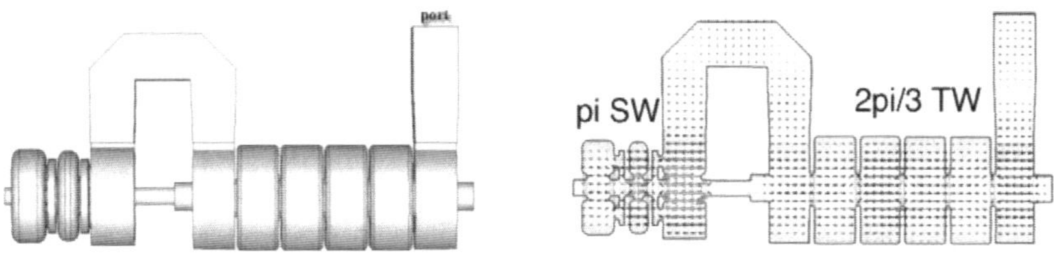


a) b)

**Figure 4.** Geometry and fields in the hybrid linac

a) b) c)

**Figure 5.** Transient processes in the hybrid linac

**Table 1.** Output beam parameters

| Maximum energy, MeV | Average energy, MeV | Beam radius, mm | Capture coefficient, % | Efficiency, % | Beam current, A |
|---|---|---|---|---|---|
| 13.1 | 10 | 3 | 58 | 57 | 0.3 |

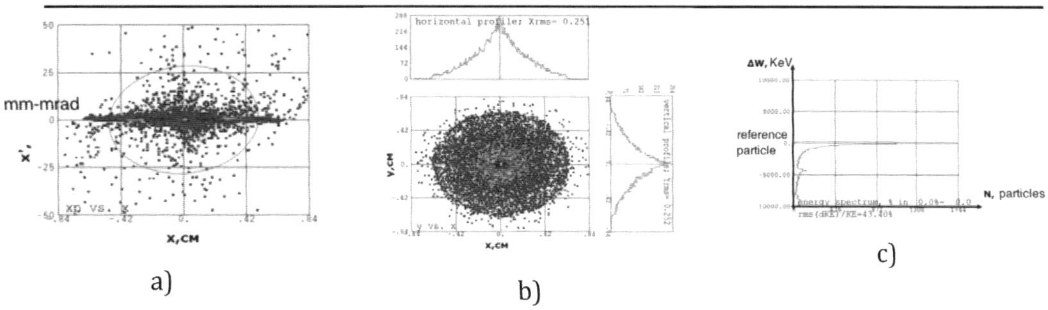


a) b) c)

**Figure 6.** (a) Beam transverse emittance; (b) beam cross section; (c) energy spectrum

## 4. Conclusion

Hybrid accelerating structure with magnetic coupled DLS-M cells has been investigated. SW and TW parts of the accelerating structure were tuned to achieve high efficiency and captivity coefficients. Beam dynamics has been simulated. In result the possibility of developing an accelerating structure without the dumping loads was provided.

**Acknowledgement**

The research has been made under support of the Ministry of Education and Science grant 3.245.2014 /r

## References

[1] S. Kutsaev et al., "Design of hybrid electron linac with standing wave buncher and travelling wave structure," Nucl. Instr. Meth. Phys. Res., pp. 13–30, 2011.

[2] S. Kutsaev et al., "Magnetic coupled accelerating structure," Problems Of Atomic Science And Technology, 2009.Jain S C, Willander M, Narayan J and van Overstraeten R 2000 *J. Appl. Phys.* **87** 965

[3] E. Savin and N. Sobenin, "Biperiodic accelerating structure with inner coupling cells with an increased coupling coefficient," Technical Physics, 2013

# Combined generating-accelerating buncher for compact linear accelerators

**E.A. Savin[1,3], S.V. Matsievskiy[1], N.P. Sobenin[1], I.D. Sokolov[1]**
**A.A. Zavadtsev[2]**

, 3 , , , ,
, 3 , , ,

**Abstract.** Described in the previous article [1] method of the power extraction from the modulated electron beam has been applied to the compact standing wave electron linear accelerator feeding system, which doesnt require any connection waveguides between the power source and the accelerator itself [2]. Generating and accelerating bunches meet in the hybrid accelerating cell operating at $TM020$ mode, thus the accelerating module is placed on the axis of the generating module, which consists from the pulsed high voltage electron sources and electrons dumps. This combination makes the accelerator very compact in size which is very valuable for the modern applications such as portable inspection sources. Simulations and geometry cold tests are presented.

## 1. Geometry and fields overview

Presented in this paper Generating-Accelerating Module (GAM) operates at 2856 MHz and consists of three cells (Figures 1 a,b). First two Biperiodic Accelerating Structure (BAS) cells operate at $TM_{010}$ mode [3] and the third cell is a combined generating-accelerating cell, operating at $TM_{020}$ mode.

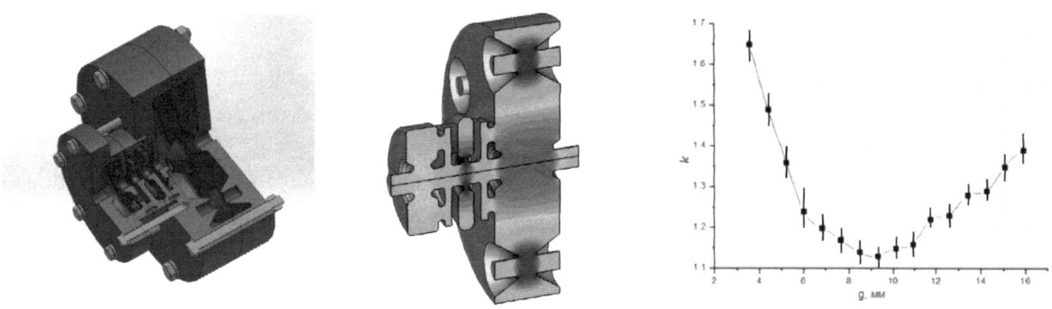

a) b) c)

**Figure 1.** (a) CAD model of GAM; (b) 2D projection of the electric field in GAM; (c) generator to accelerator maximum Ez field ratio vs generator gap

[3] To whom any correspondence should be addressed.

Six circumferential channels of the third cell are used to draw energy from driver beams. Varying the gap (g) between their noses allows tuning of generator to accelerator voltage ratio $E_{max\ gen}/E_{max\_acc}$ (Figure 1 c).
As GAM is used for electron beam bunching, phase velocities and accelerating field relations (Figure 2) were set to obtain high capture coefficient and low energy spectrum for 30 keV electron beam of specified DC electron gun.

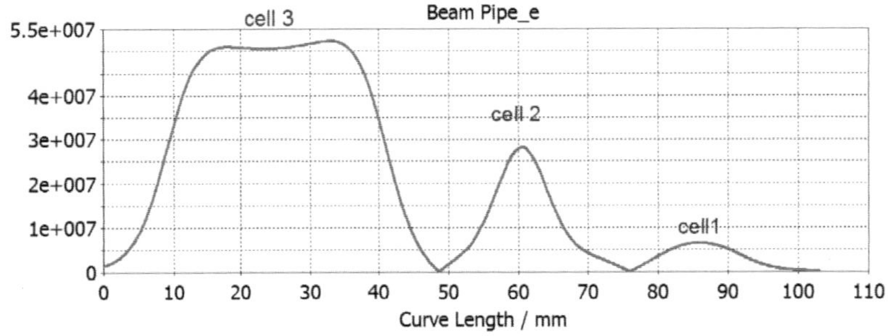


**Figure 2.** On-axis accelerating electric field in GAM

## 2. Beam dynamics simulations

Simulation of electron dynamics in GAM was carried out. Six 100 kV driver beams with 5A peak current were used for powering the structure and induce electric field of 35 MV/m in 250 ns in unloaded structure (Figure 3 a).

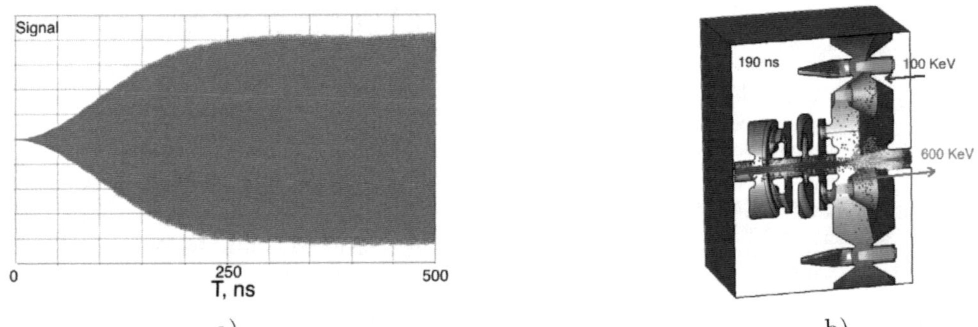


a) b)

**Figure 3.** (a) electrical field induction versus time in GAM; (b) accelerating beam dynamics in GAM

After field induction, 30 keV electron bunch acceleration in the structure was simulated(Figure 3b). Parameters of the accelerated beam are presented in the Table 1.

## 3. Measurements

The assembly has been manufactured (Figure 4) for field tests. Modes have been excited by the antenna, placed at the one of the generation channels entrance.

Operating frequency and electric field distributions in both generating and acceleration channels are corresponding to simulations (Figure 5).

**Table 1.** Parameters of the accelerated beam

| Parametres | Input | Output |
|---|---|---|
| Accelerated beam current | 0.56 A | 0.44 A |
| Accelerated beam energy | 30 keV | 600 keV |

a) b) c)

**Figure 4.** (a), (b) manufactured GAM cells; (c) part of the measurement stand.

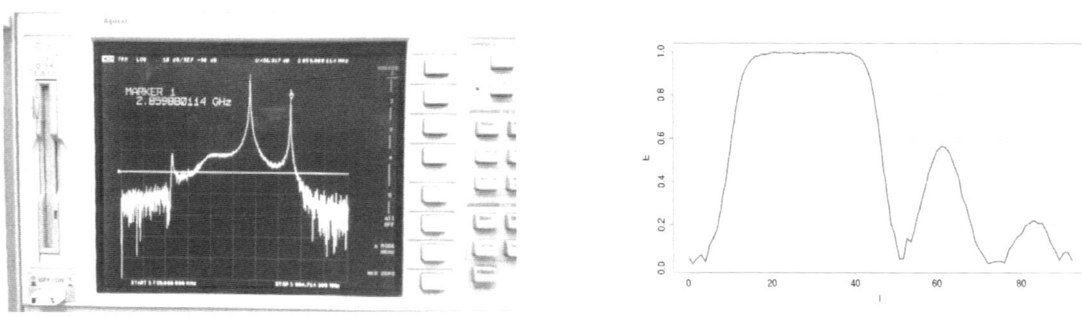


a) b)

**Figure 5.** (a) reflection scattering parameter in frequency range (b) on-axis electric field

## 4. Conclusion

Compact LINAC combined with the generation system structure has been proposed. Simulations shows that this structure works and test manufacture was successful.

Assembly is undergoing preparations for a beam test. Accordingly to the simulations, efficiency of this structure (accelerating mode power over generating beams power) exceeds 50% and depends on the generating beam parameters.

The biggest problem for the S-band GAM is the absence of the high-frequency modulated electron sources. Usually, modulation of the beam is achieved using a control grid. Unfortunately this method is limited to the L-band frequency due to the grid manufacturing difficulties. Thus, GAM scheme may be used in L-band, but needs additional research on high current electron beam sources for use of higher frequencies.

**Acknowledgement**
The research has been made under support of the Ministry of Education and Science grant 3.245.2014/r

**References**

[1] E. Savin *et al.*, "Different solutions for the generator-accelerator module," *Physics Procedia*, 2014.
[2] J. Potter *et al.*, "The klynac: an integrated klystron and linear accelerator," in *AIP*, 2013.
[3] E. Savin and N. Sobenin, "Biperiodic accelerating structure with inner coupling cells with an increased coupling coefficient," *Technical Physics*, 2013.

# Deflector for XFEL TDS BC1

**E V Volobuev[1,2,3], A A Zavadtsev[1,2,4], D A Zavadtsev[1,2], L V Kravchuk[1], V V Paramonov[1], N P Sobenin[3] and D V Churanov[1,2]**

[1]Institute for Nuclear Research of the Russian Academy of Sciences, Prospekt 60-letiya Oktyabrya 7a, Moscow 117312, Russia
[2]Nano Invest LLC, Office 017, 9 Academician V.N.Chelomey Str., Reutov, Moscow Region 143969, Russia
[3]National Research Nuclear University MEPhI (Moscow Engineering Physics Institute), Kashirskoye shosse 31, Moscow 115409, Russia

E-mail: azavadtsev@yandex.ru

**Abstract.** Deflector is the part of the Transverse Deflecting System TDS BC1 of the European X-ray Free Electron Laser (XFEL). TDS BC1 is located on the XFEL beam line at the coordinate z=206 m. This system is designed to monitor the longitudinal phase space and the emittance of the accelerated electron bunch after Bunch Compressor 1 (BC1) , where electron beam energy is 600 MeV. The deflector includes waveguide window, waveguide load, E-bend, ion pump adapters, two antennas, two ion pumps and 1.7 m long disk-loaded EH-hybrid mode deflecting structure. Operating frequency is 2997.2 MHz. Input RF power is up to 24 MW. The deflector has been manufactured, and all designed RF parameters have been obtained experimentally at low RF power level.

## 1. Introduction Design of TDS System BC1

The XFEL TDS BC1 is one of three deflecting systems in the European XFEL to monitor the longitudinal phase space and the emittance of the accelerated electron beam.
XFEL TDS BC1 is located in the XFEL accelerator tunnel, after the Bunch Compressor 1, at the coordinate z=206 m, where electron beam reaches energy of 600 MeV.

The TDS System BC1 consists of the deflector and High Power RF system (HPRF). The deflector includes all units, connected to the ultra-high vacuum volume of the XFEL accelerator: waveguide RF window, waveguide RF load, E-bend, ion pump adapters, two antennas, two ion pumps and the deflecting structure.

HPRF includes modulator, pulse transformer, klystron, solenoid, waveguide system, water cooling system, power supply distribution block, control and power supply cabinet.

Design of the TDS System BC1 (except control and power supply cabinet, which is located beside) is shown in Figure 1.

The deflector is assembled on the steel alignment girder in the clean room. The structure is installed on the girder via precise supporting units. After assembling the girder is transported from the clean room to the accelerator tunnel. The girder is installed on two concrete pillars via alignment units, allowing precise alignment of the structure to the beam axis.

[4] To whom any correspondence should be addressed.

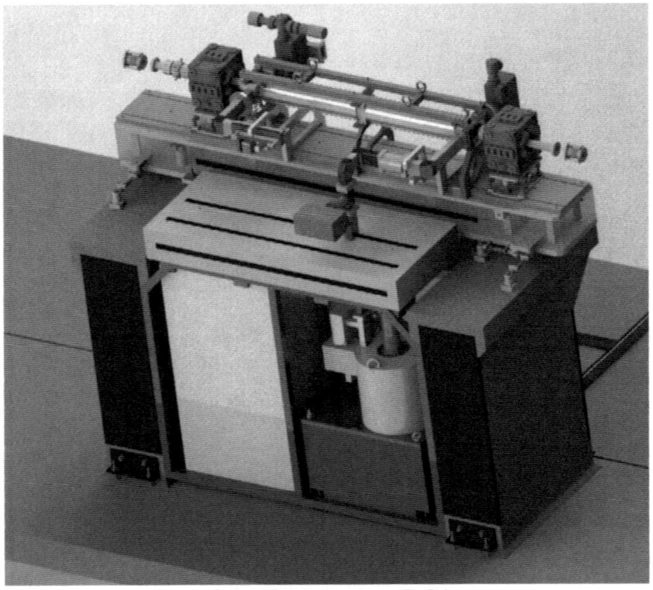

**Figure 1.** Design of the TDS System BC1.

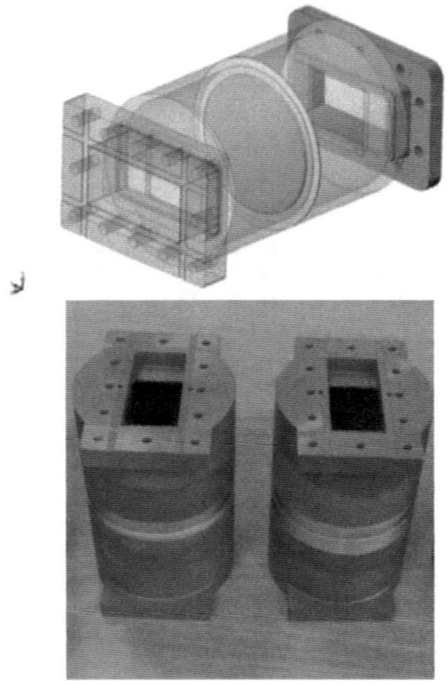

**Figure 2.** Waveguide window.

## 2. RF window

Special waveguide ceramic window has been designed for the TDS systems. This is double-mode window. 3D design and the picture of the window are shown in Figure 2. Designed and measured parameters of the window are shown in Figure 3.

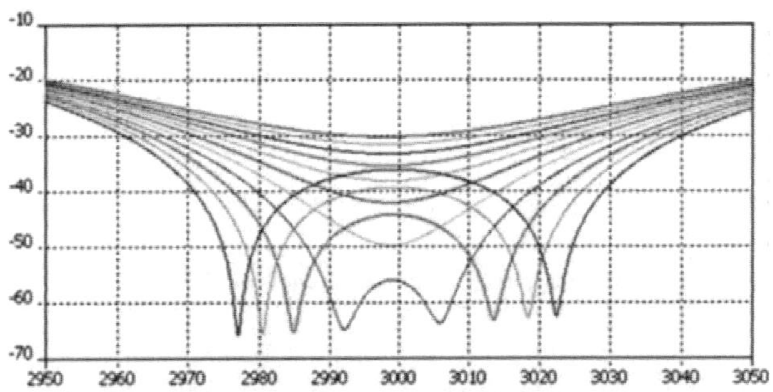


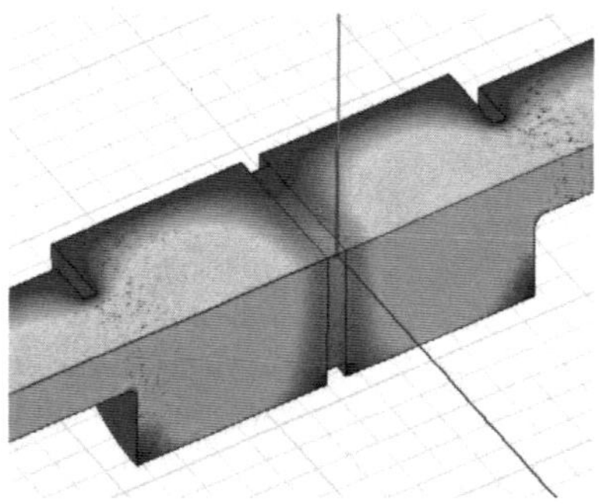

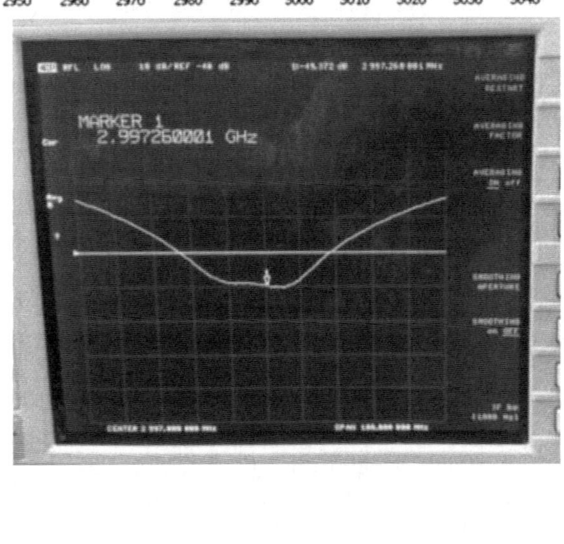


**Figure 3.** Window: reflection S11(f) for simulated tuning process (operating curve is brown), distribution of the complex amplitude of the electric field and reflection S11(f) measured in the built window. S11=-45 dB at operating frequency.

Journal of Physics: Conference Series **747** (2016) 012080 doi:10.1088/1742-6596/747/1/012080

Advantages of the double-mode window are: low electric field at the ceramic surface (4 times less than in regular waveguide), i.e. low multipacting probability and wide bandwidth (70MHz@30dB and 40MHz@40dB).

## 3. E-bend

Waveguide E-bend is to be matched and to have low overvoltage on the inner surface. Designed variant is shown in Figure 4. Max electric field on the surface is just 1.5 times higher than in regular waveguide. The bandwidth is 150MHz@40dB. Reflection measured at the produced sample is S11=-47 dB.

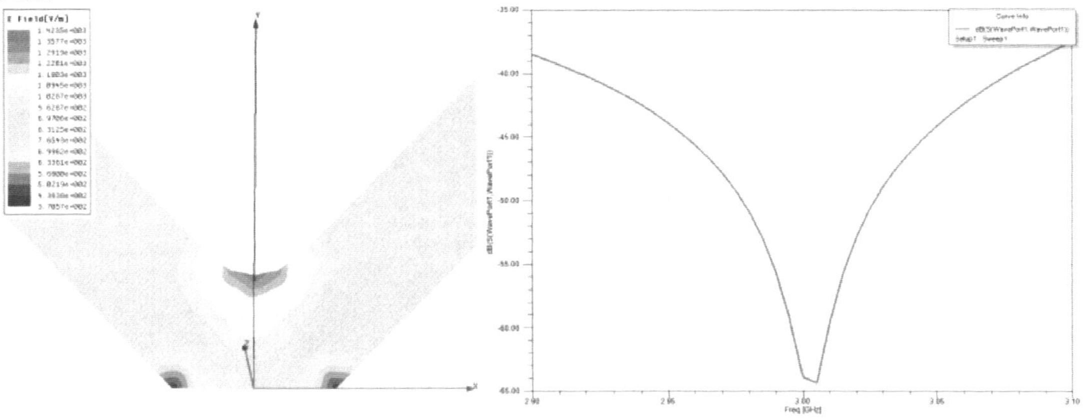

**Figure 4.** E-bend: distribution of the complex amplitude of electric field in calculation model and reflection S11(f).

## 4. RF load

The waveguide RF load has been developed specially for the XFEL TDS Systems. Main requirements for the load are a) the good matching in wide bandwidth, b) materials and connections in use to be acceptable for ultra-high vacuum volume of the XFEL accelerator, c) separation of the inner volume and water cooling volume via brazing, welding, ceramics and glass is not acceptable. It means that RF absorbing material is to be located inside the load and it has to be acceptable for ultra-high vacuum volume of the XFEL accelerator.

Sendust has been chosen for the load as an absorbing material. It is a ferromagnetic alloy, consisting of 85% Fe, 5.2-5.6 % Al, 9.4-9.8% Si, with conductivity of 0.6 μOhm·m. Special model of Sendust layer baked on the copper plate has been produced and tested in DESY vacuum group. DESY conclusion is: Sendust is acceptable for ultra-high vacuum volume of the XFEL accelerator.

0.3 mm thick Sendust layer is baked on the inner copper wall of the load. Outside surface of this copper wall is cooled with water flow.

The load is the waveguide with decreasing height on special function. Figure 5 shows calculation model with distribution of the complex amplitude of electric field and photo of load @ E-bend assembly. Measured reflection from the built load is S11=-47 dB. Prototype of the load has been built. It is tested at DESY PITZ TDS stand now [1].

## 5. Deflecting structure

S-band TW operation constant gradient disk loaded structure with $2\pi/3$ mode of hybrid EH11 wave has been chosen for XFEL TDS. Main parameters are f=2997.2 MHz, length L=1.7 m, aperture radius Ra =21.71 mm, group velocity $\beta_g$ =-0.01587, attenuation $\alpha$ =0.168 1/m, $\frac{E_d\lambda}{\sqrt{P}}$ =242 $\mathrm{Ohm}^{1/2}$.

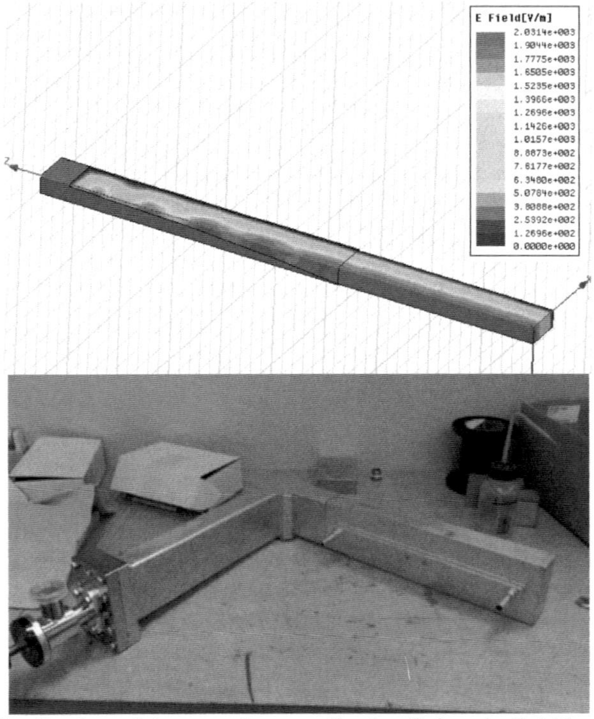


**Figure 5.** The load: distribution of the complex amplitude of electric field in calculation model and photo of load @ E-bend assembly.

The structure is analogous to the structure for XFEL TDS INJ [2]. Prototype of the structure has been built [1] and it is tested at PITZ facility now. The design model of the structure is shown in Figure 6.

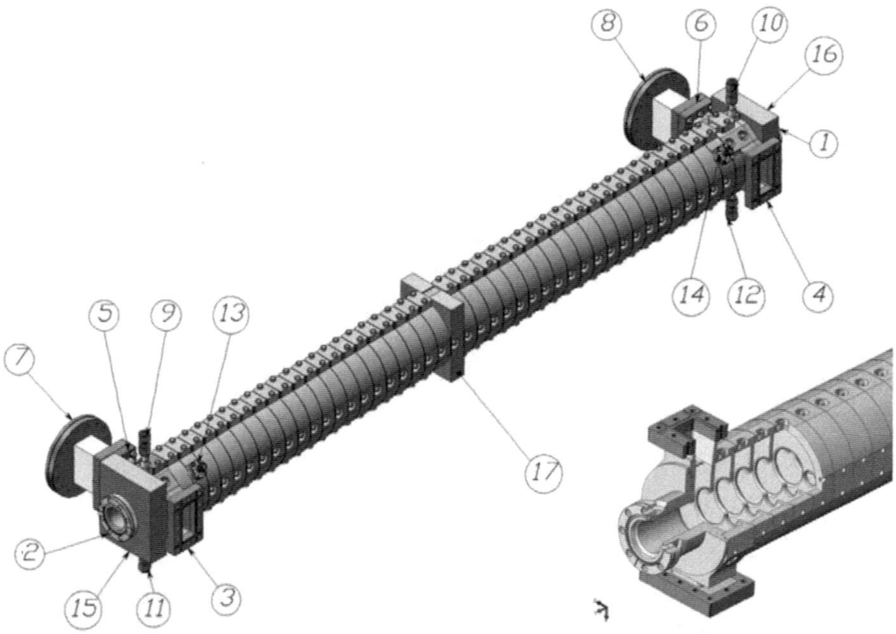


**Figure 6.** Design model of the structure.

The deflector has been manufactured and tuned. It is shown in Figure 6. The manufactured structure and measured distribution of the electric field are presented in Figure 7

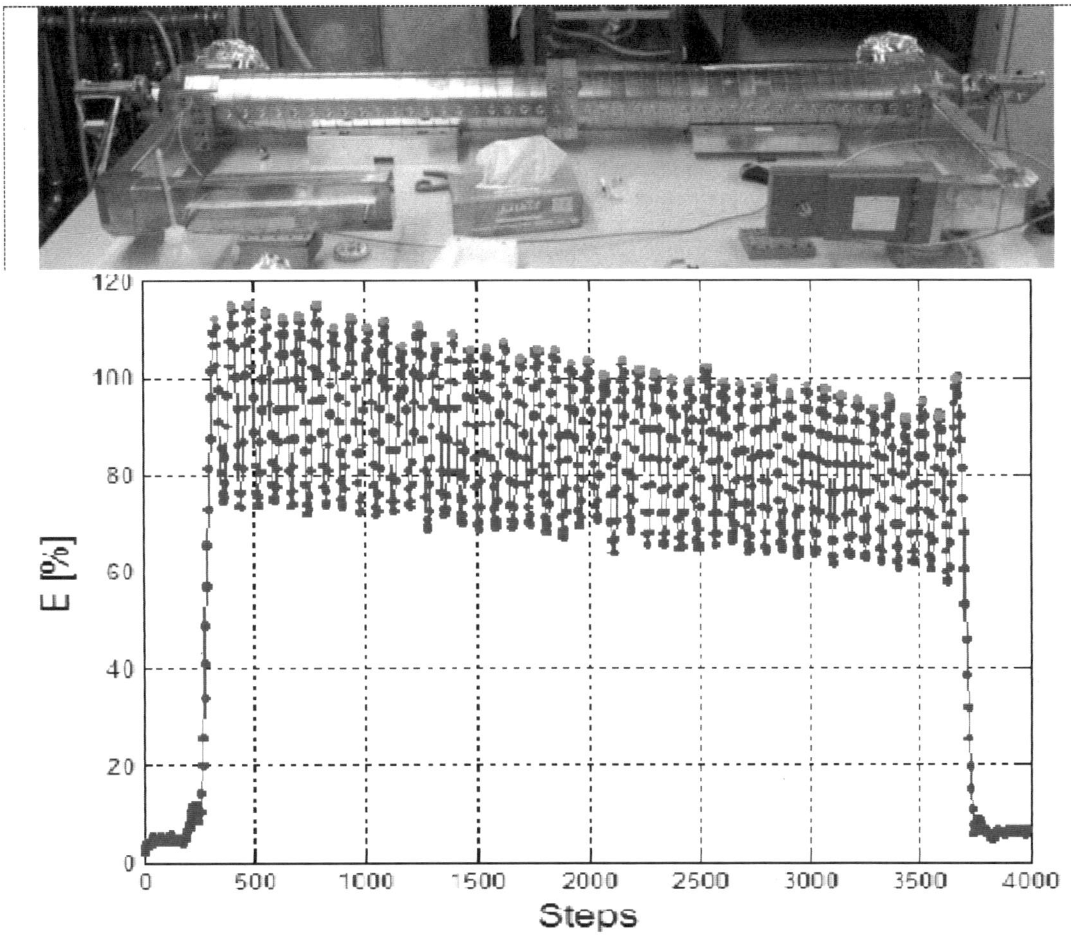


**Figure 7.** Manufactured deflector structure and measured distribution of the amplitude of the deflecting electric field.

## References

[1] Otevrel M *et al.* 2012 Diagnostics at PITZ 2.0 Beamline: Status and New Developments *Proceedings of IPAC2012 (New Orleans, Louisiana, USA)* 634-636

[2] Volobuev E, Zavadtsev A, Zavadtsev D, Kravchuk L, Paramonov V, Lalayan M, Smirnov A, Sobenin N and Churanov D 2016 Transverse deflecting structure XFEL TDS INJ *Proceedings of II International Conference «Plasma and laser research and technologies» (PLRT, Moscow, Russian Federation)* IOPscience

## Acknowledgements

The present work has been done as an in-kind contribution of Russian Federation into European XFEL project.

# High power RF system for transverse deflecting structure XFEL TDS INJ

**E N Volobuev[1,2], A A Zavadtsev[1,2,4], D A Zavadtsev[1,2], A J Smirnov[3], N P Sobenin[3] and D V Churanov[1,2]**

[1]Institute for Nuclear Research of the Russian Academy of Sciences, Prospekt 60-letiya Oktyabrya 7a, Moscow 117312, Russia
[2]Nano Invest LLC, Office 017, 9 Academician V.N.Chelomey Str., Reutov, Moscow Region 143969, Russia
[3]National Research Nuclear University MEPhI (Moscow Engineering Physics Institute), Kashirskoye shosse 31, Moscow 115409, Russia

E-mail: azavadtsev@yandex.ru

**Abstract**. The high power RF system (HPRF) is designed for RF feeding of the transverse deflecting structure of the transverse deflecting system XFEL TDS System INJ of the European X-ray Free Electron Laser. The HPRF system includes klystron, waveguide ceramic windows, directional couplers, waveguide vacuum units, spark detector and waveguide line. Operating frequency is 2997.2 MHz. Peak input power is up to 3 MW. The HPRF system has been developed, manufactured and assembled in the XFEL Injector building. The total length of the waveguide line is 55 m from the klystron at the -5 floor to the transverse deflecting structure at the -7 floor. All designed RF parameters have been obtained experimentally at low RF power level.

## 1. Design and manufacture of the HPRF system

There are three transverse deflecting systems in the X-ray Free Electron Laser (XFEL) for monitoring the longitudinal phase space and the emittance of the accelerated electron beam. The first TDS System INJ is located in the injector. The transverse deflecting structure is located in the injector tunnel at the -7 floor at longitudinal coordinate z=53 m from cathode. There is no space for the klystron and the modulator close to the structure. Therefore they are located at the -5 floor. It means the waveguide line connecting the klystron and the transverse deflecting structure is long (55 m) and quite a bent. Design of the transverse deflecting systems XFEL TDS INJ is shown in Fig. 1.

The HPRF system includes klystron, waveguide ceramic windows, directional couplers, waveguide vacuum units, spark detector and waveguide line [1, 2]. The filling of the waveguide line is following: a) Nitrogen at pressure up to 3 bar abs. from klystron to window 1; b) air at atm. pressure from window 1 to window 2 (the second option is technical vacuum in case of break-down problems); c) ultra-high vacuum from window 2 to transverse deflecting structure.

The first Nitrogen filled part of the waveguide line includes H-bend with spark detector and gas filling port, E-bend, directional coupler 1and window 1. The second air filled part of the waveguide line includes window 1, straight waveguides, E-bends, H-bends, four waveguide vacuum units for connection of the ion pumps and window 2. The third vacuum filled part includes window 2 and

[4] To whom any correspondence should be addressed.

straight waveguide. An additional waveguide part is the waveguide load connected to the output of the transverse deflecting structure.

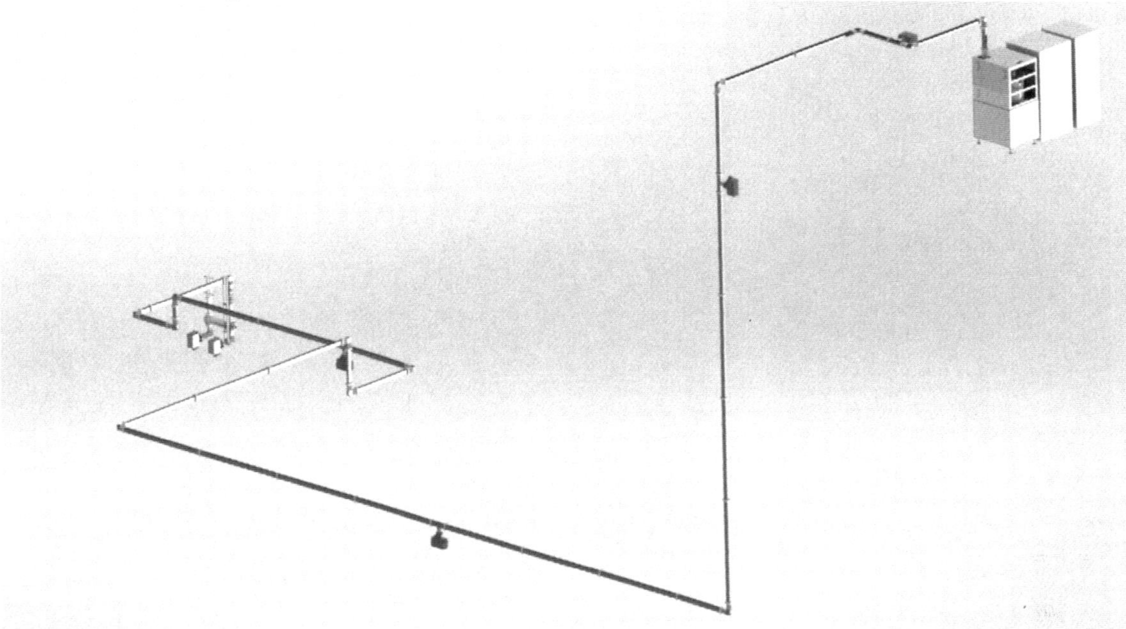

**Figure 1.** Design of the transverse deflecting systems XFEL TDS INJ.

*1.1. Klystron*

CPI klystron VKS-8262HS is used for the RF power generation (Figure 2). Main parameters of the klystron are: 3 MW peak power, 2997.2 MHz operating frequency, 110 kV voltage, 72 A current.

*1.2. Window*

Special waveguide double-mode ceramic window has been designed for the TDS systems [1]. The picture of the window is shown in Figure 3. Main parameters are a) electric field on the ceramic surface is 4 times less than in the regular waveguide, b) bandwidth is 70MHz@30dB and 40MHz@40dB.

**Figure 2.** Klystron.

**Figure 3.** Window.

Journal of Physics: Conference Series **747** (2016) 012081 doi:10.1088/1742-6596/747/1/012081

### *1.3. Directional coupler*

The directional coupler has been designed for the TDS Systems specially. The filling of the directional coupler is to be air or vacuum. Therefore it has been designed as vacuum tight unit. It includes ceramic disk brazed to the waveguide separating inner volume of the directional coupler.

The directional coupler is to monitor forward and reflected power. Main parameters of the directional coupler are: coupling -65 dB, directivity 34 dB. Directional coupler is shown in Figure 4.

**Figure 4.** Directional coupler.

**Figure 5.** Waveguide vacuum unit.

### *1.4. Waveguide vacuum unit*

The waveguide vacuum unit consists of straight waveguide with set of transverse slots on the smaller side of the waveguide and ConFlat flange for ion pump connection. The waveguide vacuum unit is shown in Figure 5.

### *1.5. Waveguide line*

The waveguide line includes straight waveguides, E-bends, H-bends (see Figure 6-8). One of the H-bend is provided with the port for connection of the spark detector and with the gas filling port. All connections of the waveguide units are realized with high vacuum DESY type rectangular flange with copper gasket.

### *1.6. Load*

Sendust coated waveguide load is used as a dummy load at the deflecting structure output [1] (Figure 9). Measured reflection is S11=-47 dB.

**Figure 6.** Straight waveguides.

**Figure 7.** H-bends and E-bends.

**Figure 8.** H-bend with spark detector and gas filling port.

## 2. Assembly and test of the HPRF system

The whole RF power supply system has been assembled and tested at low RF power level. Assembled system is shown in Figure 9. RF test at low RF power level shows that reflection from input flange of the waveguide line with the transverse deflecting structure is S11=-42 dB at operating mode (Figure 10).

**Figure 9.** Assembled and tested HPRF system.

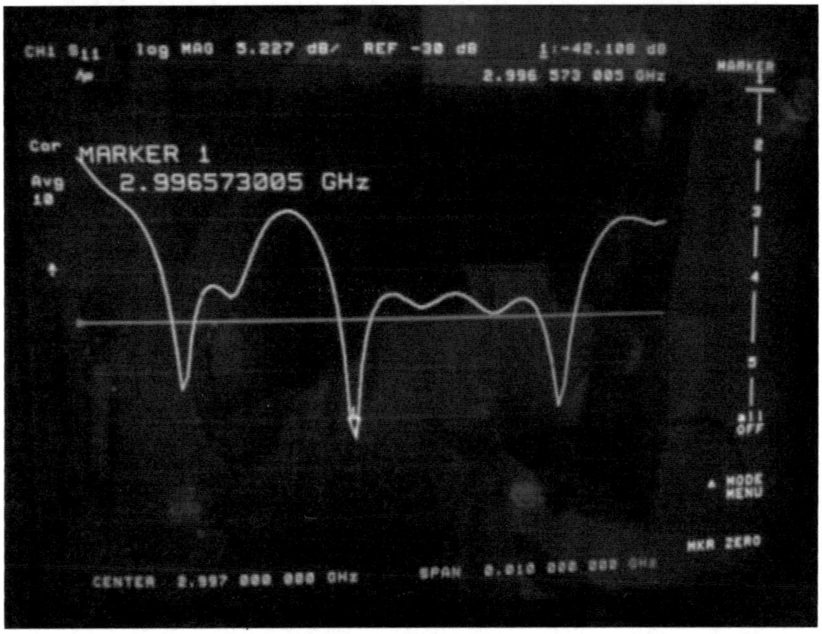


**Figure 10.** Measured reflection S11 v.s. frequency.

## References

[1] Zavadtsev A A *et al.* 2014 Components of the radio-frequency system for traveling-wave deflecting structures *Instruments and experimental techniques* **57** 706-713

[2] Volobuev E, Zavadtsev A, Zavadtsev D, Kravchuk L. Paramonov V, Sobenin N and Churanov D 2016 Deflector XFEL TDS BC1 *Proceedidngs of II International Conference «Plasma and laser research and technologies», PLRT 2016*

## Acknowledgements

The present work has been done as an in-kind contribution of Russian Federation into European XFEL project.

Journal of Physics: Conference Series **747** (2016) 012082 doi:10.1088/1742-6596/747/1/012082

# Solid state modulator for klystron power supply XFEL TDS INJ

**A A Zavadtsev[1,2,3], D A Zavadtsev[1,2], D A Zybin[1,2], D V Churanov[1,2] and P V Shemarykin[1,2]**

[1]Institute for Nuclear Research of the Russian Academy of Sciences, Prospekt 60-letiya Oktyabrya 7a, Moscow 117312, Russia
[2]Nano Invest LLC, Office 017, 9 Academician V.N.Chelomey Str., Reutov, Moscow Region 143969, Russia

E-mail: azavadtsev@yandex.ru

**Abstract**. The transverse deflecting system XFEL TDS INJ for European X-ray Free Electron Laser includes power supply for the CPI VKS-8262HS klystron. It has been designed for pulse high-voltage, cathode heating, solenoid and klystron ion pump. The klystron power supply includes solid state modulator, pulse transformer, controlled power supply for cathode heating and commercial power supplies for solenoid and ion pump. Main parameters of the modulator are 110 kV of peak voltage, 72 A peak current, and pulse length up to 6 μs. The klystron power supply has been developed, designed, manufactured, tuned, tested and installed in the XFEL building. All designed parameters are satisfied.

## 1. Introduction

Three transverse deflecting systems (XFEL TDS Systems) are developed and manufactured for the European X-ray Free Electron Laser (XFEL). These systems are designed for monitoring of the longitudinal phase space and the slice emittance of the accelerated electron beam in the Injector, after Bunch Compressor 1 and after Bunch Compressor 2. Each system includes deflecting structure, a klystron for RF drive of the structure and the power supply for the klystron. The TDS INJ System located in the Injector employes CPI VKS-8262HS klystron. The TDS Systems located after Bunch Compressors include Thales TV-2002DoD klystrons.

## 2. Modulator for PITZ TDS – prototype of modulator for XFEL TDS System INJ

The prototype of the transverse deflecting system has been developed, built and installed at DESY PITZ for the test [1]. The PITZ TDS system is an analogue of the XFEL TDS INJ System. The power supply for the klystron CPI VKS-8262HS includes solid state modulator for direct pulse high-voltage, power supply for the heater and commercial power supplies for the solenoid and the ion pump.

Schematics of the modulator is designed on base of the pulse generator invented by V Arkadjev and N Baklin in 1914. The capacitors are connected in parallel to charge and then to switched series circuit. Such a way the resulting voltage on the load is equal to sum of the voltages on each capacitor. V Arkadjev and N Baklin employed contact-mechanical switches. E Marx suggested spark gap switches instead contact-mechanical switches in 1924. IGBT are used as switches in PITZ TDS modulator.

The modulator is shown in figure 1 (lower part is the modulator in oil tank, top part is the klystron surrounded with local radiation shielding). The voltage and current pulse shapes are shown in figure 2.

[3] To whom any correspondence should be addressed.

Journal of Physics: Conference Series **747** (2016) 012082 doi:10.1088/1742-6596/747/1/012082

**Figure 1.** Modulator with klystron and local shielding.

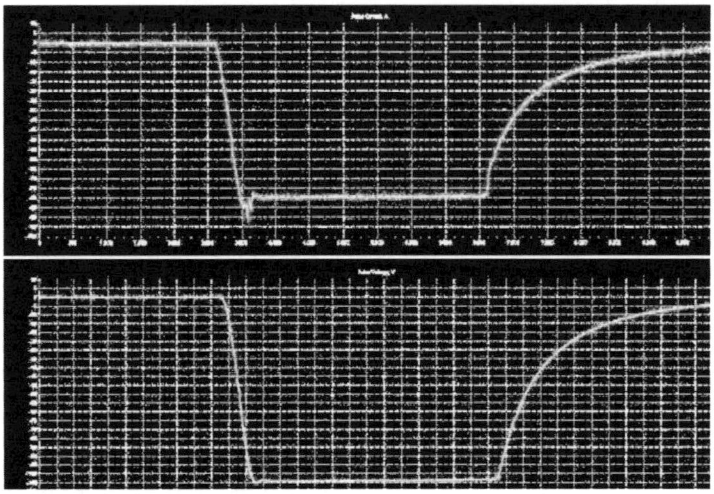

**Figure 2.** Pulse shape of the current (top) and the voltage (bottom).

The modulator is equipped with fast interlock, interrupting high-voltage pulse in case of break-down. The voltage and current pulse shapes for fast interlock event are shown in figure 3. The fast interlock leads to the voltage interruption and limitation of current.

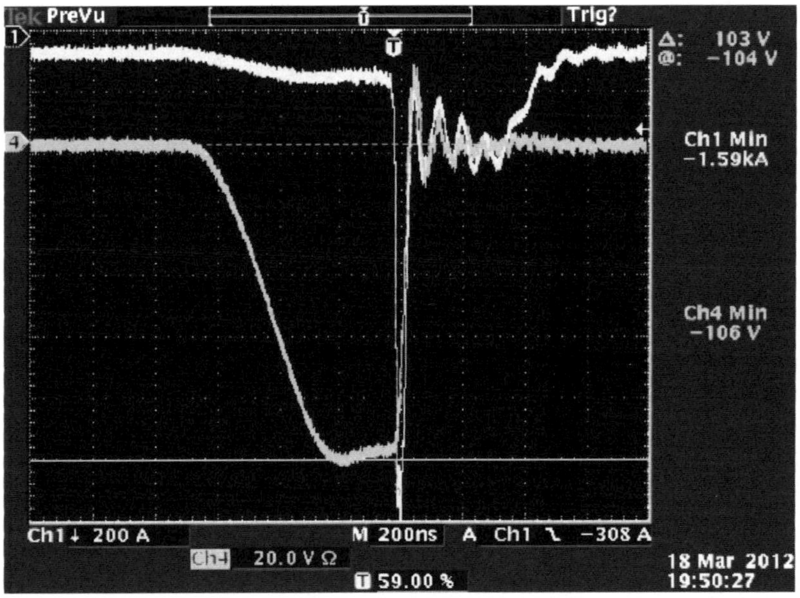


**Figure 3.** Measured pulse shape of the current (yellow) and the voltage (green) in case of fast interlock event.

Main modulator parameters are 110 kV voltage, 80 A current, 8 V heater voltage, 35 A heater current, pulse length up to 6 μs, 3.1 μs nominal flattop, 0.2 μs rise time, 2.0 μs fall time, voltage fluctuation (pulse-to-pulse, rms) <0.16%, 0.4% long-term stability (peak-to-peak), voltage unflatness (peak-to-peak) <1%, beam current limit at break-down 1500 A & 35 nsec and 350 A & 300 nsec, beam voltage fall time at break-down 35 ns, total charge for breakdown 160 μC.

## 3. Modulator for XFEL TDS Systems

Two modulators developed for two type klystron of XFEL TDS Systems are based on on the same principle as PITZ TDS modulator, but built with additional pulse transformer. The modulator for

Journal of Physics: Conference Series **747** (2016) 012082 doi:10.1088/1742-6596/747/1/012082

XFEL TDS INJ System has been manufactured and tested. The modulator for XFEL TDS BC1 System is under construction.

The output voltage in the modulator XFEL TDS INJ can be set in the range from 0 to 110 kV. Max output current is 166 A. The pulse length can be adjusted from 0 to 6 μs.

Figure 4 shows the pulse shapes measured with dummy resistive load (instead the klystron). The presented oscillograms show voltage pulse shapes at a) nominal pulse length 6 μs and nominal flattop 3.1 μs, b) reduced pulse length to 3 μs, c) 300 times zoomed pulses in one oscillogram to realise voltage stability, d) zoomed pulse top to see unflattness 0.8 V for 3.1 μs flattop. Achieved value for rise time is of 730 ns and for fall time is of 600 ns.

The pulse transformer is shown in figure 5. It is submerged for operation in the oil tank.

The measured current in function of the modulator voltage at operation for resistive dummy load is shown in figure 6.

Main parameters of the klystron power supplies are summarized in Table 1. The klystron power supply for the XFEL TDS System INJ is shown in figure 7.

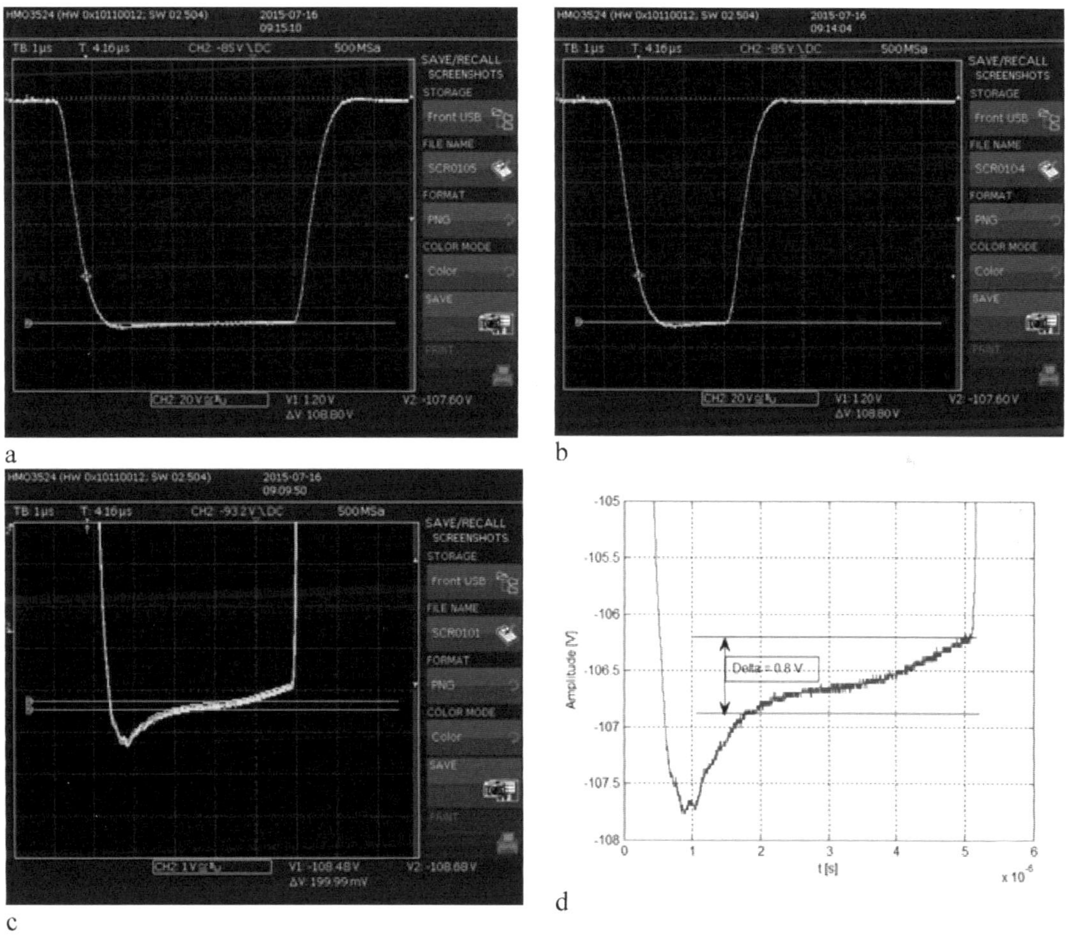


**Figure 4**. Measured pulse shapes.

**Table 1.** Main parameters of the klystron power supplies

| | Unit | XFEL TDS INJ | XFEL TDS BC1 |
|---|---|---|---|
| Number of modules in the modulator | | 6 | 2 |
| Max voltage of the module | kV | 8 | 22 |
| Max voltage of the modulator | kV | 48 | 44 |
| Max voltage in the modulator with respect to the ground | kV | 24 | 22 |
| Max current of the modulator | A | 166 | 1422 |
| Ratio of pulse transformer | | 2.3 | 5.68 |
| Max voltage of the klystron | kV | 110 | 250 |
| Nominal voltage of the klystron | kV | 101 | 230/232 |
| Max current of the klystron | A | 72 | 250 |
| Nominal current of the klystron | A | 66.4 | 214/219 |

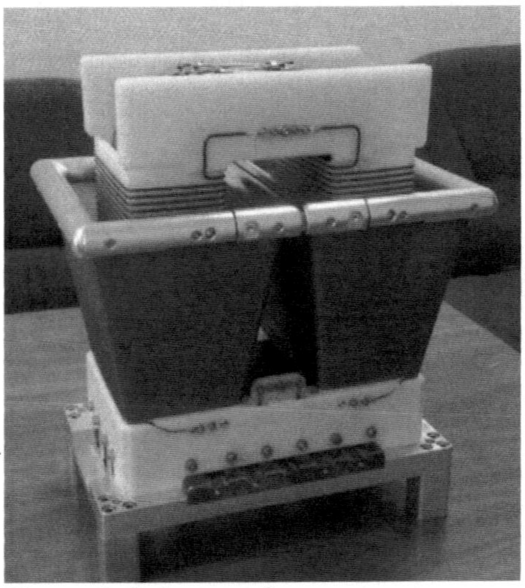

**Figure 5**. Pulse transformer.

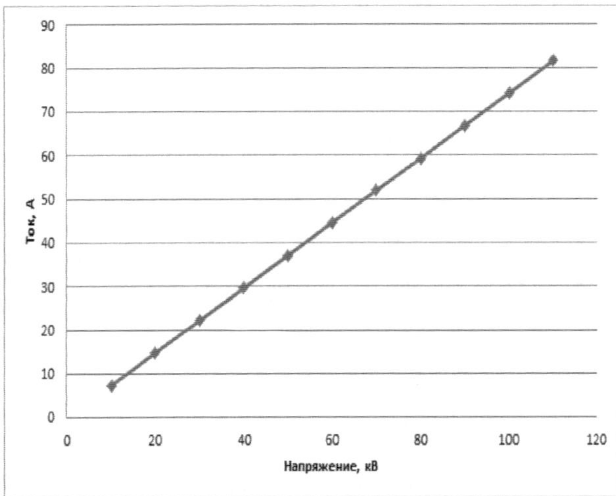


**Figure 6**. Current (A) v.s. voltage (kV) in the modulator with resistive load.

**Figure 7.** Klystron power supply XFEL TDS INJ: left – pulse transformer in oil tank with klystron and local shielding on the top; middle – modulator; right – control unit in power supply cabinet.

The transverse deflecting system XFEL TDS System INJ including the klystron power supply has been installed in the XFEL building and tested. All requirements are met.

## References

[1] Kravchuk L *et al.* 2010 Layout of the PITZ Transverse Deflecting System for Longitudinal Phase Space and Slice Emittance Measurements *Proceedings of Linear Accelerator Conference LINAC2010 (Tsukuba, Japan)* 416-418

## Acknowledgements

The present work has been done as an in-kind contribution of Russian Federation into European XFEL project.

# Transverse deflecting structure XFEL TDS INJ

**E N Volobuev[1,2], A A Zavadtsev[1,2,4], D A Zavadtsev[1,2], L V Kravchuk[1], V V Paramonov[1], M V Lalayan[3], A J Smirnov[3], N P Sobenin[3] and D V Churanov[1,2]**

[1] Institute for Nuclear Research of the Russian Academy of Sciences, Prospekt 60-letiya Oktyabrya 7a, Moscow 117312, Russia

[2] Nano Invest LLC, Office 017, 9 Academician V.N.Chelomey Str., Reutov, Moscow Region 143969, Russia

[3] National Research Nuclear University MEPhI (Moscow Engineering Physics Institute), Kashirskoye shosse 31, Moscow 115409, Russia

E-mail: azavadtsev@yandex.ru

**Abstract**. The Transverse Deflecting System XFEL TDS System INJ has been built as the part of the European X-ray Free Electron Laser for longitudinal bunch profile and slice emittance measurements. The 0.7 long Transverse Deflecting Structure (TDS), as the part of the system, has been built on base of the disk-loaded structure with hybrid EH-wave, operating frequency 2997.2 MHz and pulse input RF power up to 3 MW. The structure has been built, tuned and installed in the XFEL Injector beam line. All design parameters have been got at low RF power level. Electron beam has been passed through the whole Injector including TDS.

## 1. Structure development

For the measurement of the longitudinal bunch profile, the slice energy spread and the slice emittance three TDS Systems are foreseen in the European XFEL, located in Injector, after Bunch Compressor 1 (BC1), and after Bunch Compressor 2 (BC2), with the beam energies 130 MeV, 600 MeV and 2.5 GeV in TDS Systems correspondently.

### *1.1. TDS type and parameter definition*

TDS type and parameters are defined after detailed consideration of different conflicting demands. To allow single bunch operation, during short RF pulse (up to 3.1μs) the TDS should be filled with RF power quickly. It should provide required deflecting voltage during flat top operation and should be quickly released from RF power between RF pulses. TDS impact on regular beam due to both long range and short range wakefields should be minimized. The TDS length should be small to provide the minimal lattice perturbations. The S-band Travelling Wave (TW) constant gradient TDS, based on the disk loaded structure with 2π/3 mode of hybrid EH11 wave is the most well balanced, reasonable and realistic choice for XFEL TDS.

Considering both RF parameters and particularities of technology in structure construction for the following TDS:

- disk-loaded TW structure with two stabilizing holes [1],
- disk-loaded TW structure with two peripheral recesses [2],
- two options of disk-loaded TW structures with oval aperture [2],

[4] To whom any correspondence should be addressed.

Journal of Physics: Conference Series **747** (2016) 012083 doi:10.1088/1742-6596/747/1/012083

- disk-loaded TW structure with resonant slot,
- standing wave deflecting structure [3],

and basing on:

- practically similar RF parameters for considered TW structures,
- fulfillment of TDS requirements with well-balanced parameters combination,
- level of development both for the structure itself and for all steps in the TDS construction and tuning,
- proven experience, including high RF power operation,

the TW structure with two stabilizing holes, shown in figure 1, with backward hybrid EH11-wave has been selected for realization in XFEL TDS. This structure is well known LOLA TDS, [1], with the RF parameters optimized to meet XFEL requirements. To reduce cost of construction, parameters of cells are selected to fulfill simultaneously requirements for all three XFEL TDS and for chosen option disk inner radius is Ra =21.71 mm, relative group velocity is $\beta_g$ =-0.01587, attenuation constant is $\alpha$ =0.168 1/m, and normalized field is $\frac{E_d\lambda}{\sqrt{P}}$ =242 $Ohm^{1/2}$.

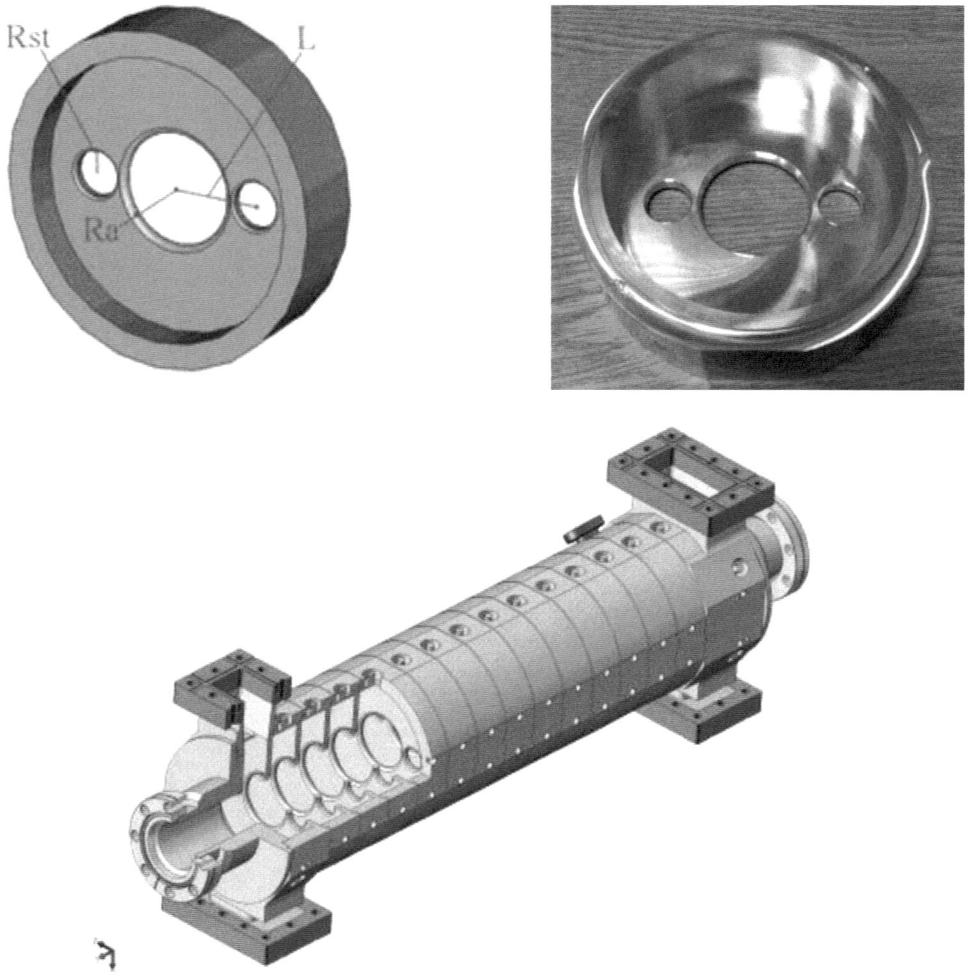


**Figure 1.** Deflecting structure. Artistic view of TDS cell, TDS cell for technical test, CAD model of TDS INJ structure.

Journal of Physics: Conference Series **747** (2016) 012083 doi:10.1088/1742-6596/747/1/012083

## 2. TDS production.

The TDS parts were produced with NC machine tool, figure 2. The regular cells were tested in the 6-cells resonant model and, if required, were tuned by removing some material from inner surface of the cell. Finally the narrow spread in cell frequencies was achieved, equivalent to ~ 3 μ deviations in the outer cell radius. The RF coupler cells have been tuned in the whole structure assembled with the press. The tuned structure has been brazed in vacuum furnace by using 72%Ag-28%Cu alloy.

**Figure 2.** The cells in the production: machine tool, produced cells and the total structure brazed in furnace.

## 3. Tuning after brazing

The final RF tuning of the structure has been performed after brazing on the low RF level stand. Results are shown in figure 3. The structure is shown at the test stand in figure 4.

Journal of Physics: Conference Series **747** (2016) 012083 doi:10.1088/1742-6596/747/1/012083

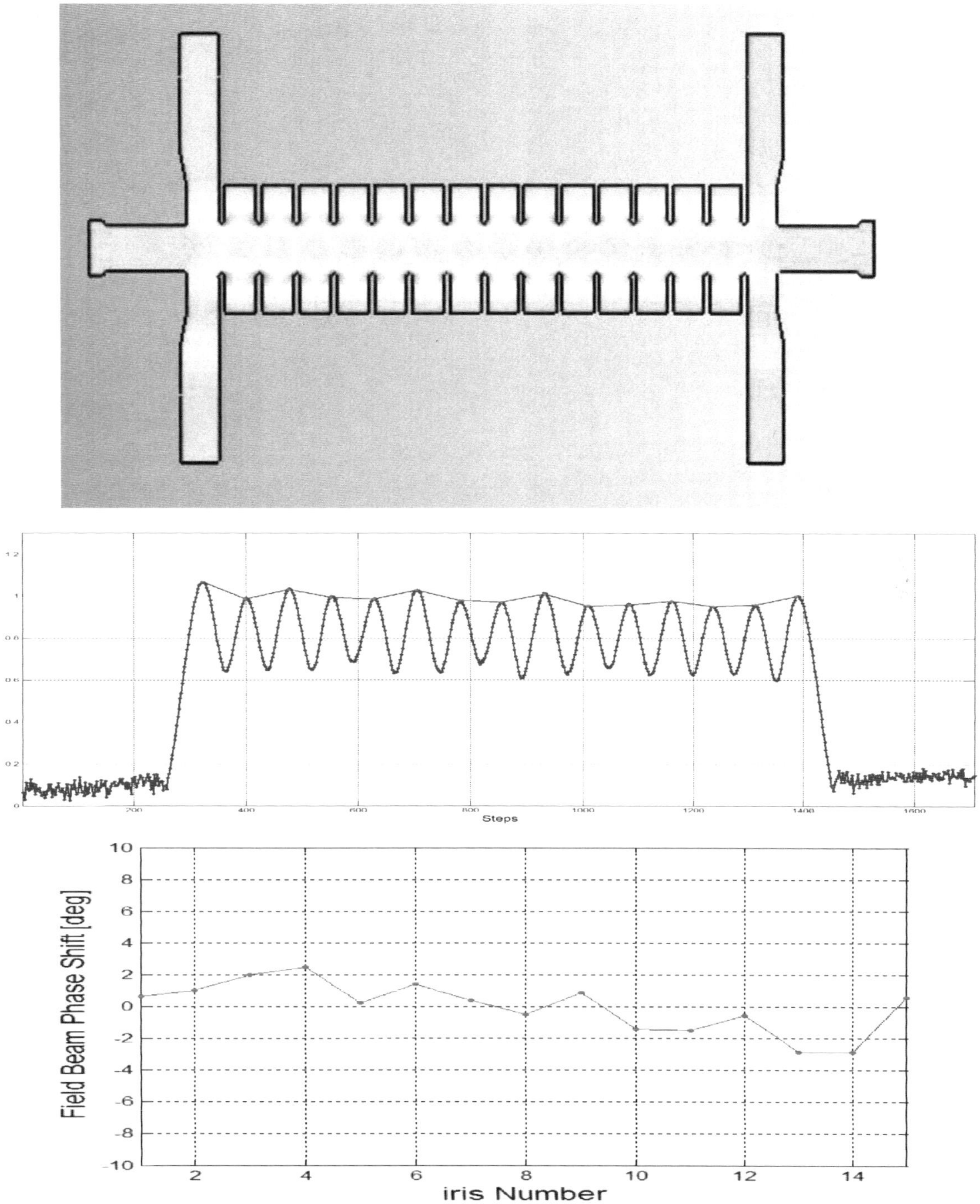


**Figure 3.** Low level RF test: complex amplitude, amplitude and phase of the electric deflecting field on the beam axis.

Journal of Physics: Conference Series **747** (2016) 012083 doi:10.1088/1742-6596/747/1/012083

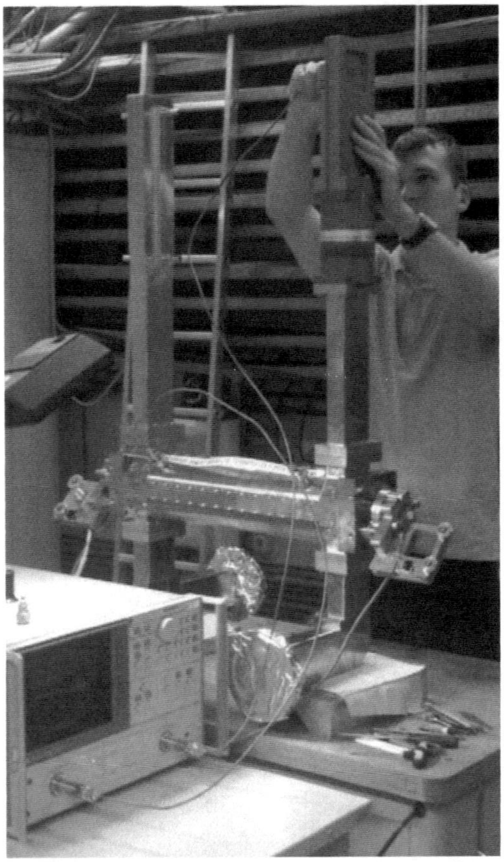

**Figure 4.** TDS structure at the test stand.

**Figure 5.** TDS structure installed in the XFEL injector tunnel.

Each TDS cell is equipped with two special hollows with the thread rod in center. This hollow has wall 2 mm thick, which can be bent either inside or outside the cell using special tools. This procedure allows us either increase or decrease cells frequency for the required structure tuning. Results, obtained in tuning of deflecting field distribution, are shown in figure 3. The measured reflection from the TDS in the frequency band is shown in figure 6.

In the TDS INJ structure RF probes for phase measurements are located in the cells #3 and #14. The measured transmissions to the antennas from input RF coupler are -66.0 dB and -66.8 dB respectively. Measured reflection from the structure with the waveguide window is S11=-41 dB at operating frequency.

## 4. Assembling of the deflector

The TDS INJ deflector includes all components connected to the ultra-high vacuum volume of the XFEL accelerator: TDS itself, waveguide RF window, two RF probes, RF load, part of straight waveguide, two ion pumps with adapters. The deflector was assembled on the steel girder. The structure is installed on the girder via two precise support units and there are special supports for the ion pumps.

After assembling in a clean room the whole deflector was tested for vacuum leaks. For transport to the injector tunnel, two aperture holes were closed with CFF blank-flanges. At the XFEL beam line, figure 5, the girder with deflector installed on two concrete pillars via special alignment units. Working with these alignment units allows us to align the structure on the beam axis.

Prototype of this structure has been built before [4] and it is tested at DESY PITZ facility now.

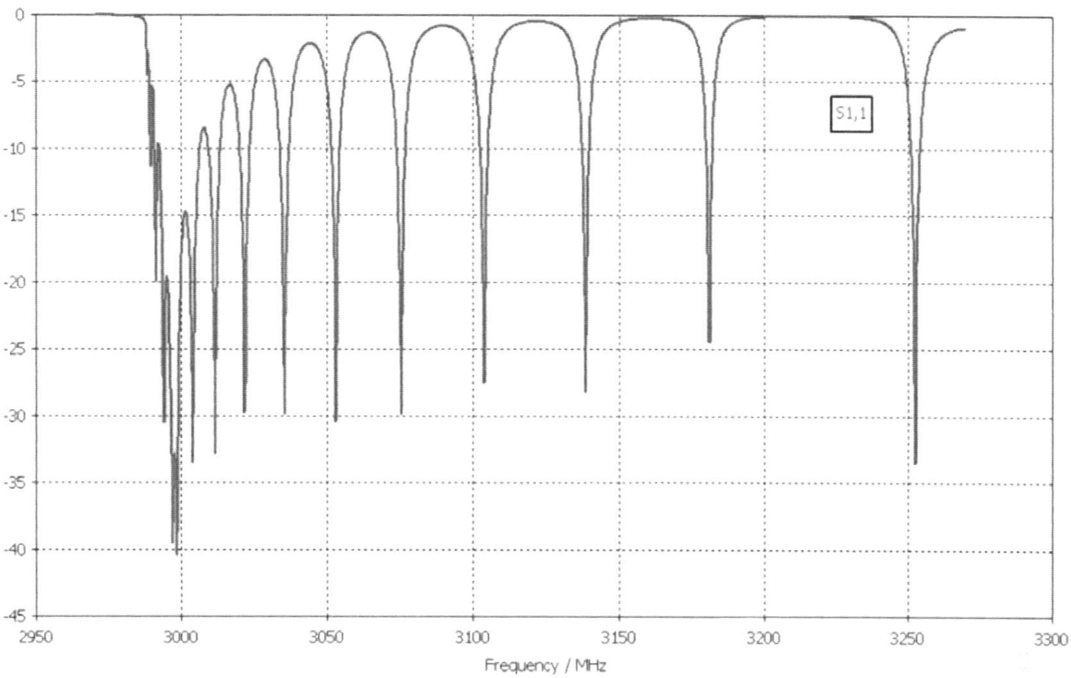


**Figure 6.** TDS structure at the test stand.

## References

[1] Altenmueller O H, Larsen R R and Loew G A 1964 Investigations of Travelling-Wave Separators for the Stanford Two-Mile Linear Accelerator *Review of Scientific Instruments* **35** 438-442

[2] Anisimov A, Isaev I, Kutsaev S, Lalayan M, Smirnov A, Sobenin N, Zavadtsev A and Zavadtsev D 2010 Stabilization of the Polarization Plane in Traveling Wave Deflectors *Proceedings of IPAC2010 (Kyoto, Japan)* 3759-3761

[3] Paramonov V, Kravchuk L and Korepanov S 2006 Effective Standing-Wave RF Structure for Charged-Particle Beam Deflection *Proc. 2006 Linac Conference* 649-651.

[4] Otevrel M *et al.* 2012 Diagnostics at PITZ 2.0 Beamline: Status and New Developments *Proceedings of IPAC2012 (New Orleans, Louisiana, USA)* 634-636

## Acknowledgements

The present work has been done as an in-kind contribution of Russian Federation into European XFEL project.

# Procedure of measuring the longitudinal emittance of electron beam

**I Yu Vladimirov**[1]

Faculty of Physics, M.V. Lomonosov Moscow State University, Moscow 119991, Russia

E-mail: timerke@mail.ru

**Abstract.** The procedure of measuring the longitudinal emittance of electron beam generated by RF gun and reconstruction of its longitudinal phase portrait is proposed. Measuring system consists of vertical deflecting RF cavity, horizontal bending dipole and screen. The beam spot on the screen is used to reconstruct the longitudinal phase portrait. In the proposed procedure an electromagnetic field of the vertical deflecting RF cavity can be approximated by the $TM_{110}$ mode of pillbox cavity. This approximation allows analytically solve the motion equations of the electron motion in the vertical deflecting RF cavity. The report contains description of the vertical deflecting RF cavity and the dipole, the formulae underlying the procedure and the results of numerical simulation.

## 1. Introduction

The spectrometer for measuring the energy of the electron beams was designed in MSU Laboratory of electron accelerators (figure 1 (a)) for GunLab [1] and bERLinPro [2] projects. The aim of the bERLinPro project is to expand the required accelerator physics and technology knowledge which is mandatory for generation of a high current (100 mA), high brilliance (norm. emittance below 1 mm mrad) electron beam. Its predecessor is GunLab project. GunLab is an research and development platform for superconducting RF photoinjectors of bERLinPro.

a b
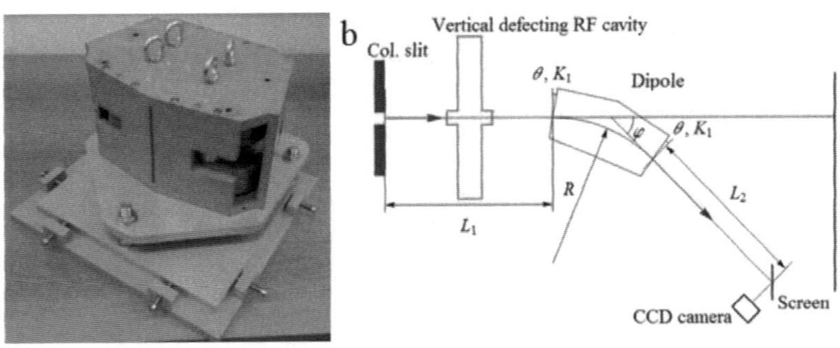


[1] To whom any correspondence should be addressed.

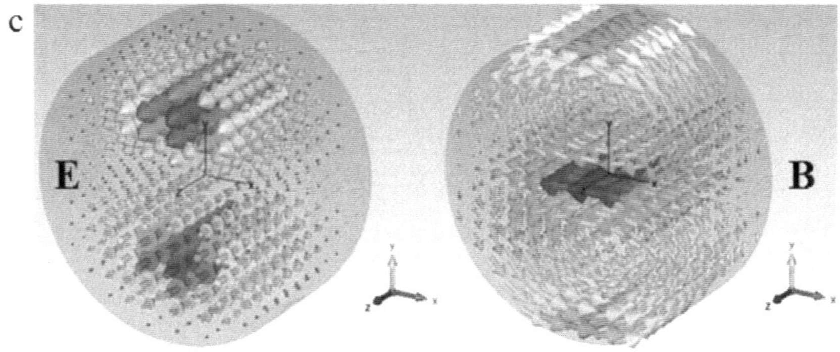


**Figure 1.** (a) The spectrometer dipole designed in MSU LEA; (b) the measuring system; (c) electromagnetic field of the $TM_{110}$ mode.

The spectrometer is a dipole magnet. The main parameters of the spectrometer are listed in the Table 1. When the vertical deflecting RF cavity is placed in front of the dipole, you get a system (figure 1 (b)) that allows to reconstruct the longitudinal phase portrait and to measure the longitudinal emittance. The vertical coordinate of the electron on the screen is connected with the time of flight to the cavity and the horizontal coordinate of the electron is related to the momentum. For GunLab and bERLinPro projects the vertical deflecting RF cavity is developed in DELTA and its status can be found at [3].

**Table 1. The main spectrometer parameters.**

| Parameter | Value | Parameter | Value | Parameter | Value | Parameter | Value |
|---|---|---|---|---|---|---|---|
| Gap | 35 mm | $L_2$ | 1 m | $\varphi$ | 45° | $K_1$ | 0.246 |
| $L_1$ | 0.8 m | $R$ | 0.25 m | $\theta$ | 7.593° | | |

## 2. Theory

We will work with a pillbox cavity instead of the real vertical deflecting RF cavity. Electromagnetic field of the $TM_{110}$ mode of the pillbox cavity can be written as [4]:

$$\mathbf{E} = \mathbf{e}_z \omega B_0 y \exp(-i(\omega t + \psi)), \mathbf{B} = i\mathbf{e}_x B_0 \exp(-i(\omega t + \psi)). \tag{1}$$

Here $\omega$ – cyclic frequency, $B_0$ – magnetic field amplitude. We suppose that the pillbox cavity length is a half of wavelength. The beam moves along the axis $z$ of the pillbox cavity. The electromagnetic field of the $TM_{110}$ mode of the pillbox cavity and coordinate system are shown on figure 1 (c).
To find the electron coordinates at the pillbox cavity output we should solve the motion equations of the electron in the electromagnetic field (1). Solutions have the form:

$$y_{\text{cav}} = y_0 - \frac{eB_0\beta\lambda^2}{2\pi^2 m_0\gamma c}\left(\cos\psi - \frac{\pi}{2}\sin\psi\right),\ y'_{\text{cav}} \equiv \frac{p_{y\text{cav}}}{p_0} = \frac{p_{y0}}{p_0} + \frac{eB_0\beta\lambda}{\pi p_0}\sin\psi, \tag{2}$$

$$p_{z\text{cav}} = p_{z0} + \frac{e^2B_0^2\beta\lambda^2}{2\pi^2 m_0\gamma c}\cos\psi\left(\cos\psi - \frac{\pi}{2}\sin\psi\right) - \left(2eB_0 y_0 + \frac{eB_0 p_{y0}\lambda}{2m_0\gamma c}\right)\cos\psi. \tag{3}$$

Here $\lambda$ – the wavelength, $\beta$ – the relative speed, $\gamma$ – the relative energy, $m_0$ – the electron rest mass, $\psi$ – the field phase when the electron is flying in the cavity, $p_0$ – the average beam momentum, $y_0$, $p_{y0}$, $p_{z0}$ – the initial electron coordinate and momentum. Assume that the momentum spread of electrons in the beam is considerably less than the average beam momentum. Suppose also that at the entrance to the measuring system the beam has zero transversal size and zero divergence. After passing the pillbox cavity beam flies rest of the drift section $L_1$, the dipole and the drift section $L_2$ (figure 1 (b)). The electron coordinates on the screen are calculated by means of transformation matrix $M$ of this system:

$$x_s = M_{16}\left(p_{z\mathrm{cav}} - p_0\right)/p_0,\; y_s = M_{33} y_{\mathrm{cav}} + M_{34} y'_{\mathrm{cav}}. \tag{4}$$

We can find approximate expression for the flight phase of the electron in the cavity by substitution (2) to (4):

$$\psi = \arcsin\left(\frac{y_s}{\sqrt{A^2+B^2}}\right) - \arcsin\left(\frac{A}{\sqrt{A^2+B^2}}\right),$$

$$A = -\frac{eB_0\beta\lambda^2}{2\pi^2 m_0\gamma c}M_{33},\; B = \frac{eB_0\beta\lambda^2}{4\pi m_0\gamma c}M_{33} + \frac{eB_0\beta\lambda}{\pi p_0}M_{34}.$$

And we can find the electron momentum by substitution (3) to (4):

$$p_{z0} = \left(1 + \frac{x_s}{M_{16}}\right)p_0 - \frac{e^2B_0^2\beta\lambda^2}{2\pi^2 m_0\gamma c}\cos\psi\left(\cos\psi - \frac{\pi}{2}\sin\psi\right).$$

Also we can estimate the average beam momentum and the momentum rms spread after the pillbox cavity using (2, 3):

$$p_z = p_{z0} + \frac{e^2B_0^2\beta\lambda^2}{4\pi^2 m_0\gamma c}\left(1 + s\left(\cos(2\psi_0) - \frac{\pi}{2}\sin(2\psi_0)\right)\right),\; s = \frac{\lambda}{2\pi\Delta}\sin\left(\frac{2\pi\Delta}{\lambda}\right), \tag{5}$$

$$\sigma_{pz} = \sqrt{\sigma_{pz0}^2 + \frac{1}{2}\left(\left(eB_0\sigma_{y0}\right)^2 + \left(\frac{eB_0\sigma_{py0}\lambda}{2m_0\gamma c}\right)^2\right)\left(1 + s\cos(2\psi_0)\right) + C},\; S = \frac{\lambda}{4\pi\Delta}\sin\left(\frac{4\pi\Delta}{\lambda}\right), \tag{6}$$

$$C = \frac{1}{2}\left(\frac{e^2B_0^2\beta\lambda^2}{4\pi^2 m_0\gamma c}\right)^2\left(\left(\frac{\pi^2}{4}+1\right)\left(1 - s^2\right) + \left(s^2 - S\right)\left(\pi\sin(4\psi_0) + \left(\frac{\pi^2}{4} - 1\right)\cos(2\psi_0)\right)\right). \tag{7}$$

Here $\Delta$ is the beam length, $\psi_0$ – the field phase when the beam center is flying in the cavity. We take into account that the beam can has nonzero transversal rms size $\sigma_{y0}$ and nonzero momentum rms spread $\sigma_{py0}$ and $\sigma_{pz0}$. But we suppose that the beam has uniform distribution of electrons.

## 3. Comparison of theory and numerical simulation

We carried out numerical simulations of electron beam dynamic in the measuring system. The simulations were performed using ASTRA code [5]. The electron beam had the following characteristics: $\sigma_{y0}$ = 0.38 mm, $\Delta$ = 11.5 mm, $p_0$ = 3.06 MeV/c, $\sigma_{py0}$ = 1.88 keV/c, $\sigma_{pz0}$ = 3.62 keV/c, $\psi_0$ = 0°. The frequency and amplitude of the electromagnetic field of the vertical deflecting RF cavity were equal to 1.3 GHz and 16.7 mT respectively.

Figure 2 (a) shows the beam spots on the screen of the measuring system. Figure 2 (b) shows the reconstructed longitudinal portraits. Portrait 1 is the true longitudinal phase portrait of the beam (rms longitudinal emittance $\varepsilon_z$ = 18.4 ps MeV/c). When the beam passes the measuring system without the horizontal collimation slit, we see on the screen spot 2 and reconstruct portrait 2 (rms longitudinal emittance $\varepsilon_z$ = 33.3 ps MeV/c). If the horizontal collimation slit (width 0.1 mm) is placed before the pillbox cavity, the beam spot on the screen will change to spot 3 and we reconstruct portrait 3 (rms longitudinal emittance $\varepsilon_z$ = 17.3 ps MeV/c). Also we simulated the dynamics of the beam with zero transversal size and zero divergence in the measuring system. Spot of this beam on the screen is spot 4 and its reconstructed longitudinal phase portrait is portrait 4 (rms longitudinal emittance $\varepsilon_z$ = 20.2 ps MeV/c).

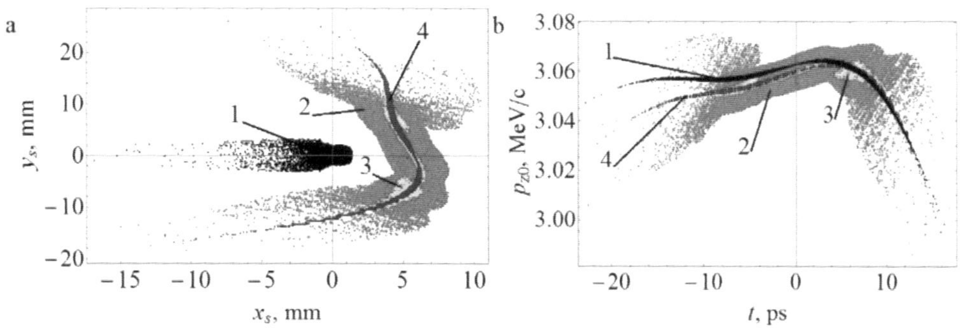


**Figure 2.** (a) The beam spots on the screen of measuring system; (b) the longitudinal phase portraits.

From figure 2 (b) we can see that the horizontal collimation slit at the entrance to the measuring system greatly improves the results of the longitudinal portrait reconstruction. This is because the reconstruction procedure assumes that the input electron beam has zero transversal size and zero divergence.

From figure 2 (a) it can be noted that the vertical deflecting RF cavity affects the average beam momentum. In figure 2 (a) spot 1 is the beam spot when the vertical deflecting RF cavity is turned off. It is evident that all the other beams, which were affected by the electromagnetic field of the mode $TM_{110}$ of the vertical deflecting RF cavity, are shifted along the horizontal axis $x_s$. Simulations show that the average momentum and the rms momentum spread of the beam which was deflected by the cavity are equal to $p_z$ = 3.082 MeV/c and $\sigma_{pz}$ = 7.14 keV/c. Theoretical estimations (5-7) give the values $p_z$ = 3.082 MeV/c and $\sigma_{pz}$ = 7.21 keV/c.

## 4. Conclusion

The report describes the reconstruction procedure of the longitudinal phase portrait of the electron beam generated by RF gun. The results of applying this procedure are in a good agreement with the true longitudinal portrait and the longitudinal emittance. Furthermore, it was found that the vertical deflecting RF cavity affects the average momentum and the momentum rms spread of the beam. There are given good estimations of their behavior in this report.

## References

[1] Völker J, *et al* 2014 *Proc. IPAC2014* 630-632
[2] Abo-Bakr M, *et al* 2014 *Proc. IPAC2014* 340-342
[3] Ferrarotto A, Riemann B, Weis T, Kamps T, Glock H-W and Völker J 2015 *Proc. IPAC2015* 827-829
[4] Burt G 2012 Transverse deflecting cavities *arxiv.org*
[5] Floettmann K 2014 A space charge tracking algorithm *DESY* http://www.desy.de/~mpyflo/Astra_manual/

# Accelerator-based electron beam technologies for modification of bipolar semiconductor devices

**Y S Pavlov[1], A M Surma[2], P B Lagov[1,3,6], Y L Fomenko[4] and E M Geifman[5]**

[1] Laboratory of Radiation Technologies, A.N. Frumkin Institute of Physical Chemistry and Electrochemistry Russian Academy of Sciences (IPCE RAS), Moscow, 119071, Russian Federation
[2] R&D Centre, JSC "Proton-Electrotex", Oryol, 302040, Russian Federation
[3] Department of Semiconductor Electronics and Semiconductor Physics, National University of Science and Technology "MISiS" (NUST "MISiS"), Moscow, 119071, Russian Federation
[4] SC "VSP-Mikron", Voronezh, 394007, Russian Federation
[5] JSC "Electrovipryamitel", Saransk, 430001, Mordovia, Russian Federation

E-mail: lagov2000@mail.ru

**Abstract.** Radiation processing technologies for static and dynamic parameters modification of silicon bipolar semiconductor devices implemented. Devices of different classes with wide range of operating currents (from a few mA to tens kA) and voltages (from a few volts to 8 kV) were processed in large scale including power diodes and thyristors, high-frequency bipolar and IGBT transistors, fast recovery diodes, pulsed switching diodes, precise temperature-compensated Zener diodes (in general more than fifty 50 device types), produced by different enterprises. The necessary changes in electrical parameters and characteristics of devices caused by formation in the device structures of electrically active and stable in the operating temperature range sub-nanoscale recombination centres. Technologies implemented in the air with high efficiency and controllability, and are an alternative to diffusion doping of Au or Pt, $\gamma$-ray, proton and low-Z ion irradiation.

## 1. Introduction

The development of physics theory and experimental equipment and methods in radiation material science during the second half of the 20th century, as well as actively seeking of new technology tools for solid-state electronics and semiconductor devices have led to the formation and application of radiation technologies in semiconductor manufacturing process due to their high efficiency. Design and creation of high-performance, high-reliable, easy-to-use and maintain electron beam accelerators with energies up to 10 MeV, fundamental improvement of individual functional units [1-6], established leading position of radiation technologies based on electron accelerators in today's semiconductor industry in comparison with other methods of radiation treatment. The effectiveness of accelerated electrons able to replace the atoms from crystal lattice sites was shown by different research teams in relation to devices based on silicon and other materials.

[6] To whom any correspondence should be addressed.

Recombination statistics based upon a single dominant level have been used to predict the relative characteristics of gold-diffused, platinum-diffused, and electron-irradiated silicon power devices [7]. The measured characteristics of power rectifiers are shown to be in good agreement with these calculations. The results are also shown to be applicable to power thyristors. In [8] the impact of the electron and proton beam density onto bipolar transistors parameters modification rate was revealed. Achievement of fast recovery characteristics for silicon power diodes and optimization of pulse power thyristors on industrial batches level was shown in [9-13]. In [14,15] shown that to fulfil the requirements of dynamic and static characteristics of bipolar static induction transistor, the irradiation/anneal processing is crucial. Defect distributions features and silicon power devices parameters modification under electron beam irradiation in a wide energy range studied in [16–18]. Extended capacitance relaxation research method for radiation defects parameters measurements in semiconductors device structures is given in [19].

## 2. Main features of electron beam technologies

### *2.1. Main features and advantages*

Main features and advantages of electron beam technology for the modification of semiconductor devices as follows:

- possibility of processing in air,
- electron flux density adjustment by beam current and distance from the output window due to scattering in air;
- providing a large processing area due to operation in the scanning mode;
- very ample opportunities for the design and construction of under-beam equipment, including beam control systems and a variety of holders, automated sample delivery systems, free-standing vacuum/environment chamber with a beam input window for special investigations;
- simultaneous pass-through processing of several semiconductor wafers, arranged one behind the other due to deep penetrating ability of accelerated electrons (range in Si about 1–2 cm);
- samples heating by beam and temperature control if necessary;
- small footprint and favorable cost of equipment, easy to use and maintain.

Mentioned advantages make it possible to do a preliminary experiment quickly and realize large volume technology to provide needed combination of electrical parameters for specific device.

### *2.2. Brief critical analysis of alternative processes*

For comparison, we present the main disadvantages of alternative methods. So gold doping process results in a negative growth of reverse currents via p-n junctions and characterizes by low concentration reproducibility from wafer to wafer and within the same wafer. The platinum doping process allows to minimize leakage currents, but does not achieve the maximum rate of recombination values and also characterizes by low reproducibility.

Sources of $\alpha$-particles provide defects formation at shallow depths (no more than 30 microns), have a small processing area and intensity that limits the range of technological tasks solved and is suitable for low volume processing of small-area wafers or some laboratory experiments.

Widely used for different tasks gamma-ray $Co^{60}$ source creates displacements in semiconductors transferring small amounts of energy; needs a long time and associates with some organizational difficulties. Most sub-nanoscale recombination centers formed have relatively low annealing temperatures and not optimum electrophysical properties.

Proton-based (several MeV-level) technology requires very complex, expensive and energy consuming equipment, which casts doubt on the cost-effectiveness of this tool today yet, due to the difficulty of high volume wafer processing as well. In addition, hydrogen doping can cause different time and current instabilities in some devices. Nowadays real application in device technology is in-air proton irradiation with an initial (in vacuum) energy of 25 MeV [20, 21] equipped with semiautomatic feeding system, which allows to process with wide (about 60 microns) struggle mechanically strong extended silicon-on-molybdenum power device structures.

## 3. Experimental results

This section contains recent results obtained by the authors for the devices of different classes that demonstrate possibilities of accelerator-based electron beam technologies. Fig. 1 shows experimental dependence of forward voltage drop $U_F$ as a function of electron fluence $\Phi_e$ (after stabilizing annealing) for silicon low voltage (6,4 V) diode structure. Volt-ampere characteristics evolution after electron irradiation is shown in Fig. 2. This structure uses as a compensated one for precise thermally compensated Zener diode. Needed value of $U_F$ = 0,54 V could be provided in preliminary Au-doped structures at fluences about $3{\cdot}10^{17}$ cm$^{-2}$.

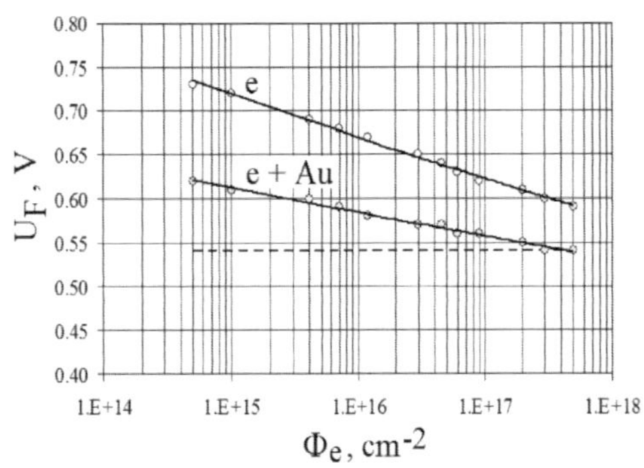


**Figure 1.** Forward voltage drop in compensated structure of precise Zener diode as function of electron fluence

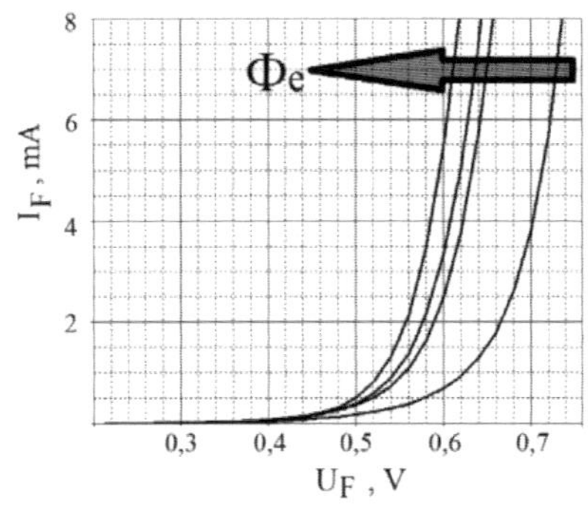


**Figure 2.** Volt-ampere characteristic evolution after electron irradiation

Regimes of electron treatment and annealing (250 kGy, 380° C) for power high frequency silicon bipolar npn- transistor (5A, 200V) with comb-like structure (Fig. 3) was developed to provide low switching time $t_s$ of 200 ns at needed gain values about 50.

Regimes of electron treatment and annealing (2000 kGy, 350° C) for pulse diode (3A, 200V) with 20 µm epitaxial p+-nn+ structure (Fig. 4) was developed to provide low reverse recovery time $t_{rr}$ of 5 ns at needed forward drop values about 1,2 V.

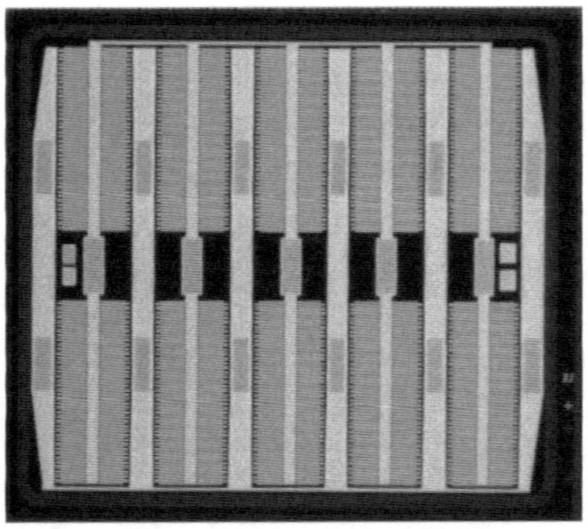

**Figure 3.** Comb-like structure of power high frequency silicon bipolar npn- transistor (5A, 200V)

**Figure 4.** Appearance of pulse diode structure (3A, 200V) with guard rings

High power electrical energy converting systems require the use of high-voltage and high-current thyristors, adapted for reliable operation in series and/or parallel connection. Thyristors for parallel and serial connection need the highest identity on-state current-voltage and reverse recovery characteristics respectively. For this purpose technology of precise parameters control of HV HC silicon thyristors by splitting semiconductor elements into specific groups with different initial parameters and control the reverse recovery time of diode-satellites during irradiation. Test diode chips and thyristor elements were made from high resistance NTD silicon with close characteristics to provide high stability of lifetime reduction coefficient values. Test diode structures were 4x4 $mm^2$ in size and were cut from Al-metallized wafer after anode/cathode diffusion that used for power diode manufacturing. During irradiation test diode structure fixed at the center of the target using contact mandrel that connected with accelerator control unit via a coaxial cable. Lifetime control scheme and control equipment for precise modification of thyristors are shown in Fig. 5 and 6. Typical results of precise modification of thyristor parameters by 8 MeV electron irradiation using the considered above lifetime control hardware are presented. Fig. 7 shows the modification of reverse recovery charge spread for silicon high-voltage thyristors (type T273-1250-44, 1250A, 4400V) after electron-beam treatment. Fig. 8 shows the statistical distribution of on-state voltage drop for the same thyristors.

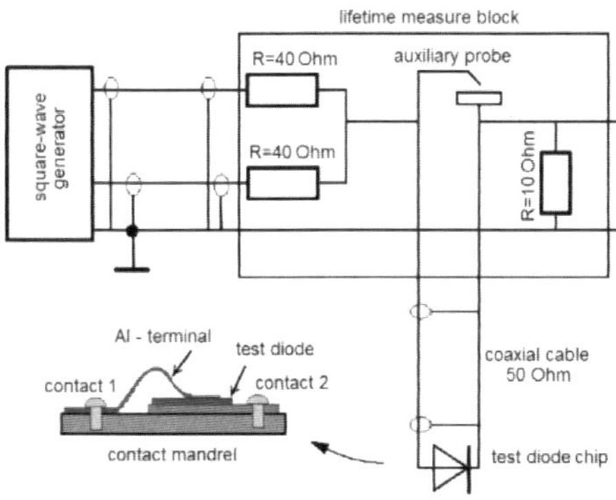


**Figure 5.** Minor charge carrier lifetime control scheme in test diode structure.

**Figure 6.** Control equipment for precise modification of high voltage thyristors.

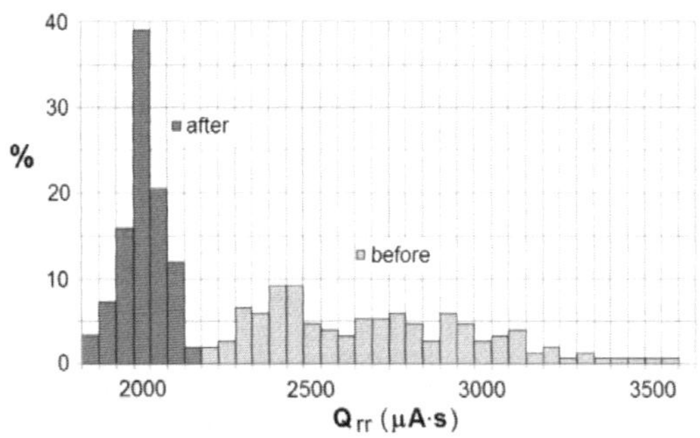


**Figure 7.** Typical distribution of reverse recovery charge for silicon HVT before and after precise modification by electron beam process

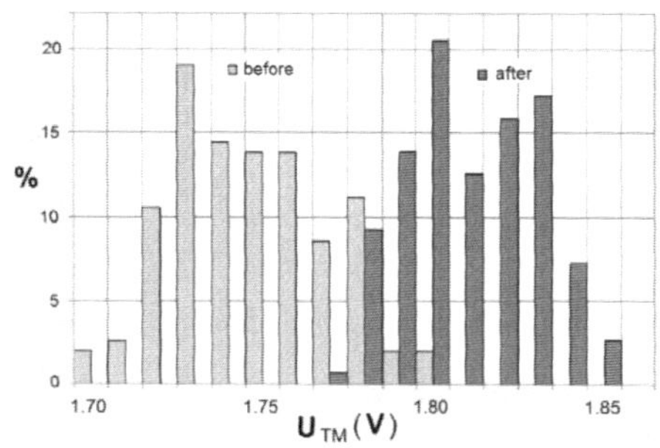


**Figure 8.** Typical distribution of on-state voltage drop for .silicon HVT before and after precise modification

Represented above and some other obtained experimental results useful for real industrial applications summarized in the Table 1.

**Table 1.** Accelerator-based electron beam technologies realized recently for different bipolar devices

| Device type | Task realized | Irradiation fluence ($cm^{-2}$) or dose (kGy) | Limiting parameter |
|---|---|---|---|
| Compensated $p^+/n^+$-structure of presice Zener diode (6.5V@7,5 mA) | $U_F$ reduction from 0.64 to 0.54 V for Au-doped structures | $3\cdot10^{17}$ $cm^{-2}$. (320°C, 2h anneal) | Irradiation time |
| Pulsed $p^+nn^+$-diode (20 μm n-epi base) (1A@200V) | $t_{rr}$ reduction to 5 ns | 2000 kGy (350°C, 1h anneal) | $U_F$ increasing |
| High frequency bipolar transistor with comb-like 'npn' structure (5A@200V) | $t_s$ reduction from 700 to 200 ns | 200–300 kGy (390°C, 1h anneal) | Gain reduction to from 60 to 50 |
| Darlington bipolar transistor | Gain, its spread and $t_s$ reduction to optimum values | 100–200 kGy (300–390°C, 1–2h anneal) | Gain reduction |
| Insulated gate transistor IGBT | Switching time reduction | 150–250 kGy (330°C, 1 h anneal) | Gain reduction |
| Fast recovery diode 50A@1700V | Reverse recovery time reduction up to 200 ns | 200 kGy (330°C, 1 h anneal) | Reverse current, softness |
| Fast recovery diode 1, 5, 15A@400, 600V | $t_{rr}$ reduction to 30–120 ns (for different types) | 400 kGy (330°C, 1 h anneal) | $U_F$ and reverse current increasing |
| Welding diode 7.1kA@400V, 600 V | $t_{rr}$ reduction to 3 us, frequency increasing up to 12 kHz | 250 kGy (250°C, 4 h anneal) | Middle softness; $U_F$ and reverse current increasing |
| HC HV Thyristor (different types) | Precise regulation parameters (spread less than 5%) | In situ lifetime control on test structures (<10 kGy) | on-state voltage drop increasing |
| mc-Silicon solar cell 15% efficiency | Radiation degradation investigation efficiency decreasing to 1 % | $1\cdot10^{12}$–$5\cdot10^{16}cm^{-2}$. | – |
| Superluminescent heterostructure diode | Culling structures with hidden defects detection | <40 (+ time to failure) | – |

Thus, the accelerator-based electron beam technologies could be effectively applied for manufacturing of competitive traditional and innovative solid state devices, elaboration new approaches in culling methods and radiation hardness investigation.

**Acknowledgements**
Some parts of this work were carried out with the financial support of the Ministry of Education and Science of Russian Federation (the unique identifier of the project RFMEFI58415X0016).

**References**

[1] Pavlov Yu S and Sharamentov S I 1977 Transistorized Frequency Divider of Microwave Pulses. *Instrum. Exp. Tech.* **20** 149-152

[2] Pavlov Yu S, Solovev N G and Tomnikov A P 1985 Syncronization Circuit for Shaping Picosecond Accelerated-Electron Pulses *Instrum. Exp. Tech.* **28** 1335-39
[3] Menshov A N, Pavlov Yu S, Shishko V I, Skvortsov A F, Lukyanov V N and Dobrokhotov V V 1989 Nanosecond Pulse Modulator with Frequency to 250 kHz in Packet Mode *Instrum. Exp. Tech.* **32** 1150-52
[4] Pikaev A K, Glazunov P Y and Pavlov Yu S 1993 Radiation Center of Institute-of-Physical Chemistry in Moscow *Radiat. Phys. Chem.* **42** 887-890
[5] Pavlov Yu S, Lagov P B 2015 Magnetic Buncher Accelerator for Radiation Hardness Research and Pulse Detector Characterization. *2015 15th European Conference on Radiation and Its Effects on Components and Systems (RADECS)* 336-338
[6] Pavlov Yu S, Dobrohotov B B, Pavlov B A, Nepomnyaschy O N and Danilichev V A 2014 Magnetic buncher accelerator UELV-10-10-T-1 for studing fluorescence and radiation physical researches *Proceedings of XXIV Russian Particle Accelerators Conference (RuPAC'2014)* Obninsk, Russia, 259-261
[7] Baliga J B and Sun E 1977 Comparison of Gold, Platinum, and Electron Irradiation for Controlling Lifetime in Power Rectifiers *IEEE Trans. Electron Devices* **24**(6) 685-688
[8] Bannikov Yu A, Gorin B M, Kozhevnikov V P, Mikhnovich V V and Gusev L I 1981 Influence of the Flux-Density on the Radiation-Damage of Bipolar Silicon Transistors by Protons and Electrons. *Soviet physics. Semiconductors* **15**(11) 1289-92
[9] Asina S S, Kuznetsov V M and Surma A M 1986 Using Radiation Technology to Produce Fast-Recovery Power Diodes. *Soviet Electrical Engineering* **57**(5) 74-78
[10] Asina S S, Surma A M, Goncharenko V P and Ivanov A A 1993 Local Electron Irradiation as a Method of Optimization of the Power Thyristor Pulse Characteristics. *Elektrotekhnika* **2** 51
[11] Ladygin E A, Lagov P B, Musalitin A M, Surma A M, Kokin S A and Uchuvatkin V V 2001 Application of fast electrons for pulse-frequency parameters improving of silicon power diodes and thyristors *Proc. APAM Fifth Topical Seminar, Materials and Processes for Power Electronics Devices* (Moscow) 44–45
[12] Pisarev A, Surma A and Chernikov A 2013 High voltage thyristor with reduced temperature dependence of reverse recovery characteristics adapted for usage in series assembly *PCIM Europe Conference Proceedings* 953–960
[13] Surma A and Kislyakov V 2013 Technology of precise adjusting of static and dynamic characteristics of high voltage thyristors *Bodo's Power Systems* **2** 18-21
[14] Usenko A Y 1993 Using Electron-Irradiation to Control the Turn-Off Time of Static Induction Transistors *Semiconductor Science and Technology* **8**(9) 1715-18
[15] Usenko A Y 1994 Irradiation-Then-Anneal Processing to Improve BSIT Performance. *IEEE Trans. Electron Devices* **41**(6) 1055-61
[16] Arinushkin V N, Berman L S, Geifman E M, Konyukhov A V, Lomasov V N, Mnatsakanov T T and Remenyuk A D 1993 Parameters of Force Silicon Diodes Irradiated with Low-Energy Electrons *Zhurnal Tekhnicheskoi Fiziki* **63**(8) 41-45
[17] Geifman E M, Serezhkin Yu N and Chibirkin V V 1997 Distribution of A centers in basic regions of semiconductor power equipment under electron irradiation *Elektrotekhnika* **11** 21-23
[18] Chibirkin V V and Geifman E M 2002 Radiative Regulation of the Lifetime of Nonequilibrium Charge Carriers in Semiconductor Power Equipment *Russian Electrical Engineering* **73**(5) 76-80
[19] Goryunov N N, Klebanov M P and Osipov R Yu 1997 Measurer of capacitance relaxation for investigating defects in semiconductors. *Pribory i Tekhnika Eksperimenta* **2** 163-164
[20] Chernikov A A, Gubarev V N, Stavtsev A V, Surma A M and Vetrov I Y 2011 One more way to increase the recovery softness of fast highvoltage diodes *Proc. of the 2011-14th European Conf. on Power Electronics and Application (EPE 2011), Birmingham, ENGLAND*
[21] Lazarev N V *et al.* 1996 30 years operation of 25 MeV proton linac I-2 in ITEP at beam current

of 200-230 mA *Proceedings of the XVIII International Linear Accelerator Conference* Ed. by Hill C, Vretenar M CERN Reports **96**(7) 542-544

# Multisectional RF compressor for the mobile electron accelerator

**G.O. Buyanov**[1], **A.P. Klachkov and A.G. Ponomarenko**
National Research Nuclear University MEPhI, 115409, Russia, Moscow, Kashirskoe highway 31

E-mail: frontlines.07@gmail.com

**Abstract.** Set out a design of the RF compression system, based on the Blumlein line principle. This particular design is based on TE01 mode in rectangular waveguide. Presented solution for TE01 excitation system with effective suppression of other modes. Presented designs 2-sectional and 6-sectional versions of compressor. Set out calculated Q factors, expected output powers and other parameters for these designs.

## 1. Introduction

RF compression systems is important part of modern high power pulsed systems. Particle accelerators is one of the areas where high RF pulsed power needed the most. We consider such systems will become important part of the future compact linear electron accelerators with high field gradient. NRNU MEPhI team worked over the RF compression system based on Blumlein line principle (BLP) [1,2,3,4].

Significant disadvantage of earlier designs [3,4] was limited frequency range. For the S-band their compression multiplier was about 100, but in X-band it goes down to 20 due to Q factor change. By the compression multiplier we mean relation of resonator $\tau$ to length of output pulse.

To increase Q factor and stored energy could be used larger waveguide with higher mode excitation. Our specific design is based on $TE_{01}$ mode. In table 1 presented comparison of designs based on $TE_{01}$ and $TE_{10}$. So for X-band (8.568 GHz to be accurate) $TE_{01}$ based design have 3 times larger stored energy and about 4 times higher output power.

**Table 1.** Comparison of $TE_{01}$ and $TE_{10}$ based designs.

| | $TE_{10}$ based 28.5×12.6 mm | $TE_{01}$ based 72×28.5 mm |
|---|---|---|
| Output pulse length, ns | 10 | |
| Input pulse length, μs | 0.2 | 0.60 |
| Q | 10700 | 32000 |
| Compression multiplier | 19.5 | 60 |

[1] To whom any correspondence should be addressed.

Journal of Physics: Conference Series **747** (2016) 012086 doi:10.1088/1742-6596/747/1/012086

| | | |
|---|---|---|
| $P_{out}$, MW | 16.6 | 70 |
| $P_{inp}$, MW | 3.3 | 1.2 |

## 2. BLP based $TE_{01}$ compressor

One of the most important parts of BLP based compressor is the output coupler. It should efficiently pass or not pass the power depending on the phase of incoming waves. It also should be optimized to withstand full output power of the compressor. 3D model of the coupler design shows on Fig. 1.

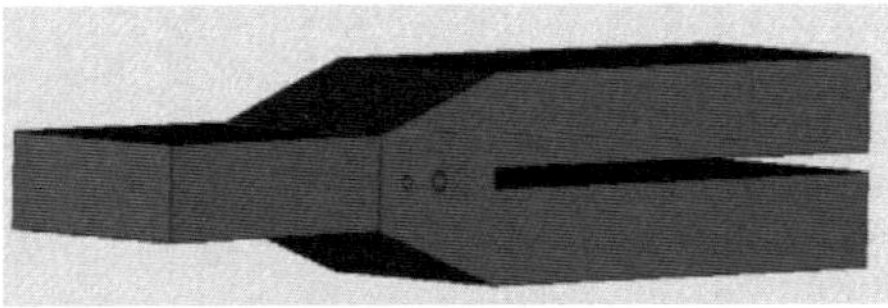

Fig. 1. Output coupler design.

Second one important part of the BLP based compressor is the input coupler. It should include mode converter from $TE_{10}$ of power source to $TE_{01}$, insure 180° phase shift and suppress $TE_{n0}$ modes. We assume to use 28.5×12.6 mm (WR-112) or 23×10 mm (WR-90) input waveguide. For the main body of the compressor was supposed to use 72×34 mm (WR-284) waveguide, but its height caused the higher and hardly suppressible modes excitation. To overcome this effect we assume to use nonstandard waveguide 72×28.5 mm. This exact size was chosen to simplify coupling with 28.5×12.6 mm input and output waveguides. On the Fig. 2 shows the input coupler designs we worked on. On Fig. 2 a,b shown smooth transition from 28.5×12.6 mm and 23×10 mm respectively. Fig. 2 c shows direct excitation with $TE_{n0}$ suppression filter. The optimal design we found in combining stepwise and smooth transition (Fig. 2 d).

Overall view of the compressor is shown on Fig. 3. Calculated electrical field distribution is shown on Fig.4. $E_y(z)$ distribution is slightly (about 2%) modulated due to hardly suppressible excitation of additional $TE_{01x}$ modes.

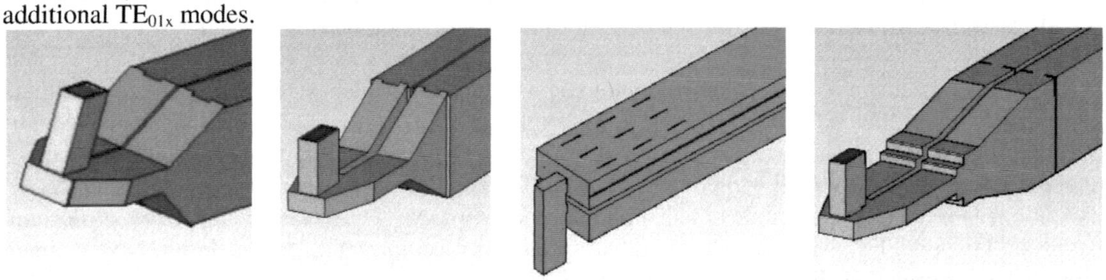

a b c d

Fig. 2. Input coupler designs.

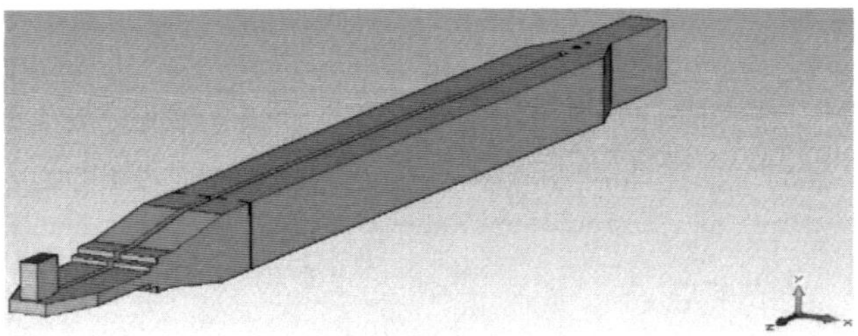

Fig. 3. 3D model of BLP compressor on 72×28.5 mm waveguides.

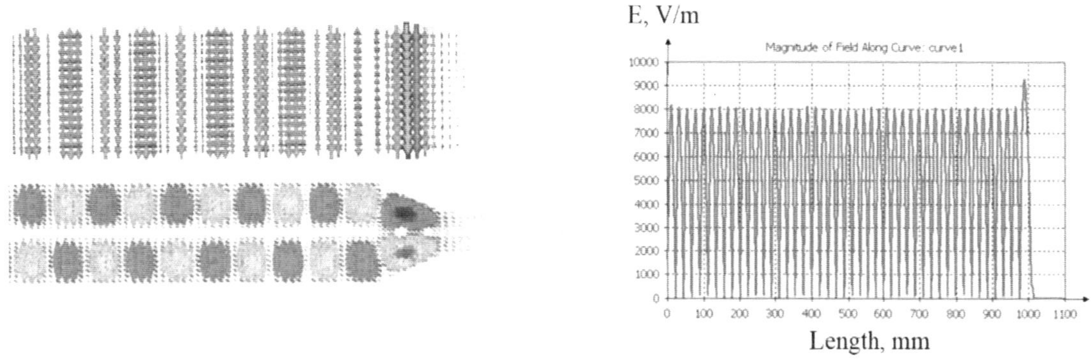


Fig. 4. Electrical field distribution in compressor.

## 3. 2-sectional $TE_{01}$ compressor

To further increase output power we suppose to use one more advantage of BLP based compressor, ability to combine several fully synchronized sections. To make it work we need joint phase shifting module, based on discharge gap for example. On the Fig. 5 a shown design of such compressor for the 2-sectional case. Excitation of both section should be synchronized. One of the possible ways to do so is using single generator and 3dB coupler (Fig. 5 b).

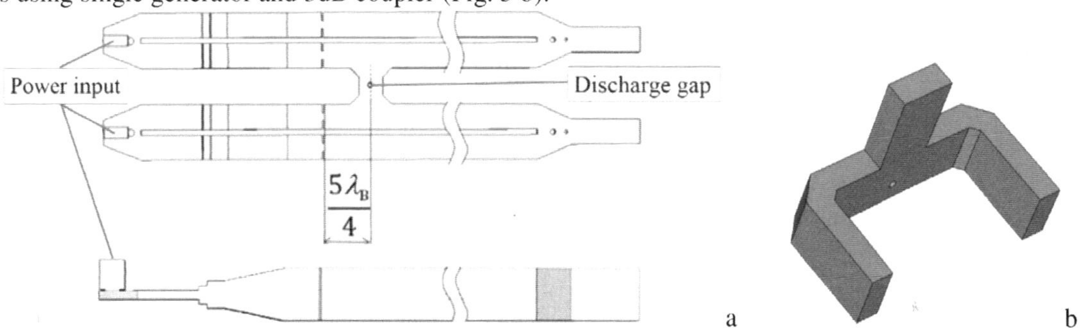


Fig. 5. 2-sectional compressor design and input 3dB coupler

### *3.1. Phase shifting module.*

Joint phase shifting module is the area with maximum EM field in 2-sectional compressor. It's design and field distribution shown in fig. 6. Electrical field in the discharge gap area about 1.4 times higher than regular field in compressor waveguides. That is unavoidable, due to stacking power flows from 2 generators with adequate field increase.

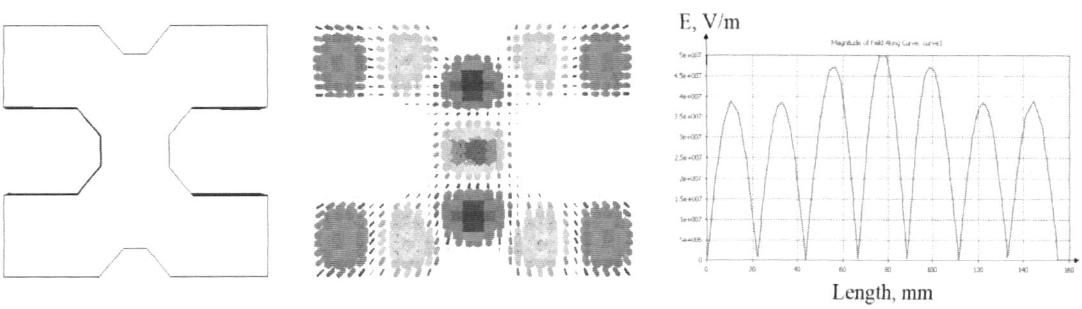


Fig. 6. Joint phase shifting module and electrical field distribution in it.

RF phase shifting is the one of the most important events in BLP based compressor. Fast (about 1ns) and lossless phase shift insures good output waveform and overall compressor operation. The good phase shifting criteria is pretty oblivious:

$$|\varphi_2 - \varphi_1| = 180°$$
$$S_{11} \sim 1$$

$\varphi_1$ – wave phase before the shift, $\varphi_2$ – wave phase after the shift, $S_{11}$ – scattering parameter (reflection coefficient).

The gap discharge itself was not in the focus of this work, therefore we analyze 2 states – "before" and "after" the phase shifting. The plasma column of discharge was represented by conductive cylinder (fig.7). Simulation results presented in the table 2. In the both cases efficient BLP based compressor operation is achievable.

Fig.7. Joint phase shifting module with conductive cylinder for plasma column imitation.

| Model № | $S_{11}$ | $\varphi_1$ | $\varphi_2$ | $R_{pc}$,mm |
|---|---|---|---|---|
| 1 | 0.9998 | 6.9° | 186.4° | 10.7 |
| 2 | 0.9834 | 207.2° | 27.7° | 2.6 |

## 4. 6-sectional $TE_{01}$ compressor

As the way to further increase output power of the BLP based compressor we considering use of 6-sectional design. The first design we tried was based on our earlier works [3, 5, 6] with $TE_{10}$ compressors (fig 8, a). The joint module itself was good, but the H-bend of the waveguide caused serious distortion in $TE_{01}$. We don't find the efficient solution for this distortion problem, but other design was offered (fig 8,b). This joint module design could be called "vortex" design because it based on the vortex electrical field. The main idea is that there is no discharge gap and phase shift should be caused by the plasma ring. That is an advantage and disadvantage in the same time. Such design could withstand extremely strong electric field, but creation the required plasma ring is difficult. One more difficulty is field line distortion in output waveguides (fig. 9, right).

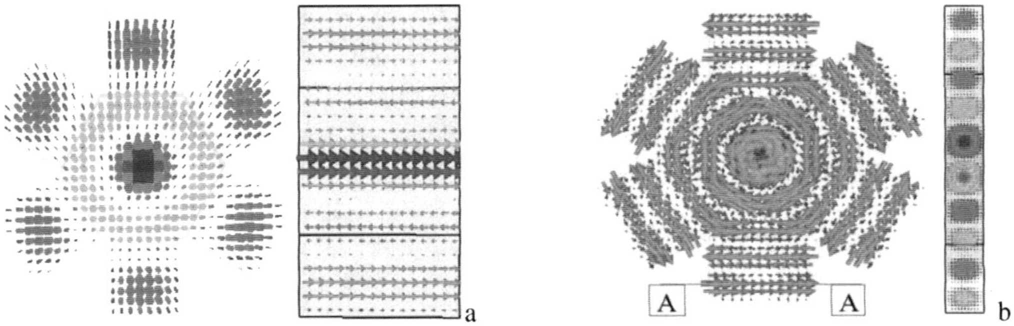


Fig. 8 Electrical field distributions for $TE_{01}$ 6-sectional compressor joint phase shifting modules.

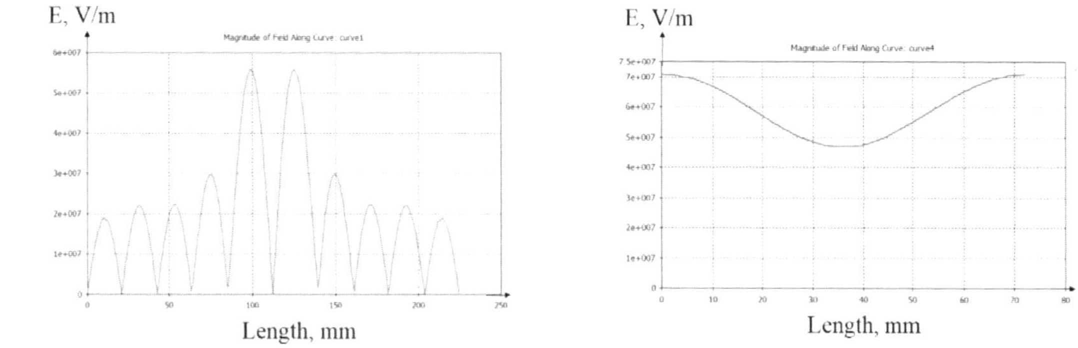


Fig. 9. $E_\varphi$ distribution in "vortex" (on the left) and in A-A section of the waveguide.

*4.1. Dielectric parts in 6-sectional compressor.*

To eliminate the field line distortion and make a surface for electrical discharge we consider partial dielectric filling of the joint module. The possible filling scheme shown in fig. 10. The calculated field distribution shown in fig. 11. Field distortion in the waveguides was reduced from 35% to less than 5% of amplitude. Fig. 12 shows electric field distribution inside of dielectric filler in the middle of joint module. The exact border of the dielectric filler should be determined by the surface discharge condition and can be moved into area of more or less intense field. The overall construction of X-band 6-sectional compressor design presented in fig.13. The waveguides length is 1m for a 10 ns output pulse.

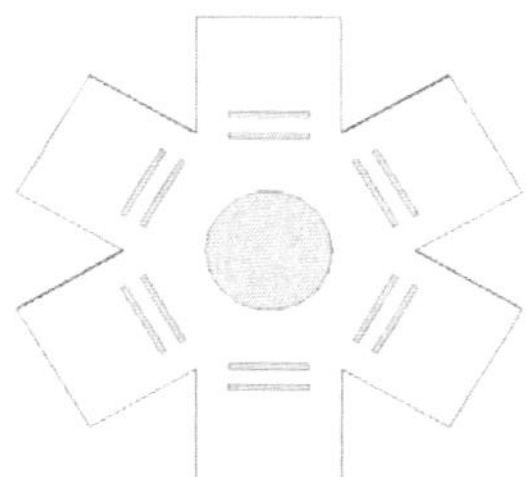

Fig. 10. 6-sectional "star" phase-shifting coupler with dielectric parts (green).

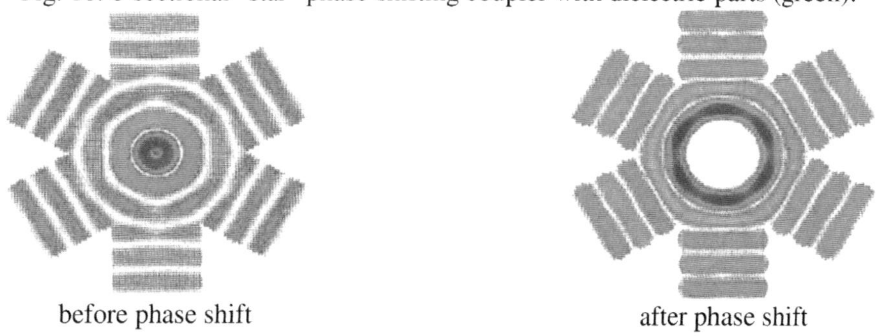


Fig. 11. Electrical field distribution for "vortex" based 6-sectional phase shifting module.

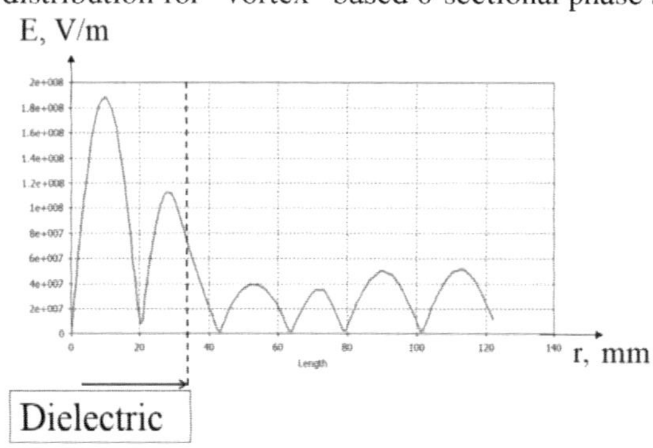

Fig. 12. Radial $E_\varphi$ distribution in "vortex" with dielectric parts installed.

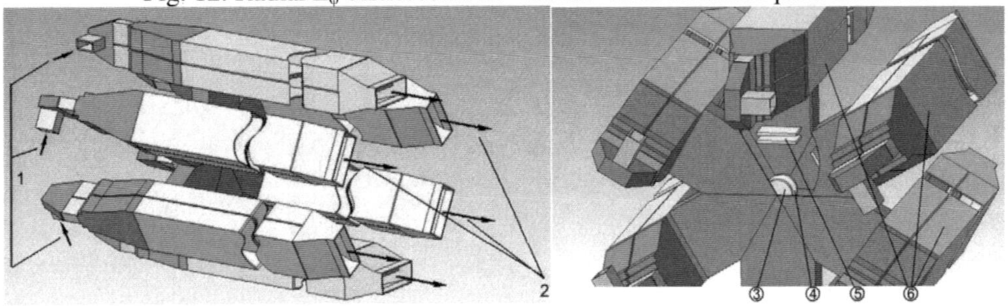


Fig. 13. 6-sectional BLP based compressor. 1- input waveguides, 2 – $TE_{01}$ power outputs, 3 central dielectric filler, 4 – surface discharge area, 5- dielectric plates for distortion suppression, 6 – BLP sections.

## 5. Conclusion

We expect the output power of the single section of BLP X-band $TE_{01}$ compressor will up to 70 MW pulsed. For the 6-section design, the total output power will about 400 MW for a 10 ns. Most important characteristics for single, dual and 6-section designs are combined in table 2. We believe that high specs of such compressor will allow to make a future generation compact particle accelerators. Authors are grateful to A.A. Osipov for a constructive criticism.

**Table 2.** X-band $TE_{01}$ compressor parameters.

| | **single** | **2-sectional** | **6-sectional** |
|---|---|---|---|
| Q | 32000 | 21500 | 18000 |
| Compression multiplier | 60 | 40 | 30 |
| Energy capacity, J | 0.75 | 1.50 | 4.50 |

## References

[1] Ponomarenko A G Microwave-pulse time compression device *Patent RU № 2293404* since 10.02. 2007.

[2] Ponomarenko A G 2007 *Moshchnaya impulsnaya tekhnika. Elementy skhem i istochniki pitaniya* (M.: MEPhI 97-111

[3] Osipov A A and Ponomarenko A G Razrabotka kompressora energii impulsa svch-pitaniya mobilnogo lineynogo uskoritelya elektronov (LUE) *Nuclear Physics and Engineering* **4** 28-32

[4] Osipov A A and Ponomarenko A G Device for Time Compression of Microwave Energy Pulses *Patent RU № 2472260* since 10.01. 2013

[5] Bogdanovich B Y, Kalyuzhny V Y, Masunov E S, Nesterovich A V, Ponomarenko A G and Senyukov V A 2009 Concept of mobile source of narrow gamma radiation *Problems of atomic science and technology* **2** 73-80

[6] Osipov A. A and Ponomarenko A G 2010 Modelirovanie perehodnogo protsessa vyvoda energii iz svch kompressora radioimpulsov na osnove DFL *Nuclear Physics and Engineering* **1** 438-449

Journal of Physics: Conference Series **747** (2016) 012087 doi:10.1088/1742-6596/747/1/012087

# Radiation Field Forming for Industrial Electron Accelerators Using Rare-Earth Magnetic Materials

**A N Ermakov, V V Khankin, N V Shvedunov, V I Shvedunov and D S Yurov**[1]

Skobeltsyn Institute of Nuclear Physics, Lomonosov Moscow State University, 1(2), Leninskie gory, Moscow 119234, Russian Federation
Laboratory of Electron Accelerators MSU Ltd, 1/77, Leninskie gory, Moscow 119992, Russian Federation

E-mail: d_yurov88@mail.ru

**Abstract**. The article describes the radiation field forming system for industrial electron accelerators, which would have uniform distribution of linear charge density at the surface of an item being irradiated perpendicular to the direction of its motion. Its main element is non-linear quadrupole lens made with the use of rare-earth magnetic materials. The proposed system has a number of advantages over traditional beam scanning systems that use electromagnets, including easier product irradiation planning, lower instantaneous local dose rate, smaller size, lower cost. Provided are the calculation results for a 10 MeV industrial electron accelerator, as well as measurement results for current distribution in the prototype build based on calculations.

## 1. Introduction

When treating materials and products with a beam of accelerated electrons at the conveyor lines, to ensure uniform distribution of the transmitted dose, electromagnets are traditionally used, which scan the beam in the direction perpendicular to the conveyor line. For pulsed linear accelerators with pulse frequency of tens to hundreds hertz and scan rate of under ten to tens hertz such an approach has a number of deficiencies. In particular, parameters such as electron beam spot size at the item being irradiated, scan width, pulse frequency, scan rate, conveyor speed, and pulse current become interdependent, which complicates irradiation planning to ensure given dose and its uniform distribution throughout the output volume. Besides, both for pulsed and continuous beams instant dose rate transmitted locally to the item being irradiated is much higher than mean dose rate during the scanning period. In some cases it requires decreasing the beam current and consequently increasing the treatment time.

The purpose of this work is to develop a nonlinear magneto-optical system on the basis of rare-earth permanent magnets, which would ensure uniform dose distribution across the entire width of the conveyor line (along the $x$ axis), to correct the above deficiencies. At that, for irradiated items moving along the $\xi$ axis, the requirement for dose uniformity is as follows:

$$Q(x) = \int q(x, \xi)\, d\xi = const, \qquad (1)$$

where $Q(x)$ is a linear charge density, $q(x, \xi)$ is a surface charge density.

[1] To whom any correspondence should be addressed.

Journal of Physics: Conference Series **747** (2016) 012087 doi:10.1088/1742-6596/747/1/012087

## 2. Principle of Radiation Field Forming

With the exception of special cases, charge distribution across beam cross-section at the accelerator output is close to being Gaussian; therefore a linear optical system fails to ensure uniform charge density at the object. The task of forming uniform charge distribution of an accelerated beam in rectangular area was addressed in publications [1-7]. The approach in the above publications is based on using nonlinear optical systems with octupole lenses. Such systems are quite versatile, and their parameters can vary widely, however, they have significant length, which is not acceptable for industrial accelerators.

We propose a new principle of forming radiation field with constant linear charge density at the object based on using nonlinear quadrupole lens, which defocuses the beam in the scanning direction with focal length increasing as distance from the axis increases, and focuses the beam in the perpendicular direction. The principle of forming uniform linear beam charge distribution out of initially non-uniform distribution is illustrated in Figure 1.

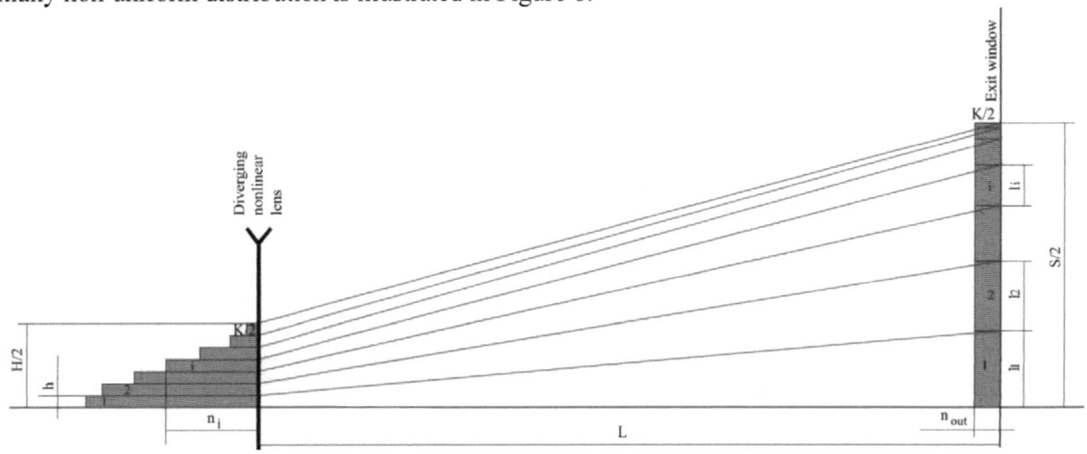


**Figure 1.** Principle of forming uniform radiation field using nonlinear lens.

Let's assume that a parallel beam with some kind of particles distribution along the transverse coordinate falls on a thin nonlinear diverging lens. Figure 1 shows part of the beam that lies in the upper half-plane and contains $N/2$ particles. Particles distribution is presented as a bar chart with a step size $h$, the number of particles at step $i$ is $n_i$, $i = 1,2, ... K/2$, at that $N/2 = h\sum_{i=1}^{K/2} n_i$. At the exit window of the scan horn we need to get uniform distribution of particles with density of $n_{out} = N/S$, where $S$ is the scan width. The condition for ensuring uniform distribution of particles is as follows: $hn_i = n_{out}l_i$, therefore $l_i = hn_i/n_{out}$. If the angle of particles to the lens axis at step $i$ of the bar chart is $x_i'$, the formula for optical power of the lens portion, which would ensure required particles distribution, is:

$$P_i = \frac{\sum_{m=1}^{i} n_m}{iLn_{out}} - \frac{1}{L} + \frac{\tan x_i'}{ih}, \tag{2}$$

where $L$ is the distance between the lens and the object being irradiated.

The optical power of quadrupole lens is determined as:

$$P_q = eGd\frac{c}{p}, \tag{3}$$

where $e$ is the electron charge, $G$ is the lens gradient, $d$ is its effective length, $c$ is the speed of light, $p$ is the beam particles momentum. Using formulas (2) and (3) one can determine dependence of the $y$ – component of magnetic induction in the median plane along the $x$-axis that is required to ensure uniform distribution of particles at the exit window:

$$B_{yi} = -\frac{p}{edc}\tan x_i' + \frac{ph}{decL}\left(\frac{1}{n_{out}}\sum\nolimits_{m=1}^{i} n_m - i\right) \tag{4}$$

Lens field nonlinearity should be formed within the beam. As a rule, the beam diameter at the linac output does not exceed several millimeters. Development of the required nonlinearity at such a small distance results in unacceptably small size of the lens aperture or high magnetic induction. Therefore, in order to increase the beam size in the scan plane at the nonlinear lens input, we use an optical arrangement shown in Figure 2, which includes an additional conventional quadrupole lens defocusing the beam in the scan plane.

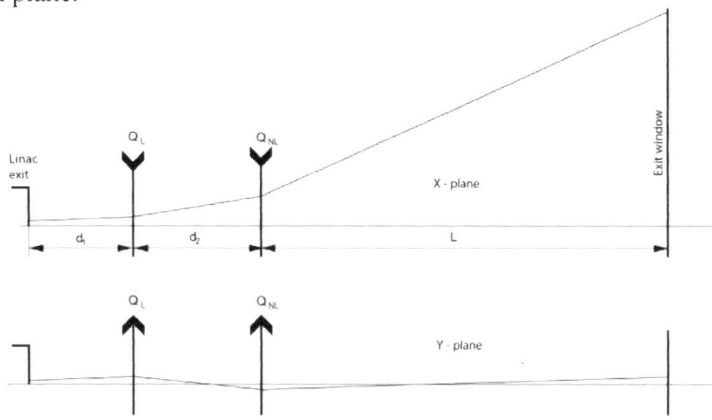


**Figure 2.** Optical arrangement for radiation field forming system. $Q_L$ is a linear and $Q_{NL}$ is a nonlinear quadrupole lens.

To ensure necessary field nonlinearity we used magnet in which the field is formed by rare-earth magnetic material. One of the possible magnet configurations is shown in Figure 3. Magnetic field is generated by two pairs of rectangular blocks made of rare-earth magnetic material magnetized as shown in the Figure. Magnetic screen is placed at some distance from the blocks. The required field distribution is achieved through selection of residual magnetization of the blocks, their parameters (*a, b, c, d*) and distances between the blocks (*h* and H).

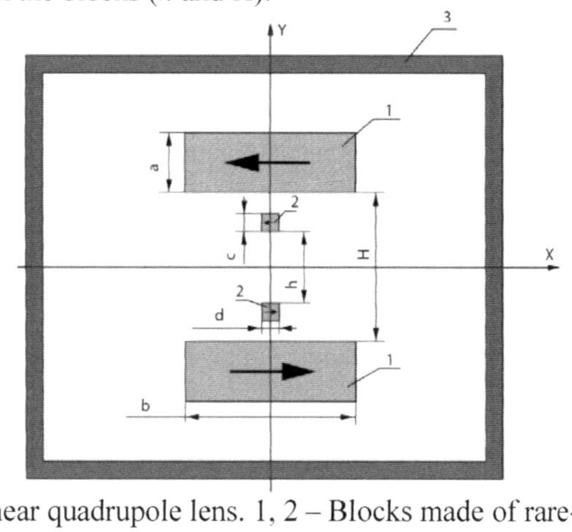


**Figure 3.** Option with nonlinear quadrupole lens. 1, 2 – Blocks made of rare-earth magnetic material, 3 – magnetic screen.

## 3. Numerical Simulation of the Radiation Field Forming System

Numerical simulation was done in 2D approximation using PANDIRA [8] code for calculating magnetic field and RTMTRACE [9] for calculating beam dynamics. The simulation was performed for 10 MeV electrons, at that calculations were done for both Gaussian distribution of particles in the beam cross-section, and for distribution obtained using PARMELA [10] code for the accelerator described in publication [11].

Preliminary estimates were performed for Gaussian beam with 2 mm root mean square radius and 1.1 millirad root mean square divergence, which is close to an actual beam at the accelerator output.

The power of quadrupole lens $Q_L$ (Figure 2) was selected in order to get a crossover in the $y$-plane at the nonlinear lens input $Q_{NL}$, which minimizes its impact on the beam in the plane.

Magnetic field distribution in median plane along the $x$-axis was calculated using formula (4) for a 100 mm long nonlinear lens and Gaussian beam at the accelerator output. Magnets configuration ensuring such distribution was found using PANDIRA. Then magnetic field and corresponding magnets configuration were modified to ensure necessary particle distribution at the exit window of the scan horn with actual beam at the accelerator output calculated using PARMELA. The resulting distribution of particles at the system output is shown in Figure 4(b).

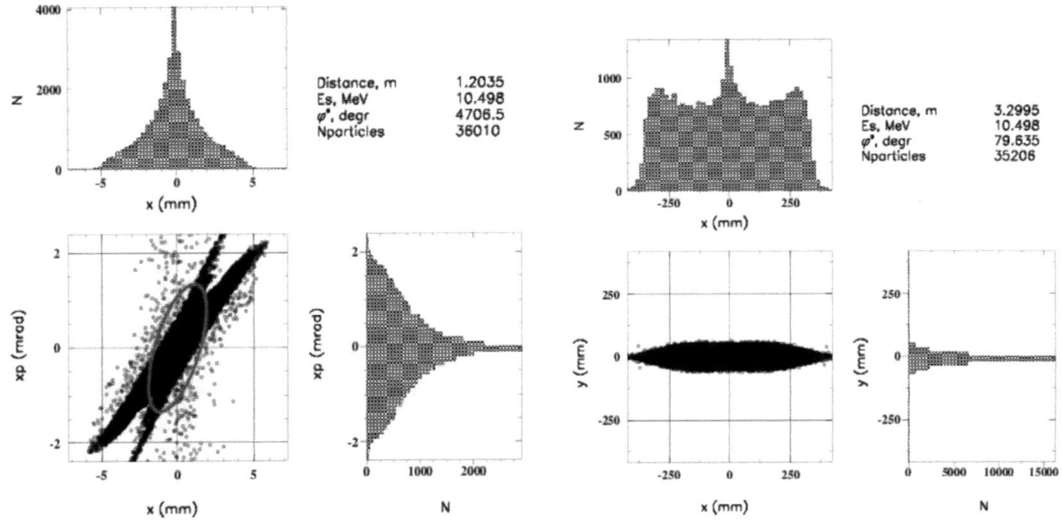


(a) (b)

**Figure 4.** a) Phase portrait of the beam at the accelerator output with 10 MeV calculated using PARMELA, b) beam portrait at the exit window of the scan horn.

## 4. Experimental Study of Radiation Field Forming System

The nonlinear lens built consistent with calculated dimensions was placed in front of a 10 MeV accelerator horn. Figure 5(a) compares this lens with a regular scanning magnet; and Figure 5(b) compares the field measured in the median plane in the central cross-section of the magnet with calculated field.

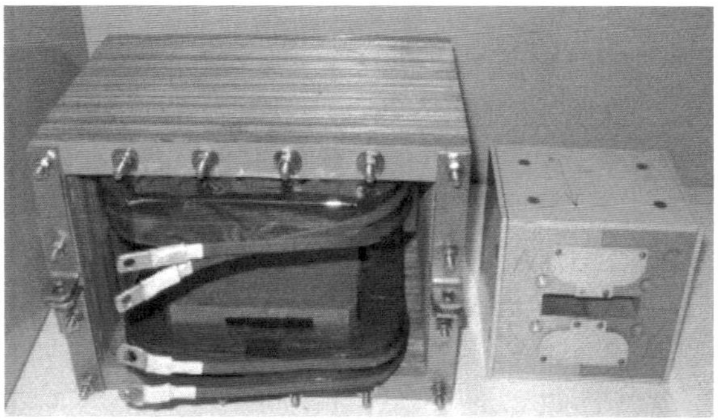

(a)

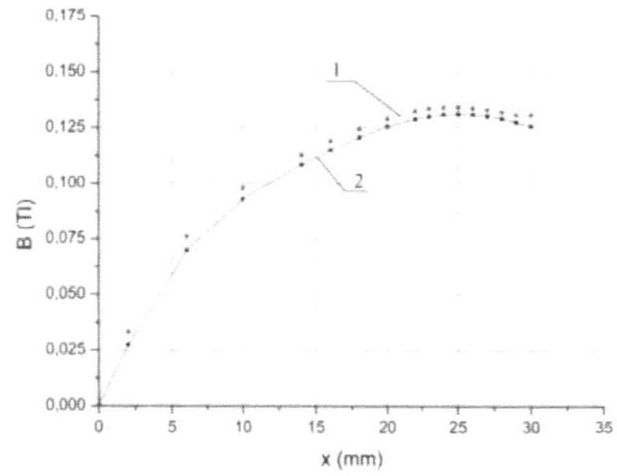


(b)

**Figure 5.** a) Comparison of the magnet (on the right) with a scanning magnet (on the left), b) comparison of calculated (Curve 2) and measured (Curve 1) field distribution.

Current distribution at the scan horn exit window was measured using relocatable 20 mm diameter copper cylinder. Uniformity of distribution was adjusted by changing the focal length of quadrupole lens placed in front of a nonlinear lens. Figure 6 shows measured current distribution after tune-up of the optical system. Scan width (about 50 cm) was determined by the size of the scan horn entrance aperture. Non-uniformity of current distribution was ±5%.

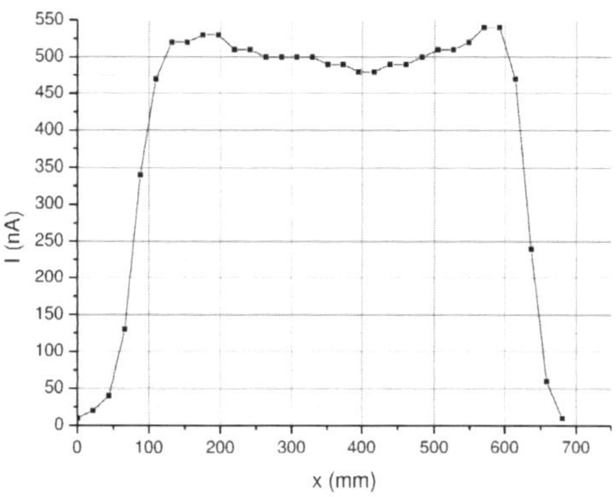


**Figure 6.** Measured copper cylinder current as a function of the coordinate along scan horn exit window.

## 5. Conclusions

As a result of this work, we have created a new type of radiation field forming system for industrial linear accelerators, which has a number of advantages over conventional beam scanning system that uses dipole magnet with variable field. The main advantages of such system are significantly easier product irradiation planning, and lower instant local dose rate. Besides, this system is much smaller, simpler and cheaper to build, and it does not require a power source.

## References

[1] Meads P F 1983 IEEE Trans. Nucl. Sci. **NS-30** 2838
[2] Kashy E and Sherrill B 1987 Nucl. Instrum. Methods Phys. Res., Sect. B **26** 610
[3] Sherrill B, Bailey J, Kashy E and Leakeas C 1989 Nucl. Instrum. Methods Phys. Res., Sect. B **40** 1004
[4] Batygin Y K 1993 Nucl. Instrum. Methods Phys. Res., Sect. B **79** 770
[5] Meot F and Aniel T 1996 Nucl. Instrum. Methods Phys. Res., Sect. A **379** 196
[6] Barlow D et al 1998 in Proceedings PAC 98, pp. 3309-3311

[7] Ji-ho Jang, Hyeok Jung Kwon and Yong Sub Cho 2011 Journal of the Korean Physical Society, **59** 604-609
[8] Halbach K and Holsinger R F 1976 *Particle Accelerators* 7 213-222
[9] Shvedunov V I et al. 1989 RTMTRACE, VINITI, N 183-B89
[10] Billen J H and Young L M 1996 PARMELA, Los Alamos National Laboratory Report, LA-UR-96-1835
[11] Alimov A S et al. 2008 in Proceedings RuPAC'2008, pp. 267-269

Journal of Physics: Conference Series **747** (2016) 012088 doi:10.1088/1742-6596/747/1/012088

# Measurement of electrodynamics characteristics of higher order modes for harmonic cavity at 2400 MHz

**Ya V Shashkov[1,4], N P Sobenin[1], M A Gusarova[1], M V Lalayan[1], D S Bazyl[1], R V Donetskiy[1], A I Orlov[1], M M Zobov[2] and A A Zavadtsev[3]**
[1] National Research Nuclear University MEPHI, Kashirskoe sh.31, Moscow, 115409, Russia
[2] LNF-INFN, Via Enrico Fermi 40, 00044 Frascati (Rome), Italy
[3] LLC « Nano Invest », Perovskaya st. 61/2, Moscow, 111394, Russia

E-mail: shashkovyv@mail.ru

**Abstract.** In the frameworks of the High Luminosity Large Hadron Collider (HL-LHC) upgrade program an application of additional superconducting harmonic cavities operating at 800 MHz is currently under discussion. As a possible candidate, an assembly of two cavities with grooved beam pipes connected by a drift tube and housed in a common cryomodule, was proposed. In this article we discuss measurements of loaded Q-factors of higher order modes (HOM) performed on a scaled aluminium single cell cavity prototype with the fundamental frequency of 2400 MHz and on an array of two such cavities connected by a narrow beam pipe. The measurements were performed for the system with and without the matching load in the drift tube..

## 1. The first section in your paper

At present the project aimed at Large Hadron Collider luminosity upgrade (HL-LHC) is being developed at CERN [1]. The implementation of 800 MHz harmonic cavities should provide a possibility to vary the length of colliding bunches in LHC which can lead in several positive effects [2]. In order to supply the required harmonic voltage several single cell superconducting cavities are to be used.

One of the main goals of the cavity design is to fulfill strict Higher Order Modes (HOM) damping requirements. That is why in order to check the simulation results and to investigate eventual HOM electrodynamics characteristics [3] the scaled version of such an array of two cavities with grooved beam pipe has been manufactured and the HOM parameters have been measured.

## 2. Another section of your paper

In [4] it was shown that it is possible to reduce the loaded quality factor of the most dangerous dipole HOM (TM110 and TE111) lower 100 and below 1000 for HOM in the higher frequency range. In order to verify the results obtained with numerical simulations a scaled aluminium cavity prototype at fundamental mode resonance frequency of 2400 MHz was built. The prototype was designed in a modular form (Figure 1) so that it is possible to carry out measurements for different lengths and shapes of the drift tubes and to further carry out measurements for a chain consisting of two such

[4] To whom any correspondence should be addressed.

cavities. The prototype assembly consists of the beam pipe with a larger radius, the cavity body, beam tube with a smaller radius having and power input feed through and two shorting plates at the ends.

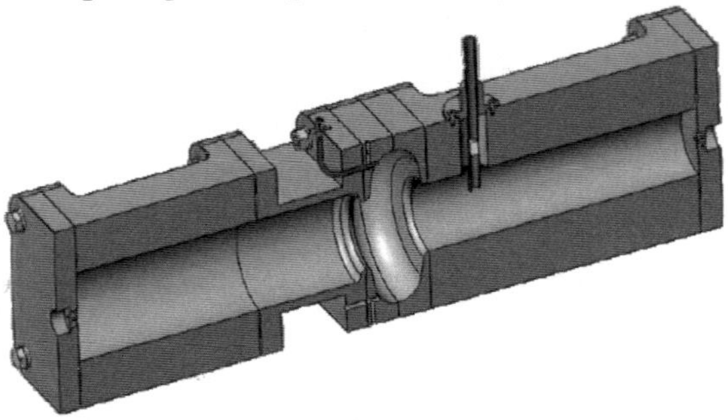

**Figure 1.** General view of single cell cavity

The measurements were performed for the length of the drift tube of 200 mm [5]. In the structure with loads in grooved beam pipe the loaded Q-factor ($Q_{load}$) of the most dangerous HOMs were significantly reduced. But it also appeared that it also affected on the operation mode. For this reason the length of the drift tube was extended up to 400 mm. After the drift tube length increase the $Q_{load}$ of the operation mode returned to its initial value while the HOM $Q_{load}$ values remained the same. The simulated $Q_0$, measured $Q_0$ (without load in beam pipe) and measured $Q_{load}$ are presented in Figure 2.

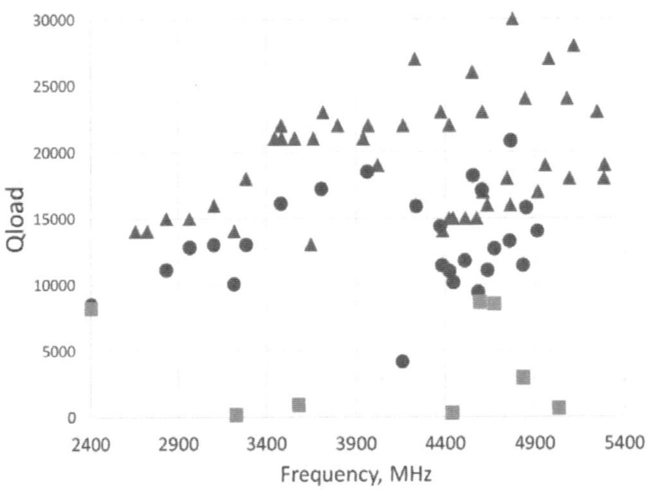


**Figure 2.** Q values of the cavity with a grooved beam pipe. Triangles – $Q_0$ calculation, circles - $Q_0$ measurements, square - $Q_{load}$ measurements

It can be seen that the calculated in CST Microwave studio and measured $Q_0$ results are in a good agreement for the modes that we were able to detect. The $Q_{load}$ values for most of the modes are lower than 1000. Other modes have significantly lower values of $Q_{load}$ and could not be identified. The modes with resonant frequencies about 4500 MHz are the quadrupole TM210 and TE211 modes with effective transverse shunt impedance Q ($R\perp_{eff}$ /Q) values of $10^{-4}$ and therefore they do not pose any effect for the beam. $Q_{load}$ for these modes are of the order of $10^4$-$10^5$.

## 3. Measurements in array of two cavities

In order to reduce the number of transitions between "cold" and "warm" parts of the accelerator vacuum chamber it has been suggested to place two harmonic cavities in a single cryostat. In order to keep the HOM parameters under control and to avoid additional HOM it has been proposed to connect the two cavities with grooved beam pipes with a pipe of a smaller radius. The respective scaled prototype (2400 MHz) has been built and HOM parameters measurements in such a structure have been performed. General view of the assembled two cell prototype is presented in Figure 3. The $Q_{load}$ values are presented in Figure 4.

**Figure 3.** General view of the assembled prototype

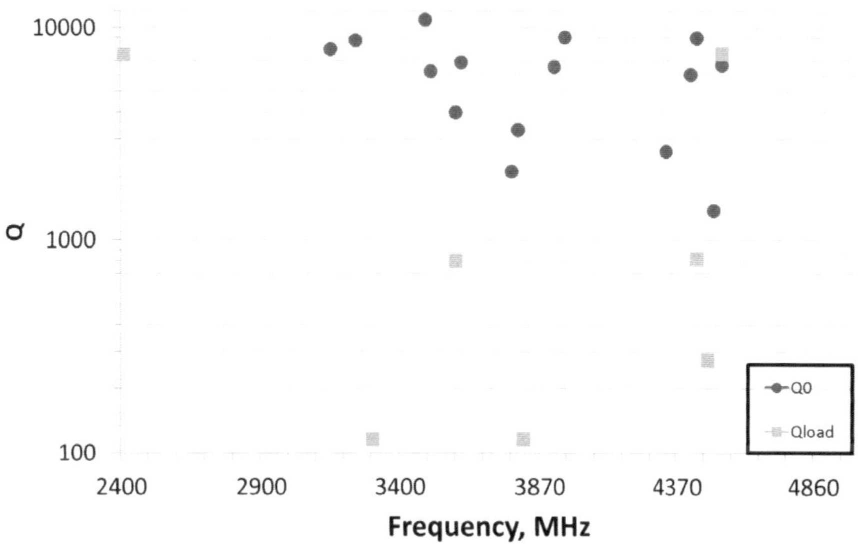


**Figure 4.** $Q_{load}$ of the array of two cavities with a grooved beam pipe. Circles - $Q_0$ measurements, square - $Q_{load}$ measurements

As it can be seen in Figure 4, similarly to the single cell case the $Q_{load}$ values of the most dangerous HOM are lower than 1000. The modes in the region around 4500 MHz are the quadrupole ones with low $R_{\perp eff}/Q$ values of the order of $10^{-4}$. Therefore they should not create any problem, neither for beam dynamics nor for the vacuum chamber heating, despite $Q_{load}$ for these modes are of the order of $10^4$-$10^5$.**Conclusion**

Thus, the performed measurements have confirmed that there are no dangerous trapped HOM both in the single cavity with grooves and in the structure composed of two such cavities. The presence of some residual HOM detected experimentally and having $Q_{load} < 1000$ is explained by the fact that the prototype was built with the shorting plates at the ends (or using the damping load only from one side) and others are the HOM with R/Q lower than $10^{-4}$. This situation is reproduced very well by simulations.

**Acknowledgments**

Work supported by Ministry of Education and Science grant 3.2458.2014/K and the EU FP7 HiLumi LHC – Grant Agreement 284404.

**References**

Journal of Physics: Conference Series **747** (2016) 012088 doi:10.1088/1742-6596/747/1/012088

[1] Rossi L 2011 *Proceedings of IPAC 2011* 908
[2] Linnecar T and Shaposhnikova E 2007 *LHC-Project-Note-394*
[3] Sobenin N and Zverev B 1999 *Electrodynamic Characteristics of Accelerating Cavities* (Gordon and Breach Science Publishers S.A.)
[4] Shashkov Ya V et al. 2014. *Nuclear Instruments and Methods in Physics Research* A767 271–280
[5] Shashkov Ya V et al. 2015 *Proceedings of SRF 2015* **G03** 1-3

Journal of Physics: Conference Series **747** (2016) 012089 doi:10.1088/1742-6596/747/1/012089

# ARIEL e-LINAC: Commissioning and Development

**R.E. Laxdal[1] and V. Zvyagintsev**
TRIUMF Canada's National Laboratory for Particle and Nuclear Physics, Vancouver, BC V6T 2A3

E-mail: lax@triumf.ca

**Abstract.** A superconducting electron Linac (e-Linac) will be a part of the ARIEL facility for the production of radioactive ion beams (RIB) at TRIUMF. The e-Linac will consist of five 1.3GHz 9-cell cavities in three cryomodules delivering a 50MeV 10mA beam. The baseline operation will be single pass but a re-circulating ring is planned to allow either energy boost or energy recovery operation. The first stage of the accelerator which consists of two cryomodules has been successfully commissioned in 2014. The paper will discuss the superconducting radio-frequency (SRF) challenges of the accelerator. Cavities , crymodules and RF system design, preparation, and performance will be presented.

## 1. Introduction

The ARIEL project [1] will allow an increase in the radioactive ion beam (RIB) hours with the addition of a new electron linac driver of 50 MeV (0.5 MW), a new proton line from the 500MeV cyclotron and new production target stations.

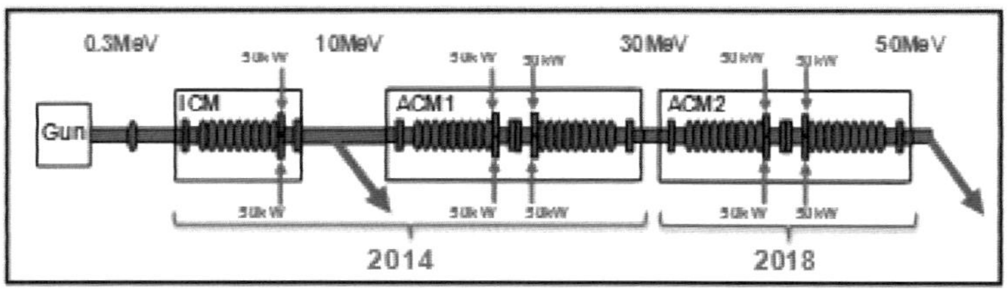


**Figure 1.** The stages of the e-Linac project.

Accelerated electrons can be used to generate RIBs via the photo-fission process [2]. The electrons are stopped in a converter to generate bremsstrahlung photons for fission in actinide target material. An electron beam intensity and energy of 10 mA and 50 MeV is required for a fission rate of $10^{13}$ fissions/sec.

The electron linac is housed in a pre-existing shielded experimental hall adjacent to the TRIUMF 500 MeV cyclotron that has been re-purposed as an accelerator vault. The e-linac is being installed in a phased way with stages shown schematically in figure 1.

A first phase consisting of a 300 kV 16 mA electron gun, an injector cryomodule, ICM, containing one 1.3 GHz nine-cell cavity and an accelerating cryomodule, ACM1, that now contains one 1.3 GHz

[1] To whom any correspondence should be addressed.

Journal of Physics: Conference Series **747** (2016) 012089 doi:10.1088/1742-6596/747/1/012089

nine-cell cavity (and eventually two cavities) plus associated beamlines is now installed and is being commissioned. This first phase is designed to accelerate cw up to 10 mA of electrons at 30 MeV. The initial beam dumps and production targets will only be compatible with 100 kW operation. A second phase, dependent on funding, will see the addition of a second accelerating module, ACM2, and a ramp up in beam intensity to the full capability of 50 MeV 0.5 MW.

## 2. ARIEL e-Linac Design

An RF frequency for accelerating cavities of 1.3 GHz is chosen to take advantage of the considerable global design effort at this frequency both for pulsed machines (ILC) but also for CW ERL applications (KEK, Cornell, BerlinPro). The linac architecture was determined by the final CW beam power of 500 kW (10 mA/50 MeV electron beam) and the available commercial CW RF couplers at 1.3 GHz. The CPI produced coupler, VWP3032, developed with Cornell for the ERL injector cryomodule is capable of operation up to 75 kW CW. In order to provide reliable operation it was decided to set a safety factor for power of 1.5 that is 50 kW CW RF power per coupler. To deliver 500 kW of RF power to the beam requires 10 such couplers. The cavity design allows two couplers per cavity arranged symmetrically around one end delivering a total of 100 kW of beam loaded power. This sets the number of cavities at 5 with a maximum gradient per cavity of 10 MV/m. It is our intention to install a future ERL ring with injection and extraction between 5-10 MeV and so a single cavity off-line injector cryomodule was chosen plus two 2-cavity accelerating modules. The electron hall is shown in figure 2 as it would appear at the end of Phase I.

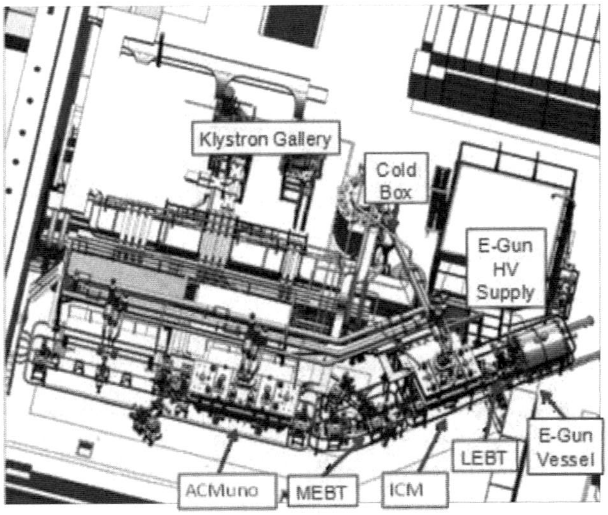


**Figure 2.** The phase I configuration of the E-Linac

### 2.1. *Electron Gun*

The electron source [3] provides electron bunches with charge up to 15.4 pC at a repetition frequency of 650 MHz. The main components of the source are a gridded dispenser cathode in a SF6 filled vessel, and an in-air high voltage power supply. The beam is bunched by superimposing a RF modulation to overcome a DC suppression voltage on the grid.

### 2.2. *Cavities*

The cavity design parameters include f=1.3 GHz, L=1.038 m, R/Q=1000, Ea=10 MV/m [4]. For $Qo=1x10^{10}$ the cavity power is $P_{cav}$=10 W at 2 K that sets the active load requirement for the cryogenics system. A rendering of the jacketed cavity is shown in figure 3.

Journal of Physics: Conference Series **747** (2016) 012089 doi:10.1088/1742-6596/747/1/012089

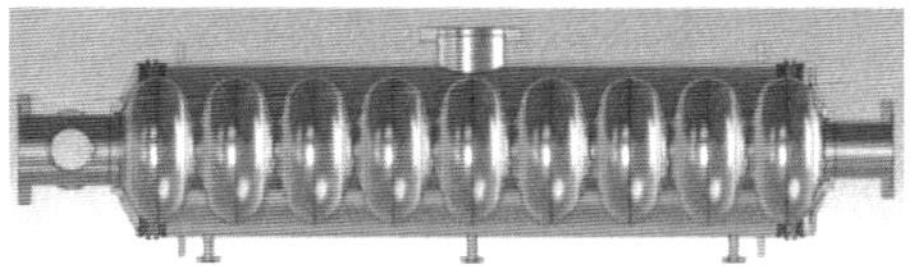

**Figure 3.** The e-Linac nine cell cavity with jacket.

The inner cells take their shape from the Tesla nine cell cavities but the end groups are modified to accept the two power couplers and to help push HOMs to dampers located on each end. On the power coupler end there is a stainless steel damping tube coaxial with the beam tube and extending into the beam pipe. On the opposite end of the cavity a coaxial CESIC tube is used [5]. Each tube is thermally anchored at 77 K and thermally isolated from the cavity by a thin walled stainless steel bellows. The dampers are sufficient to reduce the HOMs to meet the BBU criterion of $R_d/Q \bullet Q_L < 10^7$. The beam tube diameters on the coupler end and opposite end are 96 mm and 78 mm respectively. The vacuum jacket is made from Ti with a machined two convolution flexure on either end. A single 90 mm diameter chimney allows for large CW RF loads of up to 60 W per cavity assuming a conservative heat transfer of 1 W/cm$^2$.

### *2.3. Cryomodules*

A rendering of the ACM module is presented in figure 4.

The cryomodule design has been reported elsewhere [6, 7]. In brief the module is a top-loading box-like structure with a stainless steel vacuum chamber. The cold mass is suspended from the lid and includes a stainless steel strongback, a 2 K phase separator pipe, cavity support posts and the cavity hermetic unit. The hermetic unit consists of the niobium cavities, the end assemblies, an inter-cavity transition (ICT) with a stainless steel HOM damper, the power couplers (FPC) and an rf pick-up. The end assemblies include the warm-cold transition (WCT), CESIC HOM damping tubes and beam-line isolation valves. Other features include a scissor jack tuner and warm motor, LN2 cooled thermal isolation box and two layers of mu metal and alignment monitoring via a WPM diagnostic system.

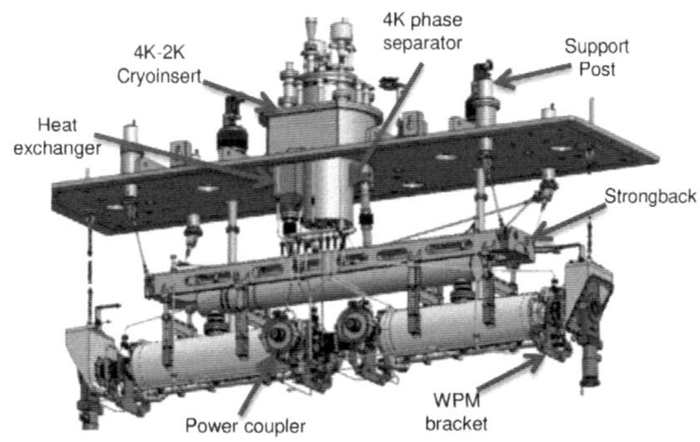


**Figure 4.** Accelerating cryomodule for ARIEL e-Linac

Each cryomodule is outfitted with an on-board 4 K to 2 K cryogenics insert. The insert consists of a 4 K phase separator, a 2.5 gm/sec heat exchanger and a JT expansion valve, a 4 K cooldown valve and a 4 K thermal intercept syphon supply and return. During cooldown the 4K valve is used to direct LHe to the bottom of the cold mass until 4 K level is reached. The level in the 4 K reservoir is regulated by the LHe supply valve, the level in the 2 K phase separator is regulated by the JT valve and the 2 K pressure is regulated by the sub-atmospheric line valve. Piping within the module delivers the syphon supply to a number of 4 K thermal intercept points (WCT, ICT and FPC) and then returns the two phase LHe back to the top of the 4 K phase separator.

Journal of Physics: Conference Series **747** (2016) 012089 doi:10.1088/1742-6596/747/1/012089

*2.4. Cryogenics System*

The design of the cryomodules allows a simplified cryogenics system. A standard commercial 4 K cold box is employed delivering 4 K liquid to a supply dewar near atmosphere. The LHe in the dewar is pushed through the cold distribution with slight overpressure (1.3 Bar) and delivered to the cryomodule 4 K reservoir with parallel feed from a common distribution trunk and cold return back from each cryomodule to the exhaust side of the trunk. The distribution has a 'keep cold' return pipe that joins the each cryomodule has an associated variable LHe supply valve. The 4 K supply and return operates as a refrigerator load. The sub-atmospheric system is independent from the cold box and operates as a liquefaction load. Each cryomodule is pumped in parallel from a common pump line while a variable valve controls the pressure. A common pump line leads between the cryomodules and the sub-atmospheric pumps in a separate building. A common valve near the sub-atmospheric pumps optimizes the operating pressure at the pumps for a given mass-flow. The return 2 K exhaust is warmed passively in a counter flow heat exchanger by thermal exchange with the helium high pressure stream going to the cold box. A simplified schematic of the system is shown in figure 5.

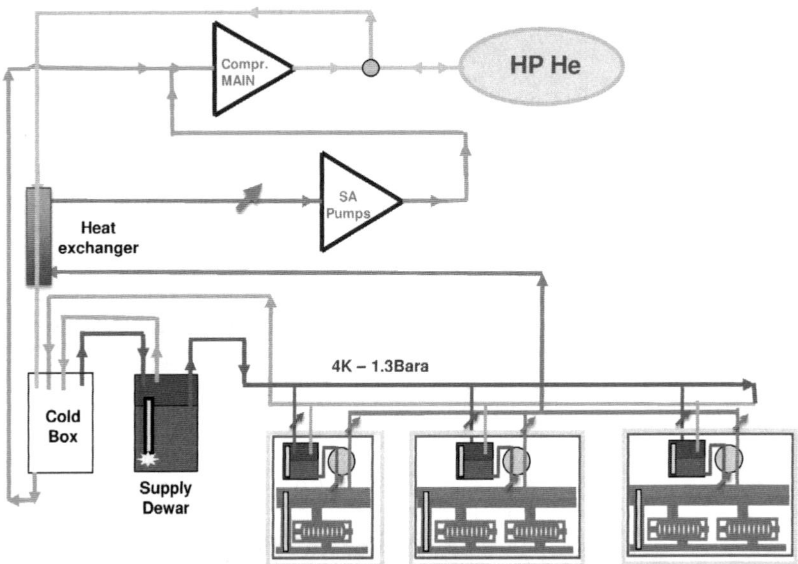


**Figure 5.** A schematic of the e-Linac cryogenic system

*2.5. RF System*

The RF system includes one high power rf source for each cryomodule [8]. In Phase I each cryomodule is driven by a dedicated 290 kW CW 1.3 GHz klystron, CPI VKL7967A (figure 6).

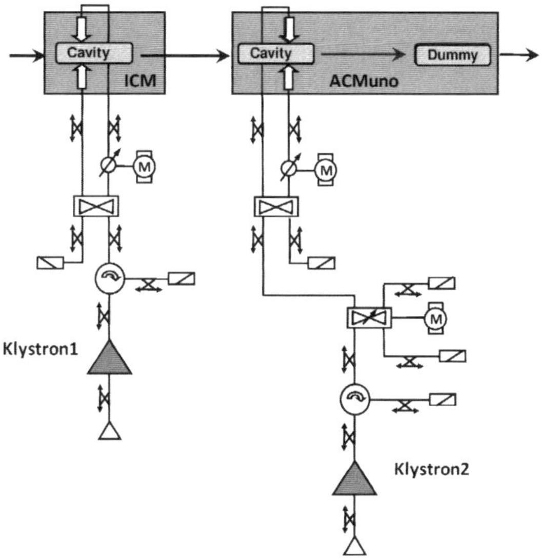


**Figure 6.** The phase-I RF System: ACMuno is ACM1 with one cavity

For Phase II one of these klystrons will drive ACM2 while the ICM will be driven by a 150 kW power source to be determined. The ACM RF power feed is split to feed each of the cavities equally. A further splitting is required to feed each of the power couplers while phase shifters in each leg are used to achieve the proper phase conditions. One LLRF system is used for each cryomodule with a vector sum compensation of voltage and phase drifts in the ACM [9].

## 3. Status and Commissioning

The first stage of the accelerator has been built and commissioned at TRIUMF. The progress has been reported elsewhere [10-13].

### *3.1. Electron Gun*

The system is installed and conditioned to 320 kV with beam extracted at 300 kV up to the full CW intensity of 10mA. RF modulation of the grid voltage is demonstrated over a wide duty factor range from 0.01% to CW. The bunch length as measured with an RF deflecting cavity in the LEBT analyzing leg at a peak current of 1.75 mA and a grid bias voltage of Ub = -160 V is ±12.3° with an energy spread of $\Delta E = \pm$ 500 eV. Estimates based on the measured transconductance ($g_{21}$ = 23 mA/V) result in a pulse length of $\varphi = \pm 10°$.

### *3.2. LEBT*

The LEBT straight section contains three solenoids to provide transverse matching and transportation. A 1.3 GHz room temperature buncher provides longitudinal matching to the ICM. A diagnostic line includes a $90^0$ bending spectrometer, diagnostic boxes and a 1.3 GHz TM011 mode RF deflector for bunch length measurements. In an initial configuration the LEBT allowed a full 3-D characterization of the beam phase spaces. The transverse emittance of the beam was measured with an Allison emittance scanner [10-12] for a peak current of 10 mA and a 1% duty factor to be $\varepsilon_{rms,norm}$ = 7.5 μm. The longitudinal phase space was mapped into the transverse space using the dispersion in the analyzing magnet to map energy spread into the horizontal plane and the vertically deflecting RF device to map time spread into the vertical plane. The LEBT is now installed and commissioned [10-12].

### *3.3. Cryogenics*

The ARIEL cryogenic system [14] includes an ALAT LL Cold Box and KAESER FSD571SFC main compressor with a mass flow rating of 112 g/s. In order to specify the required performancec the estimated static loads from the distribution and the cryomodules were multiplied by 1.5 while the active load was doubled assuming that either the $Q_0$ would be lower by a factor of two or the gradient would be increased to 14 MV/m in some modes. This resulted in a mixed mode set point with a refrigeration load of 128 W and a liquefaction load of 220 l//hr (7.6 g/s). Considering these requirements a specification of a pure refrigeration performance of 600 W and a pure liquefaction performance of 280 l/h was defined. The final commissioning produced a pure refrigeration performance of 837 W and a pure liquefaction performance of 367 l/h comfortably above the criteria. Four Busch Combi DS3010-He sub-atmospheric pumping units rated at 1.4 g/s each are installed. More can be added as the 2 K production increases in Phase II.

### *3.4. RF System*

Two CPI VKL7967A 290 kW CW 1.3 GHz klystrons and two 600 kW 65 kV klystron power supplies from AMPEGON are now installed [8]. Each klystron reached the goal specification at the factory. At TRIUMF RF tests were limited by the available power range of RF loads and RF circulators – one was operated to 250 kW CW and the other to 150 kW CW. Waveguide elements has been installed and tested [15]. The power couplers have been conditioned by two couplers at once at room temperature in a Power Coupler Test Station (PCTS) using a 30 kW IOT. Preparation procedure involved an extended bakeout (7 days) at 100C with N2 flowing to cover the ceramic and RF surfaces. RF conditioning employed both TW (up to 18 kW CW) mode with RF waveguide dummy load and SW mode (up to 10 kW in 1 ms pulse and 1% of duty cycle) with adjustable short for ~5 days [16].

### *3.5. Cavities*

The ARIEL cavities have been fabricated by PAVAC [17]. To date four cavities have been received. The cavities are tuned, degreased then given a 120 μm BCP before final tuning. After the initial cold test ARIEL1 and ARIEL2 were each degassed at FNAL at 800 C for four hours. Both cavities exhibit similar test results. The cavities reach, during vertical tests, the specified gradient of 10 MV/m but at a $Q_0$ of $6{\bullet}10^9$ [18]. Since the available cryogenic power is more than enough for Phase I it was decided to accept $Q_0{>}5{\bullet}10^9$ as a Phase I specification to allow moving forward with the cryo-engineering characterization. ARIEL3 is also jacketed and ready for installation in cryomodule. ARIEL4 cavity is ready for jacketing. Cavity jacketing is done at PAVAC. Due to problems with Ti-bellows from the sub-contractor PAVAC proposed to machine Ti flexures into the jacket. These work well with no significant increase to the cavity stiffness of 1800 N/mm.

### *3.6. Cryomodules*

The cryomodule test strategy utilizes the ARIEL1 and ARIEL2 cavities to qualify the two cryomodule types. ARIEL1 is chosen for ICM production while ARIEL2 is chosen for ACM1 installation along with a `dummy' cavity that occupies the second cavity space in the cryomodule and the RF System was adapted accordingly (Fig. 6). The 'dummy' cavity contains all the interfaces to the helium system so that all helium piping surrounding the dummy will be final. In addition the 'dummy' cavity is installed with a DC heater to replicate cavity active loads and WPM brackets to permit alignment studies. The one cavity ACM variant we term 'ACMuno'. This configuration allows a full cryo-engineering characterization of the cryomodule. The ICM and ACM preparations each consist of the hermetic unit assembly in the clean room, top down assembly in the ISAC beam assembly area and installation in the vacuum tank. The ICM assembly was completed and full cold test was done in the ISAC-II clean room before installation in the e-hall. Due to the size of the ACM it was delivered direct to the e-hall after the top plate was installed and the warm couplers were added there. Both cryomodules are equipped with protection systems developed and fabricated at TRIUMF [19] for fast trip of RF drive in case of cavity quench and threshold signals for RF power, vacuum and temperature.

*3.7. 23 MeV Beam Test*

A '23 MeV Beam Test' of the front end unit is a project milestone to validate cryogenics, HLRF, LLRF, e-Gun operation, LEBT, ICM, ACMuno engineering and overall synchronization [12].

**Table 1.** Measured Cryogenics Performance for ICM.

| Parameter | Estimate | Measurement |
|---|---|---|
| 4 K static load without syphon | 2 W | 3 W |
| 4 K static load with syphon | 6 W | 6.5 W |
| 2 K static load | 5 W | 5.5 W |
| 77 K static load | 100 W | <130 W |

*3.7.1. Cryogenics characterization.* Cooldown to 4 K and production of 2 K was straightforward. The static heat loads are measured by observing the rate of falling LHe level after the supply valves are closed to the volume and noting the volume change of LHe per unit time and the heat of vaporization. The rate of 2 K production is measured by closing the 4 K supply valve while regulating the JT valve to keep the level constant in the 2 K space. In this case the falling level in the 4 K space is a combination of the static loads of the 4 K and 2 K space plus the vapour lost due to expansion from atmosphere to 31.5 mbar. The 77 K static load is measured by noting the warmed GN2 flow required at the exhaust side in order to keep the LN2 thermal shield cold. In this case the measurement is an overestimate since it was difficult to regulate the LN2 at a lower level but the thermal shield was always cold. Measured values for the ICM are shown in Table 1 compared to estimates made during the engineering phase. The 2 K production efficiency improves as a function of mass flow as the temperature of the heat exchanger and JT valve decreases. Values are 70% at 0.5 g/s, 80% at 1g/s and 86% at 1.5 g/s. The ACMuno cryogenics test with one cavity and one 'dummy' show 6.4 W of static load for 4 K and 6.5 W of static load for 2 K.

*3.7.2. RF characterization.* The test includes cavity turn on and phase/amplitude lock, tuner frequency range and tuner lock, microphonics measurements and beam acceleration. The tuner range was measured at +400 kHz – the tuner motion was very stable and the cavity frequency could be stepped very precisely over this range. Due to the excellent frequency stability and broad bandwidth phase lock could be obtained with stable forward power even without the tuner but the tuner lock was easily achieved in any case. Cavity quality factors were estimated based on calorimetric measurements. The performance is presented in figure 7 showing RF characterization results of ARIEL1 and ARIEL2 cavities installed in ICM and ACMuno cryomodules. The $Q_0$ values in the cryomodules are higher than the values measured in the vertical test. This can be due to an additional BCP of 20 μm that each cavity received after vertical test or an improved magnetic environment or both. The cavities meet ARIEL specifications of $Q_0=10^{10}$ corresponding to power dissipation of 10 W at 2 K for Ea=10 MV/m. The results indicate that the magnetic shielding is sufficient and that the HOM dampers do not load the fundamental mode. The ideal RF coupling for ARIEL cavities for 10 mA/10 MV performance at $Q_0=10^{10}$ is $Q_{ext}=10^6$. The coupling adjustment is in the range of $Q_{ext}=7 \cdot 10^5 \ldots 3 \cdot 10^6$. For the initial beam test we set the coupling to the minimum of $Q_{ext}=3 \cdot 10^6$.

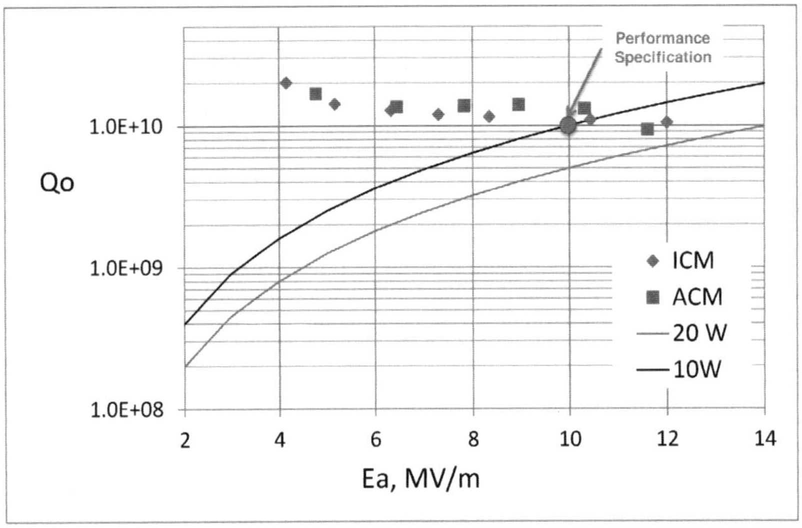


**Figure 7.** RF characterization of ARIEL1 and ARIEL2 cavities in ICM and ACM cryomodules.

*3.7.3. Beam Acceleration Test.* The beam tests were completed using the MEBT analysing leg as a beam dump. The beam energy was estimated based on the dipole setting at the maximum current intensity into the dump Faraday cup. A low duty factor unbunched beam was first aligned in the LEBT and drifted through the ICM. The ICM cavity was turned on at modest gradient and the phase scanned to achieve a good beam spot on a downstream screen. The analysing leg was then turned on and scanned until the beam was seen on the dump FC. The phase of the cavity was optimized to achieve the maximum energy for the particular cavity gradient while the buncher and optics were used to optimize the transmission. Beam simulations were done to calculate the final energy assuming a certain cavity gradient. For the beam tests a gradient of 12 MV/m is achieved for the ICM and 11 MV/m for the ACM cavity. The required forward powers are 18 kW and 14 kW CW respectively.

## 4. Conclusions

The ICM and ACMuno are assembled, installed and commissioned. Beam acceleration demonstrated that the equipment meets the performance goals. The project has now received the license to commission to higher beam intensities and the commissioning is on-going.

In spring 2016 a second ICM prepared for VECC in Kolkata will be installed and tested with beam. At the same time ACMuno will be removed and completed with the ARIEL4 cavity for re-installation in mid 2016. If funding is available a second ACM module is planned for completion in 2018 to complete the e-Linac to its full 50 MeV.

## References

[1] L Merminga, et al., "ARIEL: TRIUMF's Advanced Rare Isotope Laboratory", WEOBA001, IPAC 2011.
[2] W T Diamond NIM-A 432 (1999) pp 471-482.
[3] F Ames, et al., "The TRIUMF ARIEL RF Modulated Thermionic Electron Source WECF1133, EIC2014.
[4] V Zvyagintsev, et al., "Nine-cell Elliptical Cavity Development at TRIUMF", MOPO020, SRF2011.
[5] P Kolb, et al., "Cold Tests of HOM Absorber Material for the ARIEL eLINAC at TRIUMF", http://dx.doi.org/10.1016/j.nima.2013.05.031.
[6] R E Laxdal, et al., "The Injector Cryomodule for the ARIEL e-Linac at TRIUMF", MOPB091,

LINAC12.

[7] N Muller, et al., "ARIEL e-Linac Cryomodule - Design and Performance", THPB115, SRF2015, Whistler, Canada.

[8] A Mitra, et al., "High Power RF System for E-Linac for TRIUMF", Proc. of PAC'13, WEPHO01.

[9] Q Zheng, et al., "LLRF System for TRIUMF 1.3 GHz SRF E-Linac", LLRF Workshop 2013

[10] R E Laxdal, et al., "TRIUMF/VECC e-Linac Injector Beam test", MOPB026, LINAC12, Tel Aviv, Israel.

[11] R E Laxdal, et al., "Status of Superconducting Electron Linac Driver for Rare Ion Beam Production at TRIUMF", MOIOC01, LINAC 2014, Geneva, Switzerland.

[12] M Marchetto et al., "Commissioning and Operation of the ARIEL Electron Linac at TRIUMF", WEYC3, IPAC2015, Richmond, USA.

[13] V Zvyagintsev, et al, "Commissioning of the SRF Linac for ARIEL", TUAA02, SRF2015, Whistler, Canada

[14] A Koveshnikov, et al., "Integration and Commissioning of the ARIEL e-Linac Cryogenic System at TRIUMF", ICEC-ICMC2014

[15] Z T Ang, et al., "HPRF Waveguide Elements Study and Distributions in TRIUMF Electron Linac System", TUPB011, SRF2015, Whistler, Canada.

[16] Y Ma, et al., "High Power Coupler Test for Ariel SC Cavities", THPB103, SRF2015, Whistler, Canada.

[17] V Zvyagintsev, et al., "Production of a 1.3GHz Niobium 9-cell TRIUMF-PAVAC Cavity for the ARIEL Project", THP035, SRF2013, Paris.

[18] P Kolb, et al., "1.3 GHz Cavity Test Program for ARIEL", MOPB089, SRF2015, Whistler, Canada.

[19] Z Y Yao, et al., "Cryomodule Protection for ARIEL e-Linac", TUPB103, proc. SRF2015, Whistler, Canada.

Journal of Physics: Conference Series **747** (2016) 012090 doi:10.1088/1742-6596/747/1/012090

# Crab Waist collision scheme: a novel approach for particle colliders

**M Zobov[1] on behalf of the DAΦNE Team**

LNF-INFN, Via Enrico Fermi 40, 00044 Frascati (Rome), Italy

E-mail: mikhail.zobov@lnf.infn.it

**Abstract**. A new concept of nonlinear focusing of colliding bunches, called Crab Waist (CW) collision scheme, has been proposed at LNF INFN. It has been successfully tested at the Italian lepton collider DAΦNE in operational conditions providing luminosity for two different experimental detectors, SIDDHARTA and KLOE-2. Considering a high efficiency of the scheme for increasing collision luminosity and its relative simplicity for implementation several new collider projects have been proposed and are under development at present. These are the SuperKEKB B-factory ready to start commissioning in 2016 in Japan, the SuperC-Tau factory proposed in Novosibirsk and entered in the short list of Russian mega-science projects, the new 100-km electron-positron Future Circular Collider (FCC-ee) under design study at CERN and some others. In this paper we describe the CW collision scheme, discuss its advantages and report principal results achieved at the electron-positron Φ-factory DAΦNE.

## 1. Introduction

The crab waist collision scheme (CW) was proposed [1, 2] and have been successfully tested at the Φ-factory DAΦNE [3]. At present this scheme is considered to be most attractive for the next generation lepton factories since it holds the promise of increasing the luminosity of the storage-ring colliders by 1-2 orders of magnitude beyond the current state-of-art, without any significant increase in beam current and without reducing the bunch length. Several new collider projects seek to exploit the potential of the crab waist collision scheme. In particular, physics and accelerator communities are discussing and developing new projects which make use of the CW collision scheme: SuperB-factory in Japan (SuperKEKB [4]), SuperC-Tau factory in Novosibirsk [5], two Higgs-factories FCC-ee at CERN [6] and CEPC in China [7].

The SuperKEKB is a natural upgrade of the very successful KEKB, Japanese B-factory at KEK (Tsukuba). The design luminosity goal of the project is $0.8 \times 10^{36}$ $cm^{-2}s^{-1}$, i.e. by a factor of 40 higher than the world record luminosity of $2.1 \times 10^{34}$ $cm^{-2}s^{-1}$ achieved at KEKB. The commissioning of SuperKEKB started in the very beginning of 2016 and the first beams have been already accumulated.

The Budker Institute of Nuclear Physics (Novosibirsk, Russia) is promoting the project of a new generation SuperC-Tau factory. The crab waist concept should allow reaching the project luminosity of $(1\text{-}2) \times 10^{35}$ $cm^{-2}s^{-1}$ that by more than 2 orders of magnitude higher than the luminosity of $0.8 \times 10^{33}$ $cm^{-2}s^{-1}$ presently achieved at the operating τ-charm factory BEPCII in Beijing [8]. SuperC-Tau has entered in the list of 6 most important Russian mega-science projects. In December 2015 the injection complex of the new collider has been successfully commissioned.

[1] To whom any correspondence should be addressed.

Journal of Physics: Conference Series **747** (2016) 012090 doi:10.1088/1742-6596/747/1/012090

In 2014 CERN launched the Future Circular Collider (FCC) study aimed at the design of a 100-km proton-proton collider with the collision energy of 100 TeV. As an intermediate step, the electron-positron collider (FCC-ee) hosted in the same tunnel and covering the energy range between 90 GeV and 350 GeV is also under consideration and its intensive design study is ongoing. The CW scheme has been proposed [9] and recently chosen as the baseline option for the FCC-ee design. This scheme provides a substantially higher luminosity with respect to the traditional head-on collisions at low energies and approximately the same luminosity at higher energies (> 240 GeV) at much relaxed beam optics parameters. A CW lattice solution with twice higher vertical beta function at the interaction point, good dynamic aperture and energy acceptance and manageable photon energies has been found.

Presently China is considering to build a 54-km long Higgs factory CEPC (Circular Electron Positron Collider). Along with the head-on collision option the collider design team has started seriously evaluating a possibility to build a local double ring option with the Crab Waist interaction region [10].

In order to complete the picture, we should mention other collider projects that were considering an application of the CW scheme such as SuperB [11] and SuperTau-Charm [12] factories in Italy, a 500 GeV e+e- collider in a 233-km tunnel [13] in USA, and an upgrade option of the LHC based on collisions of very flat bunches [14, 15].

Since DAΦNE is the only collider exploiting the CW collisions, its operational experience is of great importance for all the new collider projects. In this paper we overview the CW concept and discuss principal results achieved during collider operation with SIDDHARTA and KLOE-2 experimental detectors.

## 2. Crab waist collision concept

The CW scheme can substantially increase collider luminosity since it combines several potentially advantageous ideas. Let us consider two bunches with the horizontal $\sigma_x$, vertical $\sigma_y$ and longitudinal $\sigma_z$ sizes colliding with a horizontal crossing angle $\theta$ (as shown in Fig. 1a).

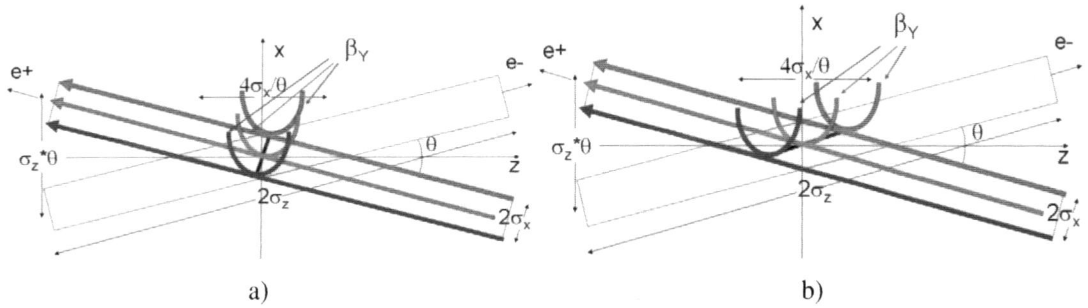


Figure 1. Crab Waist collision scheme: a) crab sextupoles OFF; b) crab sextupoles ON.

Then, the CW principle can be explained in the three basic steps. The **first one** is a large Piwinski angle $\Phi=(\sigma_z/\sigma_x)tg(\theta/2)>>1$. In the CW scheme the Piwinski angle is increased by decreasing the horizontal beam size and increasing the crossing angle. In this way we can gain in luminosity and the horizontal tune shift decreases; parasitic collisions (PC) become negligible since with higher crossing angle and smaller horizontal beam size the beam separation at the PC is larger in terms of $\sigma_x$. But the most important effect is that the overlap area of the colliding bunches is reduced, since it is proportional to $\sigma_x/\theta$ (see Fig. 1). As the **second step**, the vertical beta function $\beta_y$ can be made comparable to the overlap area size (i.e. much smaller than the bunch length):

$$\beta_y^* \approx \frac{2\sigma_x}{\theta} \cong \frac{\sigma_z}{\Phi} << \sigma_z$$

So, reducing $\beta_y^*$ at the IP gives us several advantages:

- Luminosity increase with the same bunch current.
- Possibility of the bunch current increase (if it is limited by $\xi_y$), thus farther increasing the luminosity.
- Suppression of the vertical synchrobetatron resonances [16].

Besides, there are additional advantages in such a collision scheme: there is no need in decreasing the bunch length to increase the luminosity as required in standard collision schemes. This will certainly help solving the problems of HOM heating, coherent synchrotron radiation of short bunches, excessive power consumption, etc.

However, implementation of these two steps introduces new beam-beam resonances which may strongly limit the maximum achievable tune shifts. At this point the crab waist transformation [2, 17] enters the game boosting the luminosity. This is the **third step**. As it is seen in Fig. 1b, the beta function waist of one beam is oriented along the central trajectory of the other one. In practice the CW vertical beta function rotation is provided by sextupole magnets placed on both sides of the IP in phase with the IP in the horizontal plane and at $\pi/2$ in the vertical one (as shown in Fig. 2).

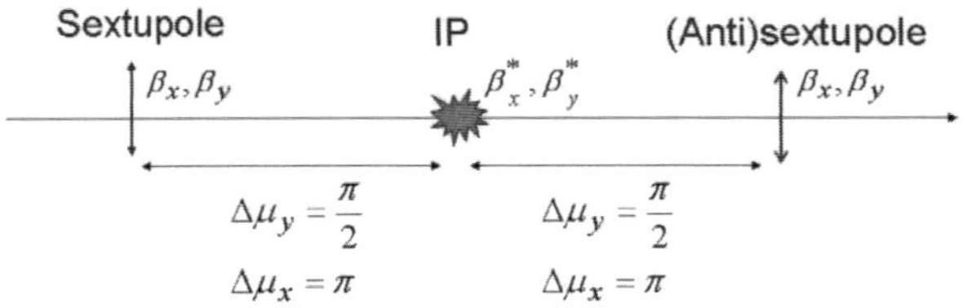


Figure 2. Crab sextupole locations with respect to the interaction point (IP).

The crab sextupole strength should satisfy the following condition depending on the crossing angle and the beta functions at the IP and the sextupole locations:

$$K = \frac{1}{\theta} \frac{1}{\beta_y^* \beta_y} \sqrt{\frac{\beta_x^*}{\beta_x}}$$

The crab waist transformation gives a small geometric luminosity gain due to the vertical beta function redistribution along the overlap area. It is estimated to be of the order of several percent. However, the dominating effect comes from the suppression of betatron (and synchrobetatron) resonances arising (in collisions without CW) due to the vertical motion modulation by the horizontal betatron oscillations [17]. Fig. 3 demonstrates the resonance suppression applying the frequency map analysis (FMA) for the beam-beam interaction in DAΦNE in CW collisions [18].

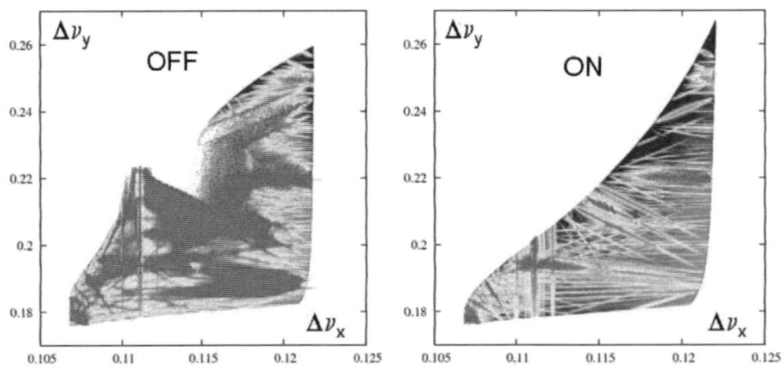


a) b)

Figure 3. DAΦNE beam -beam footprint with crab sextupoles OFF (a) and ON (b) obtained by FMA.

## 3. DAΦNE operation with crab waist collision scheme

DAΦNE is an electron-positron collider working at the energy of the Φ resonance (1.02 GeV c.m.) aimed at providing a high rate of K mesons production [19, 20]. DAΦNE complex consists of two main rings and an injection system composed of a full energy linear accelerator, a damping/accumulator ring and transfer lines. After relevant collider modifications [21, 22] the crab waist collision scheme was successfully implemented and tested at DAΦNE providing luminosity increase by a factor of 3 for the SIDDHARTA experiment [3], in a good agreement with numerical simulations [23].

Figure 4 shows the effect of the crab sextupoles on the beam-beam blow up and the distribution tails in DAΦNE. The main results achieved during the experimental run in terms of peak, specific and integrated luminosity are listed in Table 1.

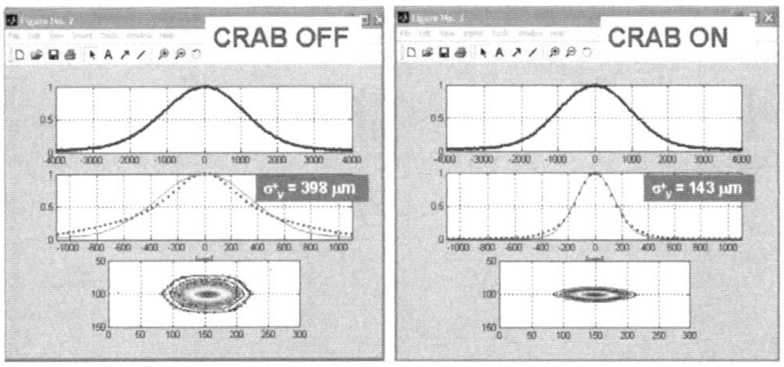


Figure 4. Transverse beam profiles with crab sextupoles OFF (left) and ON (right).

After completion of the SIDDHARTA experimental run and the collider consolidation program [24] the new, more complex, KLOE-2 experimental detector was installed at DAΦNE. Integrating the CW scheme with the large detector having a strongly perturbing solenoidal field posed new challenging issues concerning machine layout, beam acceptance and coupling correction. A new interaction region has been designed and has been equipped with a transverse betatron correction mechanism based on rotated quadrupoles and anti-solenoids independent for the two beams [25].

Table 1. Best luminosity performances during SIDDHARTA and KLOE-2 experimental runs

| Parameter | SIDDHARTA | KLOE-2 |
|---|---|---|
| Peak luminosity [$cm^{-2}s^{-1}$] | $4.53x10^{32}$ | $2.13x10^{32}$ |
| Electron beam current [A] | 1.52 | 1.13 |
| Positron beam current [A] | 1.00 | 0.88 |
| Number of bunches | 105 | 105 |
| Specific luminosity [$cm^{-2}s^{-1}mA^{-2}$/bunch] | $3.13x10^{28}$ | $2.25x10^{28}$ |
| Integrated luminosity [$pb^{-1}$/day] | 14.98 | 14.03 |

Despite the presence of the rotated quadrupoles, strong detector solenoid with nonlinear fringing fields and anti-solenoids between the crab sextupoles the effectiveness of the CW scheme has been proven both numerically [26] and experimentally [24]. As it is seen in Table 1, the presently achieved daily integrated luminosity is comparable to that obtained during the run with a simpler SIDDHARTA interaction region. However, the maximum peak luminosity is still by about a factor of 2 lower than in the previous SIDDHARTA run. Luminosity limiting factors have been identified and a work is in progress to eliminate them in order to fully exploit the CW potential. In particular, the following measures are foreseen to improve the DAΦNE performance: improving CW sextupoles alignment and their strength optimization; refining the transverse coupling correction; pushing the microwave instability threshold towards higher single bunch currents by using optics with a higher momentum compaction factor and a higher chromaticity; vacuum conditioning and scrubbing to diminish electron cloud effects; further feedback noise reduction etc.

## 4. Conclusions

Experimental measurements at DAΦNE have proven that the Crab Waist is a reliable and robust technique for the luminosity increase. It has been very successful in collisions with SIDDHARTA interaction region. The obtained results are in a very good agreement with numerical predictions and simulations. The work is now in progress to fully exploit the potential of the Crab Waist in collisions for the KLOE-2 experimental detector. At present several projects of large scale colliders based on the Crab Waist Scheme are under development.

## References

[1] Raimondi P 2006 Status on SuperB effort *in 2nd SuperB Workshop*, Frascati, Italy, 16-18 March 2006

[2] Raimondi P, Shatilov D and Zobov M 2007 Beam-beam issues for colliding schemes with large Piwinski angle and crabbed waist *Preprint* physics/0702033

[3] Zobov M *et al* 2010 Test of crab-wast collisions at DAΦNE Φ-factory *Phys. Rev. Lett.* **104** 174801

[4] Ohnishi Y *et al* 2013 Accelerator design at SuperKEKB *Prog. Theor. Exp. Phys.* **2013** 03A011

[5] Bondar A *et al* 2013 Project of Super Charm-Tau factory at the Budker Institute of Nuclear Physics *Phys. Atom. Nucl.* **76** 1072-1085.

[6] Benedikt M and Zimmermann F 2015 Outline and status of the FCC-ee design study *ICFA Beam Dyn. Newslett.* **67** 86-93

[7] CEPC web page 2016, http://cepc.ihep.ac.ch

[8] BEPCII web page 2016, http://english.ihep.cas.cn/rs/fs/bepc/index.html

[9] Bogomyagkov *et al* 2014 Beam-beam effects investigation and parameters optimization for a circular e+e- collider TLEP *Phys. Rev. ST Accel. Beams* **17** 041004

[10] Wang D *et al* 2016 CEPC partial double ring scheme and crab-waist parameters *in IAS Program on High Energy Physics* 18-21 January 2016, Hong Kong

[11] Bona M *et al* 2007 SuperB: a high-luminosity asymmetric e+e- super flavor factory. Conceptual design report *Preprint* INFN-AE-07-02 (arXiv:0709.0451)

[12] Biagini M *et al* 2013 Tau/Charm factory accelerator report *Preprint* arXiv:1310.6944

[13] Summers D J *et al* 2012 A 233 km tunnel for lepton and hadron colliders *AIP Conf. Proc.* **1507** 860-865

[14] Abelleira J L, Russenschuck S, Zimmermann F, Milardi C and Zobov M 2012 Local chromatic correction scheme and crab-waist collisions for an ultra-low beta *Conf Proc.* **C1205201** 118

[15] Abelleira J L, Russenschuck S, Tomas R, Zimmermann F, Ohmi K, Milardi C, Zobov M and Shatilov D 2013 LHC optics with crab-waist collisions and local chromatic correction *Preprint* CERN-ACC-2013-0147

[16] Pestrikov D V 1993 Vertical synchrobetatron resonances due to beam-beam interaction with horizontal crossing *Nucl. Instrum. Meth.* A **336** 427-437

[17] Raimondi P, Shatilov D and Zobov M 2008 Suppression of beam-beam resonances in crab waist collisions *Conf. Proc.* **C0806233** WEPP045

[18] Shatilov D, Levichev E, Simonov E and Zobov M 2011 Application of frequency map analysis to beam-beam effects study in crab waist collision scheme *Phys. Rev. ST Accel. Beams* **14** 014001

[19] Vignola G *et al* 1993 DAΦNE, the Frascati Φ-factory *Conf. Proc.* **C930517** 1993-1997
[20] Vignola G *et al* 1996 DAΦNE, the first Φ-factory *Conf Proc* **C960610** 22-26
[21] Zobov M *et al* 2008 DAΦNE status and upgrade plans *Phys. Part. Nucl. Lett.* **5** 560-565
[22] Milardi C *et al* 2009 Present status of the DAΦNE and perspectives *Int. J. Mod. Phys.* A **24** 360-368
[23] Zobov M 2010 Beam-beam interaction in novel, very high luminosity parameter regimes *Conf. Proc.* **C100523** 3639-3643
[24] Milardi C *et al* 2015 DAΦNE consolidation program and operation with KLOE-2 detector *ICFA Beam Dyn. Newslett.* **67** pp. 9-27
[25] Milardi C *et al* 2012 High luminosity interaction region design for collisions inside high field detector solenoid *Journal of Instrumentation* **7** T03002
[26] Zobov M *et al* 2015 Simulations of crab waist collisions in DAΦNE with KLOE-2 interaction region *Preprint* arXiv:1506.07559

**Institute of Physics**
Dirac House, Temple Back
Bristol BS1 6BE UK

ISSN: 1742-6588
ISBN 978-1-5108-3058-5

CURRAN ASSOCIATES INC.
proceedings.com